A HISTORY OF GREEK FIRE
AND GUNPOWDER

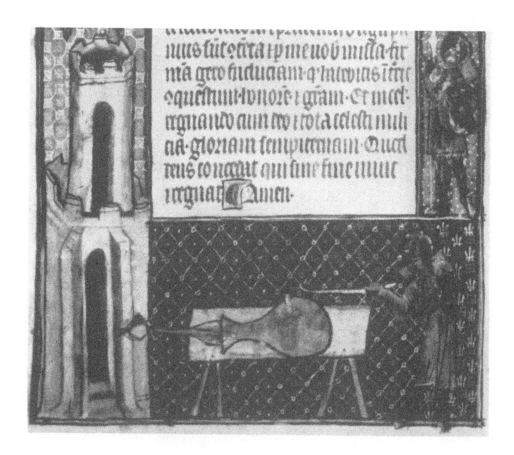

The Earliest Representation of a Gun

From the manuscript *De Notabilitatibus, Sapientis, et Prudentia* (1326) of Walter de Milemete. Christ Church, Oxford, MS. 92, fol. 70 *v*.   Original size.   (By permission of the Governing Body of Christ Church, Oxford).

A HISTORY OF
GREEK FIRE
AND
GUNPOWDER

J. R. PARTINGTON

Foreword by
LIEUT.-GEN. SIR FREDERICK MORGAN, K.C.B.

With a new introduction by
BERT S. HALL

The Johns Hopkins University Press
Baltimore and London

Hardcover edition first published 1960 by W. Heffer & Sons, Ltd., Cambridge
© 1960 Professor J. R. Partington
© 1999 The Johns Hopkins University Press
All rights reserved
Printed in the United States of America on acid-free paper

Johns Hopkins Paperbacks edition, 1999
9 8 7 6 5 4 3 2 1

The Johns Hopkins University Press
2715 North Charles Street
Baltimore, Maryland 21218-4363
www.press.jhu.edu

Library of Congress Cataloging-in-Publication Data

Partington, J. R. (James Riddick), 1886–1965.
    A history of Greek fire and gunpowder / by J. R. Partington ; forward by Frederick Morgan ;
with a new introduction by Bert S. Hall.
        p.   cm.
    "First published 1960 by W. Heffer & Sons, Ltd., Cambridge"—T.p. verso.
    Includes bibliographical references (p.   –   ) and index.
    1. Greek fire—History. 2. Gunpowder—History. 3. Incendiary weapons—History. I. Title.
TP268.P3   1999
623.4´52—dc21   98-11738
                                                                                                      CIP

A catalog record for this book is available from the British Library.

Contents

Contents

List of Illustrations

Foreword

WHEN, years ago, the author of this scholarly compendium began his painstaking work of collation he can have had little idea that he was in fact compiling what is virtually an obituary. Here is the story of the evolution of chemical explosives which, for almost precisely 600 years, from Crécy in 1346 to Hiroshima in 1945, rules as the ultimate arbiters of human affairs, the acme of war-frightfulness.

From the study of this book one can appreciate afresh the dreadful acceleration with which technical development is now taking place around us. It is only in the last half-century that so-called high explosive (already referred to as "conventional" high explosive) came to take the place of Roger Bacon's discovery of gunpowder all those years ago. And now the infernal art has been taken over from the chemists by the physicists to whom applied science has given the means of misapplying some small fraction of the power that animates our universe.

We read here how, for five hundred years and more, quite minor evolution took place in destructive science. Well within the memory of living men there were used, in the First World War, of 1914-1918, both apparatus and materials that would have seemed familiar to our remote ancestors. The early "trench mortars" bore close resemblance to the "gonnes" of Crécy, on our side the Bethune Drainpipe being but a length of cast-iron tube blocked at one end and the German mine-thrower being made of wood bound with wire. Our field guns were at first unprovided with high explosive shell, their only projectile being shrapnel of which the bursting charge consisted of a few ounces of gunpowder, a refined version of that same gunpowder that had first sullied the knightly battlefield at Crécy.

Professor Partington's research has taken him into the history of many lands, the intricacies of many languages, classic and archaic. From all of it comes confirmation of the persistence of the human dilemma. From time immemorial it is but the means that have changed in the struggle for power, for domination. Over early centuries it was a matter of acquiring first the "know-how" of Greek fire, preservation of the secret of which set an uncommonly high standard of "security." Was it, maybe, that the ancients knew our petrol which they misused otherwise than do we? Then a nation that would be a nation must have its guns and gunpowder. We, in our turn, have our "Nuclear Club" which those must join who would be heard in the world's councils.

To be sure, gunpowder had always its peaceful uses for the making of fireworks as a public spectacle and a symbol of rejoicing. We expatiate upon the peaceful uses of atomic energy. But we are worse off to the extent that our "peaceful uses" produce explosive as their by-product instead of burning it harmlessly in squib or cracker.

We, in our time, have experienced the catastrophic shock of "scientific break-through" in the explosive field. Even now, I fancy, it is little appreciated that one small projectile can be made capable of destroying at a blow the whole great city of London. Yet directly comparable in its total effect on human affairs must have been first appearance of Greek fire, of gunpowder. It is intriguing to contemplate the treatment that would have been given to the Crécy salvoes had there existed at the time our modern means of misinformation and hysterogeny. It was not just that there had suddenly appeared a new weapon of war, but this was a weapon generating unearthly din, flame and smoke, all attributes of the devil himself, and the devil was very real to our ancestors.

This book tells of the gradual spread of the horror throughout all mankind, first with hardly perceptible advance and then, as the world shrank, with increasing speed until today's Gadarene rush.

My own experience epitomises the twentieth century leap towards obliteration. Less than fifty years ago my charge was of two "modern" field guns that could deliver, by means of smokeless propellant, to a range of some four miles, projectiles of 18 lb. weight containing bursting charges of gunpowder sufficient to liberate the bullets from the body of the shell. My last official duty before retirement a few years back was to concern myself with the fabrication of a "Hydrogen Bomb" that should have explosive effect equal to that of a mass, if such can be imagined, of ten million tons of "conventional" or "orthodox" high explosive. This at a range of thousands of miles.

How now and where next?

This book tells of past centuries of evolution in the explosive field and, in so doing, inevitably gives rise to thought of the future—if there is to be a human future. Unhappily, in this context, there is as yet no known means of curbing the play of the human mind. What is here involved is a matter not of invention but of discovery. The search for knowledge is unceasing but knowledge, when found, does not of necessity bring happiness or wisdom.

To me it emerges so clearly from this great history how futile are the attempts to halt the tide of discovery by means of treaty, pact, agreement or any form of words. As futile as the comminatory epitaph, herein quoted, on the fabulous Black Berthold, legendary inventor of artillery,

> "most abominable of inhumans who, by his art, has made
> miserable all the rest of humanity."

Should we, in our turn, likewise condemn Rutherford and Cockcroft?

Professor Partington has given us, in this great compendium of knowledge, not only a masterwork of historical research but rich food for thought on what may lie ahead.

F. E. MORGAN.

Preface

ὅσα δύνασαι ποιῆσαι, ποίησον. Hermes Chemista.

Non oportet nos adhærere omnibus quæ audimus et legibus, sed examinare debemus districtissime sententias majorem, ut addamus quæ eis defuerunt et corrigamus quæ errata sunt, cum omni tamen modestia et excusatione.

Roger Bacon.

THE history of gunpowder and firearms has attracted many authors with varying interests. The general historian must take account of major inventions effecting revolutions in the life of nations. The historian of science is concerned mostly with the invention of gunpowder. The historian of technology examines the development in the manufacture of explosives and weapons, and the way in which gunpowder has found applications in the peaceful arts. The military historian deals mainly with the use of gunpowder as an explosive and a propellant (two things which are sometimes confused), and the development of firearms and their use in warfare.

No recent book in English (or for that matter in any language) has attempted a concise survey of the subject. Col. Hime's book, *The Origin of Artillery* (1915), has long been out of print and much newer information has made parts of it obsolete. Among older books, the volumes of *Études sur le passé et l'avenir de l'Artillerie*, of the Emperor Napoleon III and Col. Favé, written by two experts on the basis of original documents, have a permanent value. Not many general historians have excelled Gibbon in this field; very few sources of information known in his time managed to slip through his net and his treatment of scientific matters is based on careful study and the advice of contemporary experts.

In writing my book I first attempted to attain a mastery of the older and newer material and then to arrange the result in a continuous narrative. My sources, given in numbered references at the ends of the chapters, are detailed and extensive. From them a reader wishing to follow up topics of interest to him will be able to make a start in the literature. Mere lists of books or other publications, without page references to particular items are, in my opinion, of little value. In many cases the original texts are given, accompanied by translations or abstracts.

The first chapter, after a summary of the earlier use of incendiaries in warfare, deals with Greek fire. At least six different suggestions as to its nature have been proposed, which are all examined. I believe that the theory that Greek fire was a distilled petroleum fraction combined with other specified ingredients, but not containing saltpetre, is the only one which agrees with the descriptions of its nature and use. The use of Greek fire in the sieges of Constantinople and in the Crusades is dealt with.

xi

In entering the study of the origin of gunpowder I discuss first a Latin *Book of Fires* attributed to Mark the Greek, which, in a part not much earlier than 1300, gives a recipe for gunpowder. I show that this is probably not based on a Greek work but was written by a Spanish Jew. It is one of those recipe books which I consider in detail in a later chapter, compiled from various sources of very different dates. These books include the *Bellifortis* of Kyeser, the *Feuerwerkbuch*, and treatises by Fontana, Valturio and others, none of which has previously been analysed in English. This chapter also contains a history of infernal machines and mines.

Gunpowder was known to Roger Bacon and Albertus Magnus about 1250, but after a detailed discussion I conclude that both obtained a knowledge of it from Arabic sources. Before passing to these I deal with Berthold Schwartz, a legendary discoverer of gunpowder or inventor of cannon who is shown to be mythical, and with the invention of cannon. This is a very controversial subject. It is shown that the first known picture of a gun is in an English manuscript and the earliest mention of guns is in an Italian manuscript, both of date 1326. The claim that cannon were a German invention is shown to be doubtful. In this chapter some of the earliest cannon and other firearms, including some very large old cannon, such as were used in the capture of Constantinople in 1453, are described.

In a study of the invention and use of firearms and explosives some military knowledge is necessary, and I have made use of my training as an infantry and engineer officer in the war of 1914-18, some experience of active service, and such study of the Army manuals as conditions then allowed. This limited knowledge has been of considerable use to me.

The latest detailed survey of the history of artillery known to me is a book in German by the late Lieut.-Gen. von Rathgen, not generally available in England, in which a claim is made that cannon were a German invention. Although I have the greatest respect for such a distinguished officer and have often followed his guidance in technical matters concerned with artillery, I give reasons for dissenting from his thesis.

The chapter on gunpowder and firearms in Muslim lands first gives a sketch of general history and then deals with the early use of gunpowder. Some Muslim fire books of the same type as the European ones described in an earlier chapter are then analysed, and the early use of gunpowder as a propellant is dealt with. This is followed by a general account of Muslim artillery and firearms, the use of gunpowder and firearms in India (including the Portuguese attacks), and the artillery of the Mughal Emperors. Indian sources are in general not closely dated, but too much has been made of this by experts in other fields. Indian technical information is usually conveyed in a concise and dry manner, without emphasis on originality, and this can be misleading.

The chapter on China uses new information. Some recent short publications on Chinese gunpowder and firearms are misleading, using wrong technical terms and reporting military impossibilities, but I have attempted to make some sense of the material. In this chapter I have had valuable assistance from Dr. J. Needham, and I believe that it is the first general survey of the field which makes adequate use of modern information. If the dates of the texts are correct, the discovery of the use of saltpetre in explosives and the development of gunpowder are to be sought in China from the eleventh century.

The history of gunpowder is associated with that of saltpetre, no comprehensive account of which was available. The last chapter breaks the ground for further study.

My book, which quotes original sources, uses technical names freely, and gives extensive references, is primarily intended for readers with a scientific background. I hope that it may interest military and naval officers, who will have more knowledge of many topics in it than I have. I have tried to make the text as attractive as possible and to arrange it so that quotations and technicalities can be passed over and what remains will still form a continuous narrative. It would perhaps be unfair to me to criticise it as too heavy or technical for the general reader; it may not be, and in any case it should provide material for those better qualified than I am to produce what the French call an "œuvre de vulgarisation."

In the preparation of the book I have consulted books and manuscripts in several libraries, the staffs in which have been unfailingly courteous and helpful. On many points of detail I have had assistance from experts and I thank Professor T. Burrow, Dr. E. J. Holmyard, Mr. A. Z. Iskandar, Miss H. W. Mitchell, Mr. G. Morrison, Dr. J. Needham and Mr. A. G. Woodhead for valuable help. The publishers gave me encouragement and assistance and their skill in the production of the book deserves high praise. In a book which contains so much detail I am afraid that some errors are inevitable but I have spent much trouble in trying to reduce them to a minimum. In writing this last page of my book I take leave of my subject, and I hope that I shall convey to my readers some of the interest which it had for me.

J. R. PARTINGTON.

Cambridge, May, 1959.

Introduction, 1999

A Retrospective

JAMES RIDDICK PARTINGTON (1886–1965) was a chemist. He studied chemistry at Manchester, and he continued postgraduate work in Berlin with the famous physical chemist and thermodynamicist, Hermann Nernst. By 1913, Partington was back on the faculty of his alma mater and a recognized expert on the specific heats of gases. During the First World War he worked for the Ministry of Munitions, for which services he was later awarded an M.B.E. He was appointed chair of chemistry at Queen Mary College, University of London, in 1919, a post he held until his retirement in 1951. Partington was also a prolific writer. His earliest publication, *Higher Mathematics for Chemistry Students,* appeared in 1911, and it was followed by a string of titles that came to an end only five years after his death with his monumental (and unfortunately never finished) *History of Chemistry* (4 vols. 1960–70).

Along with his deep interest in the world of chemical theory, Partington concerned himself with educating chemists and the general public as well. Even during the war he found time to publish *The Alkali Industry* (1918), a study of industrial chemistry and its effects, and during the 1920s he became justly famous for his textbooks, including *Inorganic Chemistry for University Students* (1921; 6th ed. 1961). Its successor, *General and Inorganic Chemistry for University Students,* went through four editions between 1946 and 1966. His success as a textbook writer did not deter him from publishing at a more advanced level; his massive (1.5 million words) *Advanced Treatise on Physical Chemistry* appeared in four volumes between 1949 and 1953 to the unanimous praise of his colleagues.

Yet, inevitably, the importance of such works is bound to fade as the field advances, and it is for his historical works that Partington is chiefly remembered today. He published *Origins and Development of Applied Chemistry* in 1935, and he devoted most of his retirement years to the history of chemistry. His magnum opus, the four-volume *History of Chemistry,* occupied much of this period, but he also found time to publish *The Life and Work of William Higgins, Chemist (1763–1825)* in 1960 as well as the work reprinted here, *A History of Greek Fire and Gunpowder.*

As a human being, Partington was invariably described as "intensely reserved," with a certain "asperity of manner." A widower from 1940 to the end of his life twenty-five years later, Partington moved after his retirement from London to Cambridge, the better to pursue his historical interests. Joseph Needham, who admired Partington's work immensely, admitted that the chemist became something of a recluse in Cambridge, working in a house "literally filled with books from roof to cellar" and only rarely making an appearance at Caius High Table.[1] This dry and cautious manner is apparent in his writing, as well. Partington is no storyteller; his was not the craft of weaving narrative. "His method," notes Harold Hartley in the *Dictionary of Scientific Biography*, "consisted of summarizing the successive accomplishments of contributors to chemistry, rather than organizing . . . a sequence of themes or topics" (Hartley 1974, 329).

This same approach, with its emphasis on the individual writers who make up a scientific tradition, and, of course, Partington's magisterial knowledge of chemistry, characterizes *A History of Greek Fire and Gunpowder*. Like any single approach to a complex topic, Partington's method presupposes many things. Chief among these is the belief that Greek fire and gunpowder represent premodern forms of "scientific" knowledge. That is, these were subjects with a certain scientific content that could be contained in textual form, and these texts in turn can be studied by the modern scholar with the object of forming judgments about how nearly correct this or that ancient text writer might have been. Partington's second presumptuous belief is that the history of Greek fire and gunpowder is primarily to be understood through chemistry, that the formulas for various incendiaries and explosives were the most important things we could learn about them. These beliefs were what justified Partington's biographical approach; he was interested in those who wrote texts and in what was written in them, mainly recipes and formulas. Perhaps the most familiar practitioner of this method is the founder of the modern history of science, George Sarton, whose *Introduction to the History of Science* (1927–47) Partington's work closely resembles.

The whole text-and-author approach is somewhat removed from late-twentieth-century historiographic fashion, which emphasizes practice and experience, and which sees the text as a highly imperfect witness to the craft knowledge and skills of the early practitioner. Certainly any modern treatment of the subject would give more attention to their nonchemical aspects than Partington did. On the other hand, it must be admitted that Partington was a masterful practitioner of the older approach. His method forced him to look closely and critically at a mass of textual witnesses, testing the claims of each against the standard of twentieth-century chemical knowledge. In effect, Partington imposed a probability test on each document he scrutinized. Within the limits of certainty imposed by the language of the text, is it chemically possible that what a given author claims to be true is indeed so? It is hard to imagine anyone working in such a manner today, but paradoxically, Partington's old-fashioned method is what gives his work its enduring value. His is hardly the last word, but it is certainly a necessary starting point for any further work in the field.

PROBLEMS WITH THE HISTORY OF FIREARMS

Anyone approaching the early history of explosives and firearms should do so with a sense of caution verging on outright trepidation. Few subjects

[1] Obituary, "Prof. J. R. Partington: The History of Chemistry," *The Times* (London), 11 October, 1965, p. 12; J. Needham, Letter, "Prof. J. R. Partington," *ibid.*, 16 October, 1965.

have lent themselves to such misunderstanding as the early history of gunpowder, and few have tempted historians into such flights of fancy in the search for answers. Gunpowder is not, of course, an "invention" in the modern sense, the product of a single time and place; no individual's name can be attached to it, nor can that of any single nation or region. Fire is one of the primordial forces of nature, and incendiary weapons have had a place in armies' toolkits for almost as long as civilized states have made war. Anything that will burn fiercely can be used for military purposes. In this sense, gunpowder is—in its origins, at least—less an explosive than a substance that combusts with extreme rapidity. This is not merely a figure of speech; many, if not most, early gunpowder weapons were meant to function as incendiaries, not as bombs or guns.

This fact blurs the line between *firearms* in the modern sense of the word and a much more general definition, like "arms-that-depend-on-fire." And because the lines are blurred in this way, any history of firearms must begin with a history of incendiary agents like the famous Greek fire. Any attempt to write such a history will also encounter a serious obstacle: deeply confused and confusing terminology. One of the more awkward traits that historical documentation displays in any age is a strong tendency for writers to employ words with which their readers were most familiar. New things will more often than not be labeled with old words, while new words will only gradually creep into use—if at all. When new words do make an appearance to designate new things, it is often the case that several new terms come into use simultaneously and circulate along with older terms, until finally only one or two words survive. (Unless, of course, the old terminology resists the attempt at making new words and carries on unchanged—but is then understood in a completely different sense than its original meaning.) The historian must tiptoe through such terminological minefields. It goes without saying, however, that historians bent on special pleading, or simply with axes of their own to grind, can find rich material in these terminological thickets. Incendiaries and gunpowder show these terminological problems in abundance, and the result is a history more easily confused than most.

Some of these "arms-that-depend-on-fire" may have used the key ingredient of gunpowder, "saltpeter" (any of several nitrate salts), as their oxidizing agent. Any combustible that carries its own source of oxygen is likely to burn faster and hotter than something that depends on atmospheric oxygen to sustain its burning. Adding increasing quantities of an oxidizing agent eventually tips the process of combustion over into a mild explosion. If one happens to mix finely pulverized saltpeter with equally finely pulverized charcoal and sulfur in the rough proportions six parts saltpeter to one part each of sulfur and charcoal, that mixture will burn so rapidly as to explode. Because the mixture carries ample oxygen in the $-NO_3$ or nitrate radical, it can be confined in containers that exclude atmospheric oxygen and it will still burn. When such a mixture burns in a confined space, it combusts within a few milliseconds and bursts the containment vessel. The same mixture at lower nitrate levels and in open containers will burn more slowly, making it supremely useful as an incendiary in its own right. The result, as Partington realized somewhat more clearly than many of his predecessors, is that the historian cannot make a clear distinction between explosives and incendiaries.

Because nitrates as a class of salts were the only practical sources of chemically combined oxygen available in the Middle Ages, it is tempting to focus the early history of gunpowder on the question of saltpeter: did this or that recipe employ nitrates? This turns out to be less helpful as a tactic than one would like. Once again, terminological problems abound. Early texts do not employ our chemical vocabulary, and terms like *natron* can refer to a variety of salts, only some of which are nitrates. Often, for example, *natron* means *soda* in early works. It can be very difficult to decide just what was meant by a particular expression. Partington wisely includes an entire chapter devoted to the question of saltpeter and its cognates, including the *soda* used in the soap trade. Nitrates occur in some places as geological deposits, but they occur in nature mainly as a result of the decay of organic matter through bacterial action. Deposits of guano in caves are an important natural source, one exploited throughout the nineteenth century, for example.

The story of how nitrates came to be identified and purified (isolated from other salts) is an interesting chapter in medieval chemistry, and one written largely in Chinese sources (Needham et al. 1986, 94–161). But it is important to note that there are several common nitrate salts in nature, chiefly calcium nitrate, $Ca(NO_3)_2$, also known as "lime saltpeter"; sodium saltpeter, $NaNO_3$; and potassium nitrate, KNO_3, as well as magnesium nitrate. "Black powder" was normally made in early modern Europe with saltpeter that is nearly pure potassium nitrate, KNO_3, but early gunpowder seems to have been made with a variety of different nitrate salts. Differences between these salts have an impact on the type of gunpowder made with them. "Lime saltpeter"—calcium nitrate, $Ca(NO_3)_2$—is never used in later powder-making because it is more hygroscopic than potassium nitrate, and powders made from calcium nitrate saltpeter tend to absorb moisture from the air and spoil quite readily. Sodium saltpeter was eventually used in some gunpowders in the nineteenth century, but only after adjustments in the manufacturing and storage processes had reduced the danger of spoilage. All this suggests that it may not be wise to focus too closely on the kinds of questions that suggest themselves to a historian of chemistry, questions of how nitrates were identified and purified. Gunpowder and incendiaries could be and assuredly were made from less than chemically pure nitrates, and the real issues are often those of the gunpowder manufacturing trade rather than laboratory science.

There are further difficulties with gunpowder as well. Despite its being an important part of military operations since the end of the Middle Ages, the exact mechanisms of gunpowder's combustion do not yield easily to the prying eyes of modern science. Fabricating gunpowder is at least as much an art as a science, and seemingly trivial factors—like the species of wood from which the charcoal was made that went into the gunpowder—play a significant role. To explode efficiently, gunpowder must be "incorporated," its ingredients pulverized together in a process lasting at a minimum several hours, and then "corned" by being wetted in the final pulverization stages, and pressed while wet into shaped grains of predetermined size. Both the size and the shapes of such grains have an effect on burning time and ballistic performance. These were not mere modern refinements of once primitive and more rustic processes; they were discovered in the fifteenth century and brought into practice almost immediately. Any history of early gunpowder should afford corning a major place, but such matters as wood species and grain size and shape are things that lie outside the realm of the chemist, and indeed, of the scientist. They are part of the specialized technical craft knowledge that makes up gunpowder technology.

The discovery of all this technical knowledge was the work of many individuals, often unknown to the historian, working over many centuries on different phases of the problem. Like such modern inventions as automobiles and airplanes, gunpowder was the synthesis of many developments that had gone before. Saltpeter, the key ingredient, was isolated by Chinese alchemists looking, ironically, for compounds that would bring bodily immortality. It appears in Song Dynasty formulas from the mid 800s, and these in turn are based on much older alchemical work (Needham et al. 1986, 107–17). Gunpowder proper seems to have first appeared in 1044 A.D. in China, and to have worked its way westward over the next three centuries by routes still uncertain. In the West, as in China, gunpowder was put to a variety of uses as an incendiary, for bombs and rockets, and in guns. The development of saltpeter production and refining continued to take place, as did techniques for corning gunpowder, compounding and storing it, and reviving it after spoilage.

Consider then, what the history of gunpowder involves: its roots lie buried in confusion with earlier incendiaries; it is shadowed at every turn by profound terminological difficulties; its history stretches across Eurasia and is recorded in scores of different languages, some no longer spoken, and often in fragmentary and poorly stored documents; it makes demands on chemistry and physics that are even today not fully satisfied; and it involves arcane technical operations that lie outside the realm of modern science altogether. No wonder that the foremost modern student of the subject, Joseph Needham, called gunpowder's history an "epic." No wonder too that historians whose biases were sometimes all too evident found in this mass of material ample scope for special pleading, self-serving theses, and simply bizarre ideas.

The problems represented by the older literature can hardly be exaggerated, even if we cannot possibly review them here. Few subjects have proven so intriguing, or so difficult to answer, as the origins of gunpowder. The deeply felt human need for some coherent narrative, some "creation myth," to explain the origins of this important innovation generated many different tales about gunpowder, most of them entirely untrue. Since at least the fifteenth century, people have struggled to reconstruct how it could happen that the seemingly magical combination of saltpeter, sulfur, and charcoal was discovered. Nothing could be less satisfactory than to have to admit that we know scarcely anything tangible about the real origins of gunpowder. The subject is equally compelling to the academic mind, whether the historian is of scientific thought or of military matters. In the absence of concrete facts, fanciful explanations flourished, and these so thoroughly cluttered the picture that even the best efforts of nineteenth-century scholarship could not completely clarify the matter. Moreover, nineteenth-century historiography was itself contaminated with nationalistic prejudices of the worst sort, and few subjects could have been more calculated to bring these to the fore than the question of gunpowder.

PARTINGTON'S CONTRIBUTION

When *A History of Greek Fire and Gunpowder* first appeared in 1960, it stood in refreshing contrast to the sometimes scientifically amateurish and often highly nationalistic treatments of the subject that had often been the norm earlier. Reviewers were quick to praise the book's freedom from all kinds of *ex parte* arguments. Partington's work is blessedly free from the cant of nationalist historiography. Not for him the tortured readings of texts or the assumption-laden interpretations of vague and difficult passages. Partington was the "honest broker," the skeptical inquirer whose acidic inquiry etched away the remaining layers of overlying obfuscation from the underlying evidence about gunpowder's development. To a great extent, *A History of Greek Fire and Gunpowder* shows how successful a book grounded in a skeptical methodology can be. Like a master anatomist, he cut away appearances and laid bare the bones. Quite often in the construction of biological taxonomies, knowledge of internal organs contradicts conclusions drawn from the external appearance of animals or plants. Similarly in this case, Partington's analytic insights made clear relationships that had been previously clouded.

Partington's work appears as the great and necessary corrective to many of the misinterpretations that preceded it. Mythic figures, such as the legendary Berthold Schwarz, or Black Berthold (*Bertholdus niger* in the Latin sources), the supposed inventor of gunpowder (or was it guns?), are laid to rest by the skilled and careful reading of texts in a variety of languages (Partington 1960, 91–145). The wretched question of saltpeter and gunpowder on the Indian subcontinent is reviewed, and if not settled, at least given a rational basis for discussion (Partington 1960, 211 ff). Partington's treatment of the Chinese evidence, on which he consulted with Needham, has of course been superseded by the latter's magisterial treatment, but it is apparent on close examination just how much Needham owes to Partington's lean and skeptical reading of the evidence. Partington's organization of his material along national and regional lines does not add to the work's already meager value as narrative, but it facilitates use of the book in the way most readers probably treat it—as a work of reference. *A History of Greek Fire and Gunpowder* is an indispensable handbook, the *summa historiographica* of virtually all scholarship on gunpowder's history up to 1960, and the starting point for any serious investigation in the future.

Unfortunately, however, skeptical inquiry, textual exegesis, and biographical focus carry the historian only so far in the reconstruction of the past. At some point, history requires synthesis and imagination, the ability to tell a tale. To make this claim is not to argue that all history is a form of storytelling; certainly the "thick descriptions" like those preached and practiced by members of the French *Annales* school are a valid and valuable contribution to contemporary historiography. But it is really impossible to force Partington into the *Annales* mold. His work remains outside the fold of modern historiography. His entire historiographic outlook and his whole working method made it impossible for him to take that final step toward synthesis.

He was skilled at parsing out valid evidence from the confused tangle left by earlier writers, but like many of his generation—especially scientists turned historians—Partington assumed that "the facts would speak for themselves." He felt no need to reconstruct a story of how Greek fire and gunpowder evolved. Perhaps if the evidence were indeed clear on this point, then no synthetic narrative would have been needed. But this is far from the case, and *A History of Greek Fire and Gunpowder* remains less a "history" in the conventional sense than a prolegomenon to a future history. Paradoxically once again, Partington's eschewing of any attempt at thesis-building contributed to his high posthumous reputation as an "objective" historian. But there is the risk of canonizing his opinions. Even in order to introduce a reprint of this much-sought-after work, it is necessary to highlight the areas where Partington's views have been superseded.

GREEK FIRE

An example of how Partington's skeptical conclusions have failed to command assent is found in one of the more famous chapters in the volume, that concerning the most mysterious incendiary of all, the so-called Greek fire employed by the Byzantines at sea, the weapon that shocked and even terrified the crusaders in the twelfth century. This particular weapon was first used by the Byzantine fleet at the time of the Arab attacks on Constantinople, 671–678 A.D. It was the newest in an ancient tradition of incendiaries, one that it would be more proper to call "Kallinikos fire" after Kallinikos of Heliopolis, the man who was largely responsible for building the fire-spouting naval vessels that successfully defended the city in 678.

Kallinikos fire had some distinct characteristics: first, it was a liquid, and second, it burned even on the surface of the sea. In naval engagements, it was sprayed from siphons or nozzles, and it burned with a loud roaring or whooshing noise and generated a considerable amount of smoke. Kallinikos's innovation was immensely successful, and the reputation of the "secret weapon" of the Byzantines grew with every retelling of the tale. The Byzantines did not reveal what set Kallinikos fire apart from other incendiaries, maintaining the cloak of silence until the "secret" was apparently lost in the confusion of the Fourth Crusade in 1204 (for a review of the Greek Fire question, see Roland 1992).

The chemical composition of Kallinikos fire, or Greek fire, has generated considerable speculation. Self-igniting agents like calcium phosphide or quicklime have been postulated, but the evidence against them is irrefutable. More cogent is the possibility that saltpeter might have been mixed into the liquid as an oxidizing agent. A number of French scholars argued the saltpeter hypothesis, a line of thought that culminated with the redoubtable historian and chemist, Marcellin Berthelot, whose 1893 *La chimie au moyen âge* remains the starting point for serious study. Obviously, if Greek fire contained saltpeter, then it stands as an important precursor to gunpowder, which also relied on saltpeter as an oxidizing agent; if not, then Greek fire was the culmination of an ancient tradition of incendiaries that relied on atmospheric oxygen.

Partington sought to refute the French or saltpeter theory of Greek fire. He argued instead that the "secret" lay in the use of *distilled* natural petroleum, probably thickened artificially with various resins to produce a liquid with the desired characteristics. Such a liquid would have properties that would be unlike anything found in nature; it would be rather like modern kerosene or gasoline. As to whether saltpeter was added as an oxidizing agent, Partington rendered a negative verdict. On this question, however, Partington was in the characteristically difficult position of someone who wishes to prove a negative. Saltpeter, he argued, was not a necessary ingredient if we assume the Byzantines were practicing some form of distillation that could separate lighter and heavier petroleum fractions. Moreover, he finally concluded, saltpeter could not have been added since refined saltpeter was not known in the West before 1125. In this he was relying on the opinion of E. O. von Lippmann and assuming that the same conclusion applied to the Byzantines as well (Lippmann 1923, 1: 200–210).

This is, of course, a weak argument. The Byzantines might well have known of saltpeter and kept it secret. Or, despite Partington's emphasis on refined saltpeter, Byzantine military engineers might have used impure saltpeter harvested from natural sources like guano. The question can never be answered to the satisfaction of all. What is worse, any attempt at a full resolution will always be foiled by the opacity of technological practice. Partington's assumptions, worked into the very fabric of his text, were those of a scientist, someone rooted in a culture that establishes credit through publication. In the distant past, things were different. Workshop practice was notably reticent, and Constantinople had every reason to reinforce this with a cloak of military secrecy. Faced with such challenges, Partington's effort was doomed to fail.

Although Partington's judgment on Greek fire is still frequently quoted, subsequent researchers have departed from both of his main points. Haldon and Byrne, writing in 1977, reconstructed the invention of Kallinikos as a system using *undistilled* natural petroleum kept in heated containers aboard ship. They argue that if natural petroleum seeping from the ground were collected in the cooler months and times of day, fewer of the volatile fractions would have had the chance to evaporate. They remain skeptical about whether Byzantine technicians really could have managed the distillation of petroleum. Pászthory, writing in 1986, focused on the later and more easily portable versions of the Greek fire weapon, arguing that these "fire lances" must have used a mixture with an oxidizing agent, most likely saltpeter. He cites medical texts in support of the notion that crude saltpeter was collected and stored in late antiquity and the Middle Ages, and his work points to the whooshing sound the Greek fire flame-thrower made, a sound that so impressed observers. Pászthory's account minimizes the Greek fire weapon's use as specialized naval device and presents it through the medium of the "fire lance" as a distinct precursor of the gun. Thus, the second of Partington's points, his refutation of the saltpeter theory, has also been left behind in later work. It is doubtful that the last word has been written on this subject, but clearly the argument has moved beyond the state in which Partington sought to leave it.

ROGER BACON AND ALBERTUS MAGNUS

A second instance where Partington's methodological focus led him astray concerns two the most famous medieval intellectuals, Friar Roger Bacon and Saint Albert the Great, Albertus Magnus. The problem has to do with a failure of skepticism when confronted with a subject far from the chemist's training, manuscript studies. Here the services of a philologist were needed, not those of a scientist. Both Roger Bacon and Albertus Magnus showed a lively interest in physical sciences, including some interest in "magic," and both have something of a reputation as medieval mavericks. In fact, their writings are of genuine interest as pathways into high scholasticism's views of natural law. As to whether either knew of gunpowder, and how far such knowledge might have extended, the record is more complicated. Bacon does serve as a principal primary source for gunpowder history, although he is often cited improperly in this respect. His slightly older German counterpart, Albertus Magnus, cannot be regarded as a credible witness to Western gunpowder, even though Partington seems inclined to accept him in this role.

Bacon's is the more complex case, and we should look at it first (on Bacon, see Hackett 1982, 35–42). To begin, there are undoubtedly genuine references to gunpowder in works that are indisputably by Roger Bacon, and Partington does cite them. In the so-called *Opus maius* and again in the *Opus tertium* (both probably from the 1260s), Bacon mentions children's toys produced in many other parts of the world, where something the size of a thumb bursts with a roar greater than that of thunder and a flash greater than that of lightning. As we now know that China was already producing firecrackers for religious and ceremonial purposes by the mid-thirteenth century, there can be little reason to doubt that Roger Bacon had seen firecrackers explode or at the very least that he had heard eyewitness accounts of the experience.

Beyond this secondhand knowledge lies the question of how much more Bacon might have known. Some scholars have sought to credit Bacon with a far more profound acquaintance with gunpowder, including a complete recipe for compounding it. The "evidence" for this comes from one of his last and most obscure works, the *Letter on the Secret Workings of Art and Nature, and on the Vanity of Magic*. There is little agreement among scholars on the validity of this work; most consider some parts genuine and others not, and most assign different dates to various parts of the *Letter*. This is usually enough to frighten away scholars trained in historical philology, but of course it merely makes the whole text more attractive to others who do not see the pitfalls.

The *Letter*, like the undoubtedly genuine *Opus maius* and *Opus tertium*, mentions firecrackers once again, together with some expressions of concern about how possible "greater horrors" might come if they were made larger. Elsewhere there are many obscurities, and some of these ambiguous passages appear only in manuscripts of suspiciously late date (well after the spread of gunpowder throughout Europe). These variant texts speak of *calcem* ("chalk") and the "Stone of Aristotle." Col. Henry Hime, himself an artillerist and author of several works on early artillery and gunpowder, convinced himself in 1904 that this was a genuine reference to saltpeter refining. He went on to extort from the text a formula for gunpowder by taking a series of characters probably meant as a transliteration of a Greek phrase as an anagram that concealed the formula for gunpowder.[2] Hime's version of "Roger Bacon's gunpowder formula" has continued to generate interest and controversy in the literature ever since.

It seems strange that Partington, a skeptic and a man of considerable intellectual rigor, would have bought Hime's load of rubbish, but he did. After reviewing the evidence he concludes, against the opinions of Bacon scholars and editors, that the whole of the *Letter* is genuine, and also that Hime's reconstruction is "reasonable and sensitive" (Partington 1960, 76). Anagrams should always stir the suspicion of historians, since they can be made to say almost anything the investigator wants them to say, and there can be no doubt that Hime very much wanted his fellow countryman Bacon to have detailed knowledge of gunpowder. Also, the manuscripts containing the suspect passages are all far too late to stand as credible early witnesses. It is questionable whether Partington, whose central expertise lay elsewhere, should really have challenged the judgments of trained philologists and paleographers on such points, but he did. Moreover, even if one puts aside such objections, the Hime formula yields a gunpowder far too weak in nitrates (about 41 percent) to explode at all. Both Hime the artillery expert and Partington the chemist were fully capable of recognizing this elementary difficulty, and yet both seem to have passed over it in silence.

The case of Albertus Magnus is more easily dealt with (on Albertus, see Weisheipl 1982, 126–30). Although Albert of Bollstatt was a genuine historical person and a saint who lived from 1193 to 1280, the only passages in all his voluminous writings that mention gunpowder are found in a spurious and doubtful work to which his name was attached after his death, *On the Wonders of the World*. Most scholars now doubt that Albert had anything to do with the collection of materials that circulated under this title—even though it was an

---

[2] Hime (1915) repeating and extending his original thesis of 1904.

exceptionally popular work and one that later contemporaries fully accepted as genuine. Modern speculation would assign authorship to Arnold of Villanova, who was active a generation after Albert's death, or to another Albert, Albert of Saxony, circa. 1350 or later. Even more damning is the knowledge that the recipes attributed to Albert are very similar to recipes found in another late-thirteenth-century text, the so-called *Liber ignium* by the pseudonymous "Marcus Graecus." Partington was aware of this. In fact, his discussion of *Liber ignium* is one of the most valuable parts of his book, the starting point for all future discussion. Partington even admitted that the recipes in Albertus "may have been taken from [*Liber ignium*] rather than conversely." Today, most scholars would agree with that remark, and some would wish that Partington had taken his own suggestion more to heart. The result would have been a far more coherent discussion of thirteenth-century knowledge of gunpowder (see Foley and Perry 1979).

SALTPETER

As noted above, the problem of saltpeter runs through the early history of gunpowder like a scarlet thread. One of the problems that shapes this history is rooted in the chemistry of refining saltpeter. Despite his expert knowledge of chemistry, this is also an issue to which Partington gives what seems today like an inadequate answer. Potassium nitrate was the principal oxidizing agent that gunpowder makers preferred to use throughout the era of black powder. Other nitrates, calcium, sodium, and even magnesium nitrate, would make a workable gunpowder; the invention of gunpowder did not depend on the ability to segregate these closely related salts. Potassium nitrate's chief advantage lay in the storage of gunpowder, for it was the least water-absorbing of the common nitrates, and offered the quartermaster the product with the longest possible shelf life. Keep in mind that all gunpowder eventually spoils with exposure to air; it is only a question of how long one's own powder supplies will remain useful under whatever local conditions happen to prevail. As it has been known since the pioneering work of Serge Winogradsky, who first discovered the bacteria that produce nitrates, calcium or magnesium nitrate has the greatest share of nitrate output during decomposition and microbial breakdown of organic matter (Hall 1997, 74–79).

Since all saltpeter begins as the waste product of these bacteria, it follows that virtually all crude saltpeter begins as calcium or magnesium nitrate. What steps does the saltpeter maker take to convert these nitrates to potassium nitrate? The standard answer to this question is found in the *Pirotechnia* (1540), the magnificent and reliable technical compendium by Vannoccio Biringuccio. He describes mixing nitrate salts in water with a mixture of quicklime and wood ashes (Biringuccio [1540] 1959, 406–7). The available nitrates react with potash (K_2CO_3, formed from the wood ashes) in an alkaline solution to form calcium carbonate (limestone: $CaCO_3$) or magnesium carbonate (magnesite: $MgCO_3$), both highly insoluble salts, leaving potassium nitrate (KNO_3) in solution. Biringuccio's method was the commonly accepted means of producing potassium saltpeter throughout the early modern era.

But when did this method first appear? Partington answers this by refer-ring to the late-thirteenth-century Syrian writer, Hasan al-Rammah Najm al-Din al-Ahdab, whose *Treatise on Horsemanship and Stratagems of War* contains a recipe for using a wood ash or charcoal additive in making *barud* (saltpeter). Al-Rammah's text was discussed in detail by nineteenth-century Arabists, mainly in France, and the text is available in a French translation from which Partington seems to have worked.[3] Partington's emended translation seems to have been based on what he wanted the text to say. I have not been able to learn that Partington knew Arabic, nor does he claim to have consulted the manuscripts, which are housed in Paris (Partington 1960, 200 ff.; see also Hall 1996). There can be no doubt that the *Treatise on Horsemanship* as a whole is genuine, and it is remarkable testimony to al-Rammah's knowledge of Chinese practices in regard to the preparation of gunpowder and firearms. Unfortunately the recipe for purifying saltpeter is plainly defective in the available extant copies, and its exact meaning is extremely unclear.[4]

Indeed, even in Partington's version of the text, al-Rammah's recipe would not yield potassium saltpeter. Al-Rammah specifies willow charcoal, tra-ditionally one of the best charcoals for making gunpowder, but not a wood especially high in potassium. Biringuccio, by contrast, specifies oak ash, which is rich in potassium. Moreover, nothing in al-Rammah would supply the alkaline material needed to drive the reaction, as in the Biringuccio. There may be some garbled form of "purification" of saltpeter which al-Rammah's recipe is meant to express, but it is clearly not a method for mak-ing potassium nitrate, even though Partington assumes that it is.

This is a critical point on two separate fronts. As Gerhard Kramer (1995) has pointed out, many features of the later gunpowder recipe literature in Europe make sense only if we realize that the gunpowder is made with always-problematic calcium saltpeter, or at least saltpeter that was contaminated with excessive amounts of calcium nitrate. Why would this be true if al-Rammah had recorded a universally known method for eliminating calcium nitrate? Kramer also adds that the Chinese texts translated by Needham appear to be describing lime saltpeter along with potassium nitrate, raising the possibility that Chinese practice did not, in fact, include the deliberate reaction of calci-um nitrate with potash to create potassium nitrate. Needham is of course now dead and unable to reply, but it must be pointed out that he relies on Part-ington's interpretation of the al-Rammah text as his chief witness for Chinese practice in this regard. This may be justified by al-Rammah's remarkable knowledge of Chinese practices, but it is open to criticism where there is a lack of corroborating evidence. Is it possible that the potash method of pro-ducing potassium saltpeter may have been developed only late in the Middle Ages? The entire saltpeter question is still very much up in the air, and there may be much more to say on it before the matter is at last settled.

---

[3] Quatremère 1850, 224 ff.; Reinaud and Favé 1849. See also Hime 1915, 19 ff.
[4] I have reached this conclusion with the assistance of Ms. Muna Salloum of the Institute for the History and Philosophy of Science and Technology at the University of Toronto after consulting the published versions of al-Rammah's text.

CORNING

The final issue in gunpowder's history is corning. This is more a question of development than etiology, and it is one that has more to do with physics than with chemistry. Partington, whose *History* is very much a chemist's history of gunpowder, may be forgiven for having overlooked it. When did corning gunpowder begin to be practiced? Keep in mind that the mere mixing of the chemicals that make up gunpowder will produce a weak explosive at best. Gunpowder is a mechanical mixture of the three ingredients, saltpeter, sulfur, and charcoal. In order to burn, these have to be mixed intimately, which is achieved by grinding them together to a very fine powder for a long period of time. Eighteenth-century French treatises on how to produce the highest quality sporting gunpowders stipulated some twelve hours of uninterrupted hand grinding in a mortar and pestle and noted that the resultant mix should be as fine as women's face powder, or ground talc. Again there is no reason to assume that this was merely a later refinement of gunpowder-making; fifteenth-century evidence suggests many of the same practices were already in place.

The point of this seemingly obsessive attention was to force the small particles of saltpeter and sulfur to adhere to the micropores in the pulverized charcoal. (And this is the reason that certain species of wood make better gunpowder charcoal than others: it has to do with their porosity.) Corning was invented when small amounts of water or other liquid were added to the mix during the later stages of grinding. Saltpeter, by far the most soluble of the three ingredients, would partially dissolve and coat the inner micropore surfaces of the insoluble charcoal, carrying insoluble sulfur along with it. The pasty mass of wet gunpowder could be compressed and forced through a sieve to create granules of homogeneous composition and shape.

Gunpowder that had been made into paste and dried had two characteristics that set it apart from dry-mixed or "serpentine" gunpowders. Corned powders burned faster and thus were ballistically "stronger" for any given recipe, and they also had a longer shelf life. Early fifteenth-century texts describing the corning process as it was then practiced also comment on their superior storage qualities (see Hall 1997, 68–74). The effects of corning were remarkable. On one hand, it made possible the development of the first effective portable or shoulder arms, the *Hackenbüchse* or arquebus. Corned powder was so powerful that it was perfect for small arms. Within thirty years of corned powder's appearance, that is, by the middle of the fifteenth century, the ancestors of the musket had already been invented. The arquebus is intelligible only in terms of the new powder, which made it possible to generate supersonic missile velocities in small caliber weapons.

On the other hand, large artillery was slower to benefit from corning. The new gunpowder posed a challenge to existing guns, which were too weak to contain its higher pressures safely. Exploding guns were a constant threat and corned powder made the danger even greater. Adapting to the new situation meant redesigning artillery along lines never before seen. Barrels grew longer, and they had to be made from more robust materials, usually bronze.

To work optimally, these guns fired shot made of high density cast iron rather than stone. The gun designs that appeared in the second half of the fifteenth century were the recognizable prototypes for all smooth bore artillery used to the time of Napoleon and the American Civil War. They also employed a refinement of the corning technique; they used coarsely granulated gunpowder, the grains large enough to retard the speed of the gunpowder's combustion, thus sparing the gun the greatest strain on its breech at the moment of firing. Specialized powders served specialized guns.

All this suggests that corning was an important turning point in the history of guns and gunpowder. Accordingly, one would expect some discussion of its ancestry in Partington, even though he was more interested in the chemistry of incendiaries than their physics. Partington offers very little, however, nothing more than warmed-over versions of nineteenth-century scholarship about *Feuerwerkbuch*, the fifteenth-century text that first describes corning (see Hall 1996, 87–120). In the absence of any further positive information, an interesting question emerges, which still remains open: did the Chinese invent corning, or was it a European addition to the repertoire of gunpowder techniques? Certainly Needham follows Partington in treating corning as being of only peripheral interest, but both men also seem to assume that the Chinese practiced corning from an early date. The only evidence either scholar cites for corning in East Asia, however, stems from Chinese texts composed in the mid-sixteenth and seventeenth centuries. These discuss Western guns, muskets introduced by the Portuguese, and the types of gunpowder appropriate to them, that is, corned gunpowders (Partington 1960, 252–54; Needham et al. 1986, 358–59). Once again, the problem of trying to prove a negative arises. Additional suggestive evidence in the form of later Western observers' reports that Chinese gunpowder was usually badly corned and that Chinese miners used uncorned powder for blasting in mercury mining (Partington 1960, 285; Needham et al. 1986, 543) argues that corned gunpowder was not the normal Chinese practice. Whatever the ultimate answer to the question of Chinese corned gunpowder, it is clear that, thanks to the evidence elucidated by Partington and Needham, scholarship has now reached a point where new and very interesting questions can now be posed.

CONCLUSION

The raising of further questions is, of course, very much the point of having books like Partington's *History of Greek Fire and Gunpowder* available once again. There are works which, however flawed, represent milestones in the development of scholarship and occupy an important niche. Partington carved out for himself an indispensable place in the scheme of things. He is the trustworthy guide, the Moses who led scholars out of the wilderness without actually entering the Promised Land. Partington's Cambridge confrere, Joseph Needham, whose monumental *Science and Civilisation in China* contains the definitive history of oriental gunpowder, depends on the chemist's earlier work at every turn, while adding Chinese evidence that was unavailable when Partington was writing. Needham complements Partington, making the kind of broad, sweeping, interpretive generalizations that Partington avoided. Needham and Partington together have laid the groundwork for all further effort in the early history of incendiaries, explosives, and propellants. Everyone who researches and writes on this subject will be forever in their debt.

Bert S. Hall

References

Biringuccio, Vannoccio. [1540] 1959. *The Pirotechnia.* 2d. ed. Edited and translated by C. S. Smith and M. Gnudi. New York: Basic Books.

Foley, Vernard and Keith Perry. 1979. In defense of *Liber Igneum* [*sic*]: Arab alchemy, Roger Bacon, and the introduction of gunpowder into the West. *Journal for the History of Arabic Science* 3: 200–218.

Hackett, Jeremiah M. G. 1982. Bacon, Roger. In *Dictionary of the Middle Ages.* Vol. 2. Edited by J. R. Strayer. New York: Charles Scribner's Sons.

Haldon, J. and M. Byrne. 1977. A possible solution to the problem of Greek fire. *Byzantinische Zeitschrift* 70: 91–99.

Hall, Bert S. 1996. "The corning of gunpowder and the use of firearms in the Renaissance. In *Gunpowder: The History of an International Technology,* edited by B. Buchanan. Bath: University of Bath Press.

Hall, Bert S. 1997. *Weapons and Warfare in Renaissance Europe: Gunpowder, Tactics, and Technology.* Baltimore: Johns Hopkins University Press.

Hartley, Harold. 1974. Partington, James Riddick. In *Dictionary of Scientific Biography.* Vol. 10. Edited by C. C. Gillispie *et al.* New York: Charles Scribner's Sons.

Hime, Henry W. L. 1915. *The Origin of Artillery.* London: Longmans Green.

Kramer, Gerhard W. 1995. *Berthold Schwarz: Chemie und Waffentechnik im 15. Jahrhundert.* Abhandlungen und Berichte des Deutschen Museums, N.F. Bd. 10. Munich: Oldenbourg.

Lippmann, E. O. von. 1923. *Beiträge zur Geschichte der Naturwissenschaften und der Technik.* 2 vols. Berlin: Springer.

Needham, Joseph, Ho Ping Yü, Lu Gwei-Djen, and Wang Ling. 1986. *Science and Civilisation in China.* Vol. 5: *Chemistry and Chemical Technology,* pt. 7. Cambridge: Cambridge University Press.

Partington, J. R. 1960. *A History of Greek Fire and Gunpowder.* Cambridge: W. Heffer and Sons.

Pászthory, Emmerich. 1986. Über das 'Griechische Feuer': Die Analyse eines spätantiken Waffensystem. *Antike Welt* 17, no. 2: 27–36.

Quatremère, E. M. 1850. Observation sur le feu grégois. *Journal asiatique* 4e sér. 15: 214–74.

Reinaud, Joseph Toussaint and Ildéfonce Favé. 1849. Du feu grégois, des feux de guerre, et les origines del la poudre à canon chez les Arabes, les Persans, et les Chinois. *Journal asiatique* 4e sér. 14: 257–327.

Weisheipl, James. 1982. Albertus Magnus. In *Dictionary of the Middle Ages.* Vol. 1 Edited by J. R. Strayer. New York: Charles Scribner's Sons.

Roland, Alex. 1992. Secrecy, technology, and war: Greek fire and the defense of Byzantium, 678–1204. *Technology and Culture* 33: 655–79.

Sarton, George. 1927–47. *Introduction to the History of Science.* 3 vols. Baltimore: Williams and Wilkins.

List of Abbreviations

Abhl. K. Ges. Göttingen, Phil.-hist. Kl. Abhandlungen der Königlichen Gesellschaft der Wissenschaften zu Göttingen, Philologisch-historische Klasse.*

Abhl. K. Preuss. Akad. Wiss., Phil.-hist. Kl. Abhandlungen der Königlichen Preussischen Akademie der Wissenschaften, Philosophisch-historische Klasse, Berlin.

Ann. Chim. Annales de Chimie, Paris.

Archaeological J. Archaeological Journal.

Arch. f. d. Geschichte der Naturwissenschaften und der Technik. Archiv für die Geschichte der Naturwissenschaften und der Technik.

Arch. Gesch. Math. Naturwiss. u. d. Technik. Archiv für die Geschichte der Mathematik, der Naturwissenschaften und der Technik.

Aretaios. *De diuturn. morb.* Aretaios. De Acutorum et Diuturnorum Morborum.

Aristotle: (*a*) *De Mirab. (auscult.).* De Mirabilibus Auscultationibus. (*b*) *Meteor.* Meteorologica. (*c*) *Probl.* Problemata.

Atti R. Accad. Lincei Classe Sci. Morali, Storiche e Filologische. Atti della Reale Accademia dei Lincei, Memoria della Classe di Scienze Morali, Storiche e Filologiche, Rome.

BM. British Museum Library.

BN. Bibliothèque Nationale, Paris.

Bibl. École des Chartes. Bibliothèque de l'École des Chartes, Paris.

Bodl. Bodleian Library, Oxford.

Brit. Assoc. Rep. British Association for the Advancement of Science, Annual Report.

Bull. des Études Arabes. Bulletin des Études Arabes.

Bull. des Études Orientales. Bulletin des Études Orientales.

Bull. Inst. Hist. Medicine. Bulletin of the Institute of the History of Medicine, Baltimore.

Bull. Metropolitan Museum of Art. Bulletin of the Metropolitan Museum of Art, New York.

CUL. Cambridge University Library.

Chem. News. The Chemical News.

Chem. Ztg. Chemiker-Zeitung. Cöthen.

Corpus Script. Hist. Byzant. Corpus Scriptorum Historiæ Byzantinæ, ed. Niebuhr, Bonn.

Crell's *Annalen.* Chemische Annalen, ed. L. F. F. von Crell, Helmstädt (two vols., numbered I and II, annually from 1784 to 1803).

D.N.B. Dictionary of National Biography.

Dict. Dictionary.

Dioskourides. *Mat. Med.* Dioskourides (Dioscorides), De Materia Medica (quoted in book and chapter from the edition of J. A. Saracenus, 4°, Frankfurt, 1598).

Doukas. *Hist. Byzant.* Doukas (Ducas), Historia Byzantina.

---

* Some academies and publications have changed their names. The references are to those in use at the times for which they are quoted.

Ency. Islam. Encylclopædia of Islam.
English Histor. Rev. English Historical Review.

f., ff., folio, folios; f°, folio size of book.

Galen (*a*) *De compos. med. sec. loc.* De Compositione Medicamentorum secundum Locos, (*b*) *De simpl. med. fac.* De Temperamentis et Facultatibus Simplicium Medicamentorum. (*c*) *Method. Medend.* Medendi Methodus. (Galen is quoted by volume and page of Kühn's edition, 20 vols., 1821-33).

Geological Soc. America Spec. Paper. Geological Society of America, Special Papers, New York.
Göttingen gelehrt. Anzeign. Goettingische gelehrte Anzeigen, Göttingen.

H.N. see Pliny.
Handbuch der Röm. Altert. Handbuch der römischen Alterthümer, ed. J. Marquardt and T. Mommsen, Leipzig, 1887.
Hist. J. Peiping Nat. Acad. Historical Journal of the Peiping National Academy.

Ind. Eng. Chem. Industrial and Engineering Chemistry, Easton, Pa.
Isidore of Seville. *Orig.* Isidore of Seville (*c.* A.D. 560-636), Etymologiarum sive Originum Libri XX (q. from Lindsay's ed., Oxford, 1911).

J. Asiat. Journal Asiatique, Paris.
J. Asiat. Soc. Bengal. Journal of the Asiatic Society of Bengal.
J. Chem. Educ. Journal of Chemical Education, Easton, Pa.
J. Chem. Soc. Journal of the Chemical Society, London.
J. des Savants. Journal des Savants.
J. Hellenic Studies. Journal of Hellenic Studies.
J. North China Branch Roy. Asiat. Soc. Journal of the North China Branch, Royal Asiatic Society.
J. and Proc. Asiatic Soc. Bengal. Journal and Proceedings of the Asiatic Society of Bengal.
J. Roy. Asiatic Soc. Journal of the Royal Asiatic Society.
J. Soc. Chem. Ind. Journal of the Society of Chemical Industry, London.

Kedrenos. *Compen. Hist.* Kedrenos (Cedrenus), Compendium Historiarum.
Kgl. Königliche.
Konstantinos VII. *De Administr. Imp.* Konstantinos (Constantinus Porphyrogenitus), De Administrando Imperio.

Liebig's *Annalen.* Annalen der Chemie, ed. J. Liebig.

Mem., Mém. Memoirs, Mémoires.
Mém. Acad. Inscriptions. Mémoires de l'Académie des Inscriptions, Paris.
Mém. Acad. Imp. des Sciences de St. Pétersbourg. Mémoires de l'Académie Impérial des Sciences de St. Pétersbourg (Leningrad).
Mém. Acad. Sci. Mémoires de l'Académie des Sciences, Paris (to 1798 Royale, and various titles since).
Mem. Asiatic Soc. of Bengal. Memoirs of the Asiatic Society of Bengal.

Mém. couronnées Acad. Roy. Belg. Mémoires couronnées et autres Mémoires, Académie Royale de Belgique, Brussels.

Mém. de l'Inst. d'Égypte. Mémoires de l'Institut d'Égypte, Cairo.

Mém. div. Sav. Acad. des Inscriptions et Belles-Lettres. Mémoires présentées par divers Savants, Académie des Inscriptions et des Belles-Lettres, Paris.

Mém. div. Sav. Acad. Sci. Mémoires présentées par divers Savants, Académie des Sciences, Paris.

Mem. Manchester Lit. and Phil. Soc. Memoirs of the Manchester Literary and Philosophical Society.

Nicholson's J. Journal of Natural Philosophy, Chemistry, and the Arts, ed. Nicholson, London.

Obs. Phys. Observations sur la Physique, sur l'Histoire Naturelle et sur les Arts, ed. Rozier, Paris.

Phil. Mag. The Philosophical Magazine, London.

Phil. Trans. Philosophical Transactions of the Royal Society of London.

Philologus Suppl. Philologus Supplement.

Plutarch. (*a*) *De orac. defect.* De defectu oraculorum. (*b*) *Quæst. conviv.* Quæstioniones conviviales.

Pliny, *H.N.* Pliny, Historia Naturalium. (Quoted by book and chapter according to Hardouin's edition.)

Proc. Amer. Acad. Proceedings of the American Academy of Arts and Sciences, Boston.

Proc. British Acad. Proceedings of the British Academy.

Proc. Nat. Acad. Sci. India. Proceedings of the National Academy of Sciences of India.

Proc. R. Artillery Inst. Woolwich. Proceedings of the Royal Artillery Institute, Woolwich.

Prokopios, *Bell. Goth.* Prokopios (Procopius) De Bello Italico adversus Gothos.

Quart. J. of Science, Literature and the Arts. Quarterly Journal of Science, Literature and the Arts, publ. by the Royal Institution of Great Britain.

Quellen u. Studien z. Gesch. Naturwiss. u. Med. Quellen und Studien zur Geschichte der Naturwissenschaften und der Medizin, Berlin.

r recto (right hand page).

R.I. Acad. Royal Irish Academy, Dublin.

Rev. Deux Mondes. Revue des Deux Mondes, Paris.

Rev. Scientif. Revue Scientifique, Paris.

Sammlung chem.-und chem.-techn. Vorträge. Sammlung chemischer und chemisch-technischer Vorträge, Stuttgart.

Scribonius Largus, *De compos. med.* Scribonius Largus, De Compositione Medicamentorum.

Seneca, *Quæst. Nat.* Seneca, Quæstionum Naturalium.

Sitzungsber. (Sitzb.) königl. bayer. Akad. Wiss., Philos.-philol. Classe. Sitzungsberichte der Königlichen Bayerischen Akademie der Wissenschaften, Philosophisch-philologische Classe, Munich.

Sitzungsber. (*Sitzb.*) *Phys.-Med. Soc. Erlangen.* Sitzungsberichten der physikalisch-medizinischen Societät in Erlangen.

Sussex Archaeol. Collect. Sussex Archaeological Collections, Lewes.

Theophrastos (*a*) *Caus. plant.* De Causis Plantarum. (*b*) *Hist. plant.* Historia Plantarum. (*c*) *Odor.* De Odoribus.

Trans. Amer. Phil. Soc. Transactions of the American Philosophical Society, Philadelphia.

Trans. Korea Branch Roy. Asiat. Soc. Transactions of the Korea Branch, Royal Asiatic Society.

Trans. Newcomen Soc. Transactions of the Newcomen Society for the Study of the History of Engineering and Technology.

Trans. Roy. Irish Acad. Transactions of the Royal Irish Academy, Dublin.

v verso (left hand page).

Z. angew. Chem. Zeitschrift für angewandte Chemie.

Z. f. Assyriologie. Zeitschrift für Assyriologie.

Z. classische Philologie. Zeitschrift für classische Philologie.

Z. für die gesamte Staatswissenschaft. Zeitschrift für die gesamte Staatswissenschaft.

Z. f. historische Waffen-und Kostümkunde. Zeitschrift für die historische Waffen-und Kostümkunde.

Z. Naturwiss. Zeitschrift für die Naturwissenschaften.

Chapter I

INCENDIARIES IN WARFARE

ASSYRIAN bas-reliefs show that incendiaries were used in sieges of towns in the ninth century B.C., torches, lighted tow, burning pitch, and fire-pots being thrown down on the attacking troops.[1]* The zikkim in the Old Testament[2] *may* have been incendiary arrows. Herodotos[3] says the Persians used arrows tipped with burning tow in the capture of Athens by Xerxes (480 B.C.). The first recorded use of incendiary arrows ($\pi\nu\rho\phi\acute{o}\rho\omega\iota$ $\ddot{o}\ddot{\iota}\sigma\tau\omicron\acute{\iota}$) by the Greeks was in the Peloponnesian war (429 B.C.), the wooden walls of Plataia being protected against them by skins.[4] Catapults (ballistæ) for throwing stones were said to have been invented under Dionysios in Syracuse about 400 B.C. He assembled a large number of engineers from Greece, Italy and Carthage, and gave them every encouragement in their work.[5] Fireships and resinous torches were used in the sieges of Syracuse (413 B.C.)[6] and Rhodes (304 B.C.).[7] Aineias (Æneas) (*c.* 360 B.C.)[8] describes the production of a violent fire ($\pi\hat{\nu}\rho$ $\sigma\kappa\epsilon\nu\acute{a}\zeta\epsilon\iota\nu$ $\iota\sigma\chi\nu\rho\grave{o}\nu$) by the use of pots filled with a mixture of pitch, sulphur, pine-shavings, and incense or resin ($\mu\acute{a}\nu\nu a$); this incendiary mixture could also, he says, be attached to large wooden pestles ($\ddot{\nu}\pi\epsilon\rho a$) fitted with iron hooks at both ends and thrown on the wooden decks of ships, to which they attached themselves, or on to the wooden protections of besieging troops. The fire was not extinguished by water (which would not wet the material) but only by vinegar ($\sigma\beta\epsilon\nu\nu\acute{\nu}\epsilon\iota\nu$ $\chi\rho\grave{\eta}$ $a\grave{\nu}\tau\grave{o}$ $\ddot{o}\xi\epsilon\iota$). Aineias does not mention incendiary arrows.

The word $\mu\acute{a}\nu\nu a$, besides having the meaning of "manna," was also used for the resin exuding from the cedar tree ($\kappa\acute{\epsilon}\delta\rho\iota\nu o\nu$ $\mu\acute{\epsilon}\lambda\iota$: resinarum aliæ oleo similiores, aliæ melli), as well as incense (grana seu micæ thuris concussu).[9] The word $\acute{o}$ $\ddot{\nu}\pi\epsilon\rho o\varsigma$ in Hesiod[10] means a wooden hour-glass shaped pestle, used for pounding corn in a mortar, and of length up to about 3 or 4 yards.

All modern savage races which use the bow appear to have used incendiary arrows.[11] The shooting of incendiary missiles by Greek troops became common only after the death of Alexander the Great (323 B.C.).[12] Incendiary arrows and fire-pots are mentioned as used by both sides in the siege of Rhodes (304 B.C.).[13] Arrian[14] reported that in the siege of Tyre (332 B.C.) the Phœnicians made up a fireship with two masts at the prow, each with a projecting arm, attached to either being a cauldron filled with bitumen, sulphur, and "every sort of material apt to kindle and nourish flame." In the forepart of the ship were torches, resin and other combustibles, the hold being filled with dry brushwood and other inflammable materials. They sent this ship against the mole erected by the Macedonians to join the island to the mainland, and burnt it, in spite of its defence by two movable towers, protected against torches and weapons by curtains of raw hides, and containing engines for projecting darts and stones of large size. Thucydides[15]

* For references, see p. 32.

1

says the Boiotians in the siege of Delion (424 B.C.), in destroying fortifications which were partly of wood, used a long tube made from hollow sailyard and iron, moved on wheels, carrying a vessel containing burning charcoal, sulphur and pitch, and behind this large bellows (φῦσαι μεγάλαι), the blast from which blew the flame forward: ἐς τὸν λέβητα, ἔχοντα ἄνθρακάς τε ἡμμένους καὶ θεῖον καὶ πίσσαν, φλόγα ἐποίει μεγάλην καὶ ἦψε τοῦ τείχους. A similar apparatus using powdered coal (presumably charcoal) (ἄνθραξ λεπτός) only, was described by Apollodoros[16] (b. Damascus, the architect of Hadrian, A.D. 117–38), who says the flame may be directed on stone walls, which then crack when vinegar or other acid is poured on them: πῦρ δὲ λαβὼν ὁ ἄνθραξ ἅπτεται καὶ ἐμφυσώμενος πληγὴν ὁμοίαν ἐργάζεται φλογὶ καὶ ἐπεμβαίνει τῷ λίθῳ καὶ θρύπτει ἢ ὄξους ἢ ἄλλου τινὸς τῶν δριμέων ἐπιχεομένου; if the coal is kindled and blown it acts just like an open flame; it penetrates the stone and if it is splashed with vinegar or other acid, the fire softens it.

This is perhaps the first mention of powdered-coal apparatus. Similar apparatus is described by Heron of Byzantium (tenth century A.D.).[17] The report of the use of vinegar by Hannibal in 218 B.C. in removing rocks in crossing the Alps,[18] if it is not mythical, probably refers to a similar process. Pliny,[19] who mentions the passage of the Alps by Hannibal as a prodigy, says[20] that vinegar when poured upon rocks in considerable quantity has the effect of splitting them (saxa rumpit infusam) when fire alone has no effect; he does not mention the story of Hannibal given by Livy. Hennebert[21] suggested that this vinegar (ὄξος, acetum) was some explosive composition. S. J. von Romocki[22] quotes Lucretius as saying that vinegar dissolves rocks (rupes dissolvit aceto), but gives no reference.

Incendiary arrows were well thrown to Vergil (70–19 B.C.)[23] and Livy (59 B.C.–A.D. 17).[24] Tacitus[25] speaks of fire-lances (ardentes hastæ) thrown by machines (adactæ tormentis).[26] Ammianus Marcellinus (c. A.D. 390)[27] describes fire-arrows (malleoli), not extinguished by water but made to flame more, and put out only by sand, and says they must be shot slowly, otherwise they are extinguished:

> Malleoli autem, teli genus, figurantur hac specie. Sagitta est cannea inter spiculum et arundinem multifido ferro coagmentata, quæ in muliebris coli formam, quo nentur lintea stamina, concavatur ventre subtiliter et plurifariam patens atque in alveo ipso ignem cum aliquo suscipit alimento. Et si emissa lentius arcu invalido (ictu enim rapidiore extinguitur) hæserit usquam, tenaciter cremat aquisque conspersa acriores excitat æstus incendiorum, nec remedio ullo quam superiacto pulvere consopitur.

Vegetius (c. A.D. 380–90)[28] gives the composition of the incendiary mixture contained between the envelope (tubus) and shaft (hastile) as sulphur, resin, bitumen, and tow soaked in incendiary oil (petroleum):

> Falarica autem ad modum hastæ valido præfigitur ferro; inter tubum etiam et hastile sulphure, resina, bitumine stuppisque convolaritur infusa oleo, quod incendiarium vocant; bitumen, sulphur, picem liquidam, oleum quod incendiarium vocant.

The falarica (or phalarica, named from the falæ or wooden towers used in sieges, from which it was discharged by an engine) was a large spear with an iron head up to 3 ft. in length carrying flaming pitch and tow.[29]   Aristotle[30] said that lead shots flying through air become so hot that they melt.[31]

PETROLEUM IN ANTIQUITY

Pliny[32] says that in Samosata (on the Euphrates) there is a pool discharging an inflammable mud called maltha, which adheres strongly to every solid. The people defended their walls against Lucullus with it and burned soldiers in their armour.   It is even set on fire in water and is extinguished only by earth.   Naphtha is similar, flowing like liquid bitumen.   It is so called about Babylon and in the territory of the Astaceni in Parthia.   It has a great affinity for fire, which darts on to it instantly.   In this way, it is said, Medea burned Jason's mistress, her crown taking fire as she approached the altar for sacrificing.

Prokopios (A.D. 550)[33] says Median naphtha was called by the Greeks oil of Medea : ἀγγεῖα δε θείου τε καὶ ἀσφάλτου ἐμπλησάμενοι καὶ φαρμάκου, ὅπερ Μῆδοι μὲν νάφθαν καλοῦσιν, Ἕλληνες δὲ Μηδείας ἔλαιον, πυρί τε ταῦτα ὑφάψαντες ἐπὶ τὰς μηχανὰς τῶν κριῶν ἔβαλλον; vascula sulphure, bitumine et veneno, quod Medi Naphthan, Græci Medeæ oleum vocant, plena, atque igne concepta ardentia, in arietarias machinas mittebant.   The word νάφθα, naphtha (used only by later authors such as Dioskourides and Strabo) is the Persian naft, petroleum, such derivations[34] as "de nare, nager, et phtha, feu, synonyme de Vulcain," being fanciful.   Kinnamos[35] also speaks of πῦρ Μηδικόν.

Medeia (Μήδεια) lived in Kolchis, between the Black and Caspian Seas (a petroleum region), and in Greek legend she sent to Glauke a magic garment which, when she put it on, burst into flames and consumed her.   Some knowledge of the petroleum of Baku may underlie the story.   Otherwise, Media (Μηδία), the country between the Caspian Sea and Mesopotamia, may be meant, and the petroleum in this district was well-known in antiquity.[36] The earlier Greeks may have had some knowledge of it, but in any case it was known in the time of Alexander the Great, who captured Babylon in 324 B.C.   The "oleum Mydiacon quod contra naturam in aqua et lapidibus ardet," mentioned in the Legenda aurea (thirteenth century) is "oleum Mediacon medius seu medus."[37]

Strabo[38] reports that Alexander the Great found at Ekbatana in Media a lake of naphtha (νάφθα) which kindled as soon as a flame was brought near it.   A street was sprinkled with it, and when it was set on fire at one end the flame flashed in an instant to the other end.   If a body is soaked in the naphtha of Susiana and brought near a flame it burns so fiercely that it can be smothered only by mud, vinegar, alum and glue.   A little water makes it burn more fiercely, but it is extinguished by a very large amount of water.   Strabo also mentions (from Poseidonios) the Babylonian white and black naphtha (νάφθα τὰς πηγὰς τὰς μὲν εἶναι λευκοῦ τὰς δὲ μέλανος) or liquid sulphur (θεῖον ὑγρόν), the black naphtha being asphalt (ἄσφαλτος) which is burnt in lamps (λύχνοι). Naphtha is a Persian word (see above); the derivation of asphalt (ἄσφαλτος, asphaltum) is uncertain. Pliny[39] calls white naphtha bitumen liquidum candidum, Dioskourides[40] calls it filtered asphalt (ἄσφαλτος περιήθημα). Just what this "white naphtha," "white bitumen," or "filtered asphalt" was we do not know. It might be thought to be distilled petroleum, but distillation is generally supposed to be a somewhat later invention, although it is mentioned in a rudimentary form by Dioskourides (see ref. 249).

The preparation of a "white naphtha" by a process of purification not involving distillation, and also by distillation, is mentioned in an Arabic text of 1225 (see ch. V). That mentioned by the Greek and Roman authors may have been made by filtration through a material like fullers' earth, but since they wrote at a time very near that when distillation was known, this process cannot be excluded. Hippolytos (d. A.D. 235)[41] mentions Indian naphtha (ὁ νάφθας ὁ Ἰνδικός) which "kindles at the mere sight of fire a long way off." Pliny[42] mentions a bituminous liquid like oil, used for burning in lamps, which is skimmed from the surface of salt brine in Babylon; also[43] liquid bitumen from Zakynthos, Babylon (which is white), Apollonia, and Agrigentum (used for burning in lamps); he mentions naphtha separately (see above) but says some include it among the bitumens, "but the burning properties which it possesses, and its susceptibility of igniting, render it quite unfit for use (verum ardens ejus vis ignium naturæ cognata, procul ab omni usu abest)." He has just mentioned a liquid bitumen used for burning in lamps, and goes on to specify medical uses of pissasphaltos, which he elsewhere[44] says is a natural or artificial mixture of pitch and bitumen.

Petroleum is described by Vitruvius,[45] who says when water flows through greasy earth (per pingues terræ venas) it carries oil (oleum), as in Cilicia the river Liparis covers bathers with oil. The same happens in a lake in Ethiopia, and another in India, and a well in Carthage produces oil smelling of lemon peel, which is used for anointing cows. Dioskourides[46] says the liquid bitumen swimming on water at Agrigentum in Sicily, and used for lamps, is falsely called Sicilian oil. The Peripatetic De Mirabilibus Auscultationibus (A.D. 117–38?) says[47] that mineral oils exude from the earth in Persia, Carthaginia, Macedonia Thrace and Illyria. In Media and Persia are burning fires, in Persia so large and bright that kitchens are constructed near them. (This may refer to the use of natural gas.) Thomson[48] mentioned a large bed of bitumen in Albania. Pliny[49] describes several localities (including Etna, Bactria, the plain of Babylon, the fields of Aricia near Rome, etc.) either naturally on fire or taking fire on the approach of a torch; probably in some places there were escapes of natural inflammable gas.[50] Ailian[51] says there was a spring near Apollonia which constantly emitted flames, and Benjamin of Tudela (A.D. 1173)[52] describes the petroleum wells at Pozzuoli.

The use of incendiaries of the type described continued for many centuries. A few examples will be sufficient. Herodianos (A.D. 240)[53] says that when Maximinus took the town of Aquileia, the inhabitants threw on the soldiers and the siege-engines pots filled with a burning mixture of sulphur (θεῖον), asphalt (ἄσφαλτος) and pitch, and shot arrows with metal heads and shafts covered with burning pitch (i.e. falarica).

Braziers filled with incendiaries were used in A.D. 468 by Genseric, king of the Vandals, to destroy the Roman fleet sent by the Emperor Leo and commanded by Basiliscus.[54] Proklos the Philosopher is said by Joannes Malalas[55] to have advised the Emperor Anastasios in A.D. 515 to use sulphur (θεῖον ἄπυρον) to burn the ships of Vitalianus, but Zonaras[56] says Proklos the Soothsayer (ὀνειροκρίτης) used mirrors. Incendiary mixtures of sulphur, bitumen and naphtha were used by the Persians besieged in Petra in Kolchis in A.D. 551.[57] Incendiary arrows were used by the Visigoths in the siege of Nîmes in 673.[58] They continued to be used in the English army at least until 1599, being shot both from long-bows and cross-bows, and were used by the Chinese against the French in 1860; fire-lances, tipped with incendiaries, were used (perhaps for the last time) in the siege of Bristol in 1643.[59]

ANCIENT FIRE EXTINGUISHERS

Ancient fire extinguishers were water, sand, dry or moist earth, manure, urine (which contains phosphates), and especially vinegar, which was considered to be particularly cold. Plutarch[60] said vinegar "conquers every flame" (οὐδὲν δὲ τῶν ἀσβεστηρίων ὄξους πυρὶ μαχιμώτερον), ascribing this effect to the extreme tenuity of its parts, whilst Pliny[61] says a whirlwind may be checked by sprinkling with vinegar, which is of a very cold nature (frigidissima est natura). Vinegar is practically little better than water, but the ancients may have included as vinegar salty sauces, the salt left on the surface of burning wood helping to extinguish it, and Pliny[62] says the Gauls and Germans extinguished burning wood with salt water. The Venetians used wool soaked in vinegar to protect their vessels from Greek fire in the fourteenth century; wood and leather soaked in vinegar dry more slowly than if merely wetted with water, and burning oil would tend to run off them.[63]

Alum was well-known to prevent the combustion of wood. Herodotos[64] says Amasis, king of Egypt, sent a present of 1000 talents of alum for the rebuilding of the temple at Delphi, which had been destroyed by a fire in 548 B.C., and this may have been used for fire-proofing the timber.[65] Archelaos, a general of Mithradates, fire-proofed a wooden tower with alum in the war with Rome in 87 B.C.,[66] and the Roman siege-engines were fire-proofed with alum in the war between Constantine and the Persians in A.D. 296.[67] Mixtures containing sulphur, with or without tar or thick petroleum, resins, etc., would adhere tenaciously to surfaces, could not be washed off with water, and could with difficulty be extinguished by sand.[68]

AUTOMATIC FIRE

We may now leave the use of incendiaries of the simple type, which persist in modern war in the shape of thermit and napalm bombs, and turn to a different type of incendiary, viz. a mixture of quicklime with an inflammable material such as petroleum or sulphur. Such a mixture may inflame spontaneously when wetted with water, on account of the great heat generated in the hydration of the quicklime, an effect which was thought wonderful by Pliny.[69] St. Augustine (A.D. 415)[70] said limestone on burning takes up part of the nature of fire, which on cooling is retained in a latent form; this heat is disengaged on contact with water (the enemy of fire) but not by oil (the food of fire).

In II Maccabees (i. 20–36),[71] a "thick water" called naphtha (νεφθαι), brought from Persia, poured over wood on the altar, or on large stones, inflamed when the sun shone on it. The book, written in 135–106 B.C., reports events of 169 B.C. If quicklime was mixed with the wood soaked in petroleum, and water was poured on, or a mixture of naphtha and water poured on quicklime ("large stones"), the event may have happened as described, since a mixture of petroleum and quicklime inflames if sprinkled with water.[72] The trick persisted in the yearly kindling of lamps in the Sepulchre at Jerusalem, which began in the eighth to ninth century.[73] A mixture of sulphur and quicklime also inflames when sprinkled with water.

Livy[4] also mentions torches with heads of sulphur and quicklime, which inflame when dipped into water and taken out: Matronas Baccharum habitu . . . cum ardentibus facibus decurrere ad Tiberim, demissasque in aquam faces, quia vivum sulfur cum calce insit, integra flamma efferre. The Bacchanalia, the Latin name of the Dionysiac orgies, were widespread in Southern Italy, and although they have been supposed to be of Oriental origin, they are now regarded as fundamentally of Greek, or at most Thracian, origin, thus coming from a country bordering on a petroleum district.

Pausanias (A.D. 150)[75] mentions ashes (τέφρα) of a special kind (perhaps a mixture of quicklime and incense) as placed on the altar in a Lydian temple. A magician laid dry wood on them, put on a tiara, and uttered incantations in a foreign language, whereupon the wood inflamed without contact with fire. Lydia was under strong Oriental influence and the narrative again suggests a relation with the legend of Medeia.

The name "automatic fire" (πῦρ αὐτόματον) was, apparently, first used by Athenaios of Naukratis (c. A.D. 200),[76] who reported that Xenophon the Conjurer (ὁ θαυματοποιός)—otherwise unknown, a disciple of Kratisthenes of Phlias, astonished the world by his tricks and made an artificial fire issue from himself (πῦρ αὐτόματον ἐποίει ἀναφύεσθαι, sponte exoriri ignem faciebat). This was certainly some sort of trick like that described by Hippolytos (d. A.D. 235),[77] in which a conjurer blew sparks from his mouth into which he had put smouldering tow.

The most interesting account of a really "automatic" fire is that in a work attributed to Julius Africanus (the reading Sextus is doubtful), who was born about A.D. 160–80 in Ælia Capitolina, a town founded by Hadrian at Jerusalem, the Capitol being on the site of the Jewish Temple. Jews were forbidden to enter the town, and Julius was certainly a Christian, although the story that he was bishop of Emmæus is given only by late Syriac authors.

After extensive travels, in which he sought the remains of Noah's ark on Mount Ararat, he lived at Emmæus (Nikopolis) in Palestine. He was on good terms with the Emperor Alexander Severus, perhaps as an engineer officer.[78] Besides a *Chronography* in five books, now lost but known to Eusebius, he wrote the *Kestoi* (κεστοί) in twenty-four books, so named from the magic embroidered girdle, κεστὸς ἱμάς, of Aphrodite (Homer, *Iliad*, xiv, 214–5; hence κεστός = cestus = a girdle; i.e. a variegated patchwork), dedicated to the Emperor Alexander Severus (A.D. 222–35) and written after 225, which Duchesne calls "an amazing work."[79]

Julius Africanus believed in magic and was familiar with the Hermetic Books; the papyrus fragment contains a magical incantation from bk. xviii of the *Kestoi*. Although the language of the earlier parts of the work is classical, with a tendency to use peculiar words, that of the later part (containing the recipe) is definitely Byzantine. The editor of the Paris text, Jean Boivin (1663–1726) said (p. vii) that the publisher, Melchisedech Thevenot (1621–92), had contemplated a Latin translation (as in the case of other works in the volume), but found insuperable difficulties, much of the text being incomprehensible and incapable of restoration (permagnæ obstabant difficultates, cum multa in iis ocurrant quæ vix explicari possunt nisi restituantur). An incomplete translation by Boivin exists in the Royal Library, Copenhagen (no. 1845), but no complete translation, as far as I know, has ever been published. Some turns of style, e.g. the use of the particle ἄν with the future optative, ἀποθεῖσαν as passive, and εἰς instead of ἐν, e.g. in ἔχειν εἰς πυξίδα (to have in the box), which is characteristically Byzantine,[80] attest the late date of the text. Boivin (p. 339) had recognised that such a style was quite unlike that of Julius Africanus (Quin illud ausim affirmare, postrema hujus libri capita recentioris esse scriptoris, multum scilicet abhorrentia ab incorrupto illo Hellenismo, cujus studiosissimus fuit Africanus), and it only remains to conjecture a date when this portion of the text was added.

The later chapters (including the one containing the recipe) mention Belisarius (A.D. 505–65), and could not have been written before about A.D. 550.[81] They were probably written before the death of Konstantinos VII (Porphyrogennetos) in A.D. 959, since the text of MS. Laurentianus LV4 is either the original, or a copy not more than a century later (tenth to eleventh century), of a collection of war documents made for him, and before the time of the Emperor Leo (886–911), whose famous book on Tactics is not used in it.[82] Although Romocki[83] said that "since there is otherwise no mention in the period to the fall of the West Roman empire of incendiaries (Feuersätzen) with quicklime, we can hardly go astray" if we assume "that these incendiaries are a Greek discovery made between the sixth and tenth centuries." From what has been said above there can be no doubt that such mixtures were much older and came from an Oriental source. As Jähns[84] and Hime[85] said, this part of the text contains elements separated by several centuries, which is a characteristic feature of books of recipes, and the interpolator of the text of Africanus probably drew on much older material, although the actual form in which he gives the recipe cannot be dated safely before about A.D. 550. Vieillefond[86] gave a new text on the basis of Laurentianus LV4 (which he omits to mention was used by Lami) and Escorial, Vatican and Barbarinianus (Rome) MSS., all tenth to eleventh centuries. Since in modern scholarship there is always an "unknown original," this is assumed to be ninth century. Vieillefond has followed previous editors in giving no translation, and his index is very inadequate. His text is:

Αὐτόματον πῦρ ἅψαι καὶ τῷδε τῷ συντάγματι ⟨διδάξεται⟩· σκευάζεται γοῦν οὕτως· θείου ἀπύρου, ἁλὸς ὀρυκτοῦ, κονίας, κεραυνίου λίθου πυρίτου, ἴσα λειοῦται[a] ἐν θουίᾳ μελαίνῃ, μεσουρανοῦντος ἡλίου· μίγνυταί τε συκαμίνου μελαίνης ὀποῦ[b] καὶ ἀσφάλτου ζακυνθίας ὑγρᾶς[c] καὶ αὐτορύτου ἑκάστου ἴσον, ὡς λιγνυῶδες γενέσθαι· εἶτα προσβάλλεται ἄσβεστον τιτάνου παντελῶς ὀλίγον, ἐπιμελῶς δὲ δεῖ[d] τρίβειν, μεσουρανοῦντος ἡλίου, καὶ φυλάσσειν τὸ πρόσωπον, αἰφνίδιον[e] γὰρ ἀναφθήσεται· ἀφθεῖσαν[f] δὲ χρὴ πωμάσαι χαλκῷ τινι ἀγγείῳ πρὸς τὸ ἕτοιμον οὕτως ἔχειν εἰς πυξίδα καὶ μηκέτι δεικνύναι τῷ[g] ἡλίῳ, ἀλλ' ἐν ἑσπέρᾳ, ἐὰν βούλῃ πολεμίων ὅπλα ἐμπρῆσαι, ταῦτα καταχρίσεις[h] ἢ ἕτερόν τι, λεληθότως δέ· ἡλίου γὰρ φαινομένου, πάντα καυθήσεται.

[a]. λειοῦνται.   [b]. ὀπός.   [c]. ὑγρῆς.   [d]. δεῖ.   [e]. αἰφνήδιον.
[f]. ἀποθεῖσαν.   [g]. inserts λόγῳ.   [h]. καταχρήσεις.

Apart from punctuation, Vieillefond's text (as is seen from the notes) differs only negligibly from Lami's, which he criticised as unreadable. The reading ἄσβεστον in line 5 is mine; Boivin has ἀσφέστον; Lami and Vieillefond ἀσφάλτου, Lami saying that ἄσφαλτον τίτανον is quicklime, but this meaning seems to be unknown. The translation of this text is very difficult and depends on the punctuation. The following is an attempt:

"Automatic fire (αὐτόματον πῦρ) is composed of equal parts of native sulphur (θείου ἀπύρου), rock salt (ἁλὸς ὀρυκτοῦ), incense (κονίας), thunderbolt stone (κεραυνίου λίθου) or pyrites (πυρίτου), ground in a black mortar in the mid-day sun, and mixed with equal parts of the resin (ὀπός, lit. 'juice') of the black sycamore and liquid asphalt of Zakynthos to a greasy paste. Then some quicklime (ἀσβέστον τίτανον) is added. The mass must be stirred at mid-day with care and the body protected, since the composition easily inflames. It must be kept in bronze boxes (πωμάσαι) with tight covers, protected from the rays of the sun until it is wanted. If the engines (ὅπλα) of the enemy are to be burnt, they are smeared with it in the evening, and when the sun rises all will be burnt."

The θεῖον ἄπυρον, which Pliny[87] calls "sulphur vivum, quod Græci ἄπυρον vocant," is native sulphur which has not been melted.

The ἅλς ὀρυκτός has been supposed by many authors[88] to be saltpetre. Herodotos[89] distinguished mined or rock salt ἁλός τε μέταλλον, τὸ εἶδος ὀρύσσεται, i.e. ἅλς ὀρυκτός, from salt crystallised by the heat of the sun (ἅλες αὐτόματοι) from the water at the mouth of the river Borysthenes, and there is little doubt that the true meaning is rock-salt, which Dioskourides[90] said is the more active salt (τῶν δὲ ἁλῶν ἐνεργέστερον μέν ἐστι τὸ ὀρυκτόν). It was probably added to give a strong yellow flame, looking very hot.[91]

The word κονία, literally "dust," had many meanings; quicklime, an alkaline lye filtered through wood-ashes and quicklime (causticised lye), and incense (στακτή) which had "distilled" from resinous trees[92]; it may thus be taken as equivalent to the μάννα of the older recipes. Theophrastos[93] associated κονία with sulphides of arsenic (σανδαράκη καὶ ἀρρενικόν). Mercier,[94] who says the expression πῦρ αὐτόματον was "non employée à Byzance," translated "rigoureusement" ἁλὸς ὀρυκτοῦ κονίας as "sel extrait de la poussière," and identified it with saltpetre extracted from dusty efflorescences, e.g. on walls of cellars. He says: "on peut imaginer encore, avec une vraisemblance qui touche à la certitude, un esclave éclairant son maître d'une torche dans un couloir salpêtré, et effleurant les murs de sa flamme; ils observent alors l'aptitude du salpêtre à rendre la flamme plus vive." In the time of Julius Africanus (whom he dates A.D. 200) this efflorescence is "probablement déjà raffiné par dissolution dans l'eau et par évaporation." The word ὀρυκτός, however, always means "dug" or "quarried," and the meaning "extracted" seems improbable.

The thunderbolt stone (ceraunia) is mentioned by Pliny[95] as used by magicians (magorum studiis expetitæ), and he quotes Sotacus as saying that it was used to destroy fleets and towns (urbes per illas expugnare et classes). Voss[96] thought that lapis ceraunius or pyrites meant charcoal, the missing constituent (if sal coctus is salpetre) for gunpowder. It was probably pyrites; nodular masses of pyrites or marcasite are still called "thunderbolts." Hoefer[88] for κεραυνίου λίθου πυρίτου read "de pyrite kerdonienne (sulfure d'antimoine?)" (Lami had read κερδυνίου) and the interpretation of the text depends on the punctuation. The automatic fire (πῦρ αὐτόματον) of Africanus, inflaming "of itself (sua sponte)," is quite different from that of Athenaios.[97]

The mixture described would not inflame when "the sun rises" and the preparation is actually said to be made "in the mid-day sun" (μεσουρανοῦντος ἡλίου), a description copied uncritically into later "fire-books" (such as the Liber Ignium of Marcus Græcus, c. A.D. 1225), but it might inflame sooner or later if exposed to the heavy morning dew or to light rain. Such an uncertain weapon, we may be sure, would not be viewed with favour by military commanders.

Quicklime itself would be effective (see ref. 175). Red-hot sand, which penetrated chinks in armour, and powdered quicklime are mentioned by Quintus Curtius (first century A.D.) (see Berthelot, ref. 1), and quicklime and caustic alkalis were used in warfare after the invention of gunpowder. Pots filled with lime and soft soap (a mixture containing caustic potash) to be thrown in the eyes of the enemy were hung in the rigging of ships in 1298. Francesco da Barberino, the thirteenth-century poet, specifies among the stores for his galley: Calcina, con lancioni, Pece, pietre (saltpetre), e ronconi (bill-hooks). Christine of Pisa in her Faiz du Sage Roy Charles (V of France, 1364–80) says: "Item, on doit avoir plusieurs vaisseaulx legiers à rompre, comme poz plains de chaulx en poudre, et gecter dedens; et par ce seront comme avuglez au brisier des poz. Item, on doit avoir autres poz de mol savon et gecter es nefs des adversaires, et quant les vaisseaulx brisent, le savon est glissant, si ne se peuent en piez sous tenir, et chiéent en l'eau."[98]

The mixtures of quicklime (especially powdered) with combustible materials might inflame, rather slowly, when wetted with water and could be, and doubtless were, used in pots which could be secretly introduced into places, e.g. under the roofs of buildings, where they could produce unexpected con-flagrations.   They acted spontaneously and hence they were called "automatic fire."   Such mixtures would (in spite of what Kameniata said, see ref. 137) be quite useless in active combats, and particularly in naval battles (although they are said to have been used by the Chinese in 1232), since if thrown on the sea they would at once sink and produce no effect.   Even if the mixture in a suitable container, such as a paper bag, could float for a time and finally ignite, which is decidedly improbable, the combustion would be mild, relatively noiseless, and harmless, apart from the effect of surprise.[99]   Such mixtures were not Greek fire.

Hime[100] at first thought "sea fire . . . was a sulphur-quicklime-naphtha mixture," which ignited on water and burnt on its surface, giving a considerable volume of vapour and a series of small explosions in air, being unsafe to handle after ignition and hence necessarily discharged from siphons, and he thought this "simple explanation" (first given, he says, by Romocki)[101] "sweeps away the insurmountable difficulties raised by the saltpetre theory." Hime, however, later on changed his mind, as will be seen, and proposed an even more improbable theory (see ref. 265).

A supposed mention of fireworks by Claudianus of Alexandria[102] (A.D. 399) (Spargentes ardua flammas scena rotat), which has no parallel in any other ancient author, probably refers to some optical effect produced by mirrors.[102a] Lalanne,[103] citing Claudianus, concluded that "the Romans, and probably the other peoples of antiquity, were acquainted with gunpowder," and Mercier[104] adds that this came to them from ancient Egypt, where it was known to Moses, who made the earth open and caused fire to descend from heaven (Numbers xvi. 31, 35).   Mercier traced the use of saltpetre mixtures from the ancient Egyptians through Moses, Xenophon the Conjurer, and Julius Africanus to the Middle Ages, but such fanciful speculations are of no value.   The usually sceptical de Pauw[105] also quoted Claudianus on artificial thunder and lightning in the temple of the Eleusinian Ceres, and after discussing the possible use of the theatrical machines the keraunoskopion and bronteion, concluded that the effects were produced by "some pyrical composition," the secret being lost "like the *Greek fire*, which has not been found again, as some have pretended, to create alarms in the maritime powers."   We may now look into this matter of Greek fire.[106]

GREEK FIRE

The use of incendiary mixtures of all kinds was well-known in warfare from the earliest periods.   The Byzantine and Arab chroniclers agree that from the end of the seventh century A.D. a more terrible agent of destruction came into use, which was known to the Crusaders as Greek fire (le feu grégeois). It was first used by the Byzantines of Constantinople, but was never called by them Greek fire. The Byzantines never called themselves Greeks; they claimed to be Romans ('Ρωμαῖοι), and the name "Greek" in their time and long afterwards had a bad reputation, like "Levantine," or in modern French "Grec."[107] Byzantine Constantinople, often described as "effete," was able by diplomacy and arms to withstand the attacks of Arabs, Persians, Goths, Huns, Normans, Russians and Franks; it fell at last (1453) to the Turks, a non-Semitic race which it had resisted for centuries, and the Emperor died fighting in battle. Gibbon's unfavourable opinions of the Byzantine culture were not altogether justified; it was refined and active; it preserved the heritage of the older civilisations for the West; the army was well-disciplined and resisted attacks for a very long period, and the city was a bulwark against the tide of Asiatic conquest. The rulers were mostly able, but a large part of their work was undone, and their power was crippled, not so much by "barbarians" as by Christians in the Fourth Crusade.[108] In later times, under Asiatic emperors, there was certainly a decline.

Niketas Akominatos (Choniata), who was an eye-witness of the capture of Constantinople in 1204 by the Crusaders, reported that the admiral commanding, Michael Struphnos, had sold the bolts and anchors of the ships as well as the sails, and left the Byzantine navy without a single large ship[109]: γόμφους καὶ ἄγκυρας χρυσίου ἀλλάξασθαι ἀλλὰ καὶ λαίφεσιν ἐπιθέσθαι καὶ ἐξαργυρίσαι πρότονα, ἀπαξάπαντος πλοίου μακροῦ τὰ νεώρια Ῥωμαίων ἐκένωσε: non tantum clavos et ancoras auro permutare sed vela etiam et rudentes vendere solitus, longis navibus prorsus omnibus navalia Romana spoliarat. Yet Prokopios, another Asiatic historian, tells us that in Justinian's time, much earlier (527–65), the semi-civilised Goths (who, it is true, had been roughly handled by Belisarius) affected shame in having to dispute the kingdom of Italy with "Greeks . . . a nation of tragedians, showmen and dockyard spivs."[110]

It has been well said[111] that "Byzantine history is a series of sharp and disappointing contrasts." Some Byzantine emperors in times of financial crisis melted down their own or the Church's ornamental gold and sold their jewels instead of resorting to the debasement of coinage which had been practised by the old Romans. Yet Alexios I, who became emperor in 1081, whilst enforcing payment of taxes in gold coin, issued "gold" coins which were almost entirely copper.[112] Nikephoros III (1077–81) and Isaac Angelos (1085–95) also debased the coinage, never it is true to the alarming extent of the modern British "silver" coinage, innocent of even a trace of noble metal. Gibbon, who lived in happier times, exercised his malicious ingenuity in seeking the causes of the fall of Constantinople; we are in a better position to appreciate the effects of the enormous expenditure on public professional sport and the merciless over-taxation which was later to drive some of the best elements of the population into the territories of the Turkish sultans; the income-tax men (φορόλογοι) were feared and detested.[113]

As we turn our attention to Byzantine affairs in the seventh century A.D.

we find that a new threat has appeared at the very gates of the city: the Arabs.

Theophanes, who wrote in A.D. 811–15, deals in his *Chronography* with Byzantine history from the time of Diocletian, A.D. 285, to 813. He may have derived some material from his older friend Synkellos (d. A.D. 800), who had written a chronography from the time of Adam to Diocletian, including a very interesting account of the beginning of alchemy in Egypt.[114] Theophanes says in his *Chronography* that in A.M. (annus mundi) 6165 the Emperor learnt that the Arabs had established winter quarters in Asia Minor, and he ordered the assembly of a fleet of fire-ships equipped with siphons (κακκαβοπυρφόρους καὶ δρόμωνας σιφωνοφόρους). In the spring the siege of Constantinople commenced, and this was carried out from April to September for seven years, when, at Kyzikos, partly by a storm and partly by a fire invented by the architect Kallinikos of Heliopolis in Syria, who had fled to the Romans, the Arab boats and sailors were all burnt. And afterwards the Romans were victorious by their invention of sea-fire (θαλάσσιον πῦρ):

> τότε Καλλίνικος ἀρχιτέκτων ἀπὸ Ἡλιουπόλεως Συρίας προσφυγὼν τοῖς Ῥωμαίοις πῦρ θαλάσσιον κατασκευάσας τὰ τῶν Ἀράβων σκάφη ἐνέπρησεν, καὶ σύμψυχα κατέκαυσεν. καὶ οὕτως οἱ Ῥωμαῖοι μετὰ νίκης ὑπέστρεψαν, καὶ τὸ θαλάσσιον πῦρ εὗρον.

The last sentence implies that Kallinikos was a Greek, and the whole account appears to mean, not that Kallinikos brought the secret with him, but that it was invented in Constantinople.[115] In the Latin translation in Classen's edition, ἀρχιτέκτων is "in rebus arte parandis strenuus," and the last sentence is "sicque Romani victores reversi sunt, ignemque marinum invenere"; for κακκαβοπυρφόρους, see the notes in vol. ii, p. 501; the Latin translation is "biremes etiam cacabos igne oppletas ferentes, et dromones siphonibus eundem ignem spirantes."

The date for the beginning of the battle given by Theophanes is A.M. 6165, and since Classen[116] takes A.M. 5777 as A.C. 277, this gives 665, which Gibbon (see below) says is A.D. 673. De Boor does not give an A.D. date, Romocki takes 671, and Hime, who dates the siege 674–6, puts the arrival of Kallinikos in 673, whereas Theophanes says he arrived at the end of the battles lasting seven years, viz. in 678 if they began in 671. The exact date is not of importance for our purposes.

The Emperor was Konstantinos IV (Pogonatos), A.D. 668–85, the commander of the Arab forces was Yazīd, son of the Khalif Mu'-āwiya of Syria, and the attack terminated on the death of the latter. It was renewed in 717–18. Arabic accounts[117] say that the first attack was made in 669 and was repelled by Konstantinos IV, that the second attack was "the war of seven years," 674–80, in which Greek fire was first used, the Arabs having established a naval base on the peninsula of Kyzikos, and the third attack was in 717–8. Other accounts say the third attack was in 718–20, and the city was defended by the Emperor Leo III the Isaurian, who used Greek fire against the Arabs. Gibbon[118] says: "Theophanes places the *seven* years of the siege of Constantinople in the year of *our* Christian era 673 (of the Alexandrian 665, 1st September), and the year of the peace of the Saracens four years afterwards; a glaring inconsistency! . . . Of the Arabians [who described the event], the Hegira 52 (A.D. 672, 8th January) is assigned by Elmacin, the year 48 (A.D. 668, 20th February) by Abulfeda, whose testimony I esteem the most convenient and creditable." Gibbon says "the Saracens were dismayed by the strange and prodigious effects of artificial fire" in this first attack, but deals with this in his account of the second attack in A.D. 718.

J. H. Mordtmann,[117] who says that the invasion of A.D. 667 was by a land army, agrees that Greek fire was used on the Arab fleet in A.D. 672 and it retired in 673; the chronology, he says, is very confused. The position seems to have been clarified by M. Canard[117] (correcting Brooks[117]), who says there were really five expeditions: three in the time of the Khalif Mu'-āwiya in 665, 668 and 674; one under Sulaymān ibn 'Abd al-Malik in 715; and one under al-Mahdī, commanded by his son Hārūn, in 782. Of these, only two were true sieges, viz. the one under Yazīd and Abū Ayyūb in 668, and that under Maslama ibn 'Abd al-Malik in 715, when Greek fire was used. The expedition of "seven years" (674–80) confounds sieges and simple raids; Turkish traditions have nine sieges. Theophanes and Arab historians have jumbled these events: "on les dédouble, et on fait bon marché de la chronologie." Only once an Arab army made a landing, and camped before the wall which had been erected by Theodosios II at the extremity of the city, from the Sea of Marmara to the Golden Horn. It failed to capture Constantinople, and only once again after the attacks mentioned above, in 782, did the Arabs venture within sight of Constantinople, once again turning away.[119]

It is noteworthy that Theophanes says the "fire-ships" were fitted with "siphons" in 671, and presumably the invention of Kallinikos made these more effective. If he had brought it with him, it must presumably have been known to the Arabs in Syria, which does not appear to have been the case. Constantinople was full of inventors and craftsmen. The "philosopher" Leo of Thessalonika made for the Emperor Theophilos (829–42) a golden tree, the branches of which carried artificial birds which flapped their wings and sang, a model lion which moved and roared, and a bejewelled clockwork lady who walked.[120] These mechanical toys continued the tradition represented in the treatise of Heron of Alexandria (*c.* A.D. 125), which was well-known to the Byzantines.[121] We may be sure that when Kallinikos arrived as a displaced person he found in Constantinople many native rivals. The association of his name with Greek fire may even have been accidental, since officials ignorant of science always think that foreigners, particularly refugees, are alone capable of making inventions of service to a state.

The name Kallinikos ("handsome winner") is late Greek, but then, as now, names of Orientals meant little,[122] and it is more than likely that he was a Jew. It also seems quite probable that Greek fire was really invented by the chemists in Constantinople who had inherited the discoveries of the Alexandrian chemical school, some of whose writings are contained in manuscripts palmed off as works of the emperors Justinian and Herakleios. As Finlay[123] said: "The Byzantine army was superior to every other in the art of defending fortresses. The Roman arsenals, in their best days, could probably have supplied no scientific or mechanical contrivances unknown to the corps of engineers of Leo's army, for we must recollect that the education, discipline, and practice of these engineers had been perpetuated in uninterrupted succession from the times of Trajan and Constantine." Since Greek fire was a chemical, the same would hold for the chemists. The later Emperor, Constantine Porphyrogenitus, as we shall see (ref. 190), said that the recipe for Greek fire had been revealed by an angel to Constantine the Great, and the earliest chemists called their science "the divine art ($\theta\epsilon\hat{\iota}\alpha$ $\tau\epsilon\chi\nu\eta$)."

Theophanes is regarded as a reliable author, but Nikephoros Patriarcha (758–828)[124] in his description of the siege of Constantinople by the Arabs, does not mention Greek fire. Anastasios[125] gives the same account as Theophanes. Among the Byzantine chroniclers only Kedrenos (eleventh century)[126] reported that Kallinikos came from Heliopolis in Egypt, not Syria, and this was the view taken by Gibbon,[127] who mistakenly said it was in ruins at that time, and by Butler,[128] who reports a MS. account in Sebeos saying that ships built in Alexandria for Mu'-āwiya were equipped with "fire-spouting engines." If this is so, it is curious that the Khalif did not send them to Constantinople.

Konstantinos VII (Porphyrogennitos)[129] says that Kallinikos, who fled from the town of Heliopolis to the Romans, invented the art of projecting liquid fire through siphons ($\tau\grave{o}$ $\delta\iota\grave{\alpha}$ $\tau\hat{\omega}\nu$ $\sigma\iota\phi\acute{\omega}\nu\omega\nu$ $\grave{\epsilon}\kappa\phi\epsilon\rho\acute{o}\mu\epsilon\nu o\nu$ $\pi\hat{\upsilon}\rho$ $\acute{\upsilon}\gamma\rho\grave{o}\nu$ $\kappa\alpha\tau\epsilon\sigma\kappa\epsilon\acute{\upsilon}\alpha\sigma\epsilon$: qui Græcum ignem siphonibus emitti solitum paravit), which seems to imply that the method of *using* it was the novelty, yet Theophanes (see above) says siphons were fitted to the ships before he came. Zonaras (twelfth century)[130] gives a similar account to Theophanes, as do some other compilers.[131]

Earthenware pots supposed to have been used as receptacles for Greek fire were found at Baalbek and Ḥamā in Syria.[132] Mercier[133] described and illustrated many pots which he thinks were receptacles for Greek fire, and gave analyses of the small amounts of débris they contained. The qualitative analysis[134] showed the presence of traces of nitrate, chloride and sulphate. The analyst (P. Woog of Paris) very sensibly pointed out that these came from the soil in which the pots had lain, but Mercier is of the opinion that Greek fire contained saltpetre. The use of these pots as containers for Greek fire has never been established. Mercier[135] suggested that Greek fire contained naphtha and other inflammable materials as well as sulphur and quicklime, and also saltpetre, i.e. he confounds the separate preparations which I have considered and adds saltpetre. I cannot accept his conclusions. He quotes R. Pique, *La Poudre Noire et le Service des Poudres*, 1927, to the effect that gunpowder was used by Moses, "hypothèse assez séduisante."[136]

Earthenware grenades ($\grave{o}\sigma\tau\rho\acute{\alpha}\kappa\iota\nu\alpha$ $\sigma\kappa\epsilon\acute{\upsilon}\eta$) filled with pitch, quicklime, etc., were mentioned by Ioannes Kameniata[137] as used by the Arabs in the siege of Salonika in A.D. 904: $\tau\grave{\alpha}$ $\delta\grave{\epsilon}$ $\mathring{\eta}\nu$ $\pi\acute{\iota}\sigma\sigma\alpha$ $\kappa\alpha\grave{\iota}$ $\delta\hat{\alpha}\delta\epsilon\varsigma$ $\kappa\alpha\grave{\iota}$ $\acute{\alpha}\sigma\beta\epsilon\sigma\tau o\varsigma$ $\kappa\alpha\acute{\iota}$ $\tau\iota\nu\alpha$ $\acute{\alpha}\lambda\lambda\alpha,$

οἷς τρέφονται τὸ τάχος πυρὸς φλόγες, σκεύεσιν ὀστρακίνοις ἐπιτετηδευμένα; impetum in muro picem et tædas et calcem aliaque quædam, quibus flammæ ignis celeriter innutriuntur, in vasis testaceis disposita præparare. These small hand-grenades were quite minor instruments of warfare and never played any prominent part in the use of Greek fire.

Some details on the use of Greek fire in naval warfare are given in a work on tactics and strategy (περὶ τακτικῆς καὶ στρατηγοῦ or τῶν ἐν πολέμοις τακτικῶν σύντομος παράδοσις) attributed to the Emperor Leo, identified with Leo III the Isaurian (A.D. 717–41)[138] or Leo VI the Armenian, or the Philosopher, or the Wise (A.D. 886–911).[139] Leo[140] says many valuable military engines of destruction have been invented by the ancients and moderns for attacking ships and sailors. Of this kind is the emission of fire with thunder and burning smoke thrown (emitted) by siphons, which burns ships : πολλὰ δὲ καὶ ἐπιτηδεύματα τοῖς παλαιοῖς καὶ δὴ καὶ τοῖς νεωτέροις ἐπενοήθη κατὰ τῶν πολεμικῶν πλοίων καὶ τῶν ἐν αὐτοῖς πολεμούντων. Οἷον τό τε ἐσκευασμένον πῦρ μετὰ βροντῆς καὶ καπνοῦ προπείρου διὰ τῶν σιφώνων πεμπόμενον, καὶ καπνίζον αὐτά.

The "noise like thunder (μετὰ βροντῆς)" in Leo does not mean an explosion but the rumbling noise which Joinville reported for the tub of burning fire projected from a ballista,[141] and Anna Comnena[142] had compared the flash of a blow-pipe with a fiery whirlwind; these exaggerated metaphors have misled some modern writers into thinking that gunpowder was used. Romocki[143] and Hime[144] had emphasised this point, which should always be kept in mind in considering the accounts of the effects of Greek fire.

Leo also says[145] : χρήσασθαι δὲ καὶ τῇ ἄλλῃ μεθόδῳ τῶν διὰ χειρὸς βαλλομένων μικρῶν σιφώνων ὄπισθεν τῶν σιδηρῶν σκουταρίων παρὰ τῶν στρατιωτῶν κρατουμένων, ἅπερ χειροσίφωνα λέγεται, παρὰ τῆς ἡμῶν βασιλείας ἄρτι κατεσκευασμένα. Ῥίψουσι γὰρ καὶ αὐτὰ τοῦ ἐσκευασμένου πυρὸς κατὰ τῶν προσώπων τῶν πολεμίων; the other device of the small siphons discharged by hand from behind iron shields, which are called hand-siphons and have recently been manufactured in our dominions. For these can throw the prepared fire into the faces of the enemy.

The word βαλλομένων, literally "thrown," as an adjective qualifying χειροσίφωνα, misled some readers into supposing that the "little hand-siphons (μικροὶ σιφῶνες, χειροσίφωνα)" were thrown bodily in the manner of grenades. Fortunately, the correct interpretation, which had been given by Romocki[146] and Oman[147] on the basis of a statement by Anna Comnena to be dealt with presently, viz. that some instrument held in the hands was used to project ("throw") the fire, is not now in doubt, since a picture of the actual apparatus is available. A small hand-pump projecting Greek fire is shown in an eleventh-century Vatican MS. (1605).[148] The Spanish Muslim physician Abu'-l-Qāsim Khalaf ibn 'Abbās al-Zahrāwi (Abulcasis, d. A.D. 1013) in his book on surgery, after describing and depicting a cylindrical syringe with a piston, also says[148a] : "un liquide . . . soit repoussé au loin comme il arrive avec cet tube au moyen duquel on lance le naphte dans les combats de mer."

The name σίφων (siphon) was used for the double-action force-pump or

fire-engine invented by Ktesibios and improved by Heron.[149]   In Hesychios[150]
it means a water-pump for extinguishing incendiaries.   Many examples of
the meanings of σίφων (including some now considered) are given by
Stephanus.[151]   Siphon also meant a bent tube for transferring liquids (in
use in ancient Egypt) and a pipe through which water was forced like a
fountain.[152]

Kameniata[153] says that in the siege of Thessalonika in A.D. 904 the liquid
fire was blown from the siphons by compressed air: πῦρ τε διὰ τῶν σιφώνων
τῷ ἀέρι φυσήσαντες, καί τινα ἄλλα σκεύη καὶ αὐτὰ πυρὸς ἀνάμεστα εἴσω τοῦ
τείχους ἐξακοντίσαντες; ignemque siphonibus aëri insufflantes, atque alia vasa
(ipse quoque igne plena) intra murum iacientes.

A MS. in the Bibliothèque de l'Arsenal written for Louis XI (1423–83)
shows a defender of Constantinople holding a pipe about 5 ft. long, with
flames issuing from its funnel-shaped mouth.[154]

No doubt, on ships, the pump was connected with the metal tube from
which the liquid fire was projected by some flexible tube of leather, and
this is indicated by the mention in the Tactics of Konstantinos VII
(Porphyrogennitos)[155]: πρὸς δὲ τοὺς προσφερομένους πύργους εἰς τὸ τεῖχος, ἵνα ὦσι
στρεπτὰ μετὰ λαμπροῦ καὶ συφώνια καὶ χειροσύφωνα καὶ μαγγανικά; "flexible
apparatus with (artificial) fire, siphons, hand-siphons, and manjaniks are to
be used, if at hand, against any tower that may be advanced against the wall
of a besieged town."   The Latin translation reads: "Adversus turres vero,
quæ admoventur muro, habeant strepta cum splendore (!), & syphonia, &
chirosyphonia, & manganica," but in Byzantine and modern Greek (e.g. in
Cyprus) τὸ λαμπρόν has the meaning φῶς (light) or πῦρ (fire).[156]

Lalanne[157] concluded that Greek fire was modern gunpowder, that σίφων
was a rocket (fusée de guerre, fusée volante), that Greek fire was explosive and
detonated, and that: "il a été démontré rigoureusement, au point de vue
historique et chimique, que la poudre à canon était le seul mélange qui pût
convenir à la composition des feux grégois."   Each and every one of these
statements is incorrect.   Reinaud and Favé[158] said the translation by de
Mezeroy[159] was more correct than Lalanne's, except that they thought σίφων
should be "tube" and not "siphon" as de Mezeroy gave it, his translation
being: "ces feux préparés dans des siphons, d'ou ils partent avec un bruit
de tonnerre & une fumée enflammée qui va bruler les vaisseaux sur lesquels
on les envoie . . . petits siphons à la main."   But de Mezeroy[160] supposed
that the micro-siphons were filled with gunpowder, and were concealed
behind the shields of the users (devant leur boucliers), which is completely
incorrect.

Isaac Voss[161] quoted Leo, Constantine Porphyrogenitus "and others not
published," on throwing fire, but said they are all silent on throwing stones
or metal balls with it (sed de lapidibus seu globis metallicis simul explosis
altum omnino apud omnes silentium), thus suggesting that it was not gunpowder.

Finlay[162] says the Byzantine dromon (δρόμων) was the war-galley, which
had taken the place of the triremes of the ancient Greeks and the quinqueremes

of the Romans; it had only two tiers of rowers, and the largest carried 300 men, of whom seventy were marine soldiers. The cheland (τὸ χελάνδιον) was a smaller and lighter vessel or barge, adapted for rapid movements, fitted with tubes for launching Greek fire, and their crews varied from 120 to 160 men.

Mercier[163] reproduces from the Vatican MS. 1605 a picture of a boat with only three rowers, from the prow of which projects a wide tube, apparently bell-mouthed, manipulated by two men, from which flames are issuing forward but rather diffusely, and enveloping another small boat containing three or four rowers. One of the men standing may, or may not, have been manipulating the pump throwing the fire. Unless the size varied, the ships used for throwing Greek fire in close combat were quite smaller than those described by Finlay.

Leo[164] said:

> Ἐχέτω δὲ πάντως τὸν σίφωνα κατὰ τὴν πρώραν ἔμπροσθεν χαλκῷ ἠμφιεσμένον, ὡς ἔθος, δι' οὗ τὸ ἐσκευασμένον πῦρ κατὰ τῶν ἐναντίων ἀκοντίσαι. καὶ ἄνωθεν δὲ τοῦ τοιούτου σίφωνος ψευδοπάτιον ἀπὸ σανίδων, καὶ αὐτὸ περιτετειχισμένον σανίσιν, ἐν ᾧ στήσονται ἄνδρες πολεμισταὶ τοῖς ἐπερχομένοις ἀπο τῆς πρώρας τῶν πολεμίων ἀντιμαχόμενοι, ἢ κατὰ τῆς πολεμίας νεὼς ὅλης βάλλοντες δι' ὅσων ἂν ἐπινοήσωσιν ὅπλων.

The old Latin translation is:

> In prora siphonem ære obtectum de more habeas ad ignem in hostes ejaculandum, et celse supra siphonem pseudopatium ex asseribus confectum, et asseribus circumtectum, in quo viri ad bellandum instructi sint, qui adversum hostes ex prora pugnent, vel in hostilem navem tela, aut alia ad vastandos hostes spectantia injiciant.

Before offering a translation of the very difficult Greek text, some notes on special phrases in it will first be given.

Hime[165] translated ἔμπροσθεν χαλκῷ as "cased with bronze," referring to St. Luke vii. 25: ἐν μαλακοῖς ἱματίοις ἠμφιεσμένον, "clothed in soft raiment," and thence supposed that the siphons were made of wood with an internal casing of bronze. Oman[166] translated this part of the text as "a bronze tube protected by a solid scantling of boards." The word ἔμπροσθεν, however, means simply "in front." The word ψευδοπάτιον, untranslated in the old Latin, is presumably ψευδόπατον, which means a "false floor." The other technical terms do not offer much difficulty.

The translation of ἐσκευασμένον πῦρ as "prepared fire"[167] is the obvious one, but is not certain. In Greek papyri found in Egypt, σκευή meant quicklime.[168] Stephanides[169] pointed out that in late Greek σκευή meant a cannon, and βοτάνον τῆς σκευῆς gunpowder for a cannon; Bulliaud in his edition of Doukas[170] said in a note: βοτάνης σκευασία. βοτάνη appellant Græci moderni pulverem tormentarium seu pyrium. In modern Greek ἡ σκευασία means a "mixture or composition" in the pharmaceutical sense, τὸ σκεῦος a piece of furniture, and τὰ μαγειρικὰ σκεύη kitchen utensils. In classical Greek σκευή or σκεῦος had a semi-contemptuous meaning which corresponds with "gadget," as well as "druggist's stock." By combining the last meaning with the first, the interpretation "chemically-prepared fire," i.e. distilled petroleum, which is the meaning assumed here for "Greek fire," appears to be reasonable. The repeated use of forms derived from σκευή will be noticed in all the Greek extracts given.

A paraphrase of Leo's text may thus be offered as follows (a word-for-word translation is not required):

"The front part of the ship had a bronze tube so arranged that the prepared fire could be projected forward to left or right and also made to fall from above. This tube was mounted on a false floor above the deck on which the specialist troops were accommodated and so raised above the attacking forces assembled in the prow. The fire was thrown either on the enemy's ships or in the faces of the attacking troops."

The word προπείρου used by Leo[140] means "first tested," or "previously tried." Diels,[171] from a reading in a Munich MS., changed it into προπύρου, thus making the thunder and smoke "go before the fire"; he imagined that some rocket-like projectile was shot out (raketenartig abgeschossen werden) by a charge of gunpowder in the siphon. There is little doubt that this is incorrect.

Krause[172] and E. Pears[173] argued that Greek fire was only rarely used because it was too dangerous; since Krause thought it contained phosphorus, sulphur and saltpetre, this was a reasonable conclusion from his false premise. Such a mixture could hardly be made without exploding spontaneously.

Byzantine warfare did not confine itself to the use of Greek fire. Leo[174] specifies throwing baskets containing live scorpions and serpents at the enemy, and also[175] projecting pots of powdered quicklime, which formed a dust-cloud in the air and suffocated and blinded the enemy: καὶ χύτρας δὲ ἄλλοι ἀσβέστου πλήρεις ὧν ῥιπτομένων καὶ συντριβομένων, ὁ τῆς ἀσβέστου ἀτμὸς συμπνίγει καὶ σκοτίζει τοὺς πολεμίους, καὶ μέγα ἐμπόδιον γίνεται; Ollas calce viva plenas alii injiciunt, quibus confractis, calcis vivæ pulvis dissipatus suffocat et strangulat hostes, et magno ad præliandum impedimento est (but σκοτίζει means "darkened," or "blinded," rather than "strangled").

Poison gas was not yet invented, but quicklime dust would be quite a good lachrymatory material. Poisoned wells are mentioned,[176] and the surprise of "frightfulness" has never been lacking in warfare in any period.

Greek fire was used to repel the invasion of Constantinople by Igor the Russian in A.D. 941.[177] In this battle, says Luitprand of Cremona (tenth century), whose nephew was then an ambassador in Constantinople, the Russian flotilla of several thousand ships was defeated by fifteen semifracta chelandria, which threw liquid fire on all sides, from the prow, the stern and the sides (in Russorum medio positi ignem circumcirca proiciunt), and the Russians, rather than burn, threw themselves into the sea; those weighed down by their armour were drowned, and those who were able to swim were burnt. It was extinguished only by vinegar.[178]

Anna Komnena (b. 1083), the gifted but vain daughter of the Emperor

Alexios I Komnenos, in her biography of him,[179] describes a battle between the Greeks and Pisans near the island of Rhodes in 1103. Each of the Byzantine galleys was fitted in the prow with a tube ending with the head of a lion or other beast made of brass or iron, and gilded, frightful to behold, through the open mouth of which it was arranged that fire should be projected by the soldiers through a flexible apparatus: ἐν ἑκάστῃ πρώρᾳ τῶν πλοίων διὰ χαλκῶν καὶ σιδήρων λεόντων καὶ ἀλλοίων χερσαίων ζώων κεφαλὰς μετὰ στομάτων ἀνεῳγμένων κατασκευάσας χρυσῷ τε περιστείλας αὐτά, ὡς ἐκ μόνης θέας φοβερὸν φαίνεσθαι, τὸ διὰ τῶν στρεπτῶν κατὰ τῶν πολεμίων μέλλον ἀφίεσθαι πῦρ διὰ τῶν στομάτων αὐτῶν παρεσκεύασε διιέναι, ὥστε δοκεῖν τοὺς λέοντας καὶ τἆλλα τῶν τοιούτων ζώων τοῦτο ἐξερεύγεσθαι. The translation of στρεπτῶν by "springs" is probably incorrect, since steel springs on arbalests were not used till the end of the Middle Ages[180]; the expression also occurs in the account of Konstantinos VII (ref. 129).

Anna continues in her picturesque style to tell us that the enemy ship was rammed in the stern, and the fire pumped over it. The Pisans fled, having no previous experience of this device and wondering that fire, which usually burns upwards, could be so directed downward or towards either side according to the will of the engineer who discharges it: καὶ αὐτὸς δὲ ὁ Λαντοῦλφος πρῶτος προσπελάσας ταῖς Πισσαϊκαῖς ναυσὶν ἄστοχα τὸ πῦρ ἔβαλε, καὶ οὐδέν τι πλέον εἰργάσατο, τοῦ πυρὸς σκεδασθέντος .... ἐκδειματωθέντες οἱ βάρβαροι τὸ μὲν διὰ τὸ πεμπόμενον πῦρ (οὐδὲ γὰρ ἐθάδες ἦσαν τοιούτων σκευῶν ἢ πυρὸς ἄνω μὲν φύσει τὴν φορὰν ἔχοντος, πεμπομένου δ' ἐφ' ἃ βούλεται ὁ πέμπων κατά τε τὸ πρανὲς πολλάκις καὶ ἐφ' ἑκάτερα), τὸ δὲ ὑπὸ τοῦ θαλαττίου κλύδωνος συγχυθέντες τὸν νοῦν φυγαδείας ἤψαντο.

The directions of projection agree exactly with the account in Leo (ref. 164). Apparently the projection apparatus was able to swivel, the open mouths of the fixed beasts being wide enough to allow of this. Anna[181] gives the composition of an incendiary material:

τοῦτο δὲ τὸ πῦρ ἀπὸ τοιούτων μηχανημάτων αὐτοῖς διεσκεύαστο. Ἀπὸ τῆς πεύκης καὶ ἄλλων τινῶν τοιούτων δένδρων ἀειθαλῶν συνάγεται δάκρυον εὔκαυστον. Τοῦτο μετὰ θείου τριβόμενον ἐμβάλλεταί τε εἰς αὐλίσκους καλάμων καὶ ἐμφυσᾶται παρὰ τοῦ παίζοντος λάβρῳ <λάβρᾳ> καὶ συνεχεῖ πνεύματι κᾆθ' οὕτως ὁμιλεῖ τῷ πρὸς ἄκραν πυρὶ καὶ ἐξάπτεται καὶ ὥσπερ πρηστὴρ ἐμπίπτει ταῖς ἀντιπρόσωπον ὄφεσι.

"This fire they made by the following arts. From the pine and certain such evergreen trees inflammable resin is collected. This is rubbed with sulphur and put into tubes of reed, and is blown by men using it with violent and continuous breath. Then in this manner it meets the fire on the tip and catches light and falls like a fiery whirlwind on the faces of the enemy."

This device of incendiary blowpipes, with a powder or paste projected by the breath past a kindling device on the tip (probably a small torch of resinous wood), is a new one, and (in spite of what Romocki[182] says) is quite different from Leo's "micro-siphon," which, as was shown, was a small hand-pump. The effect would be to produce a cloud of fire, like that formerly (and perhaps still) used in theatres by igniting a cloud of lycopodium. Anna[183] says it was used in the siege of Durazzo in 1108, when the Normans under Bohemond had mined under the walls, and the Byzantines had countermined until they reached the sap, when the appearance produced by these pyrotechnics would be terrifying but not very dangerous. The Normans, she said, had their beards singed but were not much injured. Joinville (see ref. 221) says Guy Malvoisin had his armour covered with the blazing fire but his soldiers soon put it out.[184]

The use of an improved incendiary mixture containing saltpetre (the recipe for which had been given by Albertus Magnus about 1250) projected by a mouth blowpipe is described in a Berlin MS. of about 1425–50[185]:

"Wilt du machen ain fliegends für des fert in die höchin vnd verbrennt was es begrifft, so nim ain tail colofonia das ist kriechisch hartz vnd ij. tail lebendigs schwebels vnd iij. tail salniter, das rib alles gar klein vnd rib es dann mit ainem linsatöl oder loröl das es darinn zergang vnd werd als ain confect vnd tu das in ain aichin ror das lang sy vnd zünd es an vnd blas in das ror, so fert es wohin du das ror kerst vnd verwüst vnd verbrennt alles das es ankumpt."

"If you will make a flying fire which rises above and burns what it encounters, take one part of colophonium, that is Greek resin, two parts of native sulphur and three parts of saltpeter. Rub all small and then rub it with one of linseed oil or laurel oil till it is taken up and becomes like a paste. And put this in a long bronze tube and kindle it and blow into the tube, when it goes to wherever you turn the tube and destroys and burns up everything it meets."

Anna[186] mentions pots of liquid pitch (ὑγρὰ πίττη) and naphtha (νάφθα) as thrown by the besieged in Tyre against the wooden engines of the Crusaders, and (apparently) igniting at sunrise (the old recipe of Julius Africanus, p. 8): ἅμα τε γὰρ ἡμέρα διέλαμπε καὶ τὸ πῦρ συνεξέλαμπεν ἀπὸ τῶν ξυλίνων χελωνῶν εἰς αἰθέρα πυργούμενον; itaque cum prima luce ignis quoque e ligneis testudinibus ætherem versus turris instar se efferens incuxit. Niketas[187] refers to the use of such incendiary pots containing "liquid fire (ὑγρὸν πῦρ)" as used to burn houses in the capture of Constantinople by the Crusaders in 1204: τὸ ὑγρὸν πῦρ, ὃ τοῖς στέγουσιν ἐφυπνοῦν σκεύεσι κατὰ τὰς ἀστραπὰς ἐξαίφνης προΐησι τὰ ἐξάλματα καὶ πιμπρᾷ καθ' ὧν διεκπίπτον ἀφίεται. Græcus ignis ædificiis ad fretum sitis injicitur, qui in vasis occultus, subito fulguris instar emicans, omnia quæ contigerit incendit.

In Anna Comnena's time Greek fire was regarded as a Byzantine state secret. Kedrenos[188] reported that in his time (eleventh century) the secret of the fire was possessed by Lampros, a descendant of Kallinikos: ἐκ τούτου κατάγεται ἡ γενεὰ τοῦ Λαμπροῦ, τοῦ νυνὶ τὸ πῦρ εὐτέχνως κατασκευάζοντος. Since λαμπρόν is a late Greek name for fire, πῦρ, it looks as if Lampros is a hero eponymos, a purely fictitious person. The state chemist who had the recipe, a state secret not put into writing, no doubt took an oath not to divulge it, and the Emperor, who lent troops and engines to his allies, reserved for himself the secret of Greek fire, and sent it ready-made as a sort of ancient hydrogen bomb to his worthy but not wholly reliable dependents.[189] Constantine Porphyrogenitus (tenth century) in his instructions to his son[190] had commanded him emphatically to keep the composition secret: ὡσαύτως χρή σε καὶ περὶ τοῦ ὑγροῦ πυρὸς τοῦ διὰ τῶν σιφώνων ἐκφερομένου μεριμνᾶν τε καὶ μελετᾶν. The secret had in the past been given by an angel to Constantine the Great, those imparting it were anathema, and one about to communicate it had been struck by lightning.

Greek fire, like our hydrogen bomb, was not a prerogative of angels or emperors; it had been discovered by a chemist, and further east, in Saracen lands, there were, even then, other chemists no less inventive.

It was thought that the secret of Greek fire had been lost (see ref. 105); a Dupré, born in Dauphiné, claimed to have rediscovered it and sold the recipe to Louis XV in 1756.[191] The secret was in reality never lost, and Greek fire only slowly made way to artillery and gunpowder[192]; in the earlier stages both Greek fire and cannon were sometimes used together.

We have now reached the period of the Crusades, and in the meantime the Saracens, who had been so surprised and perhaps terrified by Greek fire in their siege of Constantinople in 678, had themselves become thoroughly familiar with it, and used it against the Crusaders in Syria and Egypt.

THE CRUSADES

With the many and various judgments of the Crusades we are not concerned; perhaps the most ridiculous of them is that of Sarton[193] that they were largely occupied in persecuting Jews. This has always been a feature of history, and whatever its cause may be, we can be certain that it was no speciality of the Crusaders. The sack of Constantinople in 1204,[194] an incident in a regular military operation, also cannot be regarded, properly speaking, as a Crusaders' plan; it was conceived by the Venetians as a means of securing payment for services partly rendered, a purely business proposition executed under the supervision of the Doge in person.[195] The sack of Rome in 1527 was no work of Crusaders. The Crusades are traditionally numbered as: I 1097, II 1147–49, III 1189–92, IV 1202–4, V 1218–21, VI 1228–9, VII (or VIII) 1249, VIII (or IX) 1270, but the movement was more or less continuous. They began with the recovery of Jerusalem, which had been captured by the Turks in 1071, and effectively ended with its recapture by them in 1244; the later Crusades failed to achieve anything. In the Crusades, the Europeans fighting in Syria and Egypt encountered the use of Greek fire by the Muslims, which at first inspired exaggerated terror and was regarded as an invention of evil magicians.

Lalanne,[196] who mistakenly thought that Greek fire was gunpowder, said that it was not used in the first three Crusades (1097–1192), that it was not known to the sultans of Iconium in 1161, that it was not used by the Greeks in the sieges of Constantinople in 1203–4,[197] and that it was first used by the Turks in the siege of Damietta in 1218. The attitude which should be taken to these statements will become clear in the account of the facts which will now be given. It is probably true to say that the Latin Crusaders never thoroughly understood Greek fire, which the Muslims used against them effectively at the end of the twelfth century, burning the wooden towers and engines of the Crusaders by incendiaries, so that they were then protected by covering them with iron plates.[198]

The Greek fire was used by special "naphtha troops" (naffāṭūn) attached to each corps of archers in the Muslim army of the 'Abbāsid period, who wore fireproof suits and threw incendiary material. Army engineers had charge of catapults, mangonels and battering-rams, one of them, ibn-Ṣābir al-Manjanīqi, who flourished later under al-Nāṣir (1180–1225), left an unfinished book on the art of warfare in all its details.[199] The Arabs were well acquainted with petroleum. Ibn Hauqal (902–68) says Persia is rich in naphtha (nafṭ).[199a] The naphtha of Ferghāna (Transoxania) is mentioned by al-Muqaddasī, who travelled between 965 and 985.[200] The naphtha of Baku is mentioned by Abulfeda (Abu'-l-Fidā', 1273–1331).[201] The naphtha troops are said by an Arabic author, Ibn Qutaiba (822–89), to cover themselves for protection with preparations of talc.[202] These matters will be dealt with more fully later, but it may be emphasised here that the name "naphtha" confirms very decidedly the view of the nature of Greek fire which is being put forward.

Besides large instruments, the naphtha troops used small naffāṭa of copper containing the liquid, which they threw burning at the enemy (or perhaps threw the burning liquid from them—see ref. 146). Quatremère's[203] suggestion that nafṭ sometimes meant gunpowder (la poudre) is dealt with in ch. V.

Lebeau[204] had said that it was improbable that the Arabs invented Greek fire and the Byzantines learnt it from them, since the Arabs suffered from its use for a considerable time before they knew what it was; it dates only from the time of Kallinikos and it was invented in the West. Michael, the Syriac historian (1126–99), called Greek fire naphtha.[205] Abu'-l-Faraj (Bar Hebræus) (1226–86) in his Syriac chronicle says the Arabs, formerly victorious, began to lose after Greek fire was invented.[206]

The accounts of the weapons used by the Muslims up to 1225, as given both by the Crusaders and in Arabic sources, leave no doubt that up to that time they were ignorant of gunpowder, and no Arabic work before that time mentions saltpetre. In a Latin translation of an Arabic chemical work attributed to Morienus (ch. 11) it is said: "sal anatron id est sal nitri" ("soda is nitrous salt," i.e. not saltpetre).[207] The first definite mention of saltpetre in an Arabic work is that in al-Bayṭār (d. 1248), written towards the end of his life, where it is called "snow of China." Al-Bayṭār was a Spanish Arab, although he travelled a good deal and lived for a time in Egypt.

An account by Raymund de Agiles of the siege of Jerusalem in the First Crusade, which began in June, 1099, says[208] that as the Christians approached the walls with machines, the Arabs threw not only stones and arrows but also all kinds of incendiaries, including pitch, wax, sulphur and tow:

> Sed cum iam proximarent cum machinis ad muros, non solum lapides & sagittæ, verumetiam ligna & stipula proiiciebantur, et super hæc ignis; & mallei lignei involuti pice, & cera, & sulphure, & stuppa, & panniculis igne succensis proiiciebantur in machinas. Mallei inquam clauati ab omni parte, vt quaqua parte ferirent, hærerent, et hærendo inflammarent. Ligna vero & stipula ideò iaciebant, vt saltim incendia indè accensa, retardarent quos neque gladius, & alta mænia retardarentur, nec profundum vallam retinere poterat.

William of Tyre (*c.* 1130–90) in describing the siege of the Turks in Nicæa (Izniq) in the same Crusade[209] says the Saracens threw on the machines pitch, oil, fat and all kinds of things serving as incendiary material: picem quoque, oleum, & aruinam, & cætera quæ incendiis solent fomitem ministrare, & accensas faces in nostras machinas dirigentes [saraceni]; . . . immissis magnis molaribus, et igne superiniecto. . . .

Albert of Aachen (after 1121), in describing the same siege, says[210] the Turks set fire to a wooden tower and machines by incendiaries: turrim ligonibus perfodi, adipem, oleum, picemquè stuppis & facibus ardentissimis commixtam, fundebant à mœnibus, quæ instrumentum arietis & crates vimineas prorsus absumserunt . . . faculas ardentes cum pice & adipe iactant super machinam.

He also says that in the siege of Assur (1099) the Turks set fire to a tower by means of pointed iron stakes wrapped in tow soaked in oil, pitch and other combustibles, the fire being inextinguishable by water: palos ferreos & acutos, oleo, stuppis, pice, ignis fomite inuolutos, & omnino aqua inextinguibiles.

They also set fire to and destroyed a second tower in a similar way, and although men and women attempted to extinguish the machine by vessels of water, this amount of water served no purpose, since this kind of fire is inextinguishable by water and the great flame is insuperable: simili iaculatione palorum ignitorum; . . . mox ad extinguendam machinam de omni exercitu & tentoriis concurrunt viri ac mulieres, aquam singuli in singulis vasis afferentes. Sed minimè profecit tanta aquarum suffusio. Nam huius ignis aquâ erat inextinguibile, & flamma magna et insuperabilis.

In describing another siege, William of Tyre[211] mentions the same primitive incendiaries only:

> In ipsas machinas, torres incensos, tela ignita sulphure, pice, pastâ & oleo, & iis quæ incendio solent fomitem ministrare, vt eas exurerent, certatim iaculabantur. . . . Nostri verò iniectis ignibus occurrentes, aquas desuper fundebant copiosiùs, vt incendiorum comprimerent importunitatem . . . ignem incessanter in ollis fragilibus, & quibuscunque modis poterant, cum sulphure, & pice, aruina, & adipe, stupa, cæra, lignis aridis, & stipula, & quæcunque solent incendium irritare, & ignibus ministrare fomitem, iaculabantur.

The method of attack used in siege warfare consisted in filling up the ditch

or moat before the walls, e.g. by stones thrown by ballistæ, which served at
the same time to damage the walls, if not to breach them, and then pushing
forward a high wooden tower, moving on rollers, close to the walls. The
tower had at the top a hinged drawbridge, which was lowered on to the
rampart, and across this the storming party, concealed in the tower, crossed
over to engage in hand-to-hand conflict.

This method was used by the Normans under Robert Guiscard in attacking
the Byzantine army under Palæologos at Dyrrachium (1082). The Norman
tower, built from the wood of ships put out of action by the Venetians, had
inside a broad staircase and 500 picked troops in complete armour. During
the building of the tower the besieged had built on the ramparts a slender
framework of masts and yards, which excited the contempt of the Normans.
The immense tower was pushed forward on an inclined plane and wooden
tramway up to the wall. The framework on the wall descended and wedged
the drawbridge firmly against the structure, closing the tower as by a door.
At the same instant an immense quantity of incendiary material was poured
and projected from the walls over the wooden tower, which was quickly
enveloped in flames and smoke. As the tower, with its contents, collapsed,
a sortie was made and the work of destruction completed.[211a]

The famous siege of Acre ('Akkā) in the Third Crusade (1190–91) is
described both by Arabic and European authors. Yūsuf ibn Rāfi ibn Tamīn
Bahā' al-dīn (or Ibn Shaddād; the "Bohadin" or "Boha-eddin" of Gibbon
and others) says[212]:

> "Then a young man from Damascus, a founder (fondeur) by trade,
> promised to burn the towers [of the Christians] if he were admitted to
> the town. This was agreed upon, he entered Acre, and was given the
> necessary materials. He boiled naphtha and other drugs together in
> copper pots (marmites). As soon as these things were properly inflamed,
> when in a word they looked just like balls of fire, he threw them against
> a tower, which at once caught fire . . . the second tower was also inflamed,
> and then the third."

Bahā' al-dīn[213] continues:

> "As the danger increased, two arrows, like those shot from great
> machines, were taken, and fire put on the tips so that they shone like
> torches. This double-arrow was shot against a machine and fortunately
> it stuck fast. The enemy attempted in vain to put out the fire, since a
> strong wind arose."

A similar account is given by Ibn al-Alathīr[214]:

> "The man from Damascus, in order to deceive the Christians, first
> threw pots with naphtha and other things, not kindled, against one of
> the towers, which produced no effect. The Christians, full of confidence,
> climbed triumphantly to the highest stage of the tower, calling out in
> derision. The man from Damascus, waiting until the contents of the pots
> had soaked into the tower, at the right moment threw on to it a well-
> burning pot. At once fire broke out over the whole tower and it was

destroyed. The fire was so quick that the Christians had no time to climb down and they and their weapons were all burnt up. The other two towers were similarly destroyed.''

Richard I in the voyage from Cyprus to Acre captured a Saracen transport ship laden with ballistæ, bows, arrows, lances and abundance of Greek fire in bottles, which an eye-witness had seen put aboard at Beyrut:[215]

> Erat quidem qui diceret se apud Baruth extitisse quando navis illa his omnibus congestis fuerat onerata, centum videlicet camelorum sarcinis omnis generis armorum, videlicet magnis cumulis balistarum, arcuum, pilorum, & sagittarum, septem inerant *Sarracenorum* Admirati & octoginta *Turci* electi, & præterea omnem æstimationem excedentia omnium genera victualium. Habebant & ignem *Græcum* abundanter in phialis, & ducentos serpentes perniciosissimos, in exitium Christianorum paratos.

Pictures of such ships, stacked with spherical pots or bottles containing naphtha, are in the MS. of Ḥasan al-Rammāḥ (*c.* 1280).[216]

At this time large stocks of petroleum were available; it is reported that 20,000 barrels of it were used in 1168 by Shawar to burn down the city of Fusṭāṭ (Cairo) to prevent its recapture by the Franks.[217] In Arabic tales the Byzantines also had great stocks of naphtha and (with an anticipation of modern war-propaganda) they stored it in churches. The Byzantine king brought 4000 grenades of ''flying naphtha'' (? volatile naphtha). In naval battles the ''fire reached the surface of the water as poison travels inside the body.'' The Arabs also boasted of large stocks of ''3000 naphtha grenades'' of ''white volatile Syrian naphtha,'' and charges of 2 quintals of it. The Byzantines were expert in the ''cooking'' (distillation) of naphtha and in burning ships, etc.[218]

After the Third Crusade, the Venetians in Constantinople probably learnt the composition of Greek fire; it was used on both sides, and the accounts of battles make little mention of it, so that in later times it was wrongly assumed that the secret had been lost. As late as the Fifth Crusade, in the siege of Damietta (1208), an eye-witness[219] describes the use of Greek fire, extinguishable by vinegar and sand, and hence not containing saltpetre.

The siege of Manṣūra in the Nile Delta in 1249, in which King Louis IX (St. Louis) of France was taken prisoner by the Sultan al-Muʻaẓẓam, is graphically described by Joinville (1224–1319). When the French commander saw the Saracens preparing to discharge the fire he announced in panic that they were irretrievably lost (seigneurs, nous sommes perduz à jamais sans nul remède; nous sommes ars et bruslez). The fire was discharged from a large ballista in tubs, which Berthelot[220] says the Arabs called ''marmites de l'Irak.'' Joinville, who composed his chronicle in the period 1270–1319, says[221]: ''qu'ils nous getteront le feu grégeois . . . aussi gros que ung tonneau . . . il sembloit que ce fust fouldre qui cheust du ciel, et me sembloit d'un grant dragon vollant par l'air. . . . It was like a big tun and had a tail of the length of a large spear: the noise which it made resembled thunder, and it appeared like a great fiery dragon flying through the air, giving such a light that we could see in our camp as clearly as in broad day." When it fell it burst and the liquid burnt in a trail of flame.

The troops fighting the Crusaders in Egypt, called "Turks" by Joinville, were Mamlūks, originally Christian boys captured in the region round the Black and Caspian Seas and trained by the Sultan of Egypt in Arabic methods of warfare. These Circassian slaves founded the famous Mamlūk dynasty in Egypt. The Crusade of Louis IX was defeated by one of the three Muslim queens, Shajar al-durr, queen of the Mamlūks, in 1250. Another was Ābish, ruler of Fārs, and the third Razīya, queen of Delhi (1236–40).

Joinville, a gentlemanly soldier of the old school, describes the effects of Greek fire, a barbarian frightfulness wholly unknown to him, in a most amusing way. Yet when St. Louis died in Paris in 1270 an even more terrible and devastating engine of warfare had become known, gunpowder; and when Joinville himself died in 1319 cannon were round the corner. Joinville says the commander of a castle which had been liberally sprayed with Greek fire shouted to heaven: "Aidez nous, sire, ou nous sommes tous ars. Car veez-cy comme une grant haie de feu grégois, que les Sarrazins nous ont traict, qui vient droit à notre chastel." Joinville and his company were put on night duty, and having witnessed a terrifying attack during the day, when two castles were burnt up in a moment, they were overjoyed to find that they were left unmolested: "Et lors mes chevaliers et moy laüasmes Dieu. Car s'ilz eussent attendu a la nuict, nous eussions esté tous ars et bruslez." Yet when the fire fell on the clothes of one of the king's heralds, Guillaume le Bron, it could be put out with little damage: "ilz amenèrent ung autre villain Turc, qui leur gecta trois foiz le feu grégois et à l'une des foiz il print à la robbe de Guilleaume de Bron, et l'estaignit tantost, dont besoing lui fut. Car s'il se fust allumé, il fust tout brûlé." Guy Malvoisin was set on fire, but his men put him out and he was not much the worse, being a strong and tough man: "il lui tiroient le feu grégois sans fin, tellement que une foiz fut, que à grant peine de lui peurent estaindre ses gens à heure, mais nonobstant ce, tint-il fort et ferme, sans estre vaincu des Sarrazins." It is noteworthy, in fact, that amidst all this terror and religious fervour, Joinville does not produce for us a single fatal casualty due to Greek fire. The Crusaders believed that anyone struck by it was lost—"ars et bruslez," and it seems to have been regarded as a kind of old-fashioned atomic bomb.

Joinville says the fire was thrown in tubs by means of a "perrière" (engine for throwing stones) or (presumably in pots) by an "arbaleste à tour." The mobile troops probably threw it in jars or "avecques instrumens qu'ilz avoient propice," perhaps hand-pumps. When the Crusaders attacked the Saracen ships, the latter: "commancèrent à tirer à nous . . . grant foizon de pilles avec feu grégois, tant qu'il ressembloit que les estoiles cheussent du ciel," perhaps small masses (pilles) discharged from mouth blowpipes, incendiary pea-shooters, of which we heard above from Anna Comnena. Of gunpowder there is no word.

Marino Sanudo (1260–1337), in his *Liber Secretorum Fidelium Crucis*, completed in 1313, gives minute directions for equipping a new crusade (which did not materialise) but says nothing of Greek fire, gunpowder, or firearms.[222] He gives eye-witness details of the commerce of Alexandria and other Egyptian ports, mentions Indian pepper, ginger, silk and cotton, the pearls and aromatic woods of the Malabar coast and Ceylon, Aden as a trading port, and the incense of Arabia; these, and Egyptian sugar were taken by Venetian ships to Europe in exchange for timber and iron.[223]

Up to the tenth century Greek fire was mostly used in naval battles, but was also projected from ships on buildings on shore. In the siege of Acre ('Akkā) on the coast of Palestine in the Third Crusade (1191), Greek fire was used on land and sea. The author of the *Itinerarium Regis Richardi* (an eye-witness) says[224]: "oleo incendiario quod ignem Græcum vulgus nominant . . . ignis iste perniciose fœtore, flammisque livientibus, silices & ferrum consumit; & cùm aquis vinci nequeat, harenâ respersus comprimitur; aceto perfusus, sedatur"; "the incendiary oil, commonly called Greek fire, burnt with a livid flame, smelt abominably, consumed even stone and iron, and could not be extinguished by water but only by sand, though affussion of vinegar subdued it." He mentions its use by cavalry. Arabic authors also describe its use in the siege of Acre.[225] Acre, the last Christian possession in Syria, was lost to Khalīl in 1291, and the Latin principalities were never recovered.[226]

"Sea fire" went out of use about 1200, although "Greek fire" (usually called "wild fire" in English) was used by Edward I at Stirling in 1304.[227] Yule[228] states that Greek fire was probably not known to the Italians in the thirteenth century. Joinville (1224–1317) describes it as something quite new. Greek fire was used, along with gunpowder and cannon, in the siege of Constantinople in 1453, when the Turks used a heavy gun of over 3 ft. calibre throwing stone balls weighing 600 lb. and directed by a Hungarian engineer.[229] The use of Greek fire ($\pi\hat{\upsilon}\rho$ $\acute{\upsilon}\gamma\rho\acute{o}\nu$) in the siege of Constantinople in 1453 is mentioned by Phrantzes[230] (1477), who says a siege tower, protected by a triple covering of buffaloes' hides, was burnt by it, and by Doukas.[231] Bows and arrows were last used by English archers in a siege of 1627.[232]

After the terror caused to the Europeans when they first encountered Greek fire used by their Eastern enemies had subsided, its effects were found to be much less powerful than they had feared (see ref. 184). Unlike a projectile from a gun, the discharge from a flame-thrower can be side-stepped by sufficiently prompt evasive action. In its early use, Greek fire certainly caused panic; it was regarded with extreme disfavour as an unknightly weapon, but steady troops soon grew accustomed to it. The same phenomenon was seen when gunpowder, a more terrible agent of destruction than Greek fire, came upon the scene. Its effects were stupendous. It made hay with medieval siege fortifications and the armoured knight. In 1464, the Earl of Warwick knocked Bamborough castle to pieces in a week, and Philip of Hesse in 1523 flattened Landstuhl, the strongest feudal castle in the Rhineland, in a single day.[233] Some early capitulations probably resulted from fear rather than damage—a few stone cannon balls falling inside a fortified place producing panic. As time went on, trained infantry stood firm under heavy fire from large guns and mortars, and every invention intended to make the use of infantry impossible has always failed in its purpose. Shrapnel, high-explosives, Flammenwerfer (that old friend), poison gas, rockets, aeroplane bombs, and all the paraphernalia of modern war have failed to shake the foot-soldier. He may be blown out of existence, but he stays where he is until this happens. The war of the future, it appears, with atomic bombs and bacterial warfare, is to be directed against the civilian population, who, it is anticipated, will capitulate. Experience in London under fire from rockets from the stratosphere makes even this hope questionable, and whether the specialities of the atomic scientists will achieve it is yet unknown.

WHAT WAS GREEK FIRE?

The incendiary materials used in ancient warfare, it has been seen, included:

(i) Liquid petroleum or naphtha, from oil wells in 'Irāq (Hīt) or Kerkut (across the Tigris, in ancient Assyria), probably used, together with burning pitch and sulphur, by the ancient Assyrians. In Greek and later times the petroleum wells in Armenia and the shores of the Caspian Sea were also available.

(ii) Liquid pitch, used by the Greeks from about 430 B.C. in fire-cauldrons, fire-ships, etc., and on incendiary arrows.

(iii) Mixtures of pitch, resin and sulphur, used by the Greeks from 424 B.C.

(iv) A mixture of quicklime and sulphur, inflaming on contact with water, known to Livy (for 186 B.C.).

(v) A mixture of quicklime and sulphur with other inflammable materials such as bitumen, resin, naphtha, etc., inflaming on contact with water, mentioned in an interpolation (sixth century A.D.) in the *Kestoi* of Julius Africanus.

(vi) Greek fire, first used in the eighth century, different from all the above, and containing some secret ingredient.

The compositions (iv) and (v), which have often been regarded as Greek fire, were more in the nature of toys, pet ideas of inventors, and could never be effectively used in warfare either on land or sea.

Greek fire is always described as a liquid or semi-liquid. The liquid fire was also called incendiary oil and it burnt on water, so that it could be extinguished only with vinegar, sand, or urine; it burnt even stones or iron.[233a] At first, the incendiary was probably thrown in earthenware pots and kindled by incendiary arrows, although the large tubs thrown by ballistæ (as described by Joinville) were already burning.

Later, some apparatus called a "siphon" (σίφων) was used. Theophanes says the Greek ships were fitted with siphons before Greek fire was invented. The siphons were, apparently, flame-projectors, either hand-pumps or reservoirs worked by mechanical force-pumps. Berthelot[234] says pumps throwing burning petroleum were tried at Havre in 1758, and he himself saw them used against Paris by the Germans in 1870; their use by German troops in the First World War (1914–18) need not, therefore, have caused the surprise which it did.

In land and sea battles glass bottles or pottery jars, or tubs, of blazing incendiary were thrown by hand or projected from ballistæ. Incendiary arrows with heads wrapped in tow soaked in an inflammable liquid and kindled were shot, sometimes from arbalests (arcus ballista). In land battles, liquid fire was projected by small pumps carried by troops, or a powdery or pasty incendiary was blown by the mouth through a metal tube with a burning igniter at the tip, so as to produce a cloud of fire, or burning balls; or large tubs filled with blazing incendiaries were thrown by ballistæ, a method which overlaps the ancient use of incendiaries. In sieges, pots of burning material were dropped on wooden structures, or cascades of burning oil poured over them, or in naval battles incendiary pots or tubs were thrown from the masts on the decks of enemy ships. These were not properly speaking Greek fire. Pots filled with a mixture of quicklime and a combustible, particularly sulphur, were secretly put under the timbers of roofs and, when moistened by heavy dew in the early morning, the mixture burst into flame. In Chinese accounts, this mixture in paper bags was thrown on water and caused some kind of explosion. In the use of real Greek fire the larger fire-ships (δρόμωνες σιφωνοφόροι) were fitted with brass or iron animal heads, through the mouths of which an incendiary liquid was projected and inflamed as it left the apparatus. The fire was also projected from special flat-bottomed ships (chalandria, χάλανδρον = κράββατον in Hesychios, a mat or mattress, classical χαλάδριον), 200 of which were used in the tenth century in an expedition against the Arabs in Sicily, and one man sufficed to work the projector and row the boat. The projected fire (πῦρ πεμπόμενον) would be effective only at short range and on calm water,[235] but it could burn wooden ships down to the water-line.[236]

Petroleum was, of course, well-known to the Arabs, who (as will be shown in ch. V) were quite familiar with its distillation. Jacques de Vitry[237] reported a spring of a "water" in the East from which Greek fire was made. It was prized by the Arabs. It could be extinguished only by vinegar, urine or sand: fons quidem in partibus Orientis, ex cuius aquis ignis Græcus efficitur quibusdam aliis admixtis, qui postquam vehementer fuerit accensus, vix aut nunquā potest extingui nisi aceto & hominū urina, & sabulo. Prædicta autem fontis aquas magno pretio comparant Saraceni (A.D. 1180).

In the *Image du Monde* of Gossuin of Metz (1247), Greek fire is said to be made from a material from certain fountains in the East; the Saracens value this water more than good wine. Only vinegar and urine can extinguish it.[238] These accounts can be multiplied, but this is not necessary.[239]

Romocki[240] had realised that petrol would be a very effective incendiary if projected by pumps, and since Julius Africanus had spoken of "natural (αὐτόρυτος)" petroleum, there must have been an artificial (distilled) kind, but he thought it was probably mixed with solid materials. The recipe from Anna Comnena given (see ref. 181) shows that the solids were pine resin and sulphur, but the essential ingredient, petrol, she deliberately omits. Quick-lime would not be a suitable material at all, and is never mentioned or even hinted at, as a component of Greek fire. Darmstaedter[241] had assumed that the new invention of Kallinikos was the addition of quicklime, but the use of quicklime was well-known long before, as has been shown. Petrol, obtained by distillation, could be projected burning, or sprayed and then kindled by an incendiary arrow. It would float, still burning, on water. Both the effective range of projection, and the stability of the flame, would be increased by thickening the liquid, even but not necessarily to the extent of producing a paste, by dissolving in it resins or solid combustibles. In short, all the pro-perties and effects of Greek fire, and all descriptions of the methods of making and using it, agree in detail with this view of its nature.

Wiegleb[242] had suggested that Greek fire was Greek (Armenian?) or Persian petroleum or naphtha. Gibbon[243] concluded that the principal ingredient of Greek fire was naphtha; Lebeau[244] assumed that Greek fire was either petroleum "sometimes distilled," or a solution of sulphur in petroleum, the "oil of Medea" (see ref. 32). Petroleum, probably from the Caspian region, was imported to Constantinople. Georgios Kodinos, the Curopalates (ὁ κυροπαλάτης) or supervisor of the palace there, in an account of the antiquities of Constantinople probably written about 1450 but interpolated by later editors, mentions[245] two baths erected there by Septimius Severus (Emperor A.D. 193–211); the larger one, outside the town, called Kaminia (καμίνια, "furnaces"), was a vast building; 2000 persons could use the baths at once. These baths were heated by Median fire (λουτρὸν μετὰ τοῦ Μηδικοῦ πυρός), i.e. petroleum.[246]

A sixteenth-century Persian poet, Tanukhi, describes a yellow (bronze) piece (tube?) containing a viscous liquid of the same colour, and with a bell-shaped mouth; this may refer to a cannon,[247] except that cannon do not contain viscous liquids. The recoil, however, is mentioned.

True essential oils known in the Roman period were those of the lemon and laurel[248] and oil of turpentine, the latter obtained by a primitive process of distillation. Dioskourides[249] says resin oil [turpentine] (πισσέλαιον) is made from resin (ἐκ τῆς πίσσης) by separation of the watery part (ὑδατώδους αὐτῆς) which rises like cream on milk and is separated by heating the resin and hanging clean wool over it, which, when it is wetted through by the rising vapour (ἀναφερομένου ἀτμοῦ) is pressed out into a vessel, and this is done as long as the resin boils. The description, it is seen, is not quite accurate. Pliny[250] says the same: a fire is lighted under a pot containing the resin, the vapour (halitus) rises and condenses on wool spread over the opening of the pot, from which the oil, called pissinum or pisselaion, is afterwards squeezed: pice fit, quod pissinum appellant, quum coquitur velleribus supra halitum ejus expansis atque ita expressis . . . color oleo fulvus. About the same time, Scribonius Largus[251] says: florem picis autem appello, quod excipitur dum ea coquitur, lana superposita ejus vapori. It is interesting that he did not know of distilled water, but in his prescriptions specifies rain-water. The Baleriac islanders produced an oil from the turpentine tree.[252]

The extraction of pitch (πίσσαν καίειν; picem coquere) was also carried out in a meiler (σύνθεσις, calix) in a way described by Theophrastos[253] and Pliny.[254] Pliny describes the preparation of wood tar and cedar oil by heating the wood in chips in a closed oven; the first liquid running out by a pipe made for the purpose is as clear as water and is called by the Syrians cedrium, which is so powerful that it is used in Egypt for embalming dead bodies. This would be cedar oil or turpentine. The next running was thicker, and is pitch (pix) which was boiled with vinegar in copper cauldrons to make Bruttian pitch, used for caulking barrels.[255] This suggests that the oil of turpentine was principally made in Bruttium (South Italy, where distillation was afterwards much developed) and resin was made in Kolophon in Greece, hence the name colophonium for resin. The best resin was from the terebinths (terebinthi) of Syria and Cyprus, that from cedars, cypresses and pines being less esteemed. Every resin dissolves in oil (resina omnis dissolvitur oleo[256]). Pliny and Dioskourides thus describe the distillation of tar and cedar oil by crude processes in one of which the pitch or oil is boiled in a pot over which fleeces are spread to catch the vapour and the condensate is wrung out of them. In the preparation of mercury from cinnabar, the latter was heated in an iron vessel and the drops of mercury condensed on the lid,[257] this apparatus being called an ambix (ἄμβιξ). Distillation in which the vapour from a flask (λοπάς, cucurbita) was condensed in an upper part (φιάλη), and ran into a receiver through a spout, was invented by the first chemists in Alexandria, in the first to third centuries A.D., the invention being attributed to Maria the Jewess.[258] Glass still-heads of the fifth to the eighth century A.D. have been excavated in Egypt and Syria[258a] and are of the same form as those depicted in the Greek chemical manuscripts and there attributed to Maria.

Distillation is described by the Spanish-Arabic physician Abu'-l-Qāsim (Abulcasis),[259] d. A.D. 1013 or 1107, and it could easily have been adapted to making petrol. Before this, knowledge of distillation had passed from Egypt to Syria and might have been known there to Kallinikos; it was already known in Constantinople, where the alchemist Stephanos (who was familiar with it) flourished in the seventh century A.D. It seems probable that the process was used in Constantinople to make the essential constituent of the new invention. In an Abyssinian account, the Egyptians taught the Abyssinians how to make and use Greek fire, naft (= petroleum), in a war against the Arabs in 1420.[260]

It is very probable that the basis of the earliest Greek fire was liquid rectified petroleum or volatile petrol, which was projected in hand-grenades, or ordinary petroleum in tubs shot from ballistæ. In Leo's time this liquid was projected from jets to which it was pumped through flexible leather tubes by a force-pump, the jets being either fixed in brass figureheads on ships, or manipulated to turn in various directions. The burning liquid could also float and burn on water around the ships. Since the petroleum was imported, its nature could be kept a semi-secret among the Byzantine officers, but its use as an incendiary in warfare must soon have been learned from experience of it by the Arabs (if they did not know it already), who then began to use it themselves.

Petrol itself would not be very effective in flame-projectors; the projected jet dissipates too rapidly and does not carry far enough. It must be thickened, almost or quite to a jelly, by dissolving in it resinous substances of the kind specified by Anna Comnena, and sulphur may also have been added. The particular mixture (or mixtures) used, and the mechanical means of projecting it, together constituted the invention of Greek fire. It was an achievement of chemical engineering.

What were regarded as the properties of Greek fire were reproduced by Brock[260a] by a mixture of saltpetre 6, sulphur 1, powdered pitch 3¼, powdered glue ¼, powdered plumbago ¼, rendered plastic by heat and moulded into balls, which were pressed into a steel mortar. This when ignited at the muzzle first burned, then shot out the burning balls in succession to a distance of 100 yards. This is evidently quite different in composition from Greek fire, and such a compact mixture could not have been projected by siphons; it would not have floated on water or spread over wood. Although many other writers[261] all suggested that Greek fire contained saltpetre, others[262] rejected this supposition, and in the present writer's opinion rightly so. There is no evidence of the use of saltpetre in incendiary compositions in the West before about 1250, the reason being that purified saltpetre was not known there before about 1225. This will be discussed later.

Oman,[263] in an attempt to piece together the motley of the Byzantine writers, concluded that Greek fire was a "semi-liquid substance, composed of sulphur, pitch, dissolved nitre and petroleum boiled together and mixed with certain less important and more obscure substances." If the saltpetre is omitted, this is not a bad description.

Hime[264] at first thought Greek fire was a mixture of sulphur and quicklime. Later[265] he thought that the "sea-fire" ($\pi\hat{\nu}\rho\ \theta\alpha\lambda\dot{\alpha}\sigma\sigma\iota\sigma\nu$) or "wet fire" ($\pi\hat{\nu}\rho$ $\dot{\nu}\gamma\rho\dot{\sigma}\nu$), which he then regarded as different from Greek fire (incorrectly, as has been shown), contained calcium phosphide, made by Kallinikos by heating lime, bones and urine, which was ignited by water from hoses connected with pumps (siphons), when spontaneously inflammable phosphine gas would be evolved. Crude calcium phosphide, however, would sink rapidly in water, and even if a few harmless flashes of ignited gas did appear on the surface, the effect would be altogether insufficient to produce the extensive and terrifying phenomena attributed by all writers to Greek fire. Even with good calcium phosphide, the effects would be as described; with the product made from lime, bones and urine it is doubtful if any effect at all would be produced.

REFERENCES

1. Botta, *Monumen de Ninive*, Paris, 1849, i, plates 52, 61; 1851, v, 124 (text); Layard, *A Second Series of the Monuments of Niniveh*, London, 1853, plate 21 (fire), 39 (pots or stones); Maspero, *Passing of the Empires*, 1900, 10; Partington, *Origins and Development of Applied Chemistry*, 1935, 225; Berthelot, *Rev. Deux Mondes*, 1891, cvi, 786, did not think petroleum or resins were used.

2. Prov. xxvi, 18; Ps. cxx, 4; Isaiah *l*, 11 (eighth century B.C.); H. W. L. Hime, *The Origin of Artillery*, 1915, 139.
3. *History*, viii, 52.
4. Thucydides, ii, 75.
5. Diodorus Siculus, xiv, 41–3; H. Diels, *Antike Technik*, Leipzig and Berlin, 1924, 20, 94.
6. Thucydides, vii, 53.
7. Diodorus Siculus, xx, 86.
8. *Polyorketikon*, xxxiii–xxxv.
9. Stephanus, *Thesaurus Græcæ Linguæ*, ed. Hase and Dindorf, Paris, 1842–6, v, 569; manna (μάννα) is incense in Dioskourides, *Mat. Med.*, i, 83, 94.
10. *Opera et Dies*, 423.
11. Feldhaus, *Die Technik der Vorzeit*, 1914, 320.
12. Bauer, in Iwan Müller, *Handbuch der klassischen Altertumswissenschaft*, 1887, IV, i, 326.
13. Diodoros Siculus, xx, 48, 88, 96, 97.
14. *Exped. Alexander*, ii, 19; G. Rawlinson, *History of Phœnicia*, 1889, 516.
15. *History*, iv, 100–14, Teubner ed., 1930–6, 396; text in Romocki, i, 8 (see ref. 22).
16. Apollodoros, *Polyorketikon*, in Thevenot, *Veterum Mathematicorum*, Paris, 1693, 21; in R. Schneider, *Griechische Poliorketiker*, in *Abhl. K. Ges. Wiss. Göttingen, Phil.-hist. Kl.*, Berlin, 1908, x, no. 1, pp. 18–21, 20–21; see Plate iii, 10, 11.
17. *Polyorketika*, in Wescher, *Poliorcétique des Grecs*, Paris, 1867, 219, 224; for Heron of Alexandria, second century A.D., see Diels and Schramm, *Herons Belopoiika* (Greek and German), in *Abhl. Preuss. Akad. Wiss., phil.-hist. Kl.*, 1918, no. 2; *id.*, *Philons Belopoiika*, *ib.*, no. 16.
18. Livy, xxi, 37. See Berthelot, ref. 88, i, 370.
19. *H.N.*, xxxvi, 1.
20. *H.N.*, xxiii, 27.
21. Q. by W. Smith and Marinden, *Dict. of Greek and Roman Antiquities*, 1914, ii, 469.
22. *Geschichte der Explosivstoffe*, Berlin, 1895, i, 9.
23. *Aeneid*, ix, 705.
24. *History*, xxi, 8.
25. *History*, iv, 23.
26. Medieval artillery engines are described by H. Yule, *The Book of Ser Marco Polo the Venetian concerning the Kingdoms and Marvels of the East*, 2 vols., 1871, ii, 122 (who says the name "mangonel" is derived from μάγγανον, "a piece of witchcraft"), and C. W. C. Oman, *The Art of War in the Middle Ages*, 1905, 131 f., 543 f.
27. *History*, XXIII, iv, 14–15.
28. *Rei Militaris Instituta*, iv, 1–8, 18.
29. Yates and Wayte, in Smith and Marinden, ref. 21, i, 937.
30. *De Cœlo*, ii, 7, 289a; Ideler, ed. Aristotle, *Meteorologica*, Berlin, 1832, i, 359 (note); Lucretius, *De Rerum Natura*, vi, 305.
31. Romocki, ref. 22, i, 2, and Feldhaus, ref. 11, 321, report that Aristotle said that an arrow (Pfeil) shot very rapidly inflames, but I cannot find that he says anything about arrows in the text.
32. *H.N.*, ii, 108–9.
33. *De Bello Gothico*, iv, 11; in *Corpus Script. Hist. Byzant.*, Bonn, 1833, ii, 512; ed. Haury, Leipzig, Teubner, 1905, ii, 541.
34. Hoefer, *Histoire de la Chimie*, 1866, i, 301.
35. *Epitome Rerum Præclare Gestarum*, vi, 10; ed. Niebuhr, *Corpus Script. Hist. Byzant.*, Bonn, 1836, 165; ed. Ducange in Migne, *Patrologia Græca* (*PG*), 1864, cxxxiii, 657.
36. Lebeau, *Histoire du Bas-Empire*, Paris, 1827, ix, 211; L. Lalanne, *Mém. div. Sav. Acad. des Inscriptions et Belles-Lettres*, II. Sér., Paris, 1843, i, 294–363 (339); Partington, *Origins and Development of Applied Chemistry*, 1935, 269, 419, 421.
37. Seybolt, *Speculum*, 1946, xxi, 38; *Isis*, 1946, xxxvi, 235 (abstr.).
38. *Geography*, XVI, i, 15, p. 743 C.
39. *H.N.*, ii, 104–6.
40. *Mat. Med.*, i, 101.

41. *Refutationis omnium hæresium*, vii, 25; ed. Duncker and Schneidewin, Göttingen, 1859, 370; Plutarch, *Quæst. conviv.*, v, 7; *Moralia*, Paris, 1841, ii, 825; *The Philosophie, commonlie called the Morals*, tr. Holland, f°, 1603, 723, reported this for Median naphtha.

42. *H.N.*, xxxi, 39.

43. *H.N.*, xxxv, 51.

44. *H.N.*, xxiv, 25.

45. *Architectura*, viii, 3.

46. *Met. Med.*, i, 99.

47. ψ-Aristotle, *De Mirabil. Auscult.*, chs. 35, 113, 115, 127.

48. *History of Chemistry*, 1830, i, 106.

49. *H.N.*, ii, 110–11.

50. Hoefer, ref. 34, i, 182.

51. *Varia Historia*, xiii, 16.

52. *Itinerary*, text and tr., ed. Adler, 1907, 8.

53. *History*, viii, 4; ed. Stavenhagen, Leipzig (Teubner), 1922, 213–4.

54. Lebeau, ref. 36, 1827, vii, 16.

55. *Chronographia*, xvi; ed. Dindorf, Bonn, 1831, 403 f., 635; Migne, *PG* (ref. 35), 1860, xcvii, 597. Malalas (*c.* A.D. 491–578) also reports that in a riot in Constantinople in the time of Anastasios, fire was "injected" on buildings, which were set on fire and burnt to cinders (ἔβαλε πῦρ εἰς τὴν λεγομένην; ignes . . . quam vocant injecit), *Chronographia*, ed. Dindorf, Bonn, 1831, 394; Migne, *PG*, xcvii, 583; and also mentions "injecting fire" (ἐμβόλου ἕως) in Justinian's time (A.D. 482–565); *Chronographia*, ed. Dindorf, 1831, 474.

56. *Annals*, xiv; Migne, *PG* (ref. 35), 1864, xxxiv, 1217.

57. Agathias, *History*, iii, 5: ed. Niebuhr, Bonn, 1825, 147; Gibbon, *Decline and Fall of the Roman Empire*, ch. xlii, Everyman ed. (all refs. are to vol. and page of this ed.), vi, 318; Lebeau, ref. 36, 1827, ix, 211; Gibbon, who refers only to Agathias, has added the sulphur, bitumen, and naphtha from some other source, since Agathias does not name them. J. B. Bury, *A History of the Later Roman Empire*, 1889, i, 447 (with the date A.D. 550), has "oil of Medea."

58. Oman, ref. 26, 1905, 47.

59. Hime, ref. 2, 1915, 139–40.

60. *Quæst. conviv.*, iii, 5; *Moralia*, Paris, 1841, ii, 792; tr. Holland (ref. 41), 1603, 690.

61. *H.N.*, ii, 49.

62. *H.N.*, xxxi, 39.

63. Favé, in Napoleon III, *Études sur le passé et l'avenir de l'Artillerie*, Paris, Libraire Militaire, J. Dumaine, Libraire-Éditeur de l'Empereur, 1862, iii, 4; Berthelot, ref. 1, 1891, cvi, 795–6.

64. *History*, ii, 180.

65. Lenz, *Mineralogie der alten Griechen und Römer*, Gotha, 1861, 11.

66. Aulus Gellius, *Noctes Atticæ*, xv, 1.

67. Ammianus Marcellinus, *History*, XX, vi, 13 : aliæ unctæ alumine.

68. Berthelot, ref. 1, 1891, cvi, 790–1, 794.

69. *H.N.*, xxxvi, 23 (53) : mirum, aliquid, postquam arserit accendi aquis.

70. *De Civitate Dei*, xxi, 4.

71. Kautzsch, *Apokryphen und Pseudepigraphen des alten Testaments*, 1900, i, 87–8; Charles, *Apocrypha and Pseudepigrapha of the Old Testament*, Oxford, 1913, i, 128, 133. Cf. I Kings, xviii, 31–8.

72. Partington, *Nature*, 1927, cxx, 165; Mercier, *Le Feu Gregéois. Les Feux à Guerre depuis l'Antiquité. La Poudre à Canon*, Paris, 1952, 30, 124, seems to suggest that he and Woog confirmed this, but the text is obscure. A. Marshall, *Explosives*, 1917, i, 12, reports that he made several attempts to set on fire a mixture of quicklime and sulphur by adding water, "but although a fairly high temperature was reached the sulphur never caught fire."

73. Eisler, *Orphisch-Dionysische Mysteriengedanken in der Christlichen Antike*, Vorträge Bibliothek Warburg, Leipzig, 1925, ii, 135–8.

74. *History*, xxxix, 13.

75. *Description of Greece*, I Elis, 27; ed. Spiro, Leipzig (Teubner), 1903, ii, 85.

76. *Deipnosophistæ*, i, 35; ed. Schweighauser, Strassburg (9 vols., 8°, 1801–7), 1801, i, 73.

77. *Refutationis*, ref. 41, iv, 28–41; 1859, 89–107.

78. Salmon, in W. Smith and Wace, *Dict. Christian Biography*, 1877, i, 53; Grenfell and Hunt, *Oxyrhynchus Papyri*, 1903, iii, 36, no. 412; Duchesne, *Early History of the Christian Church*, 1909, i, 333, 411; 1912, ii, 63; Preisendanz *et al.*, *Papyri Græcæ Magicæ*, Leipzig, 1931, ii, 150; J. R. Vieillefond, *Jules Africain. Fragments des Cestes provenant de la Collection des Tacticiens Grecs*, Paris, 1932; Th. Martin, *Mém. div. Sav. Acad. Inscriptions*, 1854, iv, 337 f.

79. Text in Thevenot, *Veterum Mathematicorum Opera*, f°, Paris, 1693, 275–316 (pp. 280–9 are always missing), the passage is on p. 303 (ch. 44), and notes by Boivin, pp. 340–59, see the note on p. 353; the text is from a sixteenth-century Paris MS. old 2706, now 2439, with alternative readings in the notes from MSS. Colbert 1996 (now 2441) and Royal 2173 (now 2445). A better text, ed. by Lami, in Meursius, *Opera*, Florence, 1746, vii, 899–984 (passage in text, p. 954), is based on MS. Laurentianus LV 4 (Florence, parchment, tenth century) collated with a Leyden and several Paris MSS. See Isaac Voss, *Variarum Observationum Liber*, 4°, London, 1685, 86; Fabricius, *Bibliotheca Græca*, Hamburg, 1712, v, 268; 1795, iv, 241; Lalanne, ref. 36, 1843, 360 (text); Hoefer, ref. 34, 1866, i, 303 (from BN 2437 grec); Romocki, ref. 22, 1895, i, 9–12 (text); Vieillefond, ref. 78, 62.

80. Stephanus, ref. 9, 1835, iii, 292 : frequentissimum autem εἰς temere pro ἐν positum scriptoribus ævi Byzantini. . . . Exempla apud Byzantinos omnes omnibus paginis reperienda omitto reperere.

81. Boivin's note in Thevenot's ed., ref. 79, 357; Gelzer, *Sextus Julius Africanus*, 2 vols., Leipzig, 1880–98, i, 13.

82. Jähns, *Geschichte der Kriegswissenschaften in Deutschland*, in *Geschichte der Wissenschaften in Deutschland*, Munich and Leipzig, 1889–91, xxi, pts. 1–3; 1889, i, 5 f., 103; Krumbacher, *Geschichte der byzantinischen Litteratur*, 1891, 349–50.

83. Ref. 22, 1895, i, 13.

84. Ref. 82, i, 103.

85. Ref. 2, 1915, 33.

86. Ref. 78, p. 62.

87. *H.N.*, xxxv, 15 (50).

88. Hoefer, ref. 34, 1866, i, 303; Berthelot, *Histoire des Sciences, La Chimie au Moyen Âge*, Paris, 1893, i, 95.

89. *History*, iv, 53, 185.

90. *Mat. Med*, v, 126.

91. Romocki, ref. 22, i, 9.

92. Stephanus, ref. 9, 1841, iv. 1799.

93. *On Stones*, vii, 40.

94. Ref. 72, 1952, 33, 34, 36, 37, 124 f.

95. *H.N.*, xxxvii, 9 (51).

96. Isaac Voss, ref. 79, 1685, 86–94 : De Origine & Progressu Pulveris Bellici apud Europæos; 87, nitrum seu sal fossilis; Ceraunius seu Pyrites erat loco carbonum, nec desunt hodie qui illo ipso lapide vicem carbonum supplere fuerint conati, quo successu nescio.

97. Jähns, ref. 82, 1889, i, 104; Romocki, ref. 22, i, 10–11.

98. Yule, ref. 26, 1871, I, lxvi, lxxiii.

99. Zenghelis, *Byzantion*, 1932, vii, 265.

100. *Gunpowder and Ammunition, their Origin and Progress*, 1904, 41 f.

101. Ref. 22, i, 11.

102. *De Malii Theodori Consulatu*, 326; *Opera* (Delphin), London, 1821, i, 456.

102a. Berthelot, ref. 1, 1891, cvi, 809.

103. Ref. 36, 343.

104. Ref. 72, 11.

105. *Philosophical Dissertations on the Egyptians and Chinese*, London, 1795, i, 306–8.

106. For a summary of the names, uses, and properties of Greek fire, see Lebeau, ref. 36, 1830, xi, 421–2; Ducange, *Glossarium Scriptorum mediæ et infimæ Græcitatis*, 2 vols., f°, Paris,

1688, i, 1275; *id.*, *Glossarium mediæ et infimæ Latinitatis*, ed. Henshel, 1844, iii, 757; Lalanne, ref. 36, 300–3; Reinaud and Favé, *Du feu grégeois*, 1845, 202 f.; Favé, in Napoleon III, ref. 63, 1862, iii, 47 f.; Romocki, ref. 22, i, 13; Mercier, ref. 72, is mostly compiled from Lalanne; see also Hime, ref. 2.

107.   Romocki, ref. 22, i, 7; Hime, ref. 2, 1915, 60.   J. B. Bury, ref. 57, i, pref. v, refused to use the name "Greek" or "Byzantine" for them, and insisted that they should be called "Romans."

108.   *Shorter Cambridge Medieval History*, Cambridge, 1952, i, 245; Finlay, *History of Greece*, 7 vols., Oxford, 1877, ii, 294, 299; Diehl, *Byzance*, Paris, 1919; *Histoire de l'Empire Byzantine*, Paris, 1920.

109.   Niketas, *Historia*, Isaac Angelus, iii; ed. Bekker, Bonn, 1835, 716.

110.   Prokopios, *Bell. Goth.*, i, 18; Bonn, 1833, ii, 93 :   Γραικοὺς . . . ἐξ ὧν τὰ πρότερα οὐδένα ἐς 'Ιταλίαν ἥκοντα εἶδον, ὅτι μὴ τραγῳδοὺς καὶ μίμους καὶ ναύτας λωποδύτας.

111.   H. A. L. Fisher, *A History of Europe*, 1949, i, 218.

112.   Finlay, ref. 108, 1877, iii, 63.

113.   Finlay, ref. 108, iii, 88.

114.   Synkellos, ed. Bonn, 1829, i, 23; Scaliger, *Eusebii Chronicorum Canonum*, Leyden, 1606, Animadv., 243.   J. B. Bury, ref. 57, ii, 310 f., who refers to the confused chronology, thought the "seven years" was a mistake; since ships equipped with "siphons" were assembled before Greek fire was used, Bury thought these ships were used to hold incendiaries such as sulphur and naphtha.

115.   Theophanes, ed. Classen, in *Corpus Script. Hist. Byzant.*, ed. Niebuhr, Bonn, 1839, i, 540–2; 1841, ii, 178; De Boor's ed. of Theophanes, Leipzig (Teubner), 1883–5, i, 353–4, has no change of text of any consequence; see Lebeau, ref. 36, 1830, xi, 419; Lalanne, ref. 36, 355–6; Romocki, ref. 22, i, 5; Hime, ref. 2, 1915, 27.

116.   In the Bonn ed., ii, 297.

117.   Hitti, *History of the Arabs*, 1940 (or 1956), 201; approved by Setton, *Speculum*, 1954, xxix, 310; J. H. Mordtmann, *Encyclopaedia of Islam*, 1912, i, 867–76 (868); E. W. Brooks, *J. Hellenic Studies*, 1899, xix, 19–33; M. Canard, "Les expéditions des Arabes contre Constantinople dans l'histoire et dans la légende, *J. Asiatique*, 1926, ccviii, 61–121 (104).

118.   Ref. 57, Ch. lii; v, 385.

119.   For some Arabic accounts of sieges of Constantinople, see Mercier, ref. 72, 1952, 61–8; for the expedition of 717–8, Brooks and Canard, ref. 117.

120.   Constantine VII (Porphyrogenitus), *De Ceremoniis*, ii, 15; ed. Niebuhr, Bonn, 1829, i, 569; 1830, ii, 642 (Reiske's note); confirmed by Liutprand (the spelling is sometimes Liutprand), *Historiarum*, vi, 2, in Muratori, *Rerum Italicarum Scriptores*, Milan, 1723, II, i, 469–70; see J. H. Krause, *Die Byzantinen des Mittelalters nach den byzantinischen Quellen*, Halle, 1869, 56; G. Brett, *Speculum*, 1954, xxix, 477.

121.   Partington, *Nature*, 1947, clix, 784.

122.   Freind, *Opera omnia Medica*, 1733, 536 : hæc gentilis quædam apud eos institutio est, perinde ac altera quoque occupatio, quæ circa munitiones et commeatus bellicos comparandos versatur; nam in Historia Byzantina legimus, Judæus his rebus operam Imperatoribus navasse, quæ consuetudo itidem in omnibus etiamnum Europæis exercitibus obtinet.

123.   Ref. 108, 1877, ii, 18.

124.   *Opuscula Historica*, ed. de Boor, Leipzig (Teubner), 1880, 18–19.

125.   *Historia Eccles. ex Theophane*, in Theophanes, ed. Bonn, 1841, ii, 178 [p. 542 margin] : anno 664 . . . Callinicus architectus ab Heliopoli Syriæ ad Romanos profugus venit, qui marino igne confecto vasa Arabum exussit et una cum animabus prorsus incendit.   Ita Romani cum triumpho reversi sunt, igne ad invento marino.   The note on p. 503 says : ἀπὸ 'Ηλιουπόλεος Συρίας, ἀπὸ 'Ηλιουπόλεως τῆς Αἰγύπτου scribit Cedrenus, sed et Heliopolum Syriæ alteram ab Ægyptiaca memorat Ortelius; cf. Donne, in W. Smith, *Dict. Greek and Roman Geography*, 1854, i, 1036.   Heliopolis in Syria (modern Baalbek, "city of the Sun") is mentioned by Strabo, XVI, ii, 10, 753C.

126.   *Compend. Hist*, ed. Bekker, Bonn, 1838, i, 765.

127.   *Decline and Fall*, ch. lii; Everyman ed., v, 393.

128. *The Arab Conquest of Egypt*, Oxford, 1902, 114, 228.

129. *De Administr. Imp.*, xlviii; Migne, *PG* (ref. 35), cxiii, 369.

130. *Annalium*, xiv; Migne, *PG* (ref. 35), 1864, cxxxiv, 1293.

131. "Historia Miscella ab incerto Auctore . . . Quam Paulus Diaconus multis additis," in Muratori, ref. 120, 1723, I, i, 137, ed. by Canisius from an Ambrosian MS.; Lalanne, ref. 36, 355–6.

132. Sarre, *Keramik und andere Kleinfunde von Baalbek*, 1925, 21–4; Sarton, *Isis*, 1933, xix, 255; the pots are illustrated by Biringuccio, *Pirotechnia*, x, 8; Venice, 1540, 163.

133. Ref. 72.

134. Ref. 72, 131–2.

135. Ref. 72, 14.

136. Ref. 72, 4, 6, 35 : "Moïse aurait pu faire."

137. *De Excidio Thessalonicensi*, in *Theophanes Continuatus*, ed. Bekker, *Corpus Script. Hist. Byzant.*, Bonn, 1838, 534.

138. Finlay, ref. 108, 1877, ii, 17; Oman, ref. 26, 546; Diels, *Antike Technik*, Leipzig and Berlin, 1924, 109; Vasiliev, *History of the Byzantine Empire*, Univ. of Wisconsin Studies, Madison, Wis., no. 13, i, 288; *id.*, *Histoire de l'Empire Byzantin*, Paris, 1932, i, 314.

139. Platte, in W. Smith, *Dict. of Greek and Roman Biography and Mythology*, London, 1849, ii, 740.

140. *Tactica*, xix, § 51; Migne, *PG* (ref. 35), 1863, cvii, 1007 f.; ed. Lami, in Meursius, *Opera*, Florence, 1745, vi, 841; text in Lalanne, ref. 36, 1843, 357, and Romocki, ref. 22, i, 14; a Latin translation by Sir John Cheke, from an English MS., is *Leonis Imperatoris De Bellico Apparatv Liber, e græco in latinum conuersus, Ioan. Checo Cantabrigensi*, sm. 8°, Basel, 1554, unpaged, ded. to King Henry VIII (Cambridge Univ. Libr. Kkk. 615); French translation by Joly de Mezeroy, *Institutions Militaires de l'Empereur Léon le Philosophe*, 2 vols., Paris, 1771, with plates, ii, 137 (sea combats), 159–60 (siphons), 271–90 (notes "sur le Feu Grégeois" by de Mezeroy).

141. See ref. 221.

142. See ref. 181.

143. Ref. 22, i, 16.

144. Ref. 2, 1915, 30.

145. *Tactica*, xix, 57; Migne, *PG* (ref. 35), 1863, cvii, 1007; ed. Lami, in Meursius, *Opera*, Florence, 1745, vi, 844.

146. Ref. 22, i, 22.

147. Ref. 26, 546.

148. Feldhaus, ref. 11, 1914, 303, fig. 200; *id.*, *Die Technik der Antike*, 1931, 232 (eleventh-century Vatican MS. 1605).

148a. *La Chirurgie d'Abulcasis*, tr. L. Leclerc, Paris, 1861, bk. ii, ch. 59, p. 148, with ref. to Channing's ed., Oxford, 1778, and note : hinc patet ignem græcum tandem tempore Albucasis inter Arabes et Mauros vulgò notum fuisse, pariterque instrumentum cujus ope.

149. Heron, *Spiritalia*, in Thevenot, ref. 79, 1693, 181; Vitruvius, ix, 9 (Ctesibica machina); Pliny, *Hist. Nat.*, vii, 38; Isidore, *Orig.*, xx, 6; Rich, *Dict. Roman and Greek Antiquities*, 1874, 220, 607 (specimen from Castrum Novum, nr. Civita Vecchia); Romocki, ref. 22, i, 16; Feldhaus, ref. 148 (1931), 185 (Roman specimen with barrels, from Metz).

150. *Hesichii Alexandrini Lexicon*, ed. Albert and Schmidt, Jena, 1862, iv, 36, where a reference to Heron's σίφων is given : ὄργανόν τι εἰς πρόεσιν ὑδάτων ἐν τοῖς ἐμπρησμοῖς.

151. Ref. 9, 1848–54, vii, 303; see also Beckmann, *History of Inventions*, 1846, ii, 245.

152. Partington, ref. 36, 1935, 82, 198, 314; Seneca, *Quæst. Nat.*, ii, 16; Pliny, ii, 66; Rich, ref. 149, 607.

153. Ref. 137, 534, 536; Finlay, ref. 108, 1877, ii, 271–3.

154. Byron, *The Byzantine Achievement*, 1929, 280. The Greek and Roman trumpet (σάλπιγξ, *tuba*) was a long, straight, slightly tapering metal tube with a bell-shaped mouth (κώδων) and was probably of Etruscan (and hence Eastern) origin; W. Smith and Marinden, ref. 21, ii, 901.

155. In Meursius, *Opera*, 1745, vi, 1348.

156. Sophocles, *Greek Lexicon of the Roman and Byzantine Period*, New York, 1887, 705; Zenghelis, *Byzantion*, 1932, vii, 278.

157. Ref. 36, 336, 340, 342, 349.

158. Ref. 106, 1845, 103.

159. Ref. 140, 1771, ii, 159 f.

160. Ref. 140, ii, 286.

161. Ref. 96, xv, 86.

162. Finlay, ref. 108, 1877, ii, 316; Sophocles, ref. 156, 1887, 1175.

163. Ref. 72, 28, plate. There is a picture of the use of Greek fire from a boat in a fourteenth-century MS. (5–3 N–2, f. 34 *v.*) in the Biblioteca Nacional, Madrid; *The Shorter Cambridge Medieval History*, Cambridge, 1952, 214.

164. *Tactica*, xix, 6; Meursius, *Opera*, 1745, vi, 828; Migne, *PG* (ref. 35), 1863, cvii, 992.

165. Ref. 2, 1915, 36.

166. Ref. 26, 1905, i, 546.

167. Oman, ref. 26, 546; Diels, ref. 138, 1924, 109.

168. Brugsch, *Ägyptologie*, 1891, 397.

169. *Revue des Études Grecques*, 1922, xxxv, 296–320 (311).

170. *Historia Byzantina*, Paris, 1649; ed. Bekker, Bonn, 1834, 596.

171. Ref. 138, 109.

172. Ref. 120, 1869, 153, 271, 273.

173. *The Destruction of the Greek Empire and the Story of the Capture of Constantinople by the Turks*, 1903, 263.

174. *Tactica*, xix, 53.

175. *Tactica*, xix, 54; Migne, *PG* (ref. 35), cvii, 1007.

176. Leo, *Tactica*, xvii, 68.

177. Lalanne, ref. 36, 308; Berthelot, ref. 1, 801 (both giving A.D. 936); Romocki, ref. 22, i, 15; Kadlec, *Cambridge Medieval History*, iv, 205.

178. Luitprand, *Historia ejusque legatio ad Nicephorum Phocam*, v, 6; in Muratori, ref. 120, 1723, II, i, 463; in *ib.*, iii, 6, p. 447, he says: nullo præterquam aceti liquore extinguitur. For similar accounts by the Russian historian Nestor, by Leo the Deacon, and (for a Greek attack on Hungarian ships) Ditmar (A.D. 936) and Thurocz, see Lalanne, ref. 36, 307–8.

179. *Alexias*, xi, 10; ed. Schoepen, Bonn, 1878, ii, 116; ed. Rifferscheid, Leipzig (Teubner), 1884, 133–4.

180. Berthelot, ref. 1, 803.

181. *Alexias*, xiii, 3–4; ed. Schoepen, Bonn, 1878, ii, 189–93; ed. Rifferscheid, Leipzig (Teubner), 1884, 182–3; Lebeau, ref. 36, xi, 420; Lalanne, ref. 36, 357; Buckler, *Anna Comnena*, 1929, 379, 404; *Isis*, 1931, xv, 207–8.

182. Ref. 22, i, 22.

183. *Alexias*, xiii, 3; ed. Bonn, 1878, ii, 189.

184. Berthelot, ref. 1, 801.

185. Romocki, ref. 22, i, 199.

186. *Alexias*, xiv, 2; ed. Bonn, 1878, ii, 260–1.

187. Ref. 109, Isaac Angelus I, 1835, 510, cf. 496; Lebeau, ref. 36, xi, 421.

188. Kedrenos, ed. Bekker, Bonn, 1838, i, 765; Lebeau, ref. 36, xi, 422.

189. Gibbon, ref. 57, ch. lii, v, 394–5; Lebeau, ref. 36, xi, 423.

190. Constantine Porphyrogenitus, *De Administrando Imperio*, xiii, xlviii; ed. Bekker, Bonn, 1840, iii, 84, 216; Migne, *PG* (ref. 35), 1854, cxiii, 183, 367; Romocki, ref. 22, i, 14–15.

191. Lebeau, ref. 36, xi, 422. In another account, P. in *Nouvelle Biographie Générale*, 1862, xl, 603, the secret was offered in 1702 to Louis XIV by an Italian chemist, Martino Poli, who was given a pension to forget it.

192. Lalanne, ref. 36, 333–5.

193. *Introduction to the History of Science*, 1931, ii, 163 (Sarton was not a Jew); for some literature on the Crusades, see Michaud, *Histoire des Croisades*, 7 vols., Paris, 1819–22; Wilken, *Geschichte der Kreuzzüge*, 7 vols., Leipzig, 1807–32; Reinaud, *Extraits des Histoires Arabes relatifs aux Guerres des Croisades*, 1829; *Recueil des historiens des Croisades*, 17 vols., f°, Paris, 1841–1906

(5 vols. occidentaux, 5 vols. orientaux, 2 vols. grecs, 2 vols. armeniens); *Documents relatif à l'histoire des Croisades*, 1946–, both published by the Académie des Inscriptions, Paris; Thorndike, *Medieval Europe*, 1920, 296–310; Thompson, *An Economic and Social History of the Middle Ages*, New York, 1928, 401 f., 417; *Gesta Dei per Francos sive Orientalivm Expedetionvm, Et Regni Francorvm Hierosolimitani Historia*, ed. Jacques Bongars, 2 vols., fº, Hannover, 1611. J. C. S. Runciman, *A History of the Crusades*, 3 vols., Cambridge, 1951–4; i, 285; iii, 27; has hardly a word on the use of Greek fire, which is not in the index of the book.

194. Finlay, ref. 108, 1877, iii, 257 f., correcting Gibbon.

195. H. A. L. Fisher, *A History of Europe*, 1949, i, 266 f.

196. Ref. 36, 358–60.

197. Gibbon, ref. 57, ch. lx; vi, 168, had remarked that Baldwin and Villehardouin do not "observe any peculiar properties in the Greek fire," and implies that only fire-ships were ineffectually launched by the defenders among the Venetian ships.

198. Gibbon, ref. 57, ch. lii; v, 393; Oman, ref. 26, 545–9.

199. Quatremère, *J. Asiat.*, 1850, xv, 214; Kremer, *Culturgeschichte des Orients under den Chalifen*, Vienna, 1875, i, 223, 237, 249; Hitti, *History of the Arabs*, 1940 (or 1956), 327.

199a. Abulfeda, *Géographie*, tr. Reinaud, 1883, II, ii, 214.

200. Wiedemann, *Sitzb. Phys.-Med. Soc. Erlangen*, 1912, xliv, 252.

201. Carra de Vaux, *Penseurs de l'Islam*, 1921, ii, 362.

202. Wiedemann, ref. 200, 1915, xlvii, 101 (117).

203. Ref. 199, 256, 259.

204. Ref. 36, 1830, xi, 420, 422.

205. Lebeau, ref. 36, xi, 420.

206. Lebeau, ref. 36, xi, 419; Bar Hebræus, *Chronicon Syriacum*, text and tr., ed. P. I. Bruns and G. W. Kirsch, 2 vols., 4º, Leipzig, 1789; text and tr., Budge, *The Chronology of Abû'l Faraj, Bar Hebræus*, 2 vols., Oxford, 1932, i, 101; it is a history of the world from the creation till 1286; a different work is the *Chronicon Ecclesiasticum*, text and tr., ed. J. B. Abbeloos and T. J. Lamy, 3 vols., 8º, Louvain, 1872–7.

207. Morienus, in Manget, *Bibliotheca Chemica Curiosa*, Geneva, 1702, i, 514.

208. *Gesta Dei*, ref. 193, 178.

209. *Historia rerum in partibus transmarinis gestarum*, bk. ii, ch. 9; *Gesta Dei*, ref. 193, 670–1.

210. *Gesta Dei*, ref. 193, 294–5; Favé, in Napoleon, ref. 63, 1862, iii, 52: "Albert d'Aix paraît attribuer la propriété d'être inextinguible par l'eau non seulement à l'artifice, mais l'incendie qu'il produisait," which increased the terror.

211. *Gesta Dei*, ref. 193, 756–7.

211a. Finlay, ref. 108, 1876, ii, 76. On the tower (πύργος, *turris*) used in warfare, see W. Smith and Marinden, ref. 21, ii, 907.

212. *Recueil*, ref. 193, *Histoire Orientaux*, Paris, 1884, iii, 155 (Anecdotes etc. de la vie du Sultan Youssef (= Saladin)); on p. 221 a great tower is said to have been built in four stages, of wood, lead, iron, and copper, and was burnt by naphtha; Behâ ed-Dîn, *The Life of Saladin*, Palestine Pilgrims' Texts Society, 1897, xiii.

213. *Recueil*, ref. 193, *Historiens Orientaux*, 1884, iii, 180; cf. p. 176.

214. *Recueil*, ref. 193, *Historiens des Croisades. Historiens Orientaux*, Paris, 1876, II, ii, 19 (Extrait de la Chronique intitulée Kamel-Altevarykh par ibn-Alatyr); Favé, in Napoelon, ref. 63, 1862, iii, 52–3.

215. *Richardi Regis Itinerarium Hierosolymorum*, in Thomas Gale, *Historiæ Anglicanæ Scriptores Quinque*, Oxford, 1687, ii, 329, where the text is attributed to Gaudfridus Vinisauf (the Englishman, Geoffry de Vinsauf); it is now supposed to have been by a Richard, prior of Holy Trinity, London, about 1222, in association with a Norman poet Ambroise, who made a French poem, "L'estoire de la guerre sainte," from it; the two were together in King Richard's service; Norgate, *English Histor. Rev.*, 1910, xxv, 523; M. Amari, "Su i fuochi da guerra usati nel Mediterraneo nell' XI a XII secolo," in *Atti Reale Accad. Lincei, Mem. Classe di Sci. Morali, Storiche e filologiche*, Rome, 1876, iii, 3–16.

216. Mercier, ref. 72, 1952, 32, plate.

217. Lane Poole, *Egypt in the Middle Ages*, 1901, 184; Mercier, ref. 72, 1952, 73, quoting Maqrīzī, says 20,000 "pots" and 10,000 torches, and gives other uses on a large scale in 1169 by Saladin.

218. Mercier, ref. 72, 1952, 84–91, with some impossible interpretations.

219. *Oliveri Scholastici Historia Damiatina*, vii (Olivier l'Écolatre), in von Eckhart (Eccard), *Corpvs Historicvm Medii Ævi*, Leipzig, 1723, ii, 1404:

Ignis græcus cominus de turri fluminis, & eminus de civitate, fluminis instar, veniens, pavorem incutere potuit; sed per liquorem acetosum & sabulum ac extinctoria subventum fuit laborantibus. . . . Defensores turris extensis lanceis anteriorem partem scalæ subter unxerunt oleo, deinde ignem apposuerunt, qui erupit in flammas.   Cumque Christiani, qui in ea fuerunt, concurrerent ad ignis suffocationem, suo pondere caput scalæ depresserunt in tantum, ut pons tornatilis fronti appositus inclinaretur.

The passage is not in "Oliveri Scholastici Coloniensis de Captione Damiatæ," in *Gesta Dei*, ref. 193, which mentions Greek fire twice, on pp. 1186, 1190 (Rex igne Græco ferè combustos fuit), in a battle dated 1219.

220. Ref. 1, 794.

221. Joinville, "Historie de Saint Loys," in Petitot, *Collection complète des mémoires relatifs à l'histoire de France*, Paris, 1824, ii, 235–6; Joinville, Everyman ed., 1915, 186; *Jean Sieur de Joinville, Histoire de Saint Louis*, ed. in mod. French by W. de Wailly, Paris, 1874, 112; tr. by J. Evans, *The History of St. Louis*, 1938; Gibbon, ref. 57, ch. lix; vi, 129; Reinaud and Favé, ref. 106, 52; Lalanne, ref. 36, 329; Lane Poole, ref. 217, 234; Romocki, ref. 22, i, 29–33; Hime, ref. 2, 68; G. Paris, *Histoire littéraire de la France*, 1898, xxxii, 291–459.

222. *Gesta Dei*, ref. 193, ii, 1–316.

223. Thompson, ref. 193, 1928, 423.

224. *Gesta Dei, Historia Hierosolimitana* in ref. 193, 1167; Oman, ref. 26, 547.

225. Berthelot, ref. 1, 801–5.

226. Lane Poole, ref. 217, 1901, 286.

227. Hime, ref. 2, 1915, 41–2.

228. Ref. 26, 1871, I, lxvi.

229. Doukas, *Hist. Byzant.*, Bonn, 1834, 247; Gibbon, ref. 57, ch. lxviii; vi, 426; G. Young, *Constantinople*, 1926, 118.

230. *Annales*, iii, 3; ed. Bekker, Bonn, 1838, 244; Lalanne, ref. 36, 333.

231. Ref. 229.

232. T. Roberts, *The English Bowman, or Tracts on Archery*, London, 1801; Feldhaus, ref. 11, 790.

233. Oman, ref. 26, 553; Salzman, *English Life in the Middle Ages*, Oxford, 1926, 207 f.

233a. Lebeau, ref. 36, xi, 420 f.

234. Ref. 1, 786, 792, 800.

235. Berthelot, ref. 1, 803–4.

236. Lebeau, ref. 36, xi, 425.

237. *Historia Hierosolimitana*, i, 84, 85; iii; in *Gesta Dei*, ref. 193, 1098–9, 1133–4.

238. Langlois, *La Connaissance de la Nature et du Monde au Moyen Âge*, 1911, 62 f., 96.

239. See Lippmann, "Verwendung des Petroleums im frühen Mittelalter," in his *Beiträge zur Geschichte der Naturwissenschaften und der Technik*, Berlin, 1923, i, 136, who thought (in my opinion wrongly) that Greek fire was "a mixture of petroleum or petroleum solutions and quicklime"; he quotes very few of the sources drawn upon in the present study.

240. Ref. 22, i, 17.

241. *Handbuch zur Geschichte der Naturwissenschaften und der Technik*, Berlin, 1908, 44.

242. *Geschichte des Wachsthums und der Erfindungen in der Chemie*, 1792, i, 79; cf. Bergman, *Physical and Chemical Essays*, 1791, iii, 62; Wiegleb quotes Hanovius *Dissertatio de igne græco*, Dantzig, 1749; presumably the collection of M. Ch. Hanow, *Dantziger Erfahrungen*, monthly from 1739, quoted by Gmelin, *Geschichte der Chemie*, 1798, ii, 487.

243. Ref. 57, ch. lii; v, 393.   See also MacCulloch, *Annals of Philosophy*, 1822, iv, 390–5; *Quart. J. of Science, Literature and the Arts*, 1822–3, xiv, 22–40; J. J. Conybeare, *Annals of*

Philosophy, 1822, iv, 434–9 (mentioning MacCulloch). Conybeare concluded that Greek fire was naphtha and Kallinikos "may have revived its use or improved its composition."

244. Ref. 36, ix, 211; xi, 420.

245. *De Originibus Constantinopolitanis*, Bonn, 1843, 14; Migne, *PG* (ref. 35), 1866, clvii, 450, and note.

246. Lippmann, *Abhandlungen und Vorträge*, 1913, ii, 226, says the oil was burnt in glass or glazed lamps, and rapidly brought the water and air to the desired temperature, but where he got this information I am unable to say, since the only reference he gives, to Unger, *Quellen der byzantinischen Kunstgeschichte*, Vienna, 1878, 275, merely reproduces the passage in Kodinos, who says nothing of these things. Feldhaus, ref. 11, 753, dates the use of petroleum as 1400, i.e. the time of Kodinos, but the latter seems to imply that it was in use in much earlier times. Perhaps I have missed some source of information used but not quoted by Lippmann.

247. Mercier, ref. 72, 1952, 55.

248. Pliny, *H.N.*, xv, 7.

249. *Mat. Med.*, i, 95, 105.

250. *H.N.*, xv, 7; xxiv, 23–4.

251. *De Compos. Med.*, ch. 40.

252. ψ-Aristotle, *De Mirab. auscult.*, ch. 88.

253. *Hist. plant*, IX, iii, i; *De igne*, 67.

254. *H.N.*, xvi, 21–22; Blümmer, *Technologie und Terminologie der Gewerbe und Künste der Griechen und Römern*, Leipzig, 1879, ii, 347.

255. Pliny, *H.N.*, xiv, 25.

256. Pliny, *H.N.*, xiv, 25.

257. Pliny, *H.N.*, xxxiii, 41; Dioskourides, *Mat. Med.*, v, 110.

258. Partington, *Short History of Chemistry*, 1957, 23–5, fig. 19, from BN MSS.; Virginia Heines, *Libellus de Alchimia ascribed to Albertus Magnus*, Berkeley, 1958, plate vii.

258a. F. Sherwood Taylor, *J. Hellenic Studies*, 1930, l, 109; *id.*, *A History of Industrial Chemistry*, 1957, 155, and plate X, fig. A.

259. In Mesuë, *Opera*, f⁰, Venice, 1623, ii, 245.

260. Budge, *History of Ethiopia*, 1928, i, 302, who says, *ib.*, ii, 574, there are no scientific treatises in Ethiopian and the medical works consist of spells and native prescriptions only.

260a. *A History of Fireworks*, 1949, 232.

261. Beddoes, in J. E. Stock, *Memoirs of the Life of Thomas Beddoes*, 4⁰, 1811, Appendix, xxxi f.; MacCulloch, *Quart. J. Science, Literature and the Arts*, 1822–3, xiv, 22; Thomson, *History of Chemistry*, 1830, i, 106; Reinaud and Favé, ref. 106; Lalanne, ref. 36; Berthelot, ref. 1, 805; *id.*, *Sur la Force des Matières Explosives*, 1883, ii, 352 f.; *id.*, *La Chimie au Moyen Âge*, 1893, i, 98; Zenghelis, *Byzantion*, 1932, vii, 265; Diels, ref. 138, 108 f.; N. A. Cheronis, *J. Chem. Educ.*, 1937, xiv, 360.

262. Romocki, ref. 22, i, 7, 16, 23; Hime, ref. 2, 1915, 30; Lippmann, *Isis*, 1928, xi, 166.

263. Ref. 26, 1905, 546.

264. Ref. 100, 39 f.

265. Ref. 2, 1915, 34, 38, 40.

Chapter II

THE BOOK OF FIRES OF MARK THE GREEK

ONE of the most important documents in the history of incendiaries and gunpowder is a Book of Fires (*Liber Ignium*) attributed to Mark the Greek (Marcus Græcus). Manuscripts of the *Liber Ignium* include two in the Bibliothèque Nationale (Paris) and two in Munich:

    A. BN 7156, end of thirteenth to beginning of fourteenth century, say A.D. 1300; text on ff. 65 *r.*–66 *v.* (I found this MS. full of contractions and difficult to read).

    B. BN 7158, fifteenth century; probably copied from text of A.

    C. Munich Royal Library 267, contemporary with A.

    D. Munich 197, *c.* 1438.

Schoell[1]* had pointed out that the text of D differs markedly from that of B. These four MSS. were used by Berthelot,[2] who denoted C by M. Somewhat different from the text of A and B is that of two MSS. used by Romocki,[3] the texts of which are apparently identical:

    E. Germanische Museum Nürnberg 1481*a*.

    F. Kgl. Zeughaus Berlin 2.

Romocki also refers to Hofbibliothek Vienna 3062 and Archives of Great General Staff Berlin 117 as practically identical, and containing an abbreviated text. BM Sloane 323 and 1754 and other fourteenth century English MSS. contain it.[4]

A Vatican MS. is called *Marcus Grecus Liber Ignium*, another *Marci Greci de igne artificiale*, and there is another anonymous treatise entitled *De igne ignisque natura*; they are thirteenth century.[5] Two incendiary compositions in the Le Bègue MS. (A.D. 1431) are like those in Marcus Græcus.[6] Some MSS. of the *Liber Ignium* include "Eighty-eight Natural Experiments of Rhases" (magic tricks and chemical experiments), said to be translated from Arabic by Ferrarius,[7] and recipes included in other collections were no doubt added from various old sources. A "Marqūsh, King of Egypt" in an Arabic MS., and a "Rex Marchos" or "Marco" in Senior Zadith's *Tabula Chemica*,[8] and in a commentary on the *Turba*,[9] both derived from Arabic originals, are thought by Berthelot[10] to mean Marcus Græcus, but this is doubtful.

Although the name Marcus Græcus implies that the author of the *Liber Ignium* was a Greek (of Constantinople) the language of the original work need not have been Greek. Some words used in it, as will be seen, suggest that it was an Arabic work, or that it was written in a place where Arabic was commonly used. Hammer-Purgstall,[11] on insufficient grounds, regarded the sources of the *Liber Ignium* as Arabic, an opinion favoured by Kopp.[12] Berthelot[13] suggested an Arabic transmission of a late Greek text, with which Diels[14] agreed. Hime[15] suggested that it was compiled in Spain, and

* For references, see p. 61.

Lippmann[16] thought the Arabic words in it were of Spanish or Provençal origin. Since in MSS. A and B the name "Marcus" is written "Marchus," Reinaud and Favé[17] thought a κ in a Greek original had been mistaken for χ, but Favé later[18] thought the work had no Greek original but was written in Latin, perhaps by a Greek in Italy.

L. Dutens[19] mentioned the BN MS. and a Latin MS. owned by Dr. Mead, a copy of which was made by Samuel Jebb, who quotes from it in the preface to his edition of Roger Bacon's *Opus Majus*.[20] Jebb and Dutens (who had available the copy made by Jebb) call the work *Liber Ignium* and quote correctly from it in Latin. An erroneous idea that Mead's MS. had a Greek title, περὶ τῶν πυρῶν, seems to have originated with the Abbé Albert Fortis[21]:

> Esiste nella Real Biblioteca di Parigi un Trattato Greco ms. tutt'ora inedito, περὶ τῶν πυρῶν, il di cui autore *Marco* visse fra l'ottavo ed il nono secolo. In esso si danno precetti per fare razzi ed altri fuochi artifiziali, s'insegna ad usar del nitro per farne polvere mescolandolo collo zolfo e col carbone, ec. E un errore il creder autore di codesta terribile preparazione il Fratre Tedesco [i.e. Berthold Schwartz].

Fortis gives as natural sources of nitre mineral springs in Italy (Calabria, etc.), and caves and efflorescences in Mindanao (Molucche), Sumatra, India, Kashmīr, Tibet, Russia and abundant deposits in Peru and Chile. The last deposits (really of sodium nitrate) were not mentioned by Schelhamer.[22]

The error that the MS. had a Greek title, started by Fortis, was transmitted by a chain of copyists[23] to modern times,[24] but was corrected by the author.[25] Another error was started by Dutens,[26] who said he was not able to fix the date of Marcus Græcus, but since he is mentioned by Mesuë, c. A.D. 800, he must have lived before his time, i.e. in the eighth century. Mesuë,[27] however, merely says, in speaking of the plant Arthanita : Dixit filius Serapionis . . . Et dicit Græcus. Succus eius cum mellicrato, aut secaniabin, & proprie de quolibet partes æquales . . . est medicina experta. Mesuë's words are "the Greek says," and this has been supposed[28] to refer to Dioskourides (who does not give the information in Mesuë) or even Hippokrates.[29] Hime[30] said :

> The past tense, *dixit*, in the passage in the text would seem to show that Ibn Serapion was dead when it was written. The present tense, *dicit*, indicates similarly that "Græcus" was then living, a contemporary of Masawyah's [Mesuë's]. Yet Dutens speaks of his having lived "avant le médicin Arabe."

Hime also said the "Marcus" in Fabricius,[31] to which Dutens refers, is one mentioned by Galen, and Fabricius nowhere mentions Marcus Græcus; he says[32] the "Græcus" mentioned by Mesuë was perhaps Gereon. Although Hoefer[33] and Jähns[34] accepted Dutens' identification and date, Romocki[35] said :

> Die Behauptung Hoefers, welcher ihre apodiktische Form leider den Eingang in viele neueren Werke verschaffen hat, ist also völlig aus der Luft gegriffen; . . . so werden wir nur annehmen können, dass das Feuerbuch

des Marcus die Form, in welcher es Bacon und Albert um 1267 vorlag, erst etwas ein Jahrzehnt früher erhalten hat.

Romocki's assumption that the *Liber Ignium* was known to Roger Bacon and Albertus Magnus is, however, far from certain, as will be shown later, and his date *c.* 1257 for its composition really applies only to the later additions to it. Berthelot[36] had also corrected Hoefer's statement and pointed out that it came from Dutens' "ouvrage paradoxal"; Romocki seems to have thought it originated with Hoefer. Reinaud and Favé[37] pointed out that in a Hebrew translation of Mesuë, "Græcus" is "Iounâny," i.e. "the Greek." Jebb's mention of Marcus Græcus was noted by Benjamin Robins.[38]

The facts so far established may be summarised:

(i) The *Liber Ignium* has not in any known MS. a Greek title περὶ τῶν πυρῶν, and there is no evidence that the author was a Greek.

(ii) Marcus Græcus is not the "Marcus" mentioned by Galen, or the "Græcus" mentioned by Mesuë.

(iii) There is no evidence that he was known in the eighth century.

The text of the *Liber Ignium* has been published several times:

I. A text from BN 7156 collated with BN 7158 was published by F. J. G. La Porte Du Theil (or Dutheil) (1742–1815), Conservateur des manuscrits in the Bibliothèque Nationale: *Liber Ignium ad Comburendos Hostes, auctore Marco Græco; ou, Traité des Feux propres à détruire les Ennemies, composé par Marcus le Grec. Publié d'après deux manuscrits de la Bibliothèque Nationale,* 4°, Paris, imprim. de Delance et Lesueur, An XII (1804), vi + 18 pp. (the *Liber Ignium* is on the first three leaves, the rest consists of extracts from Cardan and Scaliger). This publication was at the instance of Napoleon, who was interested in its military significance and thought the secret of Greek fire had been lost. A copy sent by Du Theil to the Göttingen University Library was used by Beckmann.[39]

II. Hoefer[40] published a text from the same MSS. in 1842, unaware of the existence of Du Theil's publication, which he said in 1866 he had been unable to procure, "malgré nos recherches,"[41] yet there are at least six copies in the Bibliothèque Nationale.[42]

III. A text based on Du Theil's but collated with the Munich MS. was published and translated by Berthelot,[43] who divided it into paragraphs. Berthelot says Hoefer's text is "très incorrect et rempli de mauvaises lectures," and had "malheureusement servi de base à une traduction publié dans la *Revue scientifique* en 1891."[44] Hoefer, however, was expert in reading both Greek and Latin manuscripts, and his text should be regarded as an alternative version to Du Theil's and worthy of consideration. Berthelot's criticism of it is greatly exaggerated, since the differences from Du Theil's text are usually unimportant.

IV. Romocki,[45] although Berthelot's text was available to him, reproduced Hoefer's with one or two emendations and comments. He speaks of: "Der von du Theil und von Hoefer herausgegebene *Liber Ignium,* den ich, da jene

Ausgaben stellenweise bis zur Unverständlichkeit entstellt sind, hier vollständig folgen lasse"; his text (not numbered in paragraphs) is really Hoefer's.

V. Hime,[46] although he calls it "Du Theil's text," actually gives Hoefer's, as slightly amended by Romocki,[47] but numbered in paragraphs as in Berthelot. Apart from this, texts IV and V are identical, and both are essentially II.

Lippmann[48] and Diels[49] in due course severely criticised Berthelot's text and translation, but made only trifling (and probably incorrect) emendations, as will be seen. The title in MS. A is: Incipit Liber ignium a Marco Græco descriptus, cujus virtus et efficacia ad comburendos hostes tam in mari quam in terra, plurimum efficax reperitur, quorum primus hic est. Here begins the Book of Fires written by Marcus Græcus, containing the methods of proved virtue for burning the enemies as well on sea as on land, and this is the first; the first recipe following. Dutens gives the title of Mead's MS., a copy of which was given to him by Jebb, as: Incipit Liber Ignium a Marco Græco perscriptus, cujus virtus et efficacia est ad comburendum hostes tam in mari quam in terra.

Although the *Liber Ignium* is not a large work (it would fill about six pages) it will suffice for our purpose to give only some parts of it in full, and some of the different readings proposed will be noted. Du Theil's text cannot be regarded as wholly satisfactory and a modern critical edition is lacking. Berthelot's translation is defective. The paragraphs are numbered as in Berthelot; the MSS. are denoted as above; H=Hoefer's text, R=Romocki's text:

§ 1. R. sandaracæ puræ ł[a]. 1, armoniaci[b] liquidi ł. 1[c]; hæc simul pista et in vase fictili vitrato[d] et luto sapiæ[e] diligenter obturato dimitte[f]; donec liquescat ignis subponatur. Liquoris vero istius hæc sunt signa: ut ligno intromisso[g] ad modo butiri videatur.[h] Postea vero IIII libras de alkitran græco superfundas.[i] Hæc autem sub tecto fieri prohibentur,[j] quoniam[k] periculum imminerit. Cum autem in mari ex ipso operari volueris, de pelle caprina accipies utrem, et in ipsum de hujus[l] oleo l. II[m], si hostes prope fuerint intromittes[n]; si vero remoti fuerint[o] plus mittes. Postea vero utrem ad veru ferreum ligabis, lignum adversus veru grossitudinem faciens, ipsum vero inferius sepo perungens. Lignum prædictum in ripa succendens[p] et sub utre locabis. Tunc vero oleum super[q] veru et super lignum distillans accensum super aquas discurret, et quidquid[r] obviam fuerit concremabit.

Take pure sandarach resin 1 lb., liquid gum ammoniac 1 lb., grind together and put into a glazed earthenware jar closed carefully with philosophical lute. Put a fire below till [the mixture] is melted. The sign of melting is this: the product will look like butter on a piece of wood inserted in it. Then pour over it 4 lb. of Greek pitch. The performance of this operation indoors is prohibited by reason of danger [of fire]. If you wish to operate on sea, take a goatskin bag and put in it 2 lb. of this oil if the enemy is near, more if he is distant. Then attach it to an iron javelin, attach a piece of wood suited to the size of the javelin, and grease the lower part of the javelin with tallow. You will set fire to the piece of wood on the shore and put it below the bag.

The oily matter trickling over the javelin and the wood will inflame and run over the water and all it encounters will be burnt.

^a libram, H, R.   ^b amoniaci, MS. B.   ^c ana, H ("parties égales"), R.   ^d vitreato, H, R.
^e luto sapientiæ, H, R; prudentiæ, MS. C.   ^f deinde, H, R.   ^g per foramen, add. H, R;
Berthelot, add. C.   ^h buliatur, MS. B.   ⁱ infundas, H, R.   ^j prohibeantur, H, R.   ^k quum,
H, R.   ^l hoc, H, R.   ^m intromittas, add. H, R.   ⁿ minus, add. H, R.   ^o fuerunt, H.
^p succendes, H, R.   ^q sub, H, R.   ^r quiquid, H.

This is a description of an incendiary javelin, which goes back to an early period.

§ 3. Alius modus ignis ad comburendos hostes ubicunque sitos. R. Basiliscum^a al.^a balsamum, oleum Ethiopiæ, alkitran et oleum sulphuris. Hæc quidem omnia in vase fictili reposita in fimo diebus XV subfodias. Quo inde extracto, corvos eodem perungens^b ad hostilia loca super^c tentoria destinabis. Oriente enim sole, ubicunque^d id^e liquefactum fuerit accendetur. Verum^f semper ante solis ortum, aut post occasum ipsius precipimus^g esse mittendos.

Another kind of fire for burning enemies wherever they are. Take petroleum, black petroleum, liquid pitch,^h and oil of sulphur. Put all these in a pottery jar buried in horse manureⁱ for fifteen days. Take it out and smear with it crows^k which can be flown against the tents of the enemy. When the sun rises and before the heat has melted it the mixture will inflame. But we advise that it should be used before sunrise or after sunset.

^a om., H, R; in § 2, "R. balsami sive petrolei."   ^b perunguens, H, R.   ^c sive, H, R.
^d ubicumque, H, R.   ^e illud, H, R.   ^f unde, H, R.   ^g præcipimus, H, R.   ^h alkitran =
Arabic al qitrān, pix liquida.[50]   ⁱ heated by fermentation; often used by the alchemists.
^k "projectiles," Berthelot, although real crows possible. In Arabic fire books the crows are
set on fire first; such "fiery birds" appear in Chinese works (see ch. VI).

§ 4. Oleum vero sulphuris sic fit. R. sulphuris l.^a IIII quibus in marmoreo^b lapide contritis et in pulverem redactis, oleum juniperi l.^a IIII admisces; et in caldario pone, ut lento igne supposito distillare incipiat.

Oil of sulphur is made thus. Take four parts of sulphur, grind it in a stone mortar [or on a marble slab] and reduce it to powder. Mix with four parts of oil of juniper and put in a cauldron over a gentle fire until it begins to distil.^c

^a uncias, H, R.   ^b mortario, MS. B.   ^c "couler goutte à goutte," Berthelot.

§ 8. Ignis quem invenit Aristoteles quando^a cum Alexandro rege^b ad obscura loca iter ageret, volens in eo per mensem fieri illud^c quod sol in anno^d præparat, ut in spera de auricalco.^e R. æris rubicunda l. I, stagni^f et plumbi, limaturæ ferri singulorum medietatem libræ, quibus pariter liquefactis, ad modum astrolabii lamina formetur lata et rotunda, ipsam eodem igne peronunctam X diebus siccabis, XII interando; per annum namque integrum ignis idem^g succensus nullatenus deficiet. Quod si^h inunctio [hæc XIIIⁱ transcendent numerum]^k ultra annum durabit. Si vero locum quempiam inungere^l libeat, eo dessicato^m scintilla qualibet^d diffusa ardebit continue; nec aqua extingui poterit. Et hæc ist ignis prædicti compositio.

R. alkitran, colophoniam sulfuris crocei, olei ovorum sulfurinum.[o] Sulfur in marmore teratur; quo facto[p] oleum superponas. Deinde tectoris limaginem[q] ad omne pondus acceptam insimul pista et inungue.

Fire which Aristotle invented when he travelled with King Alexander in dark places, wishing to make in a month what the sun accomplishes in a year as in the brass sphere. Take red copper 1 lb., tin and lead, iron filings, ½ lb. of each. Melt them together and make a flat disc like an astrolabe. Smear it with the following fire, dry for ten days, and repeat the operation twelve times. This fire once kindled burns for a whole year without pause. If it is smeared more than thirteen times it lasts more than a year. If you smear it in any given place and let it dry, as soon as a spark falls on it it burns continually and cannot be extinguished by water. And this is the composition of the fire mentioned above.

Take pitch, colophonium, yellow sulphur, oil of eggs and sulphur. Grind the sulphur on a marble and when this is done, add the oil. Then add plasterer's tow equal to the total weight, stamp it together, and smear it.

[a] quum, H, R. [b] om., H, R. [c] id, H, R. [d] autumno, MS. B. [e] aurichalco, H, R. [f] stanni, H, R. [g] inde, MS. B. [h] quæ enim, H, R. [i] XIV, MS. B. [k] om.[ ], H, R. [l] inunguere, H, R. [m] dissiccato, H. [n] quælibet, H, R. [o] colophonii, sulphuris crocei, olei ovorum sulphurici, H, R. [p] add. universum, H, R. [q] textoris lanuginem, MS. C.

§ 9. Sequitur alia species ignis quo Aristoteles[a] domos in montibus sitas destruxit[b] incendio,[c] ut et mons ipse subsiderit.[d] R. balsami l. I, alchitran[e] l. V, oleum ovorum et calcis non extinctæ[f] ana[g] l. X. Calcem teras sum oleo, donec una fiat massa. Deinde inungas lapides ex ipso et herbas ac renascentias quaslibet in diebus canicularibus, et sub fimo ejusdem regionis sub fossa dimittes; primo[h] namque autumnalis pluviæ[i] dilapsu succendetur[k] terra[l] et indegenas comburit igne. Aristoteles namque hujus[m] ignem annis IX durare[n] asserit.

Another kind of fire with which Aristotle burnt houses situated in the mountains and burnt the mountain itself. Take petroleum 1 lb., pitch 5 lb., oil of eggs and quicklime 10 parts of each. Grind the quicklime with the oil and make one mass of them all. Then anoint [with the mixture] stones, grass, and young plants during the dog days, and bury it in horse-dung in those places underground. When the autumnal rains begin to fall the earth takes fire and its fire burns the inhabitants. Aristotle affirmed that this fire lasts nine years.

[a] Quo Alexander urbes Agarenorum in montibus, MS. C; see Berthelot's note, ref. 2, i, 107. [b] destruere, H, R. [c] ait add., H, R. [d] succenderet, MS. B. [e] alkitran, H, R (ch = Arabic q). [f] exstinctæ, H. [g] ana om., H, R. [h] postea, H, R. [i] pluvia, H. [k] succenditur, H, R. [l] terram, H, R. [m] hunc, H, R. [n] durasse, MS. B.

§ 10. Compositio inextinguibilis[a] facilis et experta. R. sulfur vivum, colofoniam,[b] aspaltum,[c] classam,[d] tartarum,[e] piculam[f] navalem, fimum ovinum aut columbinum. Hæc pulverizata[g] subtiliter dissolve[h] petroleo; post[i] in ampulla reponendo vitrea, orificio bene clauso, per dies XV in fimo calido equino subhumetur. Extracta vero ampulla, distillabis oleum in cucurbita, lento igne ac cinere mediante, calidissime et subtile,[k] in quo si bombax intincta fuerit ac incensa, omnia super quae arcu vel balista projecta fuerit, incendio concremabit.

Inextinguishable composition, easy and tested. Take native sulphur,[l] resin, asphalt, sandarach,[m] tartar, naval pitch, and sheep's or pigeon's dung. Powder these finely, dissolve in petroleum, and then put into a glass phial with the neck well stopped. Bury it for fifteen days in hot horse dung. The phial being taken out, distil the oil in a cucurbit with a gentle fire on fine burning ashes. If you soak cotton[n] [in the distillate] and set it on fire it burns up all things on which it is thrown by a bow or ballista.

[a] inexstinguibilis, H.   [b] colophonium, H, R.   [c] asphaltum, H, R.   [d] massam, H.   [e] tartari, H.   [f] piculani, H.   [g] pulveriza, H, R.   [h] om., H, R.   [i] postea, H, R.   [k] ac subtili, H, R.   [l] Pliny, H.N., xxv, 50.   The name sulphur vivum is almost always used in earlier gunpowder recipes; in French accounts it appears as soufre vif, but in one statement of 1346 soufre mort and soufre vif occur together.[51]   [m] classa: "quelque matière résineuse . . . je n'ai pu en découvrir le sens précis," Berthelot.[52]   Possibly classa is connected with "glass," or perhaps amber: Pliny, H.N., iv, 27; Aldrovandus, Musæum Metallicum, Bologna, 1648, bk. iii, ch. 18, p. 404: "Latini Succinum . . . Germani veteres . . . Glessum vocabant . . . instar vitri pelluceat"; Lippmann, Alch. (ref. 16), i, 481: "classa . . . bedeutet 'glæssa' oder 'glæssum' . . . germanische Wort für Bernstein," but Berthelot's idea is probably correct; the identification with sandarach resin is based on the mention of this in § 1.   [n] Berthelot, "un roseau"; Romocki, "Baumwolle."

§ 11. Nota quod omnis ignis inextinguibilis[a] IIII rebus extingui vel suffocari poterit, videlicet cum aceto acuto[b] aut cum urina antiqua vel arena sive filtro ter in aceto imbibito et tociens[c] dessicato, ignem jam dictum suffocat.

Note that every inextinguishable fire can be extinguished or stifled by four things, namely by strong vinegar, or by stale urine, or by sand, or by felt soaked three times in vinegar and thoroughly dried; it stifles the said fire.

[a] inexstinguibilis, H.   [b] sive forti, add. MS. B.   [c] toties, H, R.

§ 12. Nota quod ignis volatilis in aere duplex est compositio. Quorum primus est. R. partem unam colofoniæ[a] et tantum sulfuris vivi,[b] partes vero salis petrosi,[c] et in oleo lineoso[d] vel lauri,[e] quod est melius, dissolvantur bene pulverizata et oleo liquefacta. Post[f] in canna, vel ligno concavo[g] reponatur et accendatur. Evolat enim subito ad quemcunque[h] locum volueris et omnia incendio concremabit.

Note there are two compositions of fire flying in the air.[i] This is the first. Take 1 part of colophonium and as much native sulphur, 6 (?) parts of saltpetre. After finely powdering dissolve in linseed oil or in laurel oil which is better. Then put into a reed or hollow wood and light it. It flies away suddenly to whatever place you wish and burns up everything.

[a] colophonii, H, R.   [b] add. II, H, R.   [c] MS. A, VI? (Berthelot).   [d] linoso, H, R.   [e] lamii, H, R.   [f] postea, H, R.   [g] excavo, H, R.   [h] quemcunque, H, R; for different text in MS. C, see Berthelot, ref. 2, i, 109.   [i] "il y a deux compositions de fusée," Berthelot: the contrivance is a rocket.

§ 13. Secundus modus ignis volatilis hoc modo conficitur. R. Acc. l. I sulfuris vivi, l. II carbonum tiliæ[a] vel cilie,[b] VI l. salis petrosi, quæ tria subtilissime[c] terantur[d] in lapide marmoreo.[e] Postea pulverem[f] ad libitum in tunica reponatis[g] volatili vel tonitruum[h] facientem.[i] Nota[k] tunica ad volandum debet esse gracilis et longa et cum prædicto pulvere optime conculato repleta.   Tunica vero tonitruum[l] faciens debet esse brevis et grossa et prædicto pulvere semiplena et ab utraque parte fortissime[m] filo ferreo bene ligata.

The second kind of flying fire is made in this way.   Take 1 lb. of native sulphur, 2 lb. of linden or willow charcoal, 6 lb. of saltpetre, which three things are very finely powdered on a marble slab.   Then put as much powder as desired into a case to make flying fire or thunder.   Note.—The case for flying fire should be narrow and long and filled with well-pressed powder. The case for making thunder should be short and thick and half filled with the said powder and at each end strongly bound with iron wire.

[a] vitis vel salicis, H, R.   [b] om., MS. B.   [c] subtilissima, H, R.   [d] MS. B, pulverizantur. [e] MS. C, add. aut porfirico.   [f] pulvis, H, R.   [g] reponatur, H, R.   [h] tonitrum, H, R. [i] faciente, H, R.   [k] add. quod, H, R.   [l] tonitrum, H, R.   [m] MS. C, parte filo fortissimo bene clausa.   Jebb and Dutens (who quotes from p. 9 of the copy of Mead's MS.) have: "Secundus modus ignis volatilis hoc modo conficitur: lib. j sulphuris vivi; lib. ij carbonis salicis; salis petrosi vi libras," and the rest of the recipe in the form given by Hoefer, except that "ferreo" is omitted in the last line, and the order of one or two words is transposed.

§ 14.   Nota quod sal petrosum est minera terræ et reperitur in scrophulis contra lapides.[a]   Hæc terra dissolvatur in aqua bulliente, postea depurata et distila[b] per filtrum et[c] permittatur per diem et noctem integram decoqui, et invenies in fundo laminas salis conjelatas[d] cristallinas.

Note that saltpetre is a mineral of the earth and is found as an efflorescence on stones.   This earth is dissolved in boiling water, then purified and passed through a filter.   It is boiled for a day and a night and solidified, and transparent plates of salt are found at the bottom of the vessel.

[a] in scopulis et lapides, H, R.   [b] destillata, R.   [c] om., H, R.   [d] congelatas, H, R.

The first sentence offers difficulties in translation; scopula is a small brush, scopulus a pointed rock, scrofula is a swelling; efflorescence seems the only satisfactory rendering.   Berthelot[53] translated: "le saltpetre est un minéral; on le trouve sous forme d'efflorescence sur les pierres."   Beckmann[54] had translated "contra lapides" as "on walls," and Lippmann[55] inserted ". . ." after terræ (although only "et" is missing) and translated: "Salpeter ist ein Mineral *aus* dem Erd*boden*, . . . wird *aber auch* als Ausschwitzung an den *Mauern* gefunden," which is too imaginative, the words (or parts of words) italicised here being missing in the Latin.

Zenghelis[56] refers to Lippmann[57] as quoting from Pliny: "Eine Art Nitrum tritt auch als Auswitterung an feuchten Mauern auf; es wird vielfach in der Medizin gebraucht und ist ein wirksamer und treibender Dünger für zahlreiche Pflanzen, z. B. für Rettige," and thence assumes that saltpetre was known to the ancients.   None of the places in Pliny given by Lippmann mentions an efflorescence on damp walls, which seems to have been made up by Lippmann from Pliny's occasional use[58] of the name "aphronitrum" for "nitrum"; there is nothing about its use as manure for "numerous plants"; and when Pliny[59] says that in growing radishes the Egyptians "sprinkle nitrum over them (in Ægypto nitro sparguntur)" he uses the name "nitrum" and not "aphronitrum."   No Greek or Roman author, to my knowledge, ever mentions aphronitron (ἀφρόνιτρον) or aphronitrum as an efflorescence on *walls*, and the conclusion drawn by Zenghelis from an incorrect statement by Lippmann that it sometimes meant saltpetre is completely erroneous.

§§ 15 and 16 describe the preparation of inextinguishable candles from quicklime, galbanum, bile of the tortoise, cantharides and oil of zambac (oleum zambac), or from luminous worms and oil of zambac, respectively. The second is augmented when rain falls on it. Zambac (zanbāq) is the Arabic name for an essential oil, such as oil of jasmine,[60] although the usual meaning is "the white lily or the iris"[61]; "lily oil," is mentioned by Scribonius Largus (A.D. 1–50).[62]

Similar phosphorescent compositions from fireflies (noctilucæ), the biles of various animals, juices of plants, etc., are described in following paragraphs; they are often fanciful, e.g. in § 19 a light which makes a house look like silver is made by drying a black or green lizard, cutting off the tail and taking from it mercury (argentum vivum), which is used in an iron or glass lamp instead of oil. There are also some conjuring tricks with fire, and a paste of lime slaked in hot bean water (aqua fabarum), mixed with Misnian (?) earth and mallow is used to protect the hands against fire (§ 21). A "sun stone" (lapis qui dicitur petra solis) is a large round stone with white spots which shines with a light equal to that of four candles (§ 25)—perhaps a phosphorescent stone.

§ 24. Confectio vini[a] est quum[b] si aqua projecta fuerit accendetur ex toto. R. calcem vivam, eamque cum modico gummi arabici et oleo in vase candido cum sulfure confice, ex quo factum vinum[a] et aqua aspersa[c] accendetur. Hac vero confectione domus qualibet[d] adveniente pluvia accendetur.

Preparation of a wine such that if water is thrown upon it, it burns completely. Take quicklime, mix with a little gum arabic and oil in a clean vessel, also with sulphur. The wine so made inflames when sprinkled with water. This preparation put in a house takes fire on the coming of rain.

[a] visci, R.   [b] cum, H, R.   [c] add. ac, H.   [d] quælibet, H, R. The "wine" is a solid preparation analogous to the "automatic fire" of Julius Africanus (see ch. I).

§ 26. Ignem græcum tali modo facies: Recipe sulphur vivum, tartarum, sarcocollam et picolam,[a] sal coctum, oleum petroleum et oleum commune.[b] Facias bullire invicem omnia ista bene. Postea impone stupas[c] et accende. Quod si volueris extrahere[d] per embotum ut supra diximus.[e] Post illumina[f] et non extinguetur, nisi cum[g] urina vel aceto vel arena.

You will make Greek fire in this way. Take live sulphur, tartar, sarcocolla and pitch, boiled salt, petroleum oil and common oil. Boil all these well together. Then immerse in it tow and set it on fire. If you like you can pour it out through a funnel as we said above. Then kindle the fire which is not extinguished except by urine, vinegar, or sand.

[a] piceam, H; picem, R.   [b] gemmæ, H, R.   [c] stupæ, H; stupam, R.   [d] exhiberi, H, R.
[e] Since nothing is said of this previously, the recipe was taken bodily from some collection.
[f] stuppa illinita, H, R, which seems better, since it has already been said to have been kindled.
[g] om., H, R.

This recipe is one of the oldest for Greek fire, and it aroused interest in the MS. at a time when the secret of this fire was supposed to have been lost (see ch. I.). It is one of the oldest descriptions in Latin, contemporary with, if not somewhat earlier than, the one in Arabic given by Ḥasan al-Rammāḥ (c. 1280, see ch. V). This recipe was written in cipher in a thirteenth-century hand in a twelfth-century MS. in the Harvey Cushing Library, New Haven, Conn.,[63] which has "colophonium" instead of "picem," "sal tostum" instead of "sal coctum," "stuppam vel lichinen" instead of "stupas," and other small variants.

"Sal coctum" was identified with saltpetre by Berthelot,[64] but Hime[65] said (correctly) that it was common salt. In another recipe (§ 27) sal commune grossum (lumps of rock salt) is specified. Common salt was added to incendiary mixtures because it produced a yellow flame which looked very hot.[66] Sal coctum does not appear in MS. C.[67] The incorrect translation sal coctum = saltpetre refined by boiling and evaporation is repeated by Mercier.[68]

§ 27. Aquam ardentem sic facies. R. vinum nigrum spissum et vetus et in una quarta ipsius distemparatis[a] S. II[b] sulfuris vivi subtilissime pulverizati; 1 vel p. II[c] tartari extracta[d] a bono vino albo, et S. II[e] salis communis grossi; et supradicta[f] ponas in cucurbita bene plumbata, et alembico superposito[g] distillabis aquam ardentem, quam servare debes in vase vitreo clausa.[h]

You make burning water [alcohol] thus. Take black, thick, and old wine. To a quart of it add two scruples of native sulphur in very fine powder, one or two [scruples] of tartar extracted from good white wine, and two scruples of common salt in pieces. Put all into a well-luted cucurbit and with the alembic superimposed distil the burning water, which must be kept in a closed glass vessel.

[a] distemperabuntur, H, R.   [b] unciæ II, H, R.   [c] lib. II, H, R.   [d] extracti, H, R.
[e] unciæ II, H, R.   [f] subdita, H, R.   [g] supposito, H, R.   [h] clauso vitreo, H, R.

Hoefer[69] had translated una quarta as "un quart," cucurbita bene plumbata as "une cucurbite, bien plombée et lutée." Berthelot[70] translated the first as "un quart de livre" and the second as "un bon alambic de plomb," the word alembico being rendered "le chapiteau." The word quarta in early legal treatises always meant "a quarter."[71] Lippmann[72] corrected Berthelot's "a quarter of a pound" to "a quart" (i.e. Hoefer's reading), but afterwards[73] thought it was not known when "quart" was first used. Yet Hultsch[74] had pointed out that quarta (quarterius in Varro) could mean a quart of about 5 oz. measure or 136 c.c. (smaller than the modern quart). Lippmann also adopted Hoefer's translation of bene plumbata (without giving him credit for it), although the literal meaning is always connected with lead, and "plombieren" for "stopping" is a German usage.

The above is one of the old recipes for the distillation of alcohol. Berthelot[75] pointed out that it is not in MS. D as part of the text of Marcus Græcus and hence is probably a later addition to MS. A, since although D is dated 1438 it "contains older works." MS. D, however, contains another recipe (f. 78 *v*., old 75 *v*., as in Berthelot) given by Berthelot as follows and "improved" by Diels[76] by omitting the five letters in square brackets and adding the semicolon in round brackets:

Vinum in potto ardens fic hoc modo: vinum optimum rubeum vel album, in potto aliquo pone, habente caput aliquantulum elevatum cum coperculo in medio perforato. Cumque calefieri et bullire inceperit et per foramen vapor egrediatur [ac] candela accensa applica[tur] et statim vapor ille accenditur et tandiu durabit quandiu vaporis egressio(;) et est eadem cum aqua ardente.

To make wine burn in a pot as follows. Put into a pot best white or red wine, the top of the pot being high and with a lid pierced in the middle. When the wine is heated and begins to boil and vapour issues from the hole, apply a light. The vapour at once catches fire and the flame lasts as long as the vapour issues; it is the same with alcohol (aqua ardens).

The same MS. D also contains a recipe (given after the above) for the preparation of alcohol which is different from that in A and B; it is not part of the text of Marcus Græcus but is added at the end (f. 78 *v*., old 75 *v*., as in Berthelot). The text is given by Berthelot[77] but a revised text from a photostat of the MS. given by Richter[78] is reproduced here (variants B from Berthelot):

Aqua ardens ita fit: vinum antiquorum optimum cujuscunque coloris in cucurbita et alembico,[a] juncturis bene lutatis, lento igne distilla, et quod distillabitur, aqua ardens nuncupatur. Ejus virtus et proprieatus[b] ita fit, ut si pannum lini in ea madefecerit[c] et accenderis, flammam magnam prestabit,[d] qua consumpta remanebit pannus inlesus[e] integer, sicus[f] prius fuerit, si vero digitum in ea intinxeris[g] et accenderis, ardebit admodum[h] candelæ sine lesione; si vero candelam accensam sub ipsa aqua tenueris, non extinguetur. Et nota quod illa, quæ primo egreditur, est bona et ardens, postrema non est[i] utilis medicine.[k] De prima etiam mirabile fit colirium[l] ad maculam vel pannum oculorum.

Burning water (alcohol) is made so. Take old and strong wine of any colour, distil it in a cucurbit and alembic, with the joints well luted, over a gentle fire. That which distils is called burning water (alcohol). Its virtue and property is tested so. If a piece of linen cloth is wetted in it and kindled it produces a large flame. When it goes out the cloth remains intact as it was at first. If you first dip a finger in it and set it on fire it burns like a candle but produces no injury, and if you put a burning candle in this water it is not extinguished. Note that the water which comes over first is good and fiery, that later is not useful in medicine. From the first is made a wonderful collyrium for spots and films on the eyes.

[a] alembic, B.   [b] propriet, B.   [c] madefeceris, B.   [d] præstabit, B.   [e] om., B.   [f] sicut, B. [g] introduxeris, B.   [h] ad modum, B.   [i] vero est, B (the text given above contains an important correction from the MS.).   [k] medicinæ, B.   [l] collirium, B.

The last part of the recipe suggests that it was taken from a medical author. Diels[79] says the use of the word alembic shows that an Arabic source cannot be disproved (nicht widerlegt werden können). Since the recipe for the distillation of alcohol is in the appendix only of the Munich MS. D, it was probably a later addition.[80] Diels[81] thought the name aqua ardens was taken from Pliny,[82] but this is improbable.

The addition of sulphur to the wine distilled is found in a work attributed to al-Fārābī in the same Paris MS. A,[83] and also in Porta,[84] and Berthelot quotes from other treatises in MSS. A and B the idea that the great humidity of wine opposes its inflammability, so that dry things like salt, and dry and fiery things like sulphur, would overcome it; wet and dry wood were given as examples.

Some fourteenth to fifteenth century MSS. written in Irish[85] deal with the "virtues of aqua vitæ" (virtuteis aque vite in Latin titles) or usgebathadh (usqebaugh is incorrect), but say nothing of its preparation. It sharpens the intelligence, makes cheerful, improves the memory, preserves youth, and applied externally or internally it cures all kinds of ills. It gives the taste of wine to water, makes corrupted wine into good wine, weak wine into strong wine, and good wine into extra strong wine. It swims on oil and any other liquid, hardens sirica salt (? soft sugar), cooks eggs, preserves flesh and fish, draws out the virtues of every herb and spice (but not the viola). It brightens copper (only in R.I.A. MS.) and if a candle is extinguished and dipped in it, it will flame. It sets alight hair that is dipped in it but does not burn the substance.

To return to the *Liber Ignium*:

§ 28. Experimentum mirabile quod facit homines ire in igne sine læsione, vel et[a] portare ignem vel ferrum calidum in manu. R. succum bismalvæ[b] et albumen ovi et semen psilii et calcem; et pulveriza et confice cum albumine succum[c] raphani et commisce. Et ex hac confectione[d] illinias corpus tuum vel[e] manum, et dimitte desiccare,[f] et post iterum illinias; et tunc poteris audacter sustinere sine nocumento.

A wonderful experiment which allows men to walk in fire without injury or to carry fire or hot iron in the hand. Take juice of double mallow and white of egg and flea-wort[g] seed and lime and powder them. Prepare with white of egg and radish juice.[h] Anoint your body or hands with this composition and let it dry. Repeat the treatment and then you will be able boldly to undergo the test without damage.

[a] etiam, H, R. [b] bimalvæ, H, R. [c] succis, H, R. [d] commixtione, H, R. [e] et, H, R. [f] dessiccare, H, R. [g] "grain de persil (?)" (parsley seed), Berthelot, but psyllion (ψύλλιον), Pliny, *H.N.*, xxv, 90, is usually identified with flea-wort or flea-bane : he says it has a cooling nature. [h] "sève de sapin," Berthelot.

§ 30. Candela accensa quæ tantam[a] reddit flammam quod[b] crines vel vestes tenentis[c] eam comburit. R. terebentinam[d] et distilla per alembicum[e] sicut[f] aquam ardentem, quam impones in vino cui applicatur candela, et ardebit ipsa.

A lighted candle which produces a flame which burns the hair and clothes of those who hold it. Take turpentine, distil it in an alembic in the same way as burning water (alcohol), add the wine[g] and when a candle is put to it, it burns.

[a] tanta, H. [b] quæ, H, R. [c] tenentes, H, R. [d] terebinthinam, H, R. [e] alambicum, H, R. [f] om., H, R. [g] Perhaps that of § 24 ? The "wine" is also mentioned in MS. D, ch. Y (see below), where it seems to be a liquid.

§ 31.   R. coloph[onium] 1, picem græcam et ibi subtilissime tunicam proicies in ignem vel in flammam candela.   [H, R : Recipe colophonium et picem subtilissime tritam et ibi cum tunica projicies in ignem vel in flammam candelæ.]

Take colophonium and pitch, grind to a very fine powder and project this by a tube [blowpipe] into a fire or into a candle flame.

This reading, as Romocki notes, gives a method of producing a cloud of fire such as is mentioned by Anna Komnena (ch. I) and is produced in theatres by lycopodium dust.   Berthelot's text and translation are obscure, but he mentions the above use.

§ 32.   Ignis volantis in aere triplex est compositio : quorum primus fit de salepetroso[a] et sulphure et oleo lini, quibus tribus[b] insimul distemperatis et in canna positis et accensis, protinus in aere sublimetur.[c]

There are three compositions of flying fire, of which the first is made from saltpetre and sulphur and linseed oil, the three mixed together and put into a tube and lighted; it at once rises into the air.

[a] salepetro, MS. B.   [b] tritis, H, R.   [c] poterit in aerem sublevari, H; sufflari, R : "la fusée monte aussitôt en l'air," Berthelot, but the assumption that a rocket is described, rather than a Roman candle, will be examined later.

§ 33.   Alias ignis volans in aere fit ex sale petroso et sulphure vivo et ex carbonibus vitis vel salicis, quibus insimul[a] mixtis et in tenta de papyro[b] facta positis et accensis, mox in aerem volat.   Et nota quod respectu sulphuris debes ponere tres partes de carbonibus, et respectu carbonum tres partes salis petrosi.

Another fire flying in the air is made from saltpetre and sulphur and vine or willow charcoal, which are mixed together and put into a paper container and lighted and it suddenly flies into the air.   And note that for 1 part of sulphur there should be taken 3 parts of charcoal and for 1 part of charcoal 3 parts of saltpetre.

[a] om., H, R.   [b] papiro, H, R.   salpetræ, H, R.

This mixture, if the translation is correct, is 1 sulphur : 3 charcoal : 9 saltpetre.   Although *three* compositions are specified in § 32, only two are given. In § 12 it is said there are two compositions, which are given, both containing saltpetre, and in § 13 sulphur, charcoal and saltpetre in the proportions 1 : 2 : 6; the last is probably the third mentioned in § 32.

The MSS. A and B conclude with § 35, describing a lasting lamp (candela durabilis) consisting of a bronze or lead vessel containing oil and connected with a lamp by a narrow tube.   This is one of the devices supposed to explain the "perpetual lamps" to which Athanasius Kircher[86] devoted much attention.

The Munich MS. D contains chapters lettered A to Z, of which A to V are similar to those in MSS. A and B (except R, which is different); three recipes, X, Y, Z, attributed to Marcus Græcus,[87] are new. X is headed: "Ignis græcus ita componitur." It includes classa, galbanum, serapinum ("résine de sapin," B), opoponax, sulphur, kekabie (untranslated, B), naval pitch, oil of bricks, alkitran, pigeon dung, distilled oil of turpentine, oil of sulphur and liquid pitch, mixed together, and digested in a stopped glass phial in a dung-bath. The varnish-like mixture is then distilled in a well-luted alembic over a gentle charcoal fire in the same way as rose-water (veluti aquarum rosarum). The distillate is then soaked into paper wicks put into a square arrow pierced with holes, set on fire, and thrown with a bow or ballista. The fire is extinguished only by the four things (see § 11). The account concludes: In petroleo pone pulverem (?) sulfuris minutissimi intra phialam vitream et appone si vis ignem et projice ignem, quia solo motu accenditur et hic est ignis græcus. Introduce in petroleum very finely powdered sulphur in a glass bottle. Set fire to it as you wish and throw the fire. This fire burns only when in motion (?) and is Greek fire.

This recipe, containing no saltpetre, is exactly the kind of mixture said to constitute Greek fire in ch. I, and is obviously from an Arabic source.

Recipes Y and Z describe how the distilled oil of turpentine, oil of sulphur and oil of bricks specified in recipe X are prepared. In Y, turpentine is melted in a glass cucurbit, and an equal weight of linseed oil is added. The alembic (head) is luted on with a paste of wheat flour and white of egg, and an oil as white and pure as water is distilled:

et cum lento igne distilla ejus oleum clarum et purum sicut aqua. De quo si vis scintilla posueris[a] et candelam ardentem applicaveris mox flammam maximam provocabit, aut si in ampullis super vinum posueris et sic demum et accendis in tabula ignem magnum provocabis

and with gentle fire distil this oil white[b] and pure as water. If you bring near a spark and apply a lighted candle you suddenly provoke a great flame. If you put it in bottles above[c] wine and then kindle it on a board a great fire is provoked.

[a] MS. posterius.   [b] "claire," Berthelot.   [c] "décrit plus haut sous le nom de *vin*," Berthelot.

In Z, oil of sulphur is made by grinding yolks of boiled eggs in a mortar to the consistency of butter, mixing with finely powdered sulphur, and distilling to obtain what the philosophers call oil of sulphur. In another preparation, juniper oil is mixed to a paste with powdered sulphur and distilled in a cucurbit with a well-closed alembic (alambic bene clauso):

sic distillasti. At est ratio quod predicti olei distillatio prohibetur propter ejus etorem·et incendium, quia si in altum ascenderit periculum immineret.

Moreover, the reason why the said oil is so distilled [in a closed apparatus] is that distillation [in open apparatus] is prohibited because of the fœtor of it and the risk of fire, because if it boils over danger threatens.

Berthelot's translation: "à cause de l'odeur et du risque d'incendie," was criticised by Lippmann[88] who thought "etor" should be "ether," as in a quite different passage in ψ-Plato[89]; "etor" is apparently not known in Latin dictionaries (Greek ἦτορ = heart), but Albertus Magnus[90] in speaking of inflammable waters (unctuosum) says: videmus in oleo lampadis et in humido radicali ethicorum. Berthelot's reading "odeur" (fœtor would be better) is confirmed by the one in the Arabic MS. of Ḥasan al-Rammāḥ (see ch. V). The last words are translated by Berthelot: "si la flamme s'élevait, il y aurait danger."

Oil of bricks (oleum laterinum), which is mentioned in MS. BN 6514 (f. 45 v.), is in recipe Z directed to be made as follows:

Take red bricks which have not been touched by water and break them into small pieces. Heat in a strong fire and extinguish in linseed oil, nut oil, or oil of hemp seeds (cannabinum). Grind them again somewhat. Put into a large well-glazed (bene vitreata) cucurbit and distil as above. This clear and red oil is called oil of the philosophers (oleum philosophorum). If you put it on your hand and hold it up, it runs down it. Mixed with balsam of cardamoms (?) it comforts the nerves and it is good against gouty chill (contra guttam frigidam). And if a fisherman anoints himself with it, fish abound. . . . Explicit liber Ignium a Marcho Græco compositus.

Other recipes in MS. D, not attributed to Marcus Græcus, are outlined by Berthelot; it is a typical collection of recipes of the same type. There is a "strong water whitening all metals" made by dissolving sal ammoniacum comatum (which Lippmann thought was "hair salt," double iron aluminium sulphate, although sal ammoniac was made by distilling hair), common salt and alum in white vinegar and boiling in a metal pot, which became white (bullire vas metallicum et recipiet albedinem).

Recipes §§ 32, 33, 34 in A and B are missing from D, but the assumption that whatever is missing in D, which is a much later MS. than A and B, is a late addition to A and B, seems to be unsafe, although it has been made by Lippmann and others.

Hime[91] compiled an index of the *Liber Ignium*, the references being to the paragraphs as numbered by Berthelot, except that Hime joined the last four lines of Berthelot's § 6 to his § 7 to make § 7 Hime.

acetum 11, 26
aes 17, 35
  „ italicus 15
  „ rubicundus 8
æthiopiæ oleum 3
alambicum 27, 30, 34
albacarinum[a] 25
alcea[b] 22
alkitran 1, 3, 6, 8, 9
ammoniæ liquor[c] 1
anethorum oleum 2
aqua ardens 27, 30
arena 11, 26

argentum vivum 19
asphaltum 10
astrolabium 8
auricalcum 8, 16
aviculæ cerebrum 20
balsamum 2, 3, 9
bismalvæ succum 28
bombax 10
butyrum 1
calx 21, 28
  „ non extincta 9
  „ viva 15, 24
camphoræ aqua 23

cantharides 15
carbo salicis 13, 33
„ vitis 13, 33
carbunculum 34
cera 6, 7, 13
colophonium 8, 10, 12, 31
cucurbita 10, 27
cyrogaleo[d] 18
embotum 26
fabarum aqua 21
ferri limaturæ 8
ferrum 17
„ Indicum 16
filtrum 6, 11
fimum columbinum 10
„ ovinum 10
furonis[e] fel 17
galbanum 15
gummæ oleum 26
gummi Arabicum 24
juniperi oleum 4
lacertus niger 19
„ viridis 19
lamii[f] oleum 12
laterum oleum 2
laton 17
leporis marini fel 18
lini oleum 7, 12, 32
lupi aquatici fel 18
lutum sapientiæ 1
malvæ viscus 21
marrubii lignum 6
medulla cannæ ferula 2
mustelæ fel 17
noctilucæ 17, 18, 34
nux castanea 23
„ Indica 23
oleum 24, 35
„ fœtidum 15

oleum universum 8
olivarum oleum 20
ovorum albumen 22, 28
„ oleum 6, 8, 9
„ vitella 5, 22
petroleum 2, 10, 26
picula 10
pinguendo arietina 2
pix 26, 31
plumbum 8, 35
psillii semen 28
raphani succum 28
sal coctus 26
„ combustus 16
„ communis 27
„ petrosus 12, 13, 14, 32, 33
sandaraca 1, 7
sarcocolla 26
sepo 1
stannum 8
stuppa 26
sulphur 2, 4, 6, 22, 24, 29, 32, 33
„ croceum 8
„ oleum 3, 4, 5, 7, 8
„ splendidum 5
„ vivum 10, 12, 13, 26, 27, 33
tartarum 10, 26, 27
terebinthina 30
„ oleum 2
terra Messinæ 21
testudinis fel 15, 17, 18
tyriaca 18
urina 11, 26
vermes noctilucæ 16
vinum 30
„ album 27
„ nigrum 27
zembac oleum 15, 16, 17

[a] alba ceraunia? [b] alcea (ἀλκέα) is in Pliny, *H.N.*, xxvii, 6, as a plant (Malva alcea L., or Malope malachoïdes, or Hibiscus trionum?). [c] armoniaci, Hoefer; a liquid resin, not ammonia (as given by Romocki). [d] cynoglossos (the plant hound's tongue?), Pliny, *H.N.*, xxv, 41. [e] ferret, Berthelot. [f] lauri?

Beckmann[92] had pointed out the similarity between parts of the *Liber Ignium* and recipes in the *De Mirabilibus Mundi* attributed to Albertus Magnus, and in Roger Bacon, and suggested that all three came from a common source.

To these should now be added the works attributed to Michael Scot, who was still earlier. The suggestion by Fournier,[93] that the present *Liber Ignium* is part of a larger work which was available to Roger Bacon and Albertus Magnus was accepted by Romocki. Favé[94] supposed that Albertus Magnus and Roger Bacon had read the *Liber Ignium*. Albertus copied it literally, but Bacon disguised it because of his attitude that science should be concealed from the vulgar.

The only authorities named in the *Liber Ignium* are Hermes, Tholomeus (Ptolemy, presumably Ptolemy I, ruler of Egypt), Alexander (the Great) and Aristotle. These suggest an Arabic source. An old German abbreviated translation in MSS. Vienna 3062 and Berlin Great General Staff 117 (both fifteenth century)[95] attributes the work to Achilles Tabor (who is considered later) : "Das sind die Fewer die Meister Achilles Thabor geschrieben hatt." It reproduces "Armoniaci liquidi" as "czerlass Salarmoniak" (fused sal ammoniac!). The "urbes Agarenorum" (the Arabs were "Agareni" for tenth-century Byzantines) in MS. D becomes "Stett Agarrenorum"; ære ytalico (= bronze) becomes "welische Kuppher"; "ferro Judaico" is "Eysen von India"; a formula for Greek fire is different from that in Marcus Græcus and partly reproduces Kyeser's with alcohol (see ch. IV), another formula is more complicated; some recipes of Marcus Græcus are omitted.

The *Liber Ignium* is a collection of recipes given without any attempts at classification or order, and is of the well-known type of collections of "secrets" which were made not only by scholars such as Albertus Magnus, but also by technicians in the service of notabilities, and itinerant free-lances of the type of Kallinikos (ch. I). The fact that many recipes are unintelligible and some obviously worthless suggests that they were purchased as valuable secrets from some vagabond, and carefully copied out for trial in future; this type of procedure is well known among the modern business executives, ignorant of science, but highly credulous and very willing to part with other people's money. Jews in Spain, with their real knowledge of Arabic science and technology, would in the Middle Ages be particularly interested in making such collections, some of which were undoubtedly worth their price. The recipes were passed on in the course of centuries and continued to be improved. They reappear in printed works such as Wecker's *De Secretis*, the last edition of which appeared in 1753.

Leonardo da Vinci (d. 1519)[96] gives recipes for Greek fire composed of willow charcoal, saltpetre (sale nitro), alcohol (aqua vite), sulphur, pitch, frankincense and camphor, boiled together and spread over Ethiopian wool (lana etiopica). Scaliger[97] describes two incendiary mixtures, one Arabic and one Catalan, and both containing saltpetre; distillation "in seraphino" is used in both cases. It seems probable that, after the invention of gunpowder about 1225, mixtures containing the old components of Greek fire were fortified by the addition of saltpetre. Baptista Porta (1588)[98] says a mixture of sulphur and quicklime inflames when sprinkled with water, and also gives incendiary mixtures containing saltpetre. Blaise de Vigenère (end of sixteenth century) speaks of many kinds of Greek fire containing saltpetre with sulphur, bitumen, black pitch, resin, turpentine, colophony, sarcocolla, linseed oil, petroleum, laurel camphor, suet and other fats.[99]   Marcus Græcus was used and named by Porta,[100] Cardan[101] and Biringuccio.[102]

Berthelot[103] distinguished six groups of recipes in the *Liber Ignium*: (i) Those for incendiaries, going back to Julius Africanus; but oil of bricks and oil of sulphur were added by the alchemists, and alkitran is from an Arabic source.   (ii) Recipes attributed to Aristotle, from an Arabic source.   (iii) Formulæ for Greek fire.   (iv) Recipes for rockets (fusées) and petards, twelfth to thirteenth century (the oldest mention of saltpetre and gunpowder). (v) Phosphorescent materials, perhaps the oldest part, since they are mentioned in the Greek alchemical corpus (Ostanes, Maria).   (vi) Conjuring tricks (holding fire in the hand protected by a composition containing chalk, etc.; burning sulphur, with a relatively cool flame, etc.).   Berthelot refers for these to a work (Book of the Balance) of Jābir (eighth century?), but they are given by Hippolytos (see ch. I).   The alcohol recipe is a late addition to the Munich MS.   Berthelot says[104]:

> "Le *Liber ignium* . . . a été composé avec des matériaux de dates multiples, les uns remontant à l'antiquité, les autres ajoutés à diverses époques, dont les dernières étaient contemporaines, ou très voisines de celle de la transcription de chaque manuscrit."

Hime[105] pointed out that, of the thirty-five recipes, fourteen are war mixtures, six are for extinguishing incendiaries or the prevention and cure of burns, eleven are for lamps, lights, etc., and four for preparing chemicals (one, no. 14, giving the purification of saltpetre, and one, no. 27, for alcohol). The fourteen recipes for war mixtures consist of nine for various fires (nos. 1, 2, 3, 6, 7, 8, 9, 24, 26), one (no. 10) for an incendiary arrow composition, and four (nos. 12, 13, 32, 33) for rockets and Roman candles (including a "cracker"); three (nos. 9, 14, 24) contain quicklime, five (nos. 12, 13, 14, 32, 33) contain saltpetre.   Hime divided the recipes chronologically into three groups:

Earliest   750–?   nos. 1, 2, 3, 6, 7, 8, 9, 10, 15, 16, 17, 18, 19, 20, 21, 23, 25, 34

Middle   –1225?   nos. 4, 5, 11, 22, 24, 26, 27, 28, 29, 30, 31, 35

Late   1225–1300   nos. 12, 13, 14, 32, 33

The Greek names Hermes, Ptolemy, Alexander and Aristotle were common property before the *Liber Ignium* was written.   Greek words in Latinised form include alba ceraunia, asphaltum, bombax, cynoglossum, orichalcum and sarcocolla, but all these had been Latinised before; the first four are in Pliny and orichalcum is in Plautus.   Although there are three Greek names for the asphalt family (pissa, asphaltos and naphtha derived from Eastern sources), and at least three Latin words (which he actually used), pix (or picula), asphaltum and petroleum, the translator in recipe 1 used the barbarism alkitran Græcum, which is Arabic (alqitran, Spanish alquitran); so are asturlab, alambicum and zembac (zanbāq).   Sulphur splendidum and sulphur croceum are Oriental.[106]   The Spaniards did not use the word "naphtha," which does not occur in the *Liber Ignium*, where it is replaced by alkitran.

The Arabic and Spanish words occur in the "Earliest group" of recipes, the "Middle group" contains one Arabic word, alambic, and one Spanish word, petroleo.   This suggests that the earlier recipes were translated from Arabic by a Spaniard in 1182–1225, that the Middle recipes were added by other hands before 1225, and the Late recipes, mentioning saltpetre, were inserted about 1300.[107]   Hime thought that the author or translator was not a Greek or Muslim (who never used the name "Greek fire"), but a Jew or Spaniard who either did not know the Latin names for some Arabic words or thought them so familiar that they need not be translated (alkitran and zembac are untranslated; the Arabic nuhas ahmar for copper becomes æs rubicundus not cuprum, no. 9 mentions the periodic rains of the East, kharif, given as Autumnalis pluviæ dilapsu).

Hime says there are Roman candles (no. 12), rockets (no. 13) and a cracker (no. 13), but gunpowder is not used as an *explosive*, as it is in Roger Bacon, and[108] the "cracker" of recipe 13 was a toy, burst by *gradual* evolution of gas bursting a thick case.   Roger Bacon's "toy" was of thin paper only, which would have been burnt by a simple incendiary mixture, and the "bang" was produced by the direct *explosion* of the gunpowder.

Berthelot[109] and Lippmann[110] thought the recipes containing saltpetre were put in after 1300, "when it first became known in the West," and the references to ignis volans in §§ 32–3 (which are "mostly not" in the MSS. C or D) are also later interpolations.   They occur, however, in MSS. A and B, §§ 12–13.   Reinaud and Favé[111] had suggested a Chinese origin of the "feu volant," which penetrated to the Arabs and Europe with the Tartar armies about 1250.

The earliest recipes in the *Liber Ignium* are of the type described in the interpolations of the *Kestoi* of Julius Africanus (see ch. I), and others go back to the eighth century.   Probably, as Berthelot showed, some recipes are of Greek origin, but "Marcus Græcus" is a purely imaginary person, the real author (or authors) of the final collection being Jewish or Spanish of the twelfth to thirteenth century, and the gunpowder recipe is the latest of all.[112] Lippmann[113] dated it about or after 1300.   According to Hime[114] the methods given in the *Liber Ignium* and Hasan al-Rammāh (1275–95) "leave no possible doubt that in their time it had but just come into use"; the saltpetre mixtures were not true gunpowder, since they would not explode but only burn fiercely.

Reinaud and Favé,[115] who discussed Marcus Græcus on the basis of Du Theil's text (1804), refer to the two containers (cases) for gunpowder, one long and slender and the other short and thick (§§ 12, 13), and concluded that there is a description of a primitive rocket (fusée), less advanced than the Arab rockets.   Hime[116] also says the description of the rocket and its filling "is as definite and precise as many a recipe of the seventeenth century," other recipes (§§ 32–3) are as precise as those of Hasan al-Rammāh, yet the Roman candle (§ 12) and rocket (§ 13) were toys and not explosive, which Roger Bacon's was.   Reinaud and Favé[117] thought the purification of saltpetre was more primitive than Ḥasan al-Rammāḥ's. The long and slender case would be a Roman candle (chandelle romaine), with which Jähns[118] (Römerkerze) agrees; they are described as tied to the ends of sticks, lances or pikes, for use in warfare, by Biringuccio.[119] The question as to whether the "flying fire" was a rocket or Roman candle will be discussed later, and the analysis of the *Liber Ignium* will, for the present, be closed.

REFERENCES

1.   Schoell, *Geschichte der griechischen Litteratur*, Berlin, 1830, iii, 447 f.; Kopp, *Beiträge zur Geschichte der Chemie*, 1875, iii, 95.

2.   Berthelot, *La Chimie au Moyen Âge*, 1893, i, 89–135.

3.   Romocki, *Geschichte der Explosivstoffe*, Berlin, 1895, i, 123, 127.

4.   Mrs. D. W. Singer, *Catalogue of Alchemical Manuscripts in Great Britain and Ireland*, Brussels, 1930, ii, 633; cf. R. Hendrie, *An Essay upon Various Arts . . . by Theophilus*, London, 1847, p. v.

5.   Thorndike, *Isis*, 1929, xiii, 83.

6.   Mrs. Merrifield, *Original Treatises on the Art of Painting*, London, 1849, i, 73, 79; on Le Bègue, see L. Delisle, *Le Cabinet des Manuscrits de la Bibliothèque Impériale*, Paris, 1868, i, 23.

7.   L. Thorndike, *History of Magic and Experimental Science*, 1923, ii, 784.

8.   *Theatrum Chemicum*, Zetzner, Strassburg, 1660, v, 193 f., 219.

9.   Ref. 8, v, 55.

10.   Ref. 2, i, 90; *id.*, *Archéologie et Histoire des Sciences*, Paris, 1906, 261.

11.   *Jahrbücher der Litteratur*, Vienna, 1846, cxiv, 163; Reinaud and Favé, *J. Asiatique*, 1849, xiv, 257 (318).

12.   Ref. 1, iii, 96.

13.   Ref. 2, i, 98; ref. 10 (1906), 171.

14.   H. Diels, *Abhl. K. Preuss. Akad. Wiss., Phil.-hist. Cl.*, Berlin, 1913, no. iii, 18; *id.*, *Antike Technik*, 1924, 108.

15.   Hime, *Origin of Artillery*, 1915, 60.

16.   Lippmann, *Entstehung und Ausbreitung der Alchemie*, Berlin, 1919, i, 478.

17.   *Le Feu Grégeois*, Paris, 1845, 87.

18.   In Napoleon III, *Études sur le Passé et de l'Avenir de l'Artillerie*, Paris, 1862, iii, 12.

19.   Dutens, *An Enquiry into the Origin of the Discoveries attributed to the Moderns*, London, 1796, 265 f.; *id.*, *Origine des Découvertes*, 3 ed., London, 1796, 197 f.

20.   R. Bacon, *Opus Majus*, ed. Jebb, London, 1733, Pref. sig. C 1 : "quem vocant Librum ignium . . . MS. penes virum Clariss. D.Ric. Mead."

21.   Fortis, *Del Nitro Minerale*, in Amoretti and Soave, *Opuscoli scelti sulla Scienza e sulla Arti*, Milan, 1787–8, XI, iii, 145–69 (148).

22.   Schelhamer, *De Nitro*, Amsterdam, 1709, 22.

23.   Wiegleb, *Geschichte der Chemie*, 1792, ii, 140, from *Göttingen gelehrt. Anzeigen*, 1788, iii, 1948 (Stück 195)—Paris MS. of eighth century with Greek title; Gmelin, *Geschichte der Chemie*, 1797, i, 97; Kopp, *Geschichte der Chemie*, 1845, iii, 220; *id.*, ref. 1, iii, 95.

24.   Lippmann, ref. 16, i, 478; but see *id.*, *ib.*, 1954, iii, 94; Sarton, *Introduction to the History of Science*, 1927, ii, 1038.

25.   Partington, *Isis*, 1933, xix, 260.

26.   Ref. 19.

27.   Mesuë, *Opera Medica*, f°, Venice, 1562, 86 *v.*, col. 2, H.

28.   Romocki, ref. 3, i, 115.

29.   L. Lalanne, *Mém. div. Sav. Acad. des Inscriptions et Belles-Lettres*, II. Sér., Paris, 1843, i, 341.

30.   Hime, *Gunpowder and Ammunition*, 1904, 85.

31.   Fabricius, *Bibliotheca Græca*, 1726, xiii, 320.

32. Ref. 31, 1726, xiii, 172.

33. Hoefer, *Histoire de la Chimie*, 1842, i, 284; 1865, i, 304: "Les sceptiques peuvent, il est vrai, nier l'identité de notre Marcus Græcus avec celui cité par Mesué; mais ils ont encore moins de preuves pour nier, que nous pour affirmer." Hoefer mentions that Reinaud and Favé, *Le Feu Grégeois*, 1845, 97, had dated Marcus between the ninth and twelfth centuries.

34. Jähns, *Geschichte der Kriegswissenschaft*, 1889, i, 156.

35. Ref. 3, i, 115.

36. Ref. 2, i, 90.

37. Ref. 17, 89.

38. Robins, *Traité de Mathématiques*, tr. by Dupuy, Grenoble, 1771, 98; the quotation in Reinaud and Favé, ref. 17, 85, is on p. 100 of Robins.

39. Beckmann, *History of Inventions*, 1846, ii, 504; Kopp, ref. 1, iii, 95.

40. Ref. 33, 1842, i, 491; 1866, i, 517.

41. In *Nouvelle Biographie Générale*, 1859, xxix, 563, it is described as "introuvable."

42. Bibliothèque Nationale, *Catalogue Général*, 1931, cv, 847.

43. Ref. 2, i, 100–120; *J. des Savants*, 1895, 684.

44. Poisson, *Rev. Scientif.*, 1891, xlvii, 457–62, with notes.

45. Ref. 3, i, 115–23; for text from Nürnberg MS. (E), *ib.*, 123–7; cf. Berthelot, ref. 43 (1895).

46. Ref. 15, 56.

47. As Hime had previously said, ref. 30, 69.

48. Ref. 16, i, 477 f.

49. Ref. 14.

50. E. H. F. Meyer, *Geschichte der Botanik*, Königsberg, 1856, iii, 493.

51. Favé, ref. 18, 1862, iii, 82.

52. Ref. 2, i, 107.

53. Ref. 2, i, 110.

54. Ref. 39, ii, 505.

55. Ref. 16, i, 480.

56. *Byzantion*, 1932, vii, 265.

57. *Abhandlungen und Vorträge*, 1906, i, 13.

58. E.g. *H.N.*, xxii, 30.

59. *H.N.*, xix, 26.

60. Poisson, ref. 44; Ruska, *Der Islam*, 1913, iv, 320; Leclerc, *La Chirurgie d'Abulcasis*, 1861, 264.

61. Hime, ref. 15, 60.

62. *Compositiones*, ch. 156.

63. G. W. Corner, *Bull. Inst. Hist. Medicine*, 1936, iv, 745.

64. Ref. 2, i, 117.

65. Ref. 15, 11 f., 24.

66. Romocki, ref. 3, i, 9, 35.

67. Berthelot, ref. 2, i, 133.

68. *Le Feu Grégeois*, 1952, 38 f., who says the MS. was originally in Greek and eleventh to twelfth century, quoting Schoell, ref. 1.

69. Ref. 33, 1866, i, 309.

70. Ref. 2, i, 117.

71. W. Smith, *Latin-English Dictionary*, 1904, 916; Steinwenter, in Pauli-Wissowa, *Real-Encyclopädie*, xii, 2347, 2351.

72. Ref. 16, i, 480.

73. Ref. 16, 1931, ii, 175.

74. *Griechische und Römische Metrologie*, 2 ed., Berlin, 1882, 116.

75. Ref. 2, i, 138.

76. Ref. 14, 1913, 18.

77. Ref. 2, i, 142.

78. *Archiv für die Geschichte der Naturwissenschaften*, Leipzig, 1913, iv, 429 (446).

79. Ref. 14, 1913, 21.

80. Berthelot, ref. 2, i, 120, 122, 134 f.; Lippmann, ref. 16, i, 480.

81. Ref. 14, 1913, 23.

82. *H.N.*, xxxv, 50: "sentitur vis ejus et in aquis ardentibus neque alia res facilius accenditur," referring to sulphur; Bostock and Riley, *The Natural History of Pliny*, 1857, vi, 293, tr. "in hot mineral waters."

83. *Liber Alpharabii*, BN MS. Lat. 7156, f. 47 *v*.; Berthelot, ref. 2, i, 143.

84. *Magiæ Naturalis Libri IIII*, bk. ii, ch. 10; Antwerp, 1585, 113.

85. Trinity College, Dublin, 1343 (old H.3.22), pp. 107–10; 1337 (old H.3.18), p. 417; R. Irish Acad. 23.O.6, p. 36; 24.B.3, p. 94. I am indebted for the translation to Dr. (Miss) E. Knott, M.R.I.A., and for showing me the Trinity College MSS. to Dr. Sullivan. I have rearranged the order of the translation. It seems to me to be based on ch. 31 of the *Rosarius* of Arnald of Villanova, printed in Manget, *Bibliotheca Chemica Curiosa*, Geneva, 1702, i, 676, although I have not compared the two carefully.

86. Kircher, *Oedipi Aegyptiaci*, f°, Rome, 1654, iii, 548, q. an Arabic author, "Schiangia Arabs in historia memorabilium Ægypti"; Kircher, *Mundus Subterraneus*, Amsterdam, 1665, ii, 73.

87. Berthelot, ref. 2, i, 124 f.

88. Ref. 16, i, 480.

89. Berthelot, ref. 2, i, 398: "ether ist substantia lucis, vacua accidentibus," which Berthelot says "rappelle les théories de la physique moderne."

90. *De Mineralibus*, bk. III, tract. i, ch. 2.

91. Ref. 15, 56–7.

92. Ref. 39, ii, 504.

93. *Biographie Universelle*, 1852–66, xxvi, 506.

94. In Ref. 18, iii, 61–4; and ref. 17, 112, 121.

95. Romocki, ref. 3, i, 127–132; Berthelot, ref. 43 (1895), 684.

96. Richter, *The Literary Works of Leonardo da Vinci*, London, 1883, ii, 280; MacCurdy, *Leonardo da Vinci's Notebooks*, London, 1938, ii, 186, 194, 207, 217.

97. *Exotericarvm Exercitationvm*, Paris, 1557, ch. xiii, f. 30.

98. *Magia Naturalis*, bk. xii, ch. 1; Leyden, 1650, 463.

99. Berthelot, *Rev. Deux Mondes*, 1891, cvi, 786 (812).

100. Ref. 98, bk. xii, ch. 10; 1650, 479: "M. Graccho."

101. *De Subtilitate*, ii; 8° ed., 1560, 92: "Marcum Gracchum."

102. *Pirotechnia*, Venice, 1540, 163 *v*.: "Marcho Gracho."

103. Ref. 2, i, 128; ref. 43 (1895), 684.

104. Ref. 2, i, 135.

105. Ref. 15, 35, 58 f.; ref. 30, 70, 73 f.

106. Mas'ūdī, *Les Prairies d'Or*, tr. Barbier de Meynard and Pavet de Courteille, Paris, 1864, iii, 49, has "white, yellow, and other kinds of sulphur"; Rāy, *Hindu Chemistry*, Calcutta, 1902, i, 50 (white, yellow, red and black); ψ-Rāzī in Berthelot, ref. 2, i, 307 (same four).

107. Berthelot, ref. 2, i, 130, 135; Hime, refs. 15 and 30.

108. Hime, ref. 30, 87.

109. Ref. 2, i, 130, 133.

110. Ref. 16, i, 479.

111. Ref. 11, 316.

112. Hime, ref. 15, 58–60.

113. Ref. 16, i, 479.

114. Ref. 15, 15, 63.

115. Ref. 17, 79 f.

116. Ref. 15, 58, 61.

117. Reinaud and Favé, ref. 17, 84, 87; Hime, ref. 15, 61.

118. Ref. 34, i, 223.

119. Ref. 102, 162 *v*.

ROGER BACON AND ALBERTUS MAGNUS

We come now to two of the greatest scientists of the Middle Ages, Albertus Magnus (Saint Albert the Great) and Roger Bacon. They belonged to the tribe of Schoolmen, adherents to a movement known as Scholasticism. Although not infrequently an object of contempt among present-day scientists who have not the most rudimentary knowledge of it, Scholasticism in its day was a living force and some of its expositors were among the greatest men of all time. Albertus Magnus and Roger Bacon had intellects not inferior to Galileo's and Newton's, and if they had lived in the time of Galileo and Newton they would probably have been outstanding scientists.

The main impulse towards the development of Scholasticism was the study of Aristotle, a knowledge of whose principal writings reached the West in the form of Latin translations made from Arabic versions, which in turn were mostly derived from Syriac translations from the Greek. In the early Scholastic period, the Greek texts of Aristotle were unknown in the West, although they existed in Constantinople and elsewhere. Many of the Latin translations were made in Spain, one of the most celebrated translators being Gerard of Cremona, in Italy (A.D. 1114–1187), who worked in Toledo, and also translated works on alchemy and the *Canon* of Avicenna, which in the field of medicine rivalled Aristotle's writings in philosophy.

It was recognised that Aristotle's works were a great storehouse of knowledge which could not be found in any other source. It was the great contribution of Albertus Magnus to compile a system of knowledge based on the Latin translations of Aristotle but made to conform to Christian beliefs. He paraphrased nearly all the works of Aristotle, adding to them some newer knowledge and sometimes observations of his own. Albert had an excellent knowledge of Arabic works in Latin translations, although he knew no Arabic; Bacon had the advantage of knowing a little Arabic. Both men were misled into regarding as genuine works of Aristotle some treatises coming through Arabic which were really spurious; Albert, in particular (as we shall see), took one such spurious work as Aristotle's. We know very little of the lives of Albert and Bacon. We do not know with certainty when either was born; we are not certain when and where Bacon died.

Albertus Magnus, Albert the Great, or since 1931 Saint Albert the Great, was born at or near Lauingen on the Danube in Bavarian Swabia in 1193, 1195, 1205, 1206, or 1207, the dates 1193 and 1206 being probable alternatives. He belonged to the noble family of the Counts of Bollstädt. He entered the Dominican Order about 1229, and in 1233 gained his lectorate, spending the next ten years teaching in different German convents. About 1240 he was chosen to go to Paris, then the focus of intellectual life in Europe, and in 1245–6 he was admitted Doctor of Theology. As compared with the Franciscans, to which Bacon belonged, the Dominicans were short of advanced teachers, and Albert was chosen as a teacher before he had studied theology, which was very unusual. He was over forty when he gained his doctorate in Paris. Cologne, where he mainly taught, became a Studium Generale in

1248. He was Bishop of Ratisbon in 1260–62, but resigned with the consent of the Pope, and thereafter devoted his life to teaching, travelling on foot to various monasteries in Germany. He died in 1280 in the Dominican cloister in Cologne. It is said that in his youth Albert had small intellectual gifts, and that the Virgin appeared in his cell and asked him whether he wished to excel in theology or philosophy. He chose philosophy and the Virgin, disappointed, told him that he would return to his former stupidity before he died. He does seem to have become feeble-minded from about 1278, and it was said: "Albertus ex asino factus est philosophus et ex philosopho asinus." Albert's writings constitute an encyclopædia of the knowledge of the time; in the latest complete edition, by Borgnet, they fill thirty-eight volumes. Albert was a gentle, lovable man, much esteemed by all who knew him, and although he did not reach the theological eminence of his pupil, St. Thomas Aquinas, he outshone him in scientific knowledge.

Roger Bacon was born about 1214 at Ilchester in Somerset, and like Albert he belonged to a family of consequence, which had decreased its means in supporting Henry III against the barons. He studied at Oxford, probably under Grosseteste, and took the M.A. at an unknown date. Although called *Doctor Mirabilis* he was never a doctor, and perhaps not even a master, in theology, and although he wrote on it he was never a real authority on it, in the sense that Albert and Thomas were. He was thoroughly educated in scholastic philosophy. He lectured and wrote in Oxford, where the Franciscan School was part of the University organisation, and lectured on Aristotle in Paris from about 1240. Probably because he had no means and wished to study, he entered the Franciscan Order at some time in the period 1251–7. He returned to Oxford some time after 1251, and thereafter his life is almost a blank. For some reason which we do not know with certainty, he came into disfavour with his Order, and was put under restraint both in Paris and Oxford.

Whilst Guy de Foulques was still Cardinal-Bishop of Sabina, Bacon sent to him a letter concerning writings which he had not then composed. The Franciscan Order prohibited the writing of books, but the Cardinal told Bacon to proceed in secret. Guy became Pope Clement IV in 1265 and in 1266 wrote to Bacon and asked him to send him his writings. The *Opus Majus* was sent in 1268, and soon after, if not with it, the *Opus Minus* and the *Opus Tertium*, in which Bacon expands parts of the first work, and deals particularly with alchemy, for which he had a special liking and in which he probably made experiments. How these works were regarded by the Pope we do not know, and there is no indication that he ever thanked Bacon for them, or offered to defray the expenses involved in their preparation.

Pope Clement IV died in 1268 and Bacon probably returned to Oxford. The new Pope was not, apparently, interested in Bacon. Jerome of Ascoli, the Minister General of the Franciscans, met John of Vercelli, the General of the Dominicans, in 1277 to devise measures for allaying the quarrels between the two orders, and the *Franciscan Chronicle of the Twenty-four Generals*, the authenticity of which is accepted by some Catholic scholars and rejected by others, says that Jerome of Ascoli (who became Pope Nicholas IV in 1288), "by the advice of many friars, condemned and denounced the teaching of Roger Bacon of England, master of sacred theology [which he was probably not], as containing some suspected novelties, on account of which the said Roger was condemned to prison, with the order given to all the brethren that none should hold his doctrine but avoid it as reprobated by the Order. He wrote also to Pope Nicholas [III] in order that by his authority that dangerous teaching might be utterly suppressed." What the teaching was we do not know; it could hardly have been purely scientific, since there was no valid objection to this.

Bacon was loyal to the Church but dismayed at what he thought abuses. His unusual studies made him impatient with pedestrian scholarship. His lack of worldly wisdom led him to entertain the strange delusion that errors would be rectified if they were forcibly pointed out; he relied too little on others to be popular, and had too good an opinion of his own undoubted genius to fit in neatly with other members of his Order. Although deeply religious he did not sufficiently appreciate the difficulties of the higher dignitaries of the Church in steering their policies through the shoals of temporal danger, and was often unfair to them. He was inferior as a man of affairs, and too like his spiritual leader, St. Francis, to shine in such circles. He also criticised the Dominicans too freely, not clearly appreciating that their bite was at least as dangerous as their bark; the "Domini canes" were not chosen as the instrument of the Inquisition without foundation, and it fell to a Dominican to judge Joan of Arc.

Catholic writers, who are in general fair and unprejudiced, and some historians of science have, in my opinion, been too critical of Roger Bacon and have not sufficiently emphasised that his thirst for *experimental* knowledge was something quite exceptional in his time, and his criticism of authority, although restricted, was real.

After Bacon's death, although his name did not disappear, there seems to have been some justifiable timidity in too openly following him. His works were neglected, and it is only in quite recent times that they have been published; the *Opus Majus* was not published until 1733, the *Opus Tertium* and part of the *Opus Minus* in 1859; a more complete text of the last work, on the basis of a Vatican MS., was promised in 1912, but after much trouble I have been unable to find out if it has ever been published.

Albertus Magnus and Roger Bacon believed in the reality of evil. They had not the advantage of the mechanical theory of the Universe held throughout the nineteenth century, or the view of entropy and the quantum theory which replaced it in the twentieth, and they believed that the teaching of Scripture that evil was the work of the Devil was true. They did not question the reality of magic, enchantment, necromancy and the potency of animals, plants, and especially engraved gems and particular stones in magic ceremonies, which they had read of in the scientific works of Arabs which were modern in their time. They believed in the interpretation of dreams and in astrology. These were all sciences to them. Albert, however, believed that the human will and reason are free unless the will yields to sin, when it is swept away and entangled in nature, and he thought the art of images engraved on gems is evil if it imputes divinity to the stars; the effects, however, are natural and belong to the physical sciences.

A cloud of legends gathered around the memory of Albert; he made a bronze head which talked, which his pupil Thomas Aquinas broke with a hammer because its chatter disturbed his studies (the same story was told of Roger Bacon and Friar Bungay), and when the German Emperor visited Cologne in the depth of winter, Albert entertained him at table in a warm and sunny garden, but when the repast was ended uttered an incantation, when the scene disappeared in a storm of driving snow.

Unlike Bacon, Albert was less instructed in mathematics and physics than in natural history; he was a careful and accurate observer. Both had a keen interest in chemistry. Albert knew that it was penetrated by fraud, since he was less enthusiastic than Bacon, but, mistaking it as ultimately derived from Aristotle (which, in a sense, is true) he thought that there was a genuine alchemy, having the power of transmuting the base metals into gold. Both Albert and Bacon knew that the proper authorities must be consulted in particular fields. Peter of Prussia quotes Albert as saying:

> Sciendum tamen, quod Augustino in his quæ sunt de fide et moribus plus quam Philosophis [Aristotle] credendum est si dissentiunt: sed si de medicina loqueretur, plus ego crederem Galeno vel Hippocrati, et si de naturis rerum loquatur, credo Aristoteli plus quam alii experto in rerum naturis.

Both, like Roger's namesake Francis Bacon, realised that information on chemistry must be sought not in books on mathematics or astronomy but in the writings of the specialists, the alchemists. Some modern historians of science are greatly in error in believing that "science" began with mathematicians or astronomers like Galileo; chemistry did not so begin but with the much earlier alchemists, and it has its own method, quite different from that used by Galileo. Newton dabbled in alchemy and chemistry but achieved nothing of note; as compared with an alchemist like Glauber he was a mere amateur.

Bacon had followed an unusual course of study, which in his day could lead to no advancement in the Church. He studied mathematics, optics, alchemy, experimental sciences and Greek and Oriental languages such as Hebrew and Arabic. How he became interested in experimental sciences we do not know; he read translations of Arabic books and was acquainted with Peter of Maricourt, the traveller (Peregrinus), who was a soldier in the Crusade of 1269 and in that year composed a famous treatise on the magnet. Bacon esteemed him greatly.

Bacon should perhaps have done more experimental work, but a large part of his life was spent in confinement. We know that he was under some kind of restraint in the house of his Order in Paris, even if it did not amount to actual imprisonment, and that the transfer from Paris to Oxford was not voluntary on his part. No one claims to have seen him after 1278. The Franciscan historian, the Irishman Luke Wadding, says only: "The authors do not say whether he died captive or free," but that Bacon was captive for some time before his death in 1292 or 1294, Wadding, who had the records before him, did not doubt.

Modern detractors of Bacon are at pains to show, from his own words, that he had no real conception of "modern" science—whatever that may mean—although they do admit that he "sometimes comes near to being scientific." Long before Galileo, he had insisted that there was an experimental *method*, for which he used, perhaps for the first time, the name "experimental science" (scientia experimentalis), something quite different from "experiments," which many before him had made. Something more than mere experimenting, he recognised, is requisite in arriving at truth. He says that experimental science teaches how to test perfectly what can be done by nature, what by art, and what by "tricks," so that all falsity may be removed and the truth of nature and art retained. He gives some good examples of what he means, and some poor ones, but the latter cannot remove his claim to have pointed out the true way of reaching scientific truth.

Bacon recognised much more clearly than Albert did that the jumbles of "experiments" compiled into recipe books by ignorant quacks must be tested before they can be accepted, although Albert had clearly said that this is necessary in alchemy. Both accepted fallacious recipes, but neither, an ecclesiastic and not a professional scientist, could test everything. Bacon believed that *all* science could be known, but not by the unworthy, and that the coming of Christianity had revealed much that the pagans, even Aristotle, could not know. Aristotle was before us in time, and in time more and more becomes known. Argument and logic alone do not convince, but experiment does. If a man ignorant of fire had proved by logic that fire burns he would never be certain of this until he had put his hand in a fire, after which he would have no doubt.

We must beware here of two errors. First, such a statement as the last was not an obvious one. "Physics" in the sense understood by Aristotle, was not at first regarded as a proper study for Churchmen and was even regarded as sinful; it required courage to study the properties of fire. John of Wales, regent master of the Franciscans in Oxford, in 1260 quoted Alexander Neckam to prove that there was a place called Oxford, although he himself lived there. Authority was more important than observation. Secondly, we must beware of the fallacy that making experiments was a benevolently regarded commonplace in the time of Albert and Roger. Bacon tells us that his contemporaries thought the results of experiments were the work of evil spirits, theologians and canons regarding them as unworthy of a Christian; as for chemical experiments, he had omitted them altogether from his *Opus Majus* as suited only for the ears of the wisest, of whom there were not three in the world. We must remember that these words were addressed to the Pope and were not thrown off at random.

Bacon had a profound contempt for law; the canon lawyers of his time occupied choice and lucrative positions, and a knowledge of law was a sure means of advancement. He thought that legal method could equally well prove what was false, and known to be false, as demonstrate what was true. Scientific method, he thought, was better and was more sincere. From the little we know of his life, we may surely deduce that he would have done better to have studied law instead of science.

Both Albert and Roger were courageous men. Many things on which they wrote were highly suspect and regarded with much disfavour in the Church. Albert, more circumspect and calmer, overcame much of this prejudice; Bacon, rash and often violent, merely accentuated it. After them, no Churchman could or did neglect the new knowledge which they had revealed.

We will begin our study with a famous treatise by Bacon.

ROGER BACON

De secretis operibus artis et naturæ et de nullitate magiæ

A. *Latin*:

(a) In MSS. Oxford Tanner 116 (late thirteenth century, chs. i–v and beginning of vi); BM Sloane 2156 (fifteenth century, complete); Oxford Digby 164 (fifteenth century, chs. i–ix); others including Voss MS. at Leyden, sixteenth century.[1]*

(b) Printed:

(i) De his que mundo mi-/rabiliter Evenivnt: vbi / de sensum erroribus, & potentijs Anime, / ac de influentijs cælorum, F. Clau=/dij Cælestini opusculum. / De mirabili potestate ar/tis et Natvræ, vbi de / philoso-phorum lapide, F. Rogerij Ba=/chonis Anglici, libellus. / Hæc duo gratissima, & non aspernanda / Op-/uscula, Orontius F. Delph. Regius Mathe=/maticus, diligenter recognoscebat, / & in suam redigebat har=/monium, Lutetiæ / Parisiorum. / Apud Simon Colinæm, 1542. Sm. 4⁰ (CUL Syn. 7.54.33).

The editor, Oronce Fine (Briançon, 1494–Paris, 1555), was a teacher of mathematics, an astrologer, and a maker of scientific instruments, who was much admired but died in poverty.[2] The book, which is very rare, has been incorrectly dated 1541. The part by Cœlestin is on astrology.[3]

(ii) Oxford, 1594, ed. Joseph Barnes.[4]

(iii) Epistolæ Fratris Rogeri Baconis de secretis operibus artis et naturæ et de nullitate magiæ. Opera Johannis Dee Londinensis e pluribus examplaribus castigata . . . cum notis quibusdam partim ipsius Johannis Dee, partim edentis, 12⁰, Hamburgi, ex Bibliopolio Frobeniano, 1618, pp. 70 + (71 to 80) notes. Text ends: ad Guilielmum Parisiensem conscripta.[5]

* For references, see p. 79.

B. *English*:

(i) An Excellent Discourse of the Admirable Force and Efficacie of Art and Nature, by Frier Bacon, Oxford, 1597 (with the Mirror of Alchemy, Hermes and Hortulan's Emerald Tablet, etc.)

(ii) Frier Bacon his Discovery of the Miracles of Art, Natvre and Magick. Faithfully translated out of Dr. Dee's own copy by T. M. and never before in English. Printed for Simon Miller at the Starre in St. Paul's Church-yard, 12⁰, 1659. Preface quotes opinions on Bacon from Selden, Voss, Picus de Mirandulus, etc.[6]

(iii) Roger Bacon's Letter concerning the Marvelous Power of Art and of Nature and Concerning the Nullity of Magic, translated by T. L. Davis, with Notes and an Account of Bacon's Life and Work.[7]

C. *French*:

(i) Roger Bacchon, De L'admirable Povvoir Et Pvissance de l'art, & de nature, ou est traicté de la pierre philosophale, Traduit en François par Jaques Girard de Tournus. A Lyon, Par Macé Bonhomme, sm. 8⁰, 1557, Avec privilege dv Roy, 95 pp.

At the end is a letter from de Tournus to his friend Charles Fontaine, the poet. For a note on de Tournus, see Gobet, *Les Anciens Minéralogistes du Royaume de France*, 1779, ii, 603. The work is sometimes bound with another one attributed to Bacon: Des Choses merveilleuses en Nature, ou est traicté des erreurs des sens, des puissances de l'ame, & des influences des cieux, Traduit en François par Jacques Girard de Tournus, Lyon, Macé Bonhomme, 1557, 192 pp.

(ii) Lettre sur les Prodiges de la Nature et de l'Art, translated by A. Poisson, sm. 8⁰, Paris, 1893, 70 pp., with portrait of Bacon.[8]

D. *German*:

(i) Rogerius Bachon, Von der wunderbarlichen Gewalt der Kunst vnd Natur, in Morgenstern, *Turba Philosophorum*; *Das ist, Das Buch von der güldenen Kunst*, Basel, 1613, ii, 426, and Vienna, 1750, ii, 517–551.[9]

(ii) Translated by Hamberger, Hof, 1776.[10]

This is a genuine work of Bacon's. It consists of eleven[11] letters (epistola) or chapters. Charles[12] regarded chs. vii–xi as doubtfully by Bacon, because ch. vii says "I have replied to your question of the year 601 of the Arabs," i.e. A.D. 1205–6, and ch. x, "you have questioned me in the year 608 of the Arabs," i.e. A.D. 1212–13, which are before Bacon was born. The first is not in ch. vii of the text as given by the *Theatrum Chemicum*, Manget, or Brewer; the second appears as 602 corrected to 688,[13] or 621.[14] Ch. xi begins "Annis arabum 630," i.e. A.D. 1232–3, corrected to 605[15] or 603.[16]

Romocki[17] said that Bacon, an expert on calendars, would not make a mistake in the date, and since Arabic numerals were always used, long before Bacon, in stating years A.H., Bacon probably wrote 662 and 663 and not

602 and 603, the 6 being easily mistaken for 0 (as at present if carelessly written). The years 662 A.H. and 663 A.H. correspond with the period between 1st October, A.D. 1265, and 1st October, A.D. 1266, which agree with the date 1267 when the *Opus Tertium* was written. In this work, however[18] (ch. i), Bacon says: "Primo igitur in opere Secunda, secundum formam epistolæ . . . recolens me jam a decem annis exulantem," and if this refers to the present *Epistola*, they would have been written in 1257, a date which Romocki[19] elsewhere accepts. The work was supposed by Jebb[20] to have been addressed to William of Auvergne (d. 1249, when Bacon would be thirty-four years old). The Dee MS. as printed ends: "ad Gulielmum Parisiensem conscripta" (which Brewer gives without comment), but Little[21] says this explicit does not occur in any MS. he examined; he concluded[22] that the work was probably written some time before the *Opus Majus* in 1267, and Delorme[23] also dated it before 1267.

Charles[24] thought it was addressed to John of London, whom he identified with John Basingstoke (d. 1252), but in another place[25] that it was probably written before 1263, and that its correct title is *De secretis operibus artis et naturæ et de nullitate magiæ*; Little[25a] gives *De secretis operibus artis et naturæ* with the alternative *De mirabili potestate artis et naturæ*. Thorndike[26] suggested that it is later than the *Opus Majus*, *Opus Minus* and *Opus Tertium* (written in 1266–7), and is a "later compilation from these three works." This is far from certain.

Easton[27] dates the composition of Bacon's medical opuscula as 1250–60, the alchemical works as 1250–56, with the alchemical excerpts from Avicenna *c*. 1254, the *De erroribus medicorum c*. 1255, and the *De secretis operibus naturæ c*. 1260. Since a recipe for purifying saltpetre occurs in one of the alchemical works (see below), this was probably written before the *De secretis*, which would agree with Easton's dating.

Chs. vii–xi in one MS.[28] have the title "de lapide philosophorum." Davis[29] pointed out that the earliest MS.[30] has only chs. i–v and the beginning of vi; chs. vii–xi may be later additions; the reference towards the end to the "three" contraries (*sic*; the printed texts have *per* res contrarias, not *ter*)[31] indicates that the author was acquainted with mysteries which were ancient and long hidden, mysteries which were symbolised for the initiate by the "winged Babylonian sphinx"—whatever that may mean.[32]

Mrs. Singer[33] says chs. ix, x and xi of the printed text are almost certainly not by Bacon; ch. ix (although possibly by Bacon) is probably later, whilst chs. x–xi are earlier, say the beginning of the thirteenth century. She gives no reasons. Steele[34] thinks chs. i–vi were probably prompted by William of Auvergne's condemnation of magic, and considers that chs. vii–x are of a different character. Sarton[35] says chs. i–vi are supposed to have been written by Bacon in 1256–66. This all seems very arbitrary, and the position I have adopted, after a consideration of the work itself, of other works by Bacon, and of the circumstances of his life (as far as these are known), is that the whole work is Bacon's and that it was written by him soon after 1257, and that it is not later than 1265. It is earlier than the mention of

gunpowder in the earliest known text of Marcus Græcus containing the gunpowder recipe, and the *De Mirabilibus Mundi* attributed to Albertus Magnus.

The whole work is an attempt to prove that effects commonly attributed
to the work of evil magic are natural and can be imitated by experiments.
It is in response to questions addressed to Bacon on this subject, since it
begins: "Vestræ petitioni respondeo diligenter"; whether Bacon is attempting
to defend himself against a charge of being a magician is a question which
cannot be answered from the text itself. It begins by condemning conjurers,
ventriloquists and spiritualists, as using tricks of the same kind as those
described long before by Hippolytos (ch. I). It condemns the use of magic
formulæ (the efficacy of which is not denied), but agrees that exorcism as
practised in the Church is lawful. It teaches faith-healing: "The Soule
thus affected [by charms, etc.] is able to renue many things in his owne bodie,
insomuch that it may recouer his former health, through the joy and hope
it hath conceived." The elixir of life is seriously described, and there are
"predictions" of submarines, flying machines, compasses, etc. (ch. iv: De
instrumentis artificiosis mirabilibus).[36] Some predictions are startling[37]; all
are imaginary,[38] but they do credit to Bacon's scientific imagination:

(1) Currus possent fieri ut sine animali moveantur cum impetu inæstimabili—the motor car.

(2) Possunt fieri instrumenta volandi, ut homo sedens in medio instru-
menti revolvens aliquod ingenium per quod alæ artificialiter com-
positæ aerem verberent, ad modum avis volaret—the aeroplane.

(3) Possunt fieri instrumenta ambulandi in mari et in fluviis ad fundum,
sine periculo corporali—the diving bell and submarine.

(4) Pontes ultra flumina sine columna vel aliquo sustentaculo—the
suspension bridge.

(5) Naves maximæ fluviales et marinæ unico homine regente—propulsion
of ships by machines, needing only a pilot to direct them.

(6) (Ch. v: De experientiis perspectivis artificialibus.) Possunt sic
figurari perspicua, ut omnis homo ingrediens domum, videret
veraciter aurum et argentum et lapides pretiosus, et quicquid homo
vellet; quicumque festinaret ad visionis locum nihil inveniret—the
magic lantern.

The most famous passage in the work, however, is that on gunpowder
at the end of ch. xi. The Latin text as printed by Brewer[39] is:

Annis Arabum 630 . . . Item pondus totum sit 30. Sed tamen sal petræ
LURU VOPO VIR CAN VTRIET sulphuris; et sic facies tonitruum et
coruscationem, si scias artificium.

The two English translations are:

Notwithstanding, thou shalt take salt-peter, *Luro vopo vir can vtri*, and of
sulphur, and by this means make it both to thunder and to lighten.

But get of saltpetre *LURU. VOPO Vir Can Utriet* sulphuris and so you may make Thunder and Lightning if you understand the artifice (1659, 23).

The French translation (1557, 77) has:

lu, ru, vo, po, vir, can, vtri, and the margin says these "comprise the seven species of simple minerals."

R. Plot[40] says the passage is: "*componere ignem comburentem ex sale Petræ, & aliis*; which *alia*, as the Reverend and Learned D*r John Wallis* saw it in a MS. copy of the same *Roger Bacon*, in the hands of the Learned D*r Ger. Langbain*, late Provost of *Queens College*, were *Sulphur*, and *Carbonum pulvis* . . . whence t'is plain, he either invented or knew *Gunpowder*, though I think we cannot allow him less than the first, till we find out an ancienter *Author* for it, which if no Body ever does (as t'is manifold Odds they never will) in all Probability it was *invented* here at *Oxford*, where he made the rest of his affrightening Experiments."

Plot mentions Constantinus Ancklitzen of Friburg, or Bertholdus Schwartz, from Pancirollius, and thinks he got the composition from Bacon's works. Little[40a] does not list the Langbain MS., which is mentioned (from Plot?) by Lenglet du Fresnoy,[41] who quotes: "In omnem distantiam, quam volumus possumus artificialiter componere ignem comburentem ex sale petræ, et aliis (*id est* sulphure et carbonum pulvere, *ut in Ms. Germani Langbaine legitur*) præter hæc sunt alia stupenda naturæ. [R.B. in opere suo manuscripto ad Clementem IV et in Epistola ad Joannem Parisensem Episcopum cap. 6 et in tractatu de secretis artis et naturæ operibus . . . ubique edita.]"

Hime[42] by putting selected phrases from chs. ix–x of Bacon's text in brackets[43] gets out a description for purifying saltpetre:

ix.   Calcem ["chalk," for saltpetre] igitur diligenter purifica, ut fiat terra pura penitus liberata ab aliis elementis. Dissolvatur autem in aqua cum igne levi, ut decoquatur quatenus separetur pinguedo sua, donec purgetur et dealbetur . . . Iteretur destillatio [through a filter] donec rectificetur: rectificationis novissima signa sunt candor et crystallina serenitas. Ex hoc aqua materia congelatur. Lapis vero Aristotelis, qui non est lapis, ponitur in pyramide in loco calido [to dry it].

x.   Accipe igitur lapidem et calcina ipsum. Sed in fine parum commisce de aqua dulci; et medicinam laxativam [lime and wood ashes?] compone de duabus rebus quarum proportio melior est in sesquialtera proportione. Resolve [= dissolve] ad ignem et mollius calefac. Mixto igitur ex Phœnice [animal charcoal, for clarification?] adjunge, et incorpora per fortem motum; cui si in liquor calidus adhibeatur, habebis propositus ultimum. Evacuato igitur quod bonum est. Regyra cum pistillo et congrega materiam ut potes, et aquam separa paulatim.

Carefully wash the natural saltpetre and remove all [visible] impurities. Dissolve it in water over a gentle fire, and boil it until the scum ceases to rise, and it is purified and clarified. Do this repeatedly until the solution is clear and bright. Let this water deposit the stone of Aristotle which is not a stone [= lapis assius = saltpetre, Hime] in pyramids, and dry them in a warm place. Take this stone and powder it (et calcina ipsum)[44] and immerse in soft water. Make a powder of two purifying substances in the proportion of 3 to 2. Dissolve over a gentle fire. Mix with the Phœnix [= animal (?) charcoal, according to Hime] and thoroughly mix the ingredients, then pour the hot solution upon it and your object will be achieved. If the solution is good, pour it out [= decant into a crystallising vessel], stir with a pestle, collect all the crystals you can, and draw off the water [mother liquor].

Hime[45] rearranged the letters of the supposed anagram: LURU VOPO VIR CAN UTRIET, into RVIIPARTVNOUCORULVET, and combined these into the groups (interchanging u and v): R.VII.PART. V. NOV. CORUL. V. ET, which he adds to the text as follows: [sed tamen salis petræ] r(ecipe) vii part(es), v nov(ellæ) corul(i), v et (sulphuris): "but take 7 parts of saltpetre, 5 of young hazelwood ⟨charcoal⟩ , and 5 of sulphur," giving:

Saltpetre	..	..	..	41·2	1²⁄₅
Charcoal	..	..	..	29·4	1
Sulphur	..	..	..	29·4	1

and this mixture will explode "if you know the trick," viz. if you use pure saltpetre, incorporate the materials thoroughly, keep the powder dry, and do not compress it unduly. It is possible that this composition was evolved by trial from a 1 : 1 : 1 mixture.

This is amended by Davis[46] to vi parts (instead of vii) of saltpetre, "to agree with the earlier text on the preceding page of the *Theatrum Chemicum*." In the *De Mirabilibus Mundi* attributed to Albertus Magnus the materials are 6 of saltpetre, 1 of sulphur and 2 of charcoal, mixed in a stone mortar. A. Clément[47] says some letters are abbreviations: ī = is, u = ū = um, v = ꝝ = rum, and the words should read: pulveris carbonum tritorum, i.e. "of ground charcoal."[48] Steele[49] says the cipher luro vopo vir can utriet has no known manuscript authority, first appearing in the Paris 1542 edition printed from a poor manuscript by the editor Orontius Fine, and "seem to be due to an attempt to reproduce the text before him."[50] This is guesswork. Brewer[51] says his text was "collated with the Sloane MS." (which, he does not say), and in his footnotes[52] he refers to "C" and "H" for alternative readings (again omitting to say what these mean); he gives[53] as a footnote to "Luru . . . utriet" an attempt to transcribe the "cipher" as it occurs in Sloane MS. 2156, which is reproduced in facsimile below:

Hime[45] calls this "the second anagram (in Greek, Roman and Anglo-Saxon letters)," which "seems to be a note to the first." It has not been solved. The "Sloane MS." is Sloane 2156, ff. 111–6 (fifteenth century) listed by Little.[1] "C" and "H" presumably stand for Victor Cousin, who described a Douai MS. of the *Opus Tertium* in 1848, and Humphrey Hody, who quoted from the *Opus Tertium* in 1705.[54] Guttmann[55] refers to two MSS. of the *De Secretis*, viz. :

(i) BM Sloane 2156 (fifteenth century; ff. 111–6 v.), which agrees exactly with the Hamburg printed text of 1618, but in which the anagram reads : Sed tamen salpetre Kb Ka x hopospcadikis et sulphuris 5, which does not seem to be a very accurate transcription.

(ii) Oxford Bodleian Digby 164 (ff. 8–12 v.), which Guttmann said was probably of the thirteenth century, whilst it is really of the fifteenth. This contains the first six chapters, agreeing with the Hamburg text, ending "aliquibus hujus utar modis" (Mrs. Singer[56] has hiis instead of hujus), followed by "Benedicamus, etc." The whole section on the philosophical egg and the anagram are missing, and Guttmann says he did not find them in any other MS. ("Von dem ganzen philosophischen Ei, und dem anagrammatischen Satze steht darin nichts, und auch in allen übrigen Bacon-Manuscripten—ich nahm mir diese Mühe—fand sich nichts").

Romocki[57] gives as from "Gegen Ende des Cap. xi" the text : Item pondus totum fit 30. Sed tamen salis petræ LVRV NOPE CVM VBRE et sulphuris; et sic facies tonitrum et coruscationem, si scias artificium. He does not say where he found this; it is not in the three texts I have available (*Theatrum Chemicum*, Manget, Brewer; salis petræ is in *Theatrum Chemicum* and Manget; Brewer has sal petræ).

Hime[58] says "The second anagram, which is contained in the same chapter, remains unsolved," and "still defies the ingenuity of man." He[59] then quotes Romocki[60] as saying : "Et quicumque hæc reseraverit, habebit clavem quæ aperit, et nemo claudit; et quum clauserit, nemo aperit' hat dann offenbar ein Abschreiber im XIV. Jahrhundert hinzugesetzt, der seinerseits an der Auflösung der von Bacon aufgegebenen Rätsel verzweifelte." These words occur at the end of the text, after "Vale" in Brewer's text, which gives "corporis valde" as an alternative. Romocki thought his form of the anagram should read CARBONUM PULVERE, but as Hime[61] said, this means suppressing three letters and changing six others, and is bad grammar, the text requiring pulveris, not pulvere.

Bacon begins ch. vi with a description of Greek fire : possumus artificialiter componere ignem comburentem scilicet ex sale petræ et aliis. Item ex oleo petroleo rubro et aliis. Item ex maltha, et naphtha, et consimilibus. . . . His vicinus est ignis Græcus, et multa comburentia. Præterea possunt fieri luminaria perpetua et balnea ardentia sine fine. . . . He goes on to describe something different : Præter vero hæc sunt alia stupenda naturæ. Nam soni velut tonitrua possunt fieri et coruscationes in aëre, imo majori horrore quam illa quæ fiunt per naturam. Nam modica materia adaptata, scilicet ad quantitatem unius pollicis, sonum facit horribilem et coruscationem ostendit vehementem. Et hoc fit multis modis; quibus omnis civitas et exercitus destruatur, ad modum artificii Gedeonis, qui lagunculis fractis, et lampadibus, igne exsiliente cum fragore inæstimabili, infinitum Midianitarum destruxit exercitum cum trecentis hominibus.

"It is possible to compose artificially a burning fire, namely from saltpetre and other things, such as red and other petroleum, maltha, naphtha, and the like, very similar to Greek Fire and other incendiaries. . . . In addition, there can be made perpetual lamps and baths heated for all time. . . . Also there are other natural wonders. For sound like thunder and flashes in the air can be made, indeed greater horrors than those produced naturally; by means of a little of a rightly-prepared material no bigger than the thumb horrible noise and vivid flashes can be exhibited. And this is done in many ways by which a whole city and army are destroyed, as in the artifice used by Gideon, when small bottles are broken, and lamps, and fire rushes out with an immense noise, and three hundred men destroy an infinity of the Midianite army."

Bacon appears to mean that the army might be blown up, rather than put to flight by the terror caused by an explosion.[62] Hime[63] thought that Bacon was the inventor of gunpowder and that he used cryptic methods to conceal his *accidental* discovery for fear of the Inquisition, founded by Pope Gregory IX in 1233.

Thorndike[64] disagrees with Hime's date of 1248 for the *Epistola*, suggesting that it is later than 1266–7, "concocted by someone else" from Bacon's other works; he sneers at the idea that the Inquisition interfered, and it was set up in England only once in the Middle Ages; but Bacon may not have been in England when he wrote. Thorndike's discussion is not helpful or illuminating, whilst Hime's attempt to make sense of Bacon's text is at least reasonable and suggestive. Hime[65] asserted that Bacon *discovered* gunpowder; by "experimenting with some incendiary composition, prepared with pure instead of impure saltpetre, the mixture exploded unexpectedly and shattered all the chemical apparatus near it, thereby laying the foundation of the mediæval legend of the destruction of the Brazen Head."

Besides the account in the *Epistola*, Bacon refers to gunpowder in very similar terms in his other works.

I. *Opus Majus.*[66] This begins with an account of Greek fire and goes on to describe gunpowder as something superior and, at the time (1267), known in many countries:

Quædam vero solo tactu immutant et sic tollunt vitam. Nam malta, quæ est genus bituminis et est in magna copia in hoc mundo, projecta super hominem armatum comburit eum. Istud autem Romani gravi cæde perpessi sunt in expugnationibus regionum, sicut Plinius testatur 2⁰ naturalis historiæ, et historiæ certificant. Similiter oleum citrinum petroleum, id est, oriens ex petra, comburit quicquid occurrit, si rite præparetur. Nam ignis comburens fit ex eo qui cum difficultate potest extingui, nam aqua non extinguit. Quædam vero auditum perturbant in tantum, quod si subito et de nocte et artificio sufficienti fierent, nec posset civitas nec exercitus sustinere. Nullus tonitrui fragor posset talibus comparari. Quædam tantum terrorem visui incutiunt, quod coruscationes nubium longe minus et sine comparatione perturbant; quibus operibus Gideon in castris Midianitarum consimilia æstimatur fuisse operatus.

"Certain of these work by contact only and so destroy life. For malta, which is a kind of bitumen and is plentiful in this world, when projected upon a man in armour burns him up. The Romans suffered severe loss of life from this in their conquests of countries, as Pliny says in the second book of his *Natural History* and the histories attest. Similarly, yellow petroleum, i.e. oil produced from rock, if properly prepared [distilled?] burns up whatever it meets. For a consuming fire is produced by this which can be extinguished with difficulty, for water does not extinguish it. Certain inventions disturb the hearing to such a degree that if they are set off suddenly at night with sufficient skill, neither city nor army can endure them. No clap of thunder can compare with such noises. Some of these strike such terror to the sight that the coruscations of the clouds disturb it incomparably less. Gideon is thought to have employed inventions similar to these in the camp of the Midianites."

Those who have read my first chapter on Greek fire will be on familiar ground here; even my suggestion that the naphtha was distilled finds its echo in Bacon's phrase "si rite præparetur." Bacon then immediately proceeds to discuss things acting on the senses:

Et experimentum hujus rei capimus ex hoc ludicro puerili, quod fit in multis mundi partibus, scilicet ut instrumento facto ad quantitatem pollicis humani ex violentia illius salis, qui sal petræ vocatur, tam horribilis sonus nascitur in ruptura tam modicæ rei, scilicet modici pergameni, quod fortis tonitrui sentiatur excedere rugitum, et coruscationem maximam sui luminis jubar excedit.

"We have an example of these things in that children's toy which is made in many parts of the world, viz. an instrument made as large as the human thumb. From the force of the salt called saltpetre so horrible a sound is produced by the bursting of so small a thing, viz. a small piece of parchment, that we perceive it exceeds the roar of strong thunder and the flash exceeds the greatest brilliancy of the lighting."

Romocki[67] pointed out that the meaning of an *explosive* (sprengkräftigen) apparatus, as distinguished from an incendiary mixture, is quite clearly expressed. Only saltpetre is named here, and in such a way as to suggest that it was not well known when Bacon wrote.

II. *Opus Tertium.*[68] When, in 1266–8, Bacon composed his *Opus Tertium*, gunpowder had come into use in different parts of the world (in mundi partibus diversis)[69]:

Exemplum est puerile de sono et igne qui fiunt in mundi partibus diversis per pulverem salis petræ et sulphuris et carbonum salicis. Cum enim instrumentum de pergameno in quo involvitur hic pulveris, factum ad quantitatem unius digiti, tantum sonum facit quod gravat multum aures hominis. . . . Consimile fecit Gedeon in castris Madianitarum, qui ex sonitu lagenarum et ydriarum, in quibus conclusit lampades coruscantes, territi sunt Madianite et confusi.

In another fragment of this work[70] he gives his "last word" on gunpowder :

By the flash and combustion of fires, and by the horror of sounds, wonders can be wrought, and at any distance that we wish—so that a man can hardly protect himself or endure it. There is a child's toy of sound and fire made in various parts of the world with powder of saltpetre, sulphur and charcoal of hazelwood. This powder is enclosed in an instrument of parchment the size of a finger, and since this can make such a noise that it seriously distresses the ears of men, especially if one is taken unawares, and the terrible flash is also very alarming, if an instrument of large size were used, no one could stand the terror of the noise and flash. If the instrument were made of solid material the violence of the explosion would be much greater (quod si fieret instrumentum de solidis corporibus, tunc longe major fieret violentia).

Little[70] points out that Bacon was here feeling his way towards the discovery of the propulsive force of gunpowder under pressure; the loud explosion he mentions implies some kind of petard and not freely-burning powder.[71] There seems to be no reasonable doubt that Roger Bacon was very well acquainted with the preparation and properties of gunpowder. It is not certain, but probable, that he had personal experience of them. He knew well the distinction between the *deflagrative* burning of gunpowder and its *explosive* combustion in a suitable container. That he had any knowledge of its use as a *propellant* is doubtful. It is also uncertain whether Bacon himself discovered the composition of gunpowder, as Hime thought, or whether he received his knowledge from another source. What has been said of the *Liber Ignium* suggests that this may have been Arabic. If so, the origin of gunpowder is shifted back from Bacon to an earlier source, and this matter will be dealt with later.

A description of the purification of saltpetre is given in an alchemical work attributed to Bacon and probably genuine, viz. the *Breve breviarium de dono Dei*.[72] Saltpetre, it is said, grows on certain stones (cf. Marcus Græcus) and if brought in contact with glowing charcoal it at once impetuously produces fire : nascens inter quosdam lapides, quibus dicitur adhærere. Talis autem naturæ est, quod si immediate ignitos carbones tangat, statim accensum cum impetu euolat. It is purified by dissolving in water and filtering and putting the filtrate in a glass vessel, when it forms white and shining rod-shaped crystals. Bacon says it penetrates an earthen vessel, "as I have found by trial," and he also refers to other experiments with it, without here mentioning gunpowder; he also says nothing about the use of wood ashes in purifying it : Præparatio autem salis petræ est, quod resoluatur aqua, & distilletur per filtrum, & distillatum dimittetur quiescere in vitreo vase . . .

inuenies ipsum in ipsa aqua per virgulas albas & lucidas mirabiliter congelatum . . . ego probavi ex sua acuitate terrea vasa penetrat. . . . Probaui quidem ipsum; & probatum in multis laudo, quia bonum est. It melts like wax in an earthen or metal vessel over a fire but solidifies again on cooling to a white mass.

REFERENCES

1. *Roger Bacon Essays* . . . Collected and Edited by A. G. Little, Oxford, 1914, 395; E. Charles, *Roger Bacon, sa Vie, ses Ouvrages, ses Doctrines, d'après des textes inédits*, Paris (printed Bordeaux), 1861, 71.
2. A.R.D.D., in *Nouvelle Biographie Générale*, 1856, xvii, 706; L. Thorndike, *A History of Magic and Experimental Science*, New York, 1941, v, 285.
3. Thorndike, ref. 2, v, 291.
4. Mentioned by Charles, ref. 1, 57, and Little, ref. 1, 396.
5. BM 719.e.2764; CUL N*.16.41ᵃ (F). This text is reprinted in *Theatrum Chemicum*, Zetzner, Strassburg, 1600, v, 834–61; Manget, *Bibliotheca Chemica Curiosa*, fᵒ, Geneva, 1702, i, 616–24; and Brewer, *Fr. Rogeri Bacon. Opera Quædam Hactenus Inedita*, 1859, 523–51. There is also a text in *Artis Avriferæ*, Basel, 1610, ii, 327; *De mirabili potestate Artis et naturæ, Libellus* (J. Ferguson, *Bibliotheca Chemica*, Glasgow, 1906, i, 50; J. F. Gmelin, *Geschichte der Chemie*, Göttingen, 1797, i, 100, says it is in the 1572 and 1593 eds. of this work, and also gives 8ᵒ, Hamburg, 1598 and 1608); Parrot, *R. Bacon*, Paris, 1894, 62, gives Lyons, 1552 and 1612; Hoefer, *Histoire de la Chimie*, 1866, i, 395, gives Hamburg, 1508 (a mistake for 1598?) and 1608 (probably copying Gmelin). Daunou and Leclerc, *Histoire littéraire de la France*, 1842, xx, 244, give Lyons 1553 and 1612, Basel, 1593.
6. BM E.1932 (1); CUL M.18.84.
7. Chemical Publishing Co., Easton, Pa., 1923, 76 pp.; *Isis*, 1925, vii, 537.
8. Ferguson, ref. 5, i, 64; Gmelin, ref. 5, gives a French tr., Paris, 1629.
9. Ferguson, ref. 5, ii, 106–7, says this is a tr. of *Ars Aurifera* (*sic*) of 1593 or 1610.
10. Daunou and Leclerc, ref. 5.
11. Little, ref. 1, 395, says sometimes ten in MSS.
12. Ref. 1, 57.
13. *Theatrum Chemicum*, ref. 5, 859; Manget, ref. 5, 623.
14. Brewer, ref. 5, 548.
15. *Theatrum Chemicum*, ref. 5, 860; Manget, ref. 5, 623.
16. Brewer, ref. 5, 550.
17. *Geschichte der Explosivstoffe*, 1895, i, 92.
18. Brewer, ref. 5, 7.
19. Ref. 17, i, 81.
20. *Fratris Rogeri Bacon . . . Opus Majus*, 1733, pref. ei *v.*
21. Ref. 1, 395.
22. *Studies in English Franciscan History*, Manchester, 1917, 205.
23. In Vacant and Mangenot, *Dictionnaire de Théologie Catholique*, Paris, 1905, ii, 7.
24. Ref. 1, 57.
25. Ref. 1, 91.
25a. Ref. 1, 395.
26. Ref. 2, 1923, ii, 630.
27. S. Easton, *Roger Bacon and his Search for a Universal Science*, Oxford, 1952, 111.
28. BM Sloane 2629, f. 478; Charles, ref. 1, 71.
29. T. L. Davis, *Isis*, 1927, ix, 425.
30. Oxford, Tanner 116, thirteenth century; Little, ref. 1, 395.
31. Manget, ref. 5, 624; Brewer, ref. 5, 551.
32. Macdonald, *Isis*, 1928, xi, 127.
33. Mrs. D. W. Singer, *Speculum*, 1932, vii, 80 f.

34. In C. Singer, *Studies in the History and Method of Science*, Oxford, 1921, ii, 132.

35. *Introduction to the History of Science*, 1931, ii, 954.

36. Ch. 5 in English tr., 1659, 17: "Of admirable artificial instruments."

37. F. A. Pouchet, *Histoire des Sciences Naturelles au Moyen Âge*, Paris, 1853, 351 f.

38. Charles, ref. 1, 300 f.

39. Ref. 5, 551.

40. R. Plot, *Natural History of Oxfordshire*, 2 ed., 4°, Oxford, 1705, 236. Gerard Langbaine (1608–1658), from 1645 Provost of Queen's College, Oxford.

40a. Ref. 1, 395.

41. *Histoire de la Philosophie Hermétique*, Paris, 1742, i, 114.

42. In Little, ref. 1, 325 f.; *The Origin of Artillery*, 1915, 21 f., 106 f.

43. A method described by Porta, *De Occultis Literarum Notis*, 1606, 140 (modus quo una oratio sub alia bracteis perforatis occultatur); it is also described in Thackeray's *Esmond*, and is sometimes called the "Argyle cipher."

44. To calcine sometimes meant to powder in alchemical authors, cf. Hoefer, ref. 5, i, 44.

45. *Gunpowder and Ammunition*, 1904, 157; *The Origin of Artillery*, 1915, 113 f.

46. *Isis*, 1925, vii, 537; *Ind. Eng. Chem.*, 1928, xx, 772; letter from Prof. Davis, 1931.

47. "Sur l'indication de la composition de la poudre à feu chez Roger Bacon," in *Archivio di Storia della Scienza*, 1926, vii, 34; BM reprint 8902.d.13; *Isis*, 1928, x, 127.

48. H. Kopp, *Geschichte der Chemie*, 1845, iii, 226–7, had got "car on pulver" out of the anagram; an anonymous reviewer in *Nature*, 1926, cxviii, 352, says Clément's suggestion is better than Hime's. Favé, in Napoleon, *Études sur le Passé et l'Avenir de l'Artillerie*, 1862, iii, 63, had got "carvonu pulveri trito" out of "luru vopo vir can utriet."

49. *Nature*, 1928, cxxi, 208; 1928, cxxii, 563.

50. Steele, *Luru Vopo Vir can Utriet*, 1928; BM 1865.c.3 (144).

51. Ref. 5, ix.

52. Ref. 5, 523 f.

53. Ref. 5, 551.

54. *De Biblorum Textibus Originalibus*, f°, Oxford, 419–28 (from Digby MS. 218).

55. O. Guttmann, *Die Industrie der Explosivstoffe*, in P. A. Bolley, K. Birnbaum, and C. Engler, *Handbuch der chemischen Technologie*, Brunswick, 1895, VI, vi, 9–10; *Monumenta Pulveris Pyrii*, London, 1906, 2–4.

56. Mrs. D. W. Singer, *Catalogue of Latin and Vernacular Alchemical Manuscripts in Great Britain and Ireland*, Brussels, 1928, 167.

57. Ref. 17, i, 92.

58. Ref. 45, 1915, 113–4.

59. Ref. 45, 1915, 119.

60. Ref. 17, i, 93.

61. Ref. 45, 1915, 119.

62. Berthelot, *Sur la Force des Matières Explosives*, 1883, ii, 358: "plutôt pour épouvanter . . . par le bruit . . . que pour exercer une action directe"; Hime in Little, ref. 1, 333; ref. 45, 1915, 114.

63. Ref. 45, 1915, 114.

64. Ref. 2, 1923, ii, 688.

65. Ref. 45, 1904, 161.

66. Ed. Jebb, 1733, 474; ed. Bridges, 1897, ii, 217; tr. Burke, 1923, ii, 629; Charles, ref. 1, 299; Romocki, ref. 17, i, 93–4; part of the text had been given by Gmelin, ref. 5, i, 95, and before him by Borrichius, *De Ortu et Progressu Chemiæ*, Copenhagen, 1668, 126, who adds: Hic apertissimè loqvitur *Rogerius* de nitrato illo sclopetorum pulvere.

67. Ref. 17, i, 89.

68. Little, "Part of the Opus tertium," *British Society of Franciscan Studies*, Aberdeen, 1912, iv, 51; ref. 22, 1917, 206.

69. Duhem, *Un Fragment inédit de l'Opus Tertium de Roger Bacon*, Ad Claras Aquas[Quaracchi] prope Florentium, 1909, 183–4; Hime, ref. 45, 1915, 119.

70. Little, *Proc. British Academy*, 1928, xiv, 290.

71. Hime, in Little, ref. 1, 333 f.; ref. 45, 1915, 119 f.

72. In *Sanioris Medicinæ Magistri D. Rogeri Baconis Angli, de Arte Chymiæ scripta. Cvi accesservnt opuscula alia eiusdem Authoris*, 12°, Frankfurt, J. Saurius, 1603, 250.

ALBERTUS MAGNUS

De Mirabilibus Mundi

This work, of considerable interest to us, is very rare in manuscripts; it is found in fourteenth-century (St. Mark's, Venice, XIV, 40) and fifteenth-century (Florence, Paris), and possibly thirteenth-century (Wolfenbüttel) MSS.; there is no British Museum MS. In a list of Dominican works made about 1350 it is ascribed to Arnold of Liége, and it is perhaps by a pupil of Albert.[1]* It was often printed in the fifteenth century.[2] Some editions which I have seen are:

α Ferrara (Anon.), 1477. BM IA.25683 (incomplete, lacks incipit).

β Augsburg (per Johannem de Annunciata), 1478. BM IA.28670.

γ Bologna (Heydelberga), 1482. BM IA.28742.

δ London (Machlinia), 1485. BM IA.55455.

ε [Paris] (Caillout), 1490. BM IA.39554; CUL Inc.7.D.40 (old Adams 8.50.1).

ζ No place, 1490? BM IA.46455.

η Antwerp, 1491? BM IA.49921.

θ Cologne (Quentell), 1495. CUL Inc.5.A.4.24 (old Adams 6.49.1).

ι Augsburg (Schaur), 1496. BM IA.6829.

κ Antwerp, 1502. CUL F150.d.6.4.

λ Cologne, 1506? BM IA.5195.

μ Rouen, sm. 8°, s.a. BM 547.b.1.

ν Alberti co/gnomento ma/gni de secretis Mulierum Li/bellus, scholijs auctus, & / a mendis repur-/gatis. / Eivsdem de virtv-/tibus Herbarū, Lapidum, / & Animalium quo-/rum libellus. / Item de mirabilibvs / mundi, ac de quibusdam effectibus caussatis à quibusdam ant. m . . . s &. / Lvgdvni / 1566. (Unpaged.)

ξ Albertus Magnus de Secretis Mulierum. Item de Virtutibus Herbarum Lapidum et Animalium, 12°, Amstelodami, 1669, Apud H. et T. Boom. P. 204 f. contains: Michælis Scoti rerum naturalium perscrutatoris Præmium, in secreta Naturæ. Ad D. Fridericum Romanorum Imperatorem. CUL P*.13.40(G); Acton e.53.4.

ο The Secrets / of Albertus Magnus. / Of the vertues of Herbes, / Stones, and certaine / Beasts. . . . Also a Booke of the same Author, / of the merveilous things of the / world, and of certaine things / caused of certaine / Beasts. London. Printed by T. Cotes, 1637. Sm. 8° unpaged, signs. A1—H8. CUL Syn. 8.63.348. (Gaps in text and translation inaccurate.)

* For references, see p. 86.

π   Les admirables secrets d'Albert le Grand . . . tirez et traduit sur des anciens Manuscrits de l'Auteur . . . A Lion, chez les Héritiers de Beringos Fratres à l'Enseigne d'Agrippa, 12⁰, 1729, 10 ll., 306 pp., 3 ll. index; the *Des Merveilles du Monde* is contained in pp. 131–68.   This edition is much extended; it mentions (pp. 212–24) hardening and softening steel, softening crystal, soldering everything, even cold iron (fer froid), with sal ammoniac; calcined tartar and antimony; corrosive mixtures for engraving (not mineral acids); gilding by linseed oil varnish; a mixture for cleaning arms made from olive oil and lead filings kept in a pot for nine days.   As in the case of the works of Cornelius Agrippa, the imprint "Lyons, Beringos Fratres" is probably spurious, the books being published in Cologne.   Some collections published in the nineteenth century are attributed to an Albertus Parvus (Petit Albert).[3]

ρ   Cæsar Longin, Trinum Magicum Sive Secretorum Magicorum Opus, 12⁰, Frankfurt, 1614 (BM 1035.a.5.(3.)); includes *Secreta Secretorum*, 241 f., and *De Mirabilis Mundi*, 326 f.

The beginning and end of the treatise are : *inc*. Incipit opus Alberti Magni de mirabilibus mundi lege feliciter.   Postquam scivimus quod opus sapientis est facere cessare mirabilia rerum (eorum, verum); *expl*. ad faciendum vero tonitrum brevis, grossa et semiplena.   Albertus Magnus de secretis naturæ explicit (*sic* in all incunabula seen; Romocki : de mirabilibus mundi explicit).

Some editions have further text following this.   A text was given by Romocki[4] from an edition he dates 1472 (14 ll., f⁰), without place or date (Berlin, Kgl. Bibliothek, no. Le 718), collated with an edition of Strassburg, 8⁰, 1493.   The text given below is based on Romocki's, collated with the editions listed above and has no claim to be a critical one; no such critical text exists.   The work is mostly regarded as spurious,[5] but Romocki[6] regarded it as genuine, and Thorndike[7] was undecided.

Many superstitious recipes are also found in genuine works of Albert, such as the *De Animalibus* (regarded as his best scientific work),[8] including the one found in the *De Mirabilibus* that if the wax and dirt from a dog's ears are rubbed on wicks of new cotton, and these are put into a crucible of green oil and lighted, the heads of persons present will appear completely bald.[9] The language is often of poor standard, but the style of a mediæval author could change with the subject,[10] and it may have been transcribed unchanged by Albert from some Latin translations of Arabic works, which he did not fully understand and hence he did not correct.[11]   It is possible that it was written by a pupil of Albert's from Latin translations of Arabic works (mostly by Avicenna),[12] but since it is found in thirteenth-century manuscripts it may well be by Albert; it is perhaps what Roger Bacon[13] had in mind in referring to an unnamed author, probably Albert : "homo studiosissimus est, et vidit infinita, et habuit experiri [expensum, Brewer]; et ideo multa potuit colligere utilia in pelago auctorum infinito . . . quia eius scripta plena sunt falsitatibus et vanitatibus infinitis."

The recipes at the end for gunpowder, etc., are literally the same as those in the *Liber Ignium* of Marcus Græcus, as Hoefer[14] pointed out, and may have been taken from that work, rather than conversely. Wiegleb[15] drew attention to the gunpowder recipe and thought it was earlier than Bacon's.

The work begins with a scholastic discussion approximating to the form of Albert's works[16] on the validity, causes and principles of magic, and says marvels cease when they are explained, and many, as Avicenna in "VI Naturaliam"* said, are due to the power exercised over objects by the human mind, especially if aided by the stars at a favourable hour. It admits the efficacy of characters, images and incantations. Every species seeks its kind; Avicenna says that what stands in salt a long time turns into salt. Every nature has a natural friendship or enmity to another. Everything in nature is full of marvellous virtue. Specific examples are now given, the magnet of course being included. "Et alchimiste sciverunt illud veritate manifesta quomodo natura similis subingreditur et gaudet suo simili." At first authorities are cited: Hermes, Galen, filius Mesuë, Aristotle's letter to Alexander (for the experiment of killing basilisks by letting them stare themselves to death by looking in mirrors), Architas, Belbinus (= Balinus, Apollonios of Tyana), Archigenes and Cleopatra. Some of these come from Arabic sources. Ligatures and medical recipes, including parts of animals, follow. Up to this point the work is reminiscent of other treatises by Albert.

Then follow recipes of a semi-magical character, for which no authorities (except once "philosophi") are cited. This part may have been added to a complete treatise of Albertine character. Some of the recipes are for familiar purposes: to catch birds in the hand, to ward off dogs and snakes, to catch a mole, to break charms and loose bonds, to see the future in sleep, to make a chicken or other animal dance in a dish, and so on. Others deal with magic candles and lamps, which make men appear headless or with three heads, or black, or with animal or angel's faces. There is a recipe for a phosphorescent ink made from bile of a tortoise, luminous worms, etc.:

Et cum volueris ut qui sunt in palatio videantur nigri: accipe de spuma maris et calcantho et commisce ea simul; deinde humecta cum eis licinium et lumina cum eo lampadem. Quando vis, ut qui sunt in palatio videantur sine capitibus: accipe sulphur citrinum cum oleo et pone ipsum in lampade et illumina cum eo et pone in medio hominum, et videbis mirum.

"And if you would see how those in a palace look black, take sea-foam and vitriol and mix them together. Then steep wicks with it and light them as in lamps. When you wish those in a palace to seem without heads, take yellow sulphur and oil and put them into a lamp and put it in the midst of the men, and you will see something wonderful."

Ut ignis de aqua exeat. Testam ovi accipe et sulphur vivum[a]† tritum impone et calcem et claude foramen et mitte in aquam, et incenditur.

---

* This is a translation of part of Avicenna's large philosophical work, the *Shifā'*. Avicenna (Ibn Sīnā), b. A.D. 980, d. A.D. 1037.

† See notes on p. 86.

"That fire may come out of water. Take an egg-shell and put into it powdered sulphur and quicklime, and close the hole and put the shell in water and it will take fire."

Ut homines videantur habere quorumlibet animalibus capita. Accipe sulphur vivum et litargirum, et iste simul pulverisatis sparge in lampadem oleo plenam. Habeantque candelam de cera virginea, quæ permixta sit cum fæce illius animalis, cuius caput vis ut videatur habere tenens candelam accensam de lampadis igne.

"To make men anywhere look as if they have the heads of animals. Take sulphur and litharge and powder them together and sprinkle in a lamp full of oil. And have a candle of virgin wax which is mixed with the dung of that animal whose head you wish to see, and holding the burning candle light the lamp from it."

Ut videantur habere vultum canis. Suscipe adipem de aure canis, unge ex eo parum de bombicino novo, pone in lampade nova de viridi vitro, et pone lucerna inter homines et cernunt vultum canis.

"That a person may seem to have a dog's face. Take wax from a dog's ear, anoint with it a little new cotton, put it into a new lamp of green glass, and put the lamp among men, and they perceive the dog's face."

Suffumigatio alia quam cum facis vides foris virides et multiformes et mirabilia infinita quæ pro multitudine non discernuntur. Accipe cimar[b] idem vermilium et lapidem lazuli et pulegium montanum et pulverisa totum et cribella illud et confice illut cum pinguedine delphini vel equæ vel elephantis, fac grana in modum ciceris et sicca in umbra et suffumiga in eo quando volueris et fiet quod dictum est.

"Another suffumigation which when you make it will make you see green men, and many shapes, and infinite marvels which are not known to the multitude. Take cimar, which is vermilion, and lapis lazuli, and mountain pennyroyal, and powder them all and sift them. Mix with the fat of a dolphin, horse, or elephant, and make grains like chick-peas and dry them in the shade. And suffumigate with them whenever you like and what is said will be done."

A number of other recipes for magic lamps, using various materials, appear all together as if taken from some treatise on that subject. Then follow a number of pyrotechnic recipes, but as these are interrupted by some more magic lamps they probably came from the same work as the above. Some of the recipes are very similar to those in the *Liber Ignium* of Marcus Græcus (whose name is not mentioned) and some (apart from slight differences in wording) are identical. The proportions of the ingredients are, however, omitted except in the gunpowder recipe (1566, signs. Dd 5r-Dd7v; 1669, 201-3).

Si vis facere contrarium, scilicet imaginem aliquam hominis aut alterius rei, quæ quando ponitur in aqua accendetur, et si extraxeris eam, extinguetur, accipe calcem non extinctam et permisce eam cum aliquantulo ceræ et olei sesami et naptæ terra alba[c] et sulphuris et fac ex illa imaginem nam quando tu rorabis aquam accendetur ignis.

"If you will make a contrary, that is an image of a man or any other thing which when it is put in water is inflamed and when it is taken out of it is extinguished, take quicklime and mix it with a little wax, and sesame oil, and naphtha, and white earth, and make an image of it, which when you sprinkle it with water will inflame."

Si vis portare in manu ignem ut non offendat, accipe calcem dissolutam cum aqua fabarum calida et aliquantulum magrencilis et aliquantulum malvæ visci, et permisci illud cum eo bene, deinde lini cum eo palmam tuam et fac siccari et pone in ea ignem, et non nocet. Dicunt philosophi, quod talk[d] non comburitur in igne et glutum piscis salvat ab igne, et alumen Iamenum et sanguis salamandræ et fuligo furni vel lebetis, quando igitur ex istis omnibus aut quibusdam aliis fit linimentum, non offendit ignis.

"If you will carry fire in your hand so that it is not injured, take lime slaked in hot bean-water and a little mallow, and mix them well together. Then anoint the palm of your hand with it and let it dry, and put fire in it and it will not be injured. The philosophers say that talc is not burnt in fire and fish-glue preserves from fire, and Yaman alum, and blood of the salamander, and soot of the furnace or cauldron, when you make a liniment from them all or some of them, fire does not hurt."

The recipe for alcohol is given (without quantities):

Aquam ardentem sic facias. Recipe vinum nigrum, spissum potens et vetus et in una quarta ipsius distemperabis vivæ calcis, sulphuris vivi subtilissime pulverizati, tartari de bono vino et salis communis albi grossi, postea pones in curcubita bene lutata[e] et desuper posito alembico, distillabis aquam ardentem, quam servare debes in vase vitreo.

This is the same as *Liber Ignium* § 27, p. 51, which gives the quantities, but quicklime is now added (see p. 148).

Ignem græcum sic facies: Recipe <sulphur> vivum, tartarum, sarcocollam, picolum, sal coctum, petroleum et oleum commune. Fac bullire bene, et si quid imponitur in eo, accenditur, sive lignum sive ferrum, et non extinguitur nisi urina, aceto vel arena.

"Greek fire is made in this way. Take sulphur, tartar, sarcocolla, pitch, boiled salt, petroleum, and common oil. Boil them all together, and what is put into it and set on fire, whether wood or iron, is not extinguished except by urine, vinegar, or sand."

Si vis facere carbunculum vel rem lucentem in nocte, recipe noctilucas lucentes quam plurimas et ipsa contritas pone in ampullam vitream et claude. In fimo equino calido sepelias et dimitte morari per quindecim dies. Postea distillabis ex eis aquam per alembicum, quam repone in vase de crystallo aut vitro: tantam enim præstat claritatem, quod in loco obscuro quilibet potest legere et scribere. Quidam faciunt hanc aquam ex felle testudinis, felle mustelæ, felle furonis et canis aquatici, sepeliunt in fimo et destillant ex iis aquam.

"If you will make a carbuncle or thing shining in the dark, take fire-flies and crush them and put them in a glass alembic and close it. Bury it in hot horse-dung and let it remain fifteen days. Then distil from it a water in the alembic, which put into a bottle of crystal or glass. It will be very shining, so that you can read and write by it in a dark place. Some make this water from the bile of the tortoise, weasel, ferret, or water-dog, burying them in dung and distilling a water from them."

Aquam ardentem sic facies. Recipe terpentinam,[f] quam destillabis per alembicum: velut aqua ardens etiam exhibit. Misce vino aut cuivis et accenditur si appropinquas ei candelam.

"Burning water is made thus. Take turpentine, which distil by the alembic, just as described in making (the other) strong water. Mixed with wine or other thing it will inflame if you bring a candle near it."

Ignis volantis in aëre multiplex est compositio. Unus [unde] fit de sale petroso et sulphure et oleo lini, quibus insimul distemperatis et in canna[m] positis accensus ignis in aëre protinus sublimatur. Alius fit ex sale petroso, sulphure vivo et carbonibus vitis aut salicis, quæ [quibus] insimul misce [mixtis]. Sed de sulphure ibi vult esse minus, de sale aut colophonia plus.[g]

These are the recipes §§ 32 and 33 of the *Liber Ignium*, translated on p. 54.

Ignis volans. Accipe libram unam sulphuris, libras duas carbonum salicis, libras sex salis petrosi, quæ tria subtilissime terantur in lapide marmorei, postea aliquid posterius ad libitum in tunica de papyro volanti, vel tonitruum faciente, ponatur. Tunica ad volandum debet esse longa, gracilis, pulvere illo optime plena, ad faciendum vero tonitruum brevis, grossa et semiplena.

"Flying fire. Take one pound of sulphur, two pounds of willow charcoal, and six pounds of saltpetre, which three things grind finely on a marble stone. Then put as much as you wish into a paper case to make flying fire or thunder. The case for flying fire should be long and thin and well-filled with powder, that for making thunder short and thick and half-filled with powder." This is essentially recipe § 13 of the *Liber Ignium*.

There is no "definition" of saltpetre such as is given in the *Liber Ignium*, but a test to distinguish "baurach" (Arabic būraq, alkali) from "nitrum falsum" (salsum in some texts) is given in an earlier part of the text; the first salt effervesces with vinegar, which saltpetre would not:

Et quando baurat[h] et nitrum falsum[i] ponitur in vase et ponitur super ipsum acetum, bulliat fortiter absque igne.

[a] vino, Romocki, vivum ε, θ, etc. [b] cemar, Romocki, zinar, ziniar α, zimar κ, limar β. [c] om. terra alba, Romocki. [d] calx α, talc κ, talck θ, talk Romocki. [e] lutata in all texts seen, plumbeata, Romocki. [f] serpentinam in all texts seen, except ε serpentium; recte terpentinam, Romocki. [g] om. § in κ. [h] haurat θ, κ; haurit ξ. [i] salsum α, ξ.

REFERENCES

1. Thorndike, *A History of Magic and Experimental Science*, 1923, ii, 723–4; Berthelot, *La Chimie au Moyen Âge*, 1893, i, 91.

2. *Gesamtkatalog der Wiengendrucke*, Leipzig, 1925, i, nos. 691, 692 (Venice, c. 1472, 34 ll.; Venice, c. 1473, 14 ll.; the second used by Romocki, ref. 4, i, 97); Thorndike, ref. 1, ii, 721.

3. Caillet, *Manuel Bibliographique des Sciences Psychiques et Occultes*, Paris, 1913, i, 21.

4. Romocki, *Geschichte der Explosivstoffe*, 1895, i, 97–103.

5. Del Rio, *Disquisitionum Magicarum*, bk. i, ch. 3; f°, Mainz, 1603, 8; E. H. F. Meyer, *Geschichte der Botanik*, 1857, iv, 79; J. Sighart, *Leben und Wissenschaft des Albertus Magnus*, Regensburg, 1857, 298; Kopp, *Beiträge zur Geschichte der Chemie*, Brunswick, 1875, iii, 67, 81; Berthelot, ref. 1, i, 91; Strunz, *Albertus Magnus*, 1926, 96; Lippmann, *Entstehung und Ausbreitung der Alchemie*, Berlin, 1919, i, 490.

6. Ref. 4, i, 95, 103.

7. Ref. 1, ii, 720.

8. Thorndike, ref. 1, ii, 560.

9. *De animalibus*, bk. XXII, tract. ii, ch. 1, §35; *De Mirabilibus*, in Romocki, ref. 4, i, 99.

10. Thorndike, ref. 1, ii, 725.

11. Romocki, ref. 4, i, 95.

12. Wellmann, *Abhl. K. Preuss. Akad. Wiss., phil.-hist. Cl.*, 1921, no. iv, 29; *id., Philologus Suppl.*, 1934, xxvii, no. 2, 21.

13. *Opus Minus*, ed. Brewer, 1859, 327; Delorme, in Vacant and Mangenot, *Dictionnaire de Théologie Catholique*, 1905, ii, 7.

14. *Histoire de la Chimie*, 1866, i, 390.

15. *Geschichte des Wachsthums und der Erfindungen in der Chemie in der ältesten und mittlern Zeit*, Berlin and Stettin, 1792, 137.

16. Thorndike, ref. 1, ii, 730.

MICHAEL SCOT

Some interesting material from our point of view is contained in works attributed to Michael Scot (1180?–1236?), who was translating from Arabic in Toledo in 1217, was in Bologna in 1220, and after about 1227 was court astrologer and philosopher to Frederick II at Palermo.[1]*

Scot's *Liber Luminis*[2] deals with many salts, including nitrum of Alexandria (soda) and two other kinds of true nitrum (que vere nitrum salsum appellatur), a leafy, like talc (foliatum ut talcum), and a hairy (filatur et depilatur); whether the last is saltpetre is doubtful. Some supplementary chapters (in which the author speaks in the first person singular) describe, among other things, the preparation of what is apparently nitric acid or aqua regia, by distilling a mixture of sal nitrum, sal ammoniac and Roman vitriol, and sal nitrum here probably means saltpetre.

A *De Alkimia* attributed to Scot is found in different forms in manuscripts at Palermo,[3] Oxford[4] and Cambridge.[5] The Palermo text was partly published by Haskins[6] and the variants in the Cambridge text by Mrs. Singer.[7] A collation of the three manuscripts and a complete text was published by Thomson,[8] who thinks the earlier part was probably by Scot. The last part of the Oxford and Cambridge manuscripts he dates as written in the thirteenth century, say about 1275. The last parts in the three manuscripts are all different and it is doubtful if they are by Scot; they may be collections of notes, similar to those in the *De Mirabilibus* attributed to Albertus Magnus, made by pupils or younger fellow-workers in Scot's circle. The parts dealing with Greek fire, oil of turpentine, sulphur and the preparation of colours are in the Cambridge manuscript only, and these are the only ones of interest to us. Mrs. Singer[9] regards this appendix as "a compilation from various writers, including Michael Scot himself." If the recipes are due to Scot

* For references, see p. 89.

they are earlier than those in Albertus Magnus, and if they at least came
from Scot's circle, an Arabic origin is quite plausible. It has been pointed
out[10] that the earliest author cited for the distillation of alcohol is Michael
Scot, who was in Toledo in 1217 and Bologna in 1220, and used Spanish-
Arabic sources through Jewish intermediaries. The circle of alchemists in
South Italy may have developed from those of Andalusia.

Some recipes for incendiary and poisoned arrows; incendiary mixtures
containing naphtha, pitch, oils, tow and vine branches; hollow arrows, sometimes lined with copper, containing naphtha, pitch, sulphur, salt and tow;
and the protection of battering-rams from incendiaries by coverings of layers
of leather, felt, more leather, sand, linen, and finally leather, the whole well
soaked in vinegar and urine, are contained in the twelfth-century *Mappæ
Clavicula* supposed to have been written by Adelard of Bath and under Arabic
influence.[11] They are of the earlier tradition. The commentary of Bernardus
Provincialis (*c.* 1155?) in the *Salernitan Collection*[12] gives a property of what
he calls sal armoniacum, which seems to indicate that it was saltpetre.

A summary of some parts of the Cambridge Scot manuscript will now
be given. The section on salts in this manuscript[13] is different from that
in the other two. Sal nitrum de puncta is said to come from India, from
"puncta maris versus Marioth," and Alexandria. It is tested by putting it
on burning coals, and if it does not decrepitate or make a noise it is good
(quod si supra prunam diu steterit et de pruna non salierit nec stridorem
fecerit erit bonum). There is also a foliated sal nitrum somewhat long and
thick with a taste something like vinegar when touched with the tongue and
not salty, and it makes a flame over a fire (Sal nitrum foliatum est aliquantulum
longum et grossum, non nimis salsum sed quasi saporem aceti trahit quando
ipsum cum lingua tetigeris, et flammam super ignem facit). It is mentioned
in some books; it is the best for making mercury malleable (constringit
mercurium et ipsum malleabilem facit), and changes copper into the best
gold. It is found in Spain "versus Narbonem," and is exported from Aleppo.
A third kind is nitrum depilatum, from Hungary and Barbary (invenitur in
Barbaria). The apothecaries of Monte Persolano keep it. It cleans dried
pork (depuratum sicut caro porcina sicca).

There is little doubt that the sal nitrum foliatum is saltpetre, and the
observation of its deflagration on red-hot charcoal must have been made,
because this test was applied to see if the first nitrum (de puncta) decrepitated,
as it would if adulterated with common salt. The same sort of test is specified
further on for other salts:

Si flammam fecerit et videatur aerea illud est de Alexandria et est bonum,
sed si saltat de prunis non valet. Sed si illud de Armenia et de Alap vis
cognoscere ab illo de Alexandria mitte super prunas. Si diu steterit et
stridorem non fecerit bonum erit et perforat. Illud de Alexandria foditur
iuxta venam azurii: propter hoc habebit flammam. Magister Michael
Scotticus dixit quod sal nitrum nobile est inter omnia alia et melius. . . .
Si ponatur supra prunas et suffles inflammabitur et saliet de prunis, et invenitur
iuxta castrum Babilonie [Cairo?]. Saraceni aportant ipsum in Alexandriam

et in Acaton.   Saraceni appellant ipsum borach et credunt quod sit alumen.
Et in Hispania invenitur versus Argoniam in quodam monte iuxta mare, et
appellant ipsum hispani alumen acetum activum.   Of a kind from Morocco:
Si vis ipsum cognoscere pone super prunam et faciet flammam stridendi.

The appendix in the Cambridge manuscript and peculiar to it contains
recipes for incendiaries and preparing azure.[14]   The section on Greek fire
(Ignem grecum hoc modo conficies) includes sandarac (glassa), sulphur,
cacabre (amber) and naval pitch, oil of bricks, distilled turpentine, oil of
sulphur and arsenic (? alkibrit), treated in a complicated way, including
distillation in an alembic "like rose-water," and then used for incendiary
arrows tipped with cotton and fustian (fustagine).

An aqua ardens is distilled from turpentine and linseed oil; it gives a great
flame when thrown on wine and ignited.   Oil of sulphur is made from oil
of eggs or juniper oil and powdered sulphur distilled, but not under a roof
(et nota quod predicti olei distillacio sub tecto fieri prohibitur propter eius
fetorem al incendio).   The word fœtor in the *Liber Ignium* (p. 56) also occurs
here.

A fixative for gold leaf on cloth is made from almond gum and white of
egg.   A marvellous sharp water for whitening all metals to the colour of silver
is made from sal armoniac, alum, tartar and common salt, equal weights,
boiled with strong vinegar.   Gold writing is done with copper and brass
powder in gum-water.   A candle is kindled in sunlight by a mirror or a lens
(berillum ad modum ovi rotundum).   What is obviously a püstrich (see
p. 150) filled with turpentine and heated is described.

Artificial azure (azurum) is made from white earth and an infusion of
coloured flowers (qui reperitur in doliis tinctorum in persico), or from silver
plates suspended over strong vinegar in a pot of horse-dung.   There are
two kinds of azure, lombardicum and transmarinum, directions for purifying
which are given.   The Lombard azure is purified with soapy lye (per
capuellum).   Take good lye made with clavellated ashes and dissolve in it
Roman or French soap (saponens romanum sive gallicum) and in it
boil an ounce of very finely powdered azure to be cleaned (qua despumat);
pour off the lye and wash several times with clean water the flower of azure
(florem azurii) which remains in the bottom of the pot, collect the azure, dry
it, and keep it in a skin bag (in coreo).[15]

References

1.   J. W. Brown, *Michael Scot*, Edinburgh, 1897; Thorndike, *History of Magic and Experimental Science*, 1923, ii, 306 f.; Haskins, *Isis*, 1922, iv, 250; 1925, vii, 478; 1926, viii, 794; 1928, x, 350; 1930, xv, 406; *id., Studies in the History of Mediæval Science*, 1927, 5, 16, 30, 36, 67, 130, 137 f., 141 f., 153, 155 f., 242 f., 272 f., 280 f., 295, 316; *id., The Renaissance of the Twelfth Century*, 1927, 278 f., 284, 290, 301 f., 319 f., 368 f.; D. W. Singer, *Isis*, 1929, xiii, 5; 1930, xv, 212; J. Read, *Scientia*, 1938, xxxii, 190.

2.   Brown, ref. 1, 86, 249 f.

3.   Publ. Lib. 4.Qq.A.10; Speciale MS.

4.   Corpus Christi 125, ff. 116–19; fourteenth to fifteenth century.

5.   Gonville and Caius 181, ff. 19–32; thirteenth century.

6. *Isis*, 1928, x, 350–9; *id.*, *Mediæval Culture*, Oxford, 1929, 148–59; cf. *id.*, *Studies in the* *History of Mediæval Science*, 1927, 272–98.

7. Ref. 1.

8. S. H. Thomson, *Osiris*, 1938, v, 523–59.

9. Ref. 1.

10. R. Verrier, *Études sur Arnaude de Villeneuve*, 2 vols., Leyden, 1947–9, ii, 92, 156.

11. Phillips, *Archælogia*, 1847, xxxii, 183–244; recipes 264–79.

12. De Renzi, *Collectio Salernitana*, Naples, 1859, v, 309 : distempera cum aqua et scribe in carta et permitte cartem siccari cum litteris, deindes appone igni et poteris eas legere.

13. Thomson, ref. 8, 534 f.

14. Singer, ref. 1.

15. Mrs. Singer has given "capuellum [i.e. cupellation]," by an oversight, since Mrs. Merrifield, *Ancient Practice of Painting*, 1849, ii, 891, had already pointed out that the Italian capitello was caustic lye.

Chapter III

THE LEGEND OF BLACK BERTHOLD

FROM the late fifteenth century a story has been in circulation that gunpowder and cannon were invented by a mysterious alchemist, magician, or monk, Berthold Schwartz, or perhaps (as Bertholdus niger) Black Berthold, who, it is alleged, lived in the thirteenth or fourteenth century and was a German, a Dane, or a Greek. "Portraits" of Berthold Schwartz are extant,[1] and a statue of him was erected in Freiburg im Breisgau with the date 1353 as that of his "discovery." This discovery was said to be gunpowder, but since this was described by Roger Bacon about a century before, it was then said to be cannon. An attempt was also made to pre-date Berthold a century. Since the legend still regularly recurs, it is perhaps necessary to enter into some detail, in the hope that the uneasy ghost of Black Berthold may ultimately be laid to rest.

The poet Heinrich Hansjakob collected information on Berthold in 1891.[2] He says that the earliest mention of Berthold is by Felix Hemmerlin (Latin Malleolus) of Zürich (1389–1464), in a work[3] which says that a certain Black Berthold, a subtle alchemist (quidem Bertholdus niger, Alchymista subtilis; he is not called a monk) could fix mercury to a malleable metal. Wishing to kill the spirit (basilisk) of mercury, he heated this metal with sulphur and saltpetre in a closed pot, which exploded. He heated the mixture in closed metal vessels, which also exploded and blew down his laboratory walls.

In another story[4] from the same source, Berthold wishing to test the Aristotelian theory that bodies of hot and cold natures are antagonistic, mixed in a stone mortar saltpetre (cold) with sulphur (hot), together with some charcoal or linseed oil. The mortar was put over a fire, when it exploded, scattering pieces of stone. With thought (sagacitate), Berthold extended this accidental discovery and improved the vessel to "what we now call Büchsen" (i.e. guns). Hemmerlin, who "wrote about 1480 or 1450" according to Hansjakob (although the notice is not in the 1490 edition), says Berthold lived "about 200 years before," which would be about 1250, although most authors supposed that he meant 100 years and hence dated Berthold 1380. Hansjakob says[5] old French writers (e.g. de Belleforest) correctly have "Berthold le Noir"; the name Schwartz, he says,[6] was used, by reading Niger instead of niger, by Aventinus, i.e. Johannes Thurmayr or Thurnmaier of Abensburg (1466–1534), who in his annals of Bavaria[7] says for the year 1359: Berchtoldum quoque novitate inventi a se operis nobilem, eadem ætate fuisse non ignoremus. Teuto is fuit genere, religione Franciscanus, professione philosophus et magiæ, et metallariæ, artium fallacissimarum (si modo artes, et non uniuscuiusque vanissimi, ludibria dicenda sunt) consultus, invenit tormenta ærea quibus iaculamur globos ferreos et lapideos, et homines, pecudes, mœnia, muros, arces, urbes, castra, solo quasi fulmine atqui tonitru (unde bombardis nomen est) prosternimus. It will be noticed that Aventinus says nothing about "Schwartz."

91

Libri,[8] Lacabane,[9] Favé[10] and Jähns[11] quote from a sixteenth- (Favé says seventeenth) century manuscript in the Bibliothèque Nationale (Du Puy 353):

> Le XVIIe may mil trois cent cinquante-quatre [1354], ledit seigneur roy [Jean I] estant acertené de l'invention de faire artillerie trouvée en Allemagne par un moine, nommé Berthold Schwartz, ordonna aux généraux des Monnoies faire diligence d'entendre quelles quantités de cuivre estoit audit royaume de France, tant pour adviser des moyens d'iceux faire artillerie que semblablement pour empescher la vente d'iceux à estrangers et transport hors le royaume.

Lacabane says *large* guns are mentioned more frequently after 1354, and although small bronze guns were known before in France, perhaps Berthold's invention in Germany was the casting of large guns, not known before, which were imported into France from Germany; the import of large guns into France is testified in documents of 1373, 1374, 1375 and 1377. Favé was in agreement with this and said there was no reason to suspect the authenticity of the manuscript.

Köhler[12] pointed out that the word "artillerie" in this sense came into use only in the sixteenth century, as is shown by a document of 1346,[13] in which "artillery (artilharia)" meant all kinds of weapons and implements such as sieves, bows and arrows, and (it is important to notice) stones, cannon and lead (gran foyo de peyras, e canos e plomb), and hence Köhler concluded that the passage was interpolated in the manuscript (entitled "Règlement des monnoies tant de France qu'étrangères," which otherwise has nothing to do with artillery) by a copyist in the sixteenth century.

Köhler[14] also quotes from a manuscript (no. 67(148)) in the Ambraser Collection written about 1410:

> Es war ein meister von kriechenland
> Niger Berchtoldus ist er genannt,

i.e. "a Greek master called Berthold Niger." This manuscript, he said, formed the basis of the *Feuerwerkbuch* (c. 1420) to be discussed later, but it has only ten "questions of the Büchsenmeister," of which only five are found among the twelve of the *Feuerwerkbuch*. A transcription of it made at the end of the fifteenth century is in Berlin (Kgl. Bibl. Incun. 10117a, in Köhler). It is interesting that in the oldest German accounts, Berthold is a Greek. In the *Feuerwerkbuch*, printed in 1529, it is said "the art of shooting from a Büchse" was invented by "Magister niger Bertholdus," who was a Greek alchemist and "Nigromanticus," i.e. an adept in black arts (probably an interpretation of "niger").[15] Berthold is still not a monk. These records are older than that in Hemmerlin's book (c. 1495). The oldest mention, if it were authentic, would be in records for Naumburg an der Saale, which say that ignis græcus, which is gunpowder (the 4:1:1 mixture) was made in 1348 by Berthold Schwartz "aus Kriechenland."[16]

The name Constantin Anklitzen, given as the family name of Berthold, should according to Hansjakob[17] be Angelisen (corrupted to Anklitzen), which is a Freiburg name, and hence Berthold probably lived and worked in Freiburg. A Breisgau manuscript of 1245 mentions a "Meister Berthold" of Freiburg, and a Breisgau manuscript of 1371 mentions shooting from a gun (Büchsen). Hence Berthold was not mythical, gunpowder was invented in Freiburg im Breisgau about 1250, its use was spread to Italy by German soldiers (as Faber, a Swiss or German writer of Hemmerlin's time said) and later to Spain (where gunpowder was rare as late as 1484). Gunpowder and cannon appeared in Basel in 1371, which points to Freiburg as the source; an explosion of gunpowder in a store in Lübeck in 1360 (but see p. 172) and other notices of 1370, 1371, etc., attest its early use in Germany.[18]

Polydore Vergil (b. Urbino *c.* 1470, d. there *c.* 1555), Archdeacon of Wells 1508, naturalised 1510, who lived in England from 1501 until 1550 and wrote a *History of England*, compiled a history of inventions, in which he says[19] gunpowder and bombards (cannon) were invented by an unnamed German and were used by the Genoese in 1380. He derives the name bombarda from the Greek βόμβος, rumbling or buzzing like bees; he gives the legend of the discovery of gunpowder as in Hemmerlin:

> Et hæc omnia ad hominum perniciem inventa sunt . . . quod Bombardam vocant . . . cuius inventorem fuisse tradunt hominem Germanum admodum ignobilem . . . servasse aliquando domi in mortario pulverem sulphureum, cujusdam medicinæ faciendæ causa illudque texisse lapide, ac inde contigisse, ut dum ignem ex silice prope excuteret, scintilla inttò ceciderit, subitoque flamma eruperit, atque lapidem in altum tulerit . . . postea factà ferrèa fistulà, & pulvere confecto, machinam reperisse, ac ejus usum Venetis in illo bello primum ostendisse, quod ad fossam Clodiam est cum Genuensibus gestum, qui fuit annus salutis humanæ MCCCLXXX. Hæc bombarda vocatur, à bombo, id est sonitu, qui βόμβος Græcè dicitur: quidem alii tormentum æneum malunt nuncupare. . . . Sclopus n. est sonus ille, qui ex buccarum inflatione erumpit. . . . Sed vel alio nomine appellatur Arcus busius, à foramine opinior, quo ignis in pulverem fistulà contentum imittitur: nam Itali busium vulgò foramen dicunt.

Vergil's account is reproduced by Panciroli,[20] whose editor Salmuth also says guns were used by the Genoese about 1380. Sebastian Franck (d. 1542) says in his chronicle[21] that the murderous and devillish shooting from guns was invented in 1380 by a monk when there was war. Achilles Pirminius Gasser (1505–77) says in his chronicle[22] under the year 1393 that guns were first invented about this time by a certain German monk (tormenta bombardarum, hac tempestate à monacho quodam Germanico, inventa primum sunt), which Hunter[23] pointed out is long after the use of guns in the battle of Crecy.

Sebastian Munster (1489–1552), in a short history of cannon, reported that Achilles Gassius had written to him saying that cannon were used at sea by the Danes in 1354 and were invented by a chemist and monk called Berthold Schwartz:

> Achilles Gasserus medicinæ doctor & histriographus, diligentissimè

scripsit mihi, bombardos anno Christo 1354. in usu apud mare Danicum fuisse. primumqúe inuentorem & autorem extitisse Chymistam quendam nomine Bertholdū Schuuartz monachum.

Although Gasser in his earlier epitome[24] had given the date as 1393, he later[25] altered it to 1353[26]:

> von Berchtold Schwartz einem Alchimisten und Franciscaner Mönch in Teutschland ohne zweiffel auss eingebung dess leidigen Teuffels die schädliche künst das Puluer zu bereiten erfunden worden: welches auss drey theilen Salpeter / zwey theilen Lindner oder Weidener Kolen / vnd einem theil gesottenem Schwebel zugerichtet / vnnd zu den Büchsen gebrauchet wird.

The Jesuit Bernard Cæsius,[27] quoting Polydore Vergil and "Gordonius" (the Jesuit James Lesmore Gordon of Aberdeen, d. 1641),[28] gives the dates 1354 or 1380 for Berthold's invention of cannon. Athanasius Kircher[29] reported (with some picturesque details) that Berthold Schwartz was a Benedictine monk and alchemist of Goslar, who invented gunpowder in 1354, and later made it known in Italy, where it was used in the wars between Venice and Genoa. Becher[30] made Berthold live in Mainz. Theodore Janssen ab Almeloveen[31] said guns were invented in Denmark in 1354 by a monk and alchemist Barthold Schwartz. This alleged Danish origin was considered very fully by two Danish writers, Hans Gram[32] and Christian Friderich Temler,[33] who are mentioned by Hansjakob.[34]

Gram[35] reported that Antonio Cornazzano and Martin Cruse had made Berthold a native of Cologne, Becher of Mainz, Kircher of Goslar, Moreri of Freiburg im Breisgau or Freiberg in Switzerland; he had also been reported to be of Brunswick, Prague, or Jutland in Denmark. He was sometimes a Franciscan, sometimes a Dominican, and sometimes an Augustinian. Jähns[36] at first accepted Freiburg im Breisgau, later Cologne, with the approximate dates 1290–1320. Gram[37] put Berthold about 1354 in Cologne, Constantin Anklitzen being another person in Freiburg; in an appendix he discussed the old Danish names krud for gunpowder and bysse (= Büchse) for cannon.

Temler criticised Gram on some points and dealt largely with Greek fire and the early use of artillery, saying that Gram's thesis that modern gunpowder was in use in 1340 is not clearly supported by documentary evidence (which it is now known to be). Gram[38] replied to Temler's criticisms. Gram[39] had already pointed out that the date 1380 for the invention of gunpowder by Berthold Schwartz could not be correct, since there was a parchment (which he quotes fully) showing that powder was in use in Denmark in 1372. He collected all the references to Schwartz he could find, pointed out[40] that Malleolus's (Hemmerlin's) date for him, 1250, was much too early; that he was sometimes a monk, sometimes an alchemist; and that his birthplace was most variously given.

Gram,[41] who said that Achilles Gasser made Berthold a native of Jutland, rejected his Danish origin. Morhof[42] had said that gunpowder was invented by Roger Bacon, "not by that Welshman Berthold Schwartz (non Bartholdo illi Swartzio Cimbro)." Gram[43] says the Danish name for gunpowder, "Byss-Krud," is of German origin. In Doukas,[44] βοτάνη (literally plant), as the name for a gun, is the same word, from the German "Kraut," and Ducange, he says, had derived Büss or Byss from Pyxis (πυξίς, pyxis, a box). Gram[45] believed that cannon preceded small arms. For early mentions of guns, Gram[46] refers to Quesnoy (1340), Crecy (1346), Augsburg (1378; three large pieces, from Gasser), and the battle between the Genoans and Venetians (1390). He accepts[47] the reference in Petrarch[48] as genuine, as did Lalanne,[49] who adds Lübeck (1360) (see p. 172).

Temler[50] refers to the mention of the use of cannon in Italy in 1314 by Camden[51] (1551–1623) as not in the 1605 edition, but in that of 1614 published in Camden's lifetime, and says it is incorrect. Camden says the best approved authors agree that gunpowder and cannon were invented in Germany by "Berthold Swarte, a Monke skillful in . . . Alchimy." The Devil gave him a four-line Latin oracle (which Camden reproduces), on the basis of which Berthold "made a trunck of yron with learned aduice," filled it with powder and fired it. It is certain, says Camden, that Edward III used guns in the siege of Calais in 1347, "for gunnarii had their pay there, as appeareth by record. About 33. yeares before they were seene in Italy, and about that they began, as it seemeth, to be vsed in Spaine, but named by writers *Dolia igniuoma*, as fire-flashing vessels." The French, he says incorrectly, first knew of them in 1425. Benjamin Robins[52] dated Berthold about 1320.

A French author[53] says Berthold was probably born in Freiburg im Breisgau and died in prison in Venice in 1384. The tradition of the discovery of gunpowder by Berthold, which is purely legendary, goes back at least to the fifteenth century, when Crespi painted a scene now in the Uffizi Museum, Florence, showing Berthold working with others on the manufacture of powder, and a mortar has the inscription: "Pulvis excogitatus 1354. Daniâ [*sic*] Bertholdo Schwartz."[54] In 1380, Berthold went to Venice to cast for the Republic enormous cannon which were used in the siege of Chiozza and, according to the chronicle of Daniel Chinazzo, threw marble balls weighing 140–200 lb. On claiming his agreed pay he was refused it and imprisoned; to increase the power of the guns the Venetians had overloaded them with powder, which was very expensive, and the aim was found to be uncertain, so that Berthold's bill was dishonoured.

Feldaus[55] gave a summary of the manuscript and printed statements about Berthold. He rejects Hansjakob's date of 1250. He points out that Gasser had dated his invention as late as 1393 (although later he altered this to 1354— really 1353) and says that Freiburg im Breisgau is first mentioned as the place of discovery by Salmuth.[20] Köhler[56] had rejected Berthold as a purely imaginary person, as did Berthelot,[57] who pointed out that he appears first under the date 1354 when gunpowder had been known for a century and guns were used many years before 1354. Clephan[58] concluded that Berthold was an imaginary person invented by German authors wishing to claim gunpowder and cannon as discoveries of their nation.

Rathgen[59] sometimes[a] calls Berthold "legendary (sagenhafte)," sometimes[b] identifies him with an unknown German inventor of gunpowder and guns; gunpowder and rockets were known to the Tartars and Chinese in the twelfth to thirteenth century,[c] but the recipes in Albertus Magnus and Marcus Græcus (Roger Bacon is omitted) are not for gunpowder. Yet in other places[d] Berthold comes to life again; Rathgen says[e]:

> "Dieser Mönch Berthold hat in leichter Änderung der Form seines Stampfmörsers, in dem ein Pulvergemisch zufällig entzündet hatte, den ersten Schiessmörser hergestellt. Ihm gebührt das Verdienst, dass er auf diese Weise der Erfinder der Steinbüchse wurde, die die dem Pulver innewohnende Kraft zur vollen Entfaltung brachte und die bisherige geringwertige Knall- und Schreckbüchse zu der die gesamte Kriegskunst umformenden Pulverwaffe verwandelte."

The mortar was then, as "a German invention," cast in bronze by a process hardly practised abroad (kaum im Auslande geübte Verfahren) and the explosive power (Sprengkraft) and propulsive power (Treibkraft) of powder were discovered by an unknown German in 1320–30. The latest German author I know who has mentioned Berthold Schwarz rejects him as mythical; Hassenstein[60] says:

> Den bis in neuere Zeit hinein tobenden Kampf um seine Person hat die Forschung dadurch zur Ruhe gebracht, dass sie weder einen Mönch noch einen Bürger dieses Namens oder, wie man später annahm, mit Namen Anklitz oder Anklitzen in Freiburg hat feststellen können.

Black Berthold is a purely legendary figure like Robin Hood (or perhaps better, Friar Tuck); he was invented solely for the purpose of providing a German origin for gunpowder and cannon, and the Freiburg monument with the date 1353 for his discovery rests on no historical foundation.

Now that the ghost of Berthold Schwartz has been laid and the record of 1313 for Ghent and, as will be seen presently, one of 1324 for Metz, removed as forgeries, we are left in suspense. The first facts which will present themselves are that the earliest picture of a gun is in an English document, and the earliest mention of guns which is above suspicion is in an Italian document. It is surprising that German documentary evidence is, relatively, so late. The legend of the German invention of cannon is, in fact, one which requires the most careful reinvestigation. That it should be maintained by some recent German writers need cause no surprise; Diels[61] has stated, with obvious pride, that cannon were a German invention of the fourteenth century; he grouped them with the German inventions of his own time: great guns (really Austrian, from the Skoda works), Zeppelins, poison gas and submarines (deutschen Erfindungen der Riesenkanonen, Zeppeline, Gasangriffe, Tauch- boote). The first actual submarines were made by an American (Robert Fulton, 1800) in France, where submarines were first successfully used in naval warfare in 1880–5; the first steam-driven gas-filled dirigible was that of Giffard (1852) in France; we can give the Germans poison gas, since although the use of incendiary shells containing a solution of phosphorus in carbon disulphide, and poison-gas shells containing cacodyl cyanide, were proposed by Sir Lyon Playfair for use in the Crimea in 1854, they were not adopted. The Emperor Napoleon III was the first to plate warships with iron in 1855, and floating batteries plated with iron were used by the French and English in the Crimea in 1855.

Lippmann[62] said that Lt.-Gen. Bernhard Rathgen had collected information proving the leading part taken by the German fatherland in the invention and use of gunpowder and firearms (die massgebende Rolle des deutschen Vaterlandes auf diesem Gebiete), which unfortunately could not be published in existing conditions. Rathgen's book was published posthumously[16] by the Verein Deutscher Ingenieure with a preface by the President, Conrad Matschoss. In it he claimed that cannon were a German invention, first mentioned as used (or made) in Aachen in 1346 and Frankfurt in 1348. Sarton[63] says this is doubtful.

Köhler[64] had claimed to have proved, on the basis of documents, that the prevailing view that Germany had a claim to be the cradle of artillery could not be maintained. Rathgen[65] said it is the duty (Pflicht) of a modern German to refute the errors of Köhler and his copyists, and to restore Germany to its former glory, and in the interest of historical truth (der geschichtlichen Wahrheit) to make it clear that gunpowder was discovered, and firearms developed, in Germany (dass das Schiesspulver in Deutschland erfunden, dass in Deutschland die Pulverwaffe aufgekommen ist). This is all the more necessary since the gun is the source of all culture (Die Waffe ist der Ausgangspunkt aller Kultur).[66] Köhler had argued for an origin in Moorish Spain, passing by way of Italy and France to Germany, whilst Rathgen tried to show that this order should be reversed.[67]

THE EARLIEST GUNS: FLORENCE 1326 TO CRECY 1346

The earliest mention of a gun has been said to be in *De Memorial-Boek der Stad Gent*, begun in 1300 and kept up to date year by year until the fifteenth century. In an entry dated 1313, published by Lenz,[68] it is said: item, in dit jaer was aldereest ghevonden in Duutschlandt het ghebruuk der bussen van eenen mueninck (item, in this year the use of bussen was first discovered in Germany by a monk). A commercial record kept at Ghent, dated 1314, contains more than one entry of bussen met kruyt (bussen with gunpowder) despatched in this year to England.

According to Lenz, bussen were originally tubes filled with incendiary material, ignited and thrown by whatever means among the enemy, and the assumption that those mentioned were cannon is based on other material. The account is obviously of something new at the time. Van der Haeghen[69] said no such record of 1313 exists in Ghent; a sixteenth-century manuscript gives under the date 1393: "In ait jaer is ghevonden het ghebruick van het buscruit in Duytslant van eenen muninck" (in this year gunpowder was discovered in Germany by a monk), and the same manuscript contains "la fameuse phrase inscrite à l'année 1313"; the document is a forgery. Hime[70] says:

"These two passages were examined both by Sir George Greenhill and the late Mr. Oscar Guttmann a few years ago. The latter failed to find them on his first visit to Ghent, and he expressed doubts about their existence in *Kynoch's Journal*, vol. viii, no. 38; but after communicating with Sir George Greenhill, he returned to Ghent and found them. He intended to explain matters in *Kynoch's Journal*, but unhappily met his death in a motor accident before he could do so."

Van der Haeghen says he looked into the matter at the request of Guttmann and discovered the forgery. There we must leave the matter.

A cannon in the armoury of Amberg, Bavaria, has the inscription 1303.[71] If this is so, it is probably a false date, like 1322 on another gun (for 1522)[72] and Rathgen says it should probably be 1403.

Freiherr von Reitzenstein,[73] who says Berthold is fictitious, says the Klotzbüchse (*c.* 1350) in Germany were mortars, originally of wood, the powder being called Kraut or Krut. They had evolved from Roman candles, kindled at the top, by igniting the charge behind and using it to propel a shot. The name "ein Bleibuchsin" indicates that a lead ball was used.

Reitzenstein says large cannon were first used by Herzog Albrecht II of Brunswick-Grubenhagen in 1365; the first picture of a gun is in Codex 600, Munich Hof- und Stadt Bibliothek, probably of 1350 and certainly before 1400. There is, however, an earlier one of A.D. 1326 in the manuscript of Walter de Milemete in Christ Church, Oxford, showing a pear-shaped gun with a four-ended arrow, standing on a rickety table and being fired at a touch-hole by (apparently) a red-hot iron. The text of the manuscript has no reference to the gun.[74] Milemete was Prebendary of the Collegiate Church of Glaseney in Cornwall, and Chaplain of Edward III. The face of the gunner is light brown and he may be a Moor. Although it is said that the arrow is being fired at a door, it looks to me like an opening in a wall. Diels' assumption, quoted and approved by Rathgen, that Milemete's gun shot arrows by means of Greek fire, not gunpowder, is baseless; it was founded on the hypothesis of Diels that Greek fire contained saltpetre, which has already been rejected (but see ref. 113).

The Milemete gun was mentioned by Gustav Oppert in 1880; a sketch of it (printed upside-down) was published by Guttmann in 1904, and a reproduction of the picture from the manuscript by Greenhill in 1906. The Milemete MS. was published in facsimile by M. R. James[75] with part of a Holkham Hall MS. which has a picture of a "canon à quadriau," and since J. G. Mann said the miniatures in it can certainly be dated 1326–7, the doubt about the date of the other one is removed. The gun shown in the Holkham Hall MS. is much larger and it rests on stone supports. Four men are standing by it. They have now no epaulettes (shoulder-shields), but have flat-topped steel caps (Beckenhaube) and chain mail (Kettenpanzer). The gun (pot) was originally painted in silver, which has darkened. The head only of the arrow in the gun is now shown. The Holkham Hall figure is probably older, Rathgen thinks, than the Milemete figure. The size of the picture is 14 cm., the same as that given for the Milemete figure in Guttmann and Feldhaus. By comparison with the size of the men, the pot would be 2·5 m. long, with the spherical part 90 cm. diameter and the neck, which is unsupported, 1·5 m. long. Rathgen asks how this pot could have been made. The weight would be 600 kg. and this was too large for a bronze casting of the period. It was too big to have been forged. Although the armour (with shoulder-shields) is of the period 1330 or 1350 (this date seems to be too late), Rathgen concluded that the picture was copied from a German *Feuerwerkbuch* of about 1400 into a space conveniently left blank for it in the manuscript in 1326–7. Rathgen[76] says the red-hot iron (Zundeisen) was used at least as late as 1405 (it is in Kyeser's book, see ch. IV); the touch-hole appeared only in the period 1376–1401, and before its use the gun was fired, Rathgen says, by a wick (Zundfaden) from the muzzle. The match (Lunte) was used instead of the red-hot iron before 1459.[77]

I examined the Milemete MS. (Christ Church 92) on 25th July, 1956. The gun and the head and feather of the arrow are gilt (bright), the arrow shaft is red, the table top is white and the legs green. The man is dark-complexioned, the details of the face being very clear (Guttmann's reproduction is very poor). The helmet is gilt, the robe green, the shoulder-shields red, the feet are bare. The background is blue, with gilt spots. The steps to the "door" are in red and clearly show in perspective that it is not a door but an entrance to a passage. The arrow is quite out of the gun and seems to be going in front of the "door" and not through it; another figure (there are many on military subjects in the manuscript) shows an arrow from a large cross-bow flying in front of the same "door," which in both figures is surmounted by a tower with a slit in register with that in the figure below. There is a red spot over the touch-hole of the gun and the igniter is white. The manuscript is dated 1326 on the front leaf. It was presented to Christ Church in 1707 by William Carpender of Stanton, Hertfordshire.

James (pp. xix–xx) says the writing in the Milemete and Holkham MSS. is identical; the pictures, which are by more than one artist, were often put in first, since there has been difficulty in fitting in the text in the remaining spaces. His description of the gun is on pp. xxxii–xxxv, and the Holkham gun on p. lvii. The plate of the Milemete gun is on p. 181 (MS. f. 44*b*). The Milemete MS. has pictures of a machine for throwing Greek fire with four arms worked by ropes and a capstan and with flaming pots at the end of the windmill-like arms (p. 147); of beehives thrown into a fortress (p. 148); of a square box with an incendiary thrown by the point of an arrow, to which it is not attached (p. 151); and a fish-shaped kite with a large fireball suspended from it over a castle and the figures of three men pulling a rope attached to the kite (p. 154).

Post, who (incorrectly) dates the Milemete pictures 1330, since he assumed that they were added after the manuscript was written in 1326, says the gun is : "nicht nur die frühste Darstellung von Geschützen, sondern überhaupt die frühste Kunde davon" (p. 139). He shows that Rathgen's description is faulty and that his criticism of the difficulty of casting the guns is misplaced, since the insides are not bottle-shaped but have a straight bore, probably tapering to the rear, as in a Japanese cannon of similar external shape cast in 1830. The Milemete gun must be older than the manuscript, and Post thinks the original may have come from the Far East (see ch. VI).

A picture of a pot exactly like Milemete's, resting (without any support of the neck) on a board is shown (without text) in the *Wunderbuch* MS. of the Grossherzogliche Bibliothek in Weimar, written in the period 1430–1520, although some illustrations may be copied from older originals.[78] The text in the Holkham Hall MS. is: "Si uero debes impugnare incastratos. Utere instrumentis proicientibus lapides : utpote machinis Et multiplica ea iuxta modum instantis necessitatis ad hoc." This, says Rathgen[79] (who quotes it incorrectly), refers to a ballista for throwing stones, not a gun, and the blank space was afterwards filled in with "der phantastischen Zeichnung einer riesenhaften Feuerbüchse." It is true there was a "fire pot" at Rouen in 1338 (see p. 108), but this was in a "deutschen Kulturgebiet." Fire pots in Cividale in 1331 were used by "German knights" (see p. 102), and Milemete's gun was probably copied into the first manuscript not before 1330. The mention of a gun at Florence in 1326, says Rathgen, is a forgery (this is now recognised as genuine) : "Die Möglichkeit, den Ursprung, die Erfindung der Pulverwaffe in oder bei Frankfurt zu suchen, wird immer grösser." Rathgen[80] says the arrow was propelled from a gun by a wood block in the barrel, the packing being merely to hold the arrow in position. Guns shooting arrows were called spingardas and are shown in Leonardo da Vinci's *Codice Atlantico*; in Italy, however, they seem to have meant Steinbüchse in general (spinghardæ seu bombardæ).[81] The shape of Milemete's gun explains the names vasi (Italian) and pots (French) for guns[82]; there was a pot (boête) at St. Omer in 1342.[83] Pots standing on feet, with flames issuing from the mouths, and roughly of the same shape as Milemete's gun, are shown in a Berlin German Manuscript (germ.qu.1188), first half of fifteenth century,[84] but they do not seem to be intended for projecting arrows or anything else.

In 1339 there were in the Guildhall, London, "six instruments of latten [brass] called gonnes and five roleres for the same. Also pellets of lead weighing $4\frac{1}{2}$ cwt. for the same instruments. Also 32 lb. of powder for the same," and in the same year guns were used by the English in the siege of Cambrai.[85] If the illustration was added to the Milemete MS. after it was written in 1326,[86] which is a gratuitous hypothesis, it seems very probable that it cannot be much later than 1330–5. Hime's identification of "the black gunner" with Berthold Schwartz is a flight of imagination.

Guttmann[87] reproduces frescoes in the church of St. Leonardo in Lecceto, Siena, showing a cannon and two hand-guns, which were said to have been painted in 1340 by Paulo de Maestro Neri, whose receipt for payment dated 1343 is in the Siena Library. Rathgen,[88] however, says the range of the gun (Steinbüchse) shown belongs to the period 1430 at the earliest; if so, the picture must have been painted over a century after payment was made for its completion, which would seem to be a somewhat unbusinesslike procedure for Italians.

Milemete's gun threw large arrows with four-sided iron heads and metal (probably brass) feathers, i.e. cross-bow quarrels (from quatuor, four), and the gun was probably like those said by Froissart[89] to have been used at Quesnoy in 1340: "descliquèrent canons et bombardes qui jetoient grands carreaux" (arrows). The Flemish "engins appellés canons trians plonc et quarreaux" in 1349,[90] and the Burgundian "qanon à gitter garroz empennez et enferrez" in 1362 (which may have come from Flanders or the Rhine Provinces),[91] were probably of similar type. Guns throwing large arrows are shown (with old-fashioned ballistæ, etc.) in fifteenth-century manuscripts (BN Latin 7239; Munich 197); in the *Livre de Canonnerie* (see ch. IV) there is a chapter "pour tirer lances ferrées d'une bombarde, canon, ou autre baston à feu de canonnerie,"[92] and on 8th April, 1588, the Privy Council ordered "muskitts 200; arrows for the said muskittes with tamkines for eche, 1000" to be sent to Sir Francis Drake.[93]   Yet Rathgen[94] said that the last use of arrows in guns was by the French at Commines in 1382.   The English found Tournay in 1341 rich "in apparatu bellico ante hoc tempore nunquam viso."[95] Pierre of Bruges made at Tournay in September, 1346, a "tonoile" throwing an arrow attached to a 2 lb. piece of lead[96] (auquel avoit au bout devant une piece de plonch pesant ij lb.).

In an attack on Metz in 1324 the defenders are said to have had made "collevrines, arbollestres et autres traicts," and soon after, on the banks of the Moselle, William de Verey arrived in a barge provided with "serpentines et canons . . . et tirant plusieurs coptz d'artillerie . . . une serpentine qu'il fist par plusieurs fois tirer, et en tuont et blessont plusieurs."[97]   This document is a forgery; the names "serpentines" and "coulverines" were not in use at that time.[98]   The chronicles of Peter of Dusburg (Duisburg), giving detailed accounts of the wars against the Prussians in 1231 to 1326, including the capture of many castles, have no mention of cannon; the continuation of them from 1326 to 1410 mentions bombards only once, in 1410.[99]

The first definite mention of "cannon" is in an Italian document of 1326.[100]   A decree (provisione) passed by the Council of Florence on 11th February, 1326 (n.s.), appointed the priors, the gonfalonier, and twelve good men the faculty of naming two officers charged to make iron bullets and metal cannon (pilas seu palloctas ferreas et canones de metallo) for the defence of castles and villages belonging to the Republic:

> Item possint dicti domini priores artium et vexillifer justitiæ una cùm dicto officio duodecim bonorum virorum, eisque liceat nominare, eligare et deputare unum vel duos magistros in officiales et pro officialibus ad faciendum et fieri faciendum pro ipso communi pilas seu palloctas ferreas et canones de metallo, pro ipsis cannonibus et palottis habendis et operandis per ipsos magistros et officiales et alias personas in defensione communis Florentiæ et castrorum et terrarum quæ pro ipso communi tenentur, et in damnum et prejudicium inimicorum pro illo tempore et termino et cùm illis officio et salario eisdem per communi Florentiæ et de ipsius communis pecunia per camerarium camere dicti communis, solvendo

illis temporibus et terminis et cùm eâ immunitate et eo modo et forma et cum illis pactis et conditionibus, quibus ipsis prioribus et vexillifero et dicto officio XII. bonum virorum placuerit.

Other documents of the same year quoted by Davidsohn speak of payments to Rinaldo di Villamagna, a master of bombards, for making iron balls, casting cannon and obtaining gunpowder.

In an attack on Cividale in Friuli in Italy, near the Austrian border, "vasa" and "sclopi" (Italian schioppi) were used in 1331 according to Juliano[101]: Postmodum venerunt ad Pontem, et inciserunt dictum Pontem, ponentes vasa versùs Civitatem . . . et factâ die homines Civitatis, et adversarii eorum se hinc et inde fortiter balistabant, et extrinseci balistabant cum sclopo versùs Terram, et nihil nocuit Portâ apertâ existente.

A later account in the same chronicle, A.D. 1364, 12th September, gives the names of the attackers as "Dominus Franciscus de Cruspergo ac de Villalta" instead of "Johanne de Villalta," and "Zuculæ Dominorum de Spilimbergo" instead of "Domini Proëgna de Zucula," in the first account, and Romocki[102] says: "diejenigen, welche Cividale mit Geschütz angreifen, sind—was bisher ganz übersehen worden zu sein scheint—deutsche Ritter," i.e. the users of the guns were Germans, de Cruspergo being von Kreuzberg and de Spilimbergo being von Spangenberg. Rathgen[103] repeatedly refers to this use of guns by two "German knights," and says that many noble families in Friuli were of German (especially Swabian) origin, migration from Germany (Austria?) to Italy over the Isonzo being not unusual.[104] An account in Muratori[105] on the use of firearms (balistarum, sclopitarum, spingardarum, et aliorum) in Este Ferrara in 1334, was said by Rathgen to be a later addition to a work completed in 1368.

An English chamber account of 1333–4 specifies "j quart. de salpetre, 6d., iij lb. sulphuris viui, 2s. 6d., iv lb. sulphuris simplicis, 2s.," for making gunpowder.[106] A receipt in the Bibliotheque Nationale[107] dated 2nd July, 1338, specifies payment to Guillaume de Boulogne for "un pot de fer à traire garros à feu, quarant-huit garros ferrés et empanés en deux cassez, une livre de salpêtre et demi livre de souffre vif pour fare poudre pour traire les diz garros." The materials would make only 2 lb. of powder, and if this was used for forty-eight arrows, the charge for each would be only 20 g. or less than an ounce, and the "pot" (which would be like Milemete's gun) presumably weighed not more than 200 g. The use of brass and iron cannon on English warships in 1338[108] is improbable and the date should perhaps be 1379.[109]

Ducange[110] quotes Bartholomew de Drach, *Thesaurus guerrarum*, for accounts for payments "pour avoir poudres et autres choses nécessaires aux canons qui étoit devant Puy-Guillaume" (Puy-Guillem) of the date 1338 (new style 1339). In 1339, also "niewen engienen die men heet ribauds" (see p. 116) were made for a "maitre de ribaudequins" in Bruges.[111] For the defence of Cambrai against Edward III in 1339, ten cannon, five of iron and five of metal (copper), and powder for them, were made. Saltpetre and good and dry sulphur are specified for the powder, the charcoal (not mentioned) being available on the spot.   The low price of the guns, made for a wealthy man, and the quantity of powder made for them, show that they were quite small.[112]

In 1340 some kind of improved Greek fire was used at St. Omer; accounts specify 9 lb. of saltpetre, 7 lb. of sulphur, 6 lb. of colophony, 50 lb. of Greek pitch, amber powder and turpentine, a gallipot to make and a pipe (een pipe) to throw the fire.[113]   A chronicle written at Sluys in 1340 says the English ships used guns but the conditions were unfavourable.[114]   Guns (bombards) are repeatedly mentioned by Froissart as in use from 1340.[115] At Quesnoy in 1340 the besieged "desclignèrent canons et bombardes qui jetoient grands carreaux"; guns were used by the Scots in Stirling in 1341, and in the same year by the French in the defence of Hainebon.   In 1347 the King of England fortified a coastal castle with espringales, bombardes, etc., and furnished ships with bombards.   In 1356 guns, incendiary arrows and Greek fire were used by King John in the siege of the castle of Breteuil (canons jetant feu et grands carreaux; feu grégeois was thrown on a beffroi), and by the Prince of Wales in the siege of the castle of Romorantin (bombards et canons avant et à traire carreaux et feux gregéois dedans).   In the siege of Odruik (near Calais) in 1377 there were cannon which fired 140 shots, and in the same year in the siege of Ardres the French used "canons qui portoient carreaux de deux cent pesants."   In 1378 the English had 400 cannon before St. Malo.   These must have been quite small guns, but in the siege of Oudenarde in 1382 the army of Ghent:

> firent faire et ouvrer une bombarde merveilleusement grande, laquelle avait cinquante trois pouces de bec, et jetoit carreaux merveilleusement grands et gros et pesants : et quand cette bombarde descliquoit on l'ouoit par jour bien de cinq lieues loin et par nuit dix, et menoit si grands noise au décliquer que il sembloit que tous les diables d'enfer fussent au chemin.

Possibly the size of the mouth may not have been the diameter but the external circumference of the body of the gun, which Favé[116] says was often used in giving the calibre of shot.   There is, however, a gun still existing in Ghent with a calibre of 0·62 m. (15½ in.) which was used in a siege of Oudenarde in 1452 (p. 128).

It should be kept in mind that "bombard," although later used for large cannon, at first meant quite small guns; in 1364 a bombard is described as "una spanne lunghe" (a span long), a mere toy.[117]   Kervyn de Lettenhove[118] says the first large cannon or bombards, which appeared about 1360, were called "donderbussen," i.e. the German "thunder Büchsen" (cf. blunderbuss). A cannon weighing 130 lb. threw stone balls, the smaller guns of 36 lb. threw arrows, and those of 24 lb. threw lead bullets.

Petrarch,[119] in a dialogue usually dated 1344, mentions explosive bronze shells of small size fired from wooden guns:

> "I am surprised that you have not also those bronze acorns which are thrown with a jet of flame and a horrible noise of thunder (glandes æneas quæ flammis injectis horrisono tonitru jaciuntur).   It is not enough

to have the anger of an immortal God thundering in the vault of heaven but, oh cruel mixture of pride, man, sorry creature, must also have his thunder. Those thunders which Vergil thought to be inimitable, man, in his rage for destruction, has come to imitate. He throws them from an infernal machine of wood as they are thrown from the clouds (mitti solet, ligneo quidem, sed tartareo mittitur instrumento). Some attribute this invention to Archimedes. . . . This scourge was so rare that it was considered a prodigy, but now that minds are apt to invent the worst things, it is as common as other kinds of arms.''

Wooden cannon are mentioned by Biringuccio and are shown in illustrations reproduced by Jähns and Feldhaus. Even leather guns were used for a short time by Gustavus Adolphus (1625) and in Prussia (1627) and there is one in the Berlin Zeughaus; Feldhaus (1914) says they were really thin copper tubes encased in leather. Carman says the Paris Artillery Museum has a wooden gun from Cochin China, and the Turks are said to have used a cherry-wood gun. Wooden guns were used by Henry VIII in the siege of Boulogne in 1544; they are shown in wall paintings at Cowdray House as apparently 8 ft. long and 2 ft. across the bore, but they were probably dummies, a smaller metal barrel being attached to the wooden one. A description of 1588 mentions an English wooden gun ''mounted on a ship's carriage.'' Carman has a long account of leather guns. They were used in the English Civil War and made at Lambeth for Cromwell by Col. Weems, whose uncle Robert Scott offered the invention in 1628 to Gustavus Adolphus (who is wrongly given credit for the invention), but it was not accepted. An Austrian, Wurmbrandt, joined the Swedish army and tried out leather guns in 1627, but Scott's were apparently better. The specimens in England, Germany and Paris are Swedish.

A French document of 1345[120] mentions the making of ''new artillery,'' cast cannon, and large quantities of gunpowder at Cahors (canos fondus, balestas flageladas, carbo per asseiar los canos, trenta sieys liuras et meja de salpetra, vint cinq de solphre viu, que feren comprar à Tolosa per far polveryas et traire los canos). Favé, quoting original French archives from 1340 to 1381, showed that in this period guns increased in number and size; some were made from forged iron, others of bronze. The small ones threw arrows or small lead balls, and some were carried by hand; the larger ones threw stone balls or numbers of iron bullets (not solid shot); the stone balls sometimes weighed up to nearly 450 lb. Since these documents are not quoted by recent English or German writers, a brief summary of them may be given (the page references are to Favé).[121]

1340–41.   Lille, ''tuiau de tonnoire,'' a ''tonnoire'' weighing 100–125 lb., for throwing arrows (p. 76).

1342.   St. Omer (p. 77).

1345.   Toulouse, two iron cannon, 200 ''plumbatis,'' 8 lb. powder (= 25 g. per discharge) (p. 80).

1345.   Cahors, twenty-four cannon, 36 lb. saltpetre, 25 lb. sulphur
             (= 72 lb. powder, or 3 lb. per gun) (p. 80).

1346.   (September.) Tournay, a cannon (conoille) made by Peter of
             Bruges, a pewterer, throwing a *square* 2 lb. lead shot (p. 81), which
             passed through a wall and a house and then killed a man.[122]

1346.   Montauban, incendiaries, powder, "canons et du plomb" (p. 81).

1347.   Bioule, two espingales, fourteen dessaras, twenty-two canons (pp.
             81-3); Köhler, ref. 12, III, i, 229-31, has ten cannon, five iron
             and five metal (copper).

1347.   Lille, tounoile or tonnoile (last use of this word, perhaps the same
             as the Spanish truenos in Conde's translation of Arabic?) (p. 83).

1348.   Lille, cannon throwing arrows (p. 84).

1349.   Agen, cannon throwing small lead balls (p. 84).

1350.   Lille, arrows for cannon, saltpetre and sulphur (p. 87).

1356-7-8.   Laon, "large" cannon, probably small, as it cost the same as nine
             days' work of one man; 12 lb. of powder for forty-three cannon =
             70 g. per charge; twenty-four cannon on tripods (p. 88).

1357.   Chartres, cannon and espingalles (p. 91).

1358.   Ravenna (Italy), lead balls, less than 2 kg. each (p. 91).

1359-74.   Lille, two large cannon, St. Quentin, guns, saltpetre, etc. (pp.
             92-6).

1375.   Caen, iron cannon weighing 2300 lb. took forty-two days to make
             and required 2100 lb. iron and 200 lb. steel, costs; three iron
             cannon of 568 lb. each, to shoot bits of iron as well as balls; twenty-
             four cannon from metal bought from potters (probably bronze), a
             small iron cannon (24 lb.) to throw lead balls (pp. 97-101,
             detailed).[123]

1377.   Forged (iron?) cannon for Philip the Bold, Duke of Burgundy,
             made at Chalon in eighty-eight days, throwing (stone?) projectiles
             weighing 437 lb. (p. 101).

1381-2.   Lille, archives (pp. 102-5).

Accounts of Aachen of 1346[124] speak of wrought iron guns shooting arrows
(busa ferrea ad sagittandum tonitrum), saltpetre for making gunpowder, and
wood for use with the guns (pro salpetra ad sagittandum cum busa illa
7 schilde, ligneum opus ad busam). Busa is evidently the Latinised form of
Büchsen.

We have now reached the time of the battle of Crecy, fought on 26th
August, 1346. Edward III, who came to the throne in 1327, is reported
to have used guns before Crecy, viz. in the invasion of Scotland (1327), in
the siege of Cambrai (1338), at Tournay (1339) and (with wild fire) at
Quesnoy in 1339-40.[125] Archdeacon John Barbour of Aberdeen (c. 1320-95),
who visited Oxford and Paris and filled government posts in Scotland, says

in his poem "Brus" (Bruce) completed in 1375, that the engineer Crab in his defence of Berwick in 1319 had no guns ("gynis for crakys had he nane"), but that guns were used for the first time in Scotland by the English just after the accession of Edward III, when, in 1327, two novelties were seen, plumes for helmets and "crakys war of were that thai befor herd neuir eir" (the other was crakys of war which they had never before heard).[126]   Brackenbury[127] said Barbour was speaking from hearsay; no document in the Public Record Office which he consulted contained any mention of the event, which is probably fictitious.

There are four main sources for the use of guns at Crecy:

(i) A fourteenth-century manuscript of *Les Grandes Chroniques de France dit de St. Denis*[128] says the English fired three guns, whereupon the Genoese archers in the front ranks turned and fled: "lequels Anglois giettèrent trois canons, dont il avint que les Genevois arbalestriers, qui estoient au premier front, tournèront les dos, et laissèrent à traire."

(ii) The *Abrégè* of Froissart[129] says the English fired two or three guns (descliquier deux ou trois bonbardiaulx), and the account in Froissart's *Chronicle*[130] says: "Then (the Genoese) took their crossbows and artillery [= bows and arrows] to prepare to begin the battle . . . when their officers had got them into shooting formation they raised a great shouting (ils commenchièrent à huer et à juper moult hault).   But the English kept quite still and loosed off some cannon they had with them, to frighten the Genoese (li Englès demuronèrent tout quoy et descliquièrent aucuns kanons qu'il avient en le bataille, pour esbahir les Génevois)."

The earlier French or English editions of Froissart, who began his chronicle about 1360 and thus (in spite of what Lacabane says to the contrary) was reasonably near the events, are published from incomplete manuscripts, designedly expurgated by Froissart in his later years so as to avoid mentioning the use of unknightly methods of warfare by the English, a strong bias in favour of whom came over him in his later life.   They do not, for example, mention the use of guns at Crecy, so that Gibbon[131] said "the authority of John Villani [see (iv)] must be weighed against the silence of Froissart." The complete manuscript of Froissart, discovered at Amiens only in 1838 by Rigollot,[132] contains the statement given, as does the *Abrégè*, made by Froissart and genuine according to Kervyn de Lettenhove[133] ("dont l'authenticité nous semble incontestable").

The two French accounts (i) and (ii) are regarded as acceptable by French historians.   Lacabane[134] says they "ne laissent aucune doute à ce sujet," and Lalanne[135] regards the use of guns as certain.   They also rely on the contemporary French records of the making of guns which have been mentioned above.   Some Italian accounts may now be considered.

(iii) An anonymous Italian chronicle, *Istorie Pistolesi*,[136] said to have been written about the same time as Villani's (before 1348), says that in the crisis of the battle "the English knights, taking with them the Black Prince, a body of wild Welshmen (uomini salvatichi) and many bombards (con molte bombarde), advanced to meet the French army . . . they fired all the bombards at once and then the French began to flee (faccendo scoccare tutte le bombarde a uno tratto, sicchè gli Francheschi si cominciarono a mettere in fuga)." "Many bombards" is doubtless an exaggeration, but the account says that guns were used.

(iv) Giovanni Villani, a Florentine merchant, statesman and chronicler, died in 1348, two years after Crecy. His chronicle was continued to 1364 by his brother Matteo and his nephew Fillipo, but the part concerning Crecy was probably written by Giovanni and is a genuine document. In a long account of the battle, he says[137]:

> E ordinò il re d'Inghilterra i suoi arcieri, che n'havea gran quantità su per le carra, e tali di sotto, e con bombarde, che saettavano pallottole di ferro con fuoco, per impaurire e disetare i cavalli de' Francheschi. . . . Sanza i colpi delle bombarde che facieno si grande tremuto [tumulto] e romore, che parea che Iddio tonasse, con grande uccisione di gente e sfondamento di cavalli . . . in sulle carrette fediti di saette dagli arceri e dalle bombarde, onde molti ne furono fediti e morti . . . e de'fediti delle bombarde e saette, che non v'hebbe cavallo de' Francheschi che non fosso fedito, e innumerabili morti.

Muratori[138] accepted Villani's account, saying: Certamente non pochi anni avanti, cioè nell'ânno 1346 nella sanguinosa battaglia di Crecè in Francia, gl'Inglesi si servirono di Bombarde, che saetavano pallottole de ferro con fuoco.

In the main, the battle of Crecy, in which 10,000 English troops faced at least 30,000 in the French army, was decided by the English long-bow, a most deadly weapon shooting arrows which could pierce heavy defensive armour. It was the first time the French had encountered its use, and the French cross-bow was comparatively ineffective. The mercenary Genoese archers were the first to turn and run. It is probable that the noise if not the shot of the few English guns started their flight and also startled the horses of the mounted men. Added to this, the French tactics were bad. The English army moved rapidly for some distance before the battle of Crecy and had ample transport, a large waggon-park being established on the field,[139] so that there would have been no difficulty in moving cannon of moderate size if these had been used. The army also had "engineers" in 1347, and although these were probably gunners (guns being "engines")[140] there were evidently specialist troops apart from bowmen and infantry available.

Temler[141] argued that the passage in Villani was interpolated and that guns were not used at Crecy. He was unacquainted with the full text of Froissart and with the other records and information about Crecy summarised above, and his account is of no modern value. In spite of this, it has been quoted by one or two modern German historians[142] to "prove" that guns were not used at Crecy, the statement that they were being a "legend" (Hassenstein) or even a "lie" (Rathgen). The use of at least three small guns is accepted by most unprejudiced English, French, Danish and German historians.[143]

EARLY ENGLISH GUNS

Rathgen,[144] after mentioning records for Florence (1326), Metz (1324), Rouen (1338), by Bartholomew de Drach (1338—it should be 1339), Cambrai (1340), Lille (1341–2), Rihôult (1342), Artois (1343), Toulouse (1343), Cahors (1345), Montauban (1346), Tournay (1346), Crecy (1346), Bioule (1347), Lille (1347–8) and Metz (1348), all preceding his Frankfurt (1348) record, dismissed Florence as a forgery, assumed that the references for 1338 were to Greek fire, that the lead balls at Toulouse (1345), well away from German influence, came from England, that Aachen (1346) was a mere isolated incident as compared with Frankfurt, that the use of guns at Crecy was a "lie" and all the early guns described by Favé were made in Germany or by Germans. If firearms appear in any other place or at any other time they came from Germany.

"Die Wiege der Pulverwaffe hat am Ober- oder Mittelrheine gestanden. Vom Niederrhein aus hat sie weiteste Strecken zurückgelegt. Hier und in dem mit ihrem in Kulturgemeinschaft so eng gebundenen Flandern fand die Pulverwaffe besondere Pflege. Von da kam sie nach Frankreich und England, von da nach Toulouse; dort bildete sich für sie ein neuer Mittelpunkt, von dem aus sie den Weg nach Aragonien, nach Spanien, genommen haben wird."[145]

"Das was Frankfurt 1348 nachgewiesenermassen geleistet hat, ist in keinem anderen Lande vor oder zu dieser Zeit auch nur annähnerd erreicht worden. . . . Dies bezeugt, dass lange Zeit vor 1348 in Deutschland für das Ankommen der neuen Waffe ein treibender Ausgangs- und Mittelpunkt vorhanden war."

Frankfurt shows the "ureigenen deutschen Ursprung dieser Waffen."[146] From Frankfurt, artillery went to Metz, Trier, Luxembourg, Burgundian Flanders, and south to German Switzerland, until in 1450 it was everywhere.

The reader will have noticed that we have not yet reached a period when trustworthy German accounts of payments for guns can be quoted, but the English National Records show that considerable quantities of gunpowder were made in England for the use of Edward III before the army left the country, and also while the king was encamped before Calais (where he proceeded after Crecy). Payments for saltpetre (912 lb.) and sulphur (886 lb.) for the king's guns are mentioned in a writ of 10th May, 1346, i.e. three months before the battle of Crecy. Charcoal, being cheap, is not specified. Further entries show that other supplies of powder followed the army to France.[147]

A document giving the expenses of Robert de Mildenhale, keeper of the king's wardrobe, from 1344 to 1353, includes ribaldos and guns (gunnis), ten guns with handles (x. gunnis cum telar), two large (unde ij. gross), and powder for the guns (pulvere pro dictis gunnis), saltpetre (salpetre) and sulphur (sulphure vivo), charcoal (for use by smiths, and also, no doubt, for powder), small and large lead balls (pellot plumbi), 100 small engines called ribaldequin for use in France (centum minutis ingeniis vocatis Ribald, pro passagio Regis versus Normanniam), with bows, arrows, iron keys, nine coffers cum armaturis, two paribus suffocalium, ten garbis ascaris, etc., etc. On 10th May, 1346, no less than 912 lb. of saltpetre and 886 lb. of sulphur for the king's guns ("pro gunnis suis") were supplied to Thomas of Rolleston, a subordinate of Mildenhale's, and further supplies were ordered three days later from William of Stanes, "apotecaire regis" of London.[148] There is no evidence in the accounts that ribaudequin were actually sent to Crecy, but some were shipped in February, 1347, and hence were used in the siege of Calais.

In 1345 there were provided for the French campaign "gunnis" which shot lead balls and arrows (quarrels) and 100 "machines called ribaldis," i.e. clustered guns or ribaudequins (see p. 116). In 1346 ten "guns cum telariis" are mentioned, which were probably short bronze pieces with wooden handles (see ref. 163), since the repairing of the wooden shafts by helving, by a joiner, is mentioned in 1373. Hand-guns and powder (gunnarum cum pulvere) are mentioned in 1346 and 1355.[149] Accounts of this time repeatedly mention "telars" in connection with cross-bows and guns and, as stated, they were probably wooden stocks to which small guns were attached,[150] although Tout thought they were carts.[151]

For the siege of Calais which lasted from September, 1346, to August, 1347, ten guns (two heavy, "grosse"), lead balls and gunpowder were sent from London, together with ribaudequins; all the saltpetre and sulphur which could be found were ordered and 750 lb. of saltpetre and 310 lb. of sulphur were delivered in November, 1346, payment being made in September, 1347, for 2021 lb. of saltpetre and 466 lb. of sulphur. As before, the cheap charcoal is not mentioned. The English "bombards" blocked the entrance to Calais harbour and others helped to prevent the relieving army from approaching on the landward side. Between 1345–6 and 1360 there is only one reference in the privy wardrobe records, in accounts of 1353–60, for the purchase of four "gunnis de cupro," a copper mortar and an iron pestle (for making powder).[152]

In 1360 there were in the Tower of London four guns, 16 of the 16½ lb. of powder mentioned in the accounts quoted above, and the pestle and mortar. The keeper bought five guns and a small gun which the king gave to Lionel of Antwerp when he went to Ireland (probably the first gun ever seen there), but in 1365 this small gun was the only one left in the Tower. In 1365 two great guns and nine small ones went from the Tower to Queenborough Castle in Sheppey; in 1371 six guns, saltpetre and sulphur went to Dover Castle. In 1369–70 there were three great copper or brass guns and one of iron, and fifteen other guns at Calais, and a great gun and three "pots" were sent in 1372. In the expedition of 1372, twenty-nine iron guns, 1050 lb. of saltpetre and an adequate supply of lead were sent oversea.[153] In 1378 guns and material for powder were sent to Richard II at Brest.[154]

As Tout says,[153] after 1380 the accounts for saltpetre, sulphur and guns become very numerous and "firearms are so commonly mentioned that it becomes unprofitable to collect the constant references made to them."

A large gun and fifty-one smaller guns were sold to the town of Norwich by a London maker in 1384. A brass (bronze) cannon was made at Carlisle in 1385, and in 1386 sixty cannon were made for Calais, fifty being of 300 lb. weight.[155] The English bombarded Ypres with cannon in 1383.[156] From 1382–3 cannon founding became a considerable English industry and in 1396 there were nearly 4000 lb. of gunpowder and nearly 600 lb. of saltpetre in the Tower, and Hatton, the keeper, made experiments with hand-guns.[157] All persons mentioned in the records as having anything to do with the making of cannon or ammunition were men with apparently English names who worked in London.[158] In January, 1386, the Tower of London delivered to Berwick "three small cannon of brass called handgonnes" (whereas Continental writers say hand-guns were not used till 1405–10), and they were probably single barrels of ribaudequins mounted on wooden stocks for hand use.[159]

The early cast cannon were of copper or bronze and the name bronzina in Mediæval Latin meant cannon (tormentum bellicum).[160] The name cuprum (copper) was used early in England, latten (brass) appearing in 1372, and aresme (bronze) in 1397. Although "brass" guns are often mentioned, the special alloy mostly used was "gun-metal," an alloy of 9 copper and 1 tin, i.e. bronze. Bell-metal is richer in tin, 4 to 5 copper and 1 tin, and English guns were often made by bell casters, a fifteenth-century trade mark of one at Bury St. Edmunds showing a bell, a cannon and a cannon ball.[161] The Hussites in the war of 1427 melted down church bells to cast into cannon.[162]

Early cannon were almost all breech-loaders, consisting of two parts, (i) a barrel, strengthened by a number of iron rings, and (ii) a chamber or short cylinder, in which the charge of gunpowder was put. Both parts were made from iron bars arranged like the staves of a barrel and bound together by a number of iron hoops or rings. Sometimes the chamber was of cast iron. The chamber was forced against the rear end of the barrel and fastened by a stirrup or by a chock and an iron wedge driven in between this and the back of the channel of the gun-mounting in which the piece lay, or against some other retaining device. Iron cannon were made in England at least as early as 1369–71 and in considerable numbers in 1372. Double-barrelled cannon (duplices) were used in 1401, and groups of three were also in use.

The mortars, called "bombards," were short and fat muzzle-loading pieces with a core of cast iron and outer bands of wrought iron, shooting stone or iron balls up to 18 in. diameter and 225 lb. weight, but with short range and inaccurate aim. Various types of hand-guns throwing lead balls or arrows were in use during the Middle Ages, mostly miniature cannon mounted on long wooden sticks; they were used by foot soldiers or mounted men.[163] Carman mentions a hand-gun from Tannenburg, Hesse, made before 1399, which is $12\frac{1}{2}$ in. long, $\frac{7}{10}$th in. bore, weight $2\frac{3}{4}$ lb., intended for mounting on a long haft, and says a Bologna record of 1397 mentions "viii sclopes de ferro de quibus sunt tres a manibus." The fourteenth-century Swedish hand-guns were about 8 in. long and mostly octagonal.

Brackenbury[164] concluded that really large cannon were used by the English from 1378, and since guns are mentioned by Petrarch in 1344 (see ref. 119), and accounts in Muratori refer to their use in Ravenna in 1358 and Pisa in 1362, which is earlier than by the English, French and perhaps Germans (although he says the German accounts available to him were worthless), Brackenbury thought that large cannon perhaps originated in Italy. He says the use of guns by the Black Prince at Najaro in Spain in 1367 is mentioned by a fourteenth-century author. The *Chronicle* in verse of Bertrand du Guesclin (*c.* 1315–80), who was captured by the English at Najaro, says that in that battle (1367) the Black Prince had "bombardes, ars à tour, espees et espois" as well as "chars et charettes."[164a] In 1378, Richard II ordered "canons," 300 lb. of saltpetre, 100 lb. of sulphur and a cask of willow charcoal.

Gun casting was carried out in a similar way to bell founding, a clay core or "heart," strengthened with an iron bar, being surrounded by an outer mould corresponding with the cope of the bell and also bound with iron. The piece was hauled from the pit and finished internally by boring. The English, Dutch, Flemish, German, Italian and Spanish gun-casters had a very good reputation, and there is no evidence for the great superiority of German guns.[165] The guns often burst on testing and in use, so that the gunner stood ten paces behind and to the left of the gun on firing, this being the safest place to escape the flying pieces of the metal.[166] King James II of Scotland was killed by a bursting cannon in the siege of Roxburgh Castle in 1460.

Temler[167] quotes Thomas de Walshingham (*c.* 1440) as saying that in 1386 "gunnæ . . . cum . . . pulveris" were very expensive. Cannon were relatively cheap in the fourteenth century in England; the earliest recorded price, in 1355, is 13*s.* 4*d.* each, a large cross-bow costing 66*s.* 8*d.* In 1373 a brass gun and three "pots" cost 20*s.* Under Richard II a small iron cannon cost 20*s.* or 26*s.* 8*d.*, and "a great copper cannon" a little more than £3. The average price was 3½*d.* to 4*d.* per lb. At Cambrai in 1338 each of the ten cannon cost about 11*s.* to 12*s.* of English money. "Great cannon" might weigh only 150 lb., exceptionally 665–737 lb. Gunpowder, in comparison with guns, was very expensive, saltpetre in England costing 18*d.* a lb. in 1347, sulphur later 6*d.* a lb., and in 1387 powdered willow charcoal cost 18*d.* a barrel. Contrary to the usual statements of Continental authors, there is no evidence that gunpowder was imported into fourteenth-century England.[168] The following prices in £ per lb. are given for the year 1375 in England[169]: iron 2½, steel 4, lead 4, "gun-metal" 8, gunpowder 50, the corresponding prices in pence in 1865 being: 1½, 9, 2, 12, 7.

The first English iron guns are traditionally said to have been made at Buxted, ten miles from Lewes in the Sussex Weald, and there was an old mortar on Eridge Green, Sussex, with a chamber of cast iron and the other part wrought iron which was still used at the end of the eighteenth century to throw a shell over a distance of a mile for amusement.[170]   Gardner[171] illustrates a fourteenth-century cast iron grave slab at Burwash cast at Buxted and then the oldest cast iron object known, much older than the casting of iron in Siegen in Prussia in the fifteenth century which had been given[172] as the beginning of the European cast iron industry.   The works at Buxted operated for nearly three centuries, mostly in making ordnance, grave slabs and fire-backs.   The oldest cannon in the Woolwich Rotunda is also a Sussex breech-loading mortar throwing a 160 lb. stone ball, from Battle Abbey.

King Henry VIII used wrought iron and bronze guns mounted alternately on his ships[172a] and it was an innovation for Cromwell to introduce all-bronze artillery.   Bronze guns in 1559 cost £74 per ton.   A large gun bought from Spain for £18 10s. was (from its price) probably cast iron, and it probably induced Henry VIII to order from Ralphe Hogge of Buxted cast iron cannon at £10 per ton.   Henry's gun founders were Baude, a Frenchman, Arcanus de Cesena, an Italian, van Cullen, a German, and Englishmen named Owen and Johnson.   Magnificent bronze guns made by Baude, Arcanus de Cesena and the Owens are in the Tower and Woolwich.   Holinshead[173] says the first cast iron guns were made at Buxted (Bucker-steed) by Rafe Hoge and Peter Bawd (he calls both of them French) in 1543, and Stow[174] reports that Peter Bawde, a Frenchman, and another alien Peter van Collen, made in England in 1543 cast iron mortars of 11–19 in. calibre, with cast iron shells. Iron guns were used on Elizabethan warships and in 1587 Queen Elizabeth granted a licence to export 320 tons of cast iron ordnance from Sussex to Holland and Zealand.[175]   Large guns were used by Henry V in the bombardment of Harfleur in 1416; they were very effective since they shot from the surrounding heights and the lack of elevation of large guns at that time was not felt.[176]   Large English iron guns or mortars of 1400–23 were captured or abandoned on Mont St. Michel and are still there.   One is 3·65 m. long, calibre 48 cm., weight 135 kg., the other is 3·53 m. long.   They had the unusually long range of 3000 m., 1000–2000 m. being the ordinary range for contemporary German guns.[177]   Rathgen says this long range could have been achieved by an elevation of a little under 45 degrees, although mortars were little used in the fourteenth and fifteenth centuries.   The use of a heavy charge of powder is just possible, but this was not used in Germany until 1443.

At the battle of Agincourt (1415) the English had several cannon, with names,[177a] and a participant in the battle says the French there had "grant nombre de chariots & charettes, canons, & serpentines, & autres habillemens de guerre."[177b]   Edward IV in 1475 had a great brass cannon and two great "pot-guns" of brass cast for the invasion of France.[177c]

An English illustrated manuscript (BM Cotton Julius E IV) called the *Warwick Pageant*, and supposed to have been written about 1485 by John Rous of Warwick (c. 1411 or 1445–91),[177d] represents Richard, Earl of Warwick (1382–1439), giving instructions to a gunner in the battle of Caen (1417). It shows a heavy breech-loading cannon, about 6 ft. long, on a low wooden support.   It has a decorative tapering piece behind the breech which suggests that it was cast, and the barrel has four bands for strengthening. The muzzle has an upright projection cast on, which seems to have served for a sight. The breech is open at the top of the cylindrical barrel and the gunner, who is in armour but has his hands free, kneels with one knee on the side of the breech opening. He is holding a cylindrical container of the charge, which has a handle and is evidently to be put into the breech chamber. How the chamber is to be closed is not clear. The gun is to be fired into a town, where soldiers with cross-bows are shown, but no guns. Below is a picture of a boat, which is armed with a somewhat smaller cannon of the same type, projecting over the bow on the starboard side, with a gunner in armour behind it.

Mr. Carman informs me that guns with removable chambers enclosed at the side or rear were called "fowlers" or "port-pieces" in Elizabethan times. The port-piece had a long tail or handle at the end and a support to the trunnions (there are no trunnions on the Warwick gun) permitted the weapon to be placed in the wall of a ship. The Dutch East India Company used such guns in the seventeenth century, and there is a specimen in the crypt of the Royal United Services Museum in Whitehall. Favé says the guns of Charles the Bold had trunnions, and Carman that a gun from Rhodes, made in 1478 for Louis XI, had thick trunnions cast on.[177e]

EARLY FRENCH AND BURGUNDIAN GUNS

Some references to early French guns were assembled in an earlier note (p. 104). The records of Cahors for 1345[120] refer to the casting of twenty-four cannon, which Köhler thought were cast iron and Rathgen bronze. A reference to cast iron cannon at Lille in 1412 is a mistake, although they were known from 1414.[178] In 1376 a French gunmaker (magistri gallicanis) at Cologne made a gun (de tonitruis factis) costing 308 marks 4 schillings, which must, therefore, have been very large.[179]

Brackenbury,[164] who says large cannon appeared in France in 1374, refers to the detailed account[180] of the manufacture of one at Caen in 1375, and Froissart's description[115] of the large cannon used at the siege of Odruik in 1377, throwing a 200 lb. shot, and one throwing 450-lb. stone balls (corresponding with a calibre diameter of over 21 in.), made in 1377 for the Duke of Burgundy at Chalon by Jacques of Paris and Jacques and Roland of Majorca.[181] The Corporation of Dijon, in 1394, lent the Duke of Burgundy two copper guns and a large cannon.

The artillery of the Dukes of Burgundy is known from detailed records and extant guns.[182] The kingdom of Burgundy, or as it was usually called the kingdom of Arles, included Provence, Dauphiné, the valley of the Rhône, Savoy, part of Western Switzerland and the Franche-Comté, with no geographic, racial or linguistic unity. It was throughout the Middle Ages nominally attached to the German and imperial crowns, who were unable to make their claim to its government a reality, so that it came increasingly under French influence, its feudal states falling one by one into French hands; Provence (with Marseilles) in 1246, Lyons in 1313, Dauphiné in 1343, and Franche-Comté later became part of the dominions of the four Dukes of Burgundy of the house of Valois (1363–1477) : Philip the Bold (1363–1404), son of John the Good, King of France, John the Fearless (1404–19), Philip the Good (1419–67) and Charles the Bold (1467–77). Philip the Bold and John the Good were captured in the battle of Poitiers and spent the years 1357–60 as prisoners in England; Philip and his brother Charles V, King of France, were allies against the English and retook Calais in 1377.

In 1362 there is mention of "deux canons à gitter arroz achetez à Troyes de Jaquemart le serrurier, trois florins,"[183] which would be small iron guns. In 1368–9 two cannon were made by Perrenin du Pont (a good French name), with $5\frac{1}{2}$ lb. of powder, forty arrows and twelve plombées d'estain (tin), which were bullets (later references are to plombées de fer, de pierre, etc.). In 1376 a foundry for heavy guns was established in Chalon under the gunsmiths Jacques of Paris and Jacques and Roland from Majorca. An iron cannon of 384 lb., 11 in. diameter, for a 60-lb. stone ball with a charge of $1\frac{1}{2}$ lb. of powder, was made. In 1377 five larger cannon were finished, of 38, 36, 35, 24 and 20 cm. calibre, for stone balls of 130, 100, 90, 30 and 20 lb. These had a wider top (poz) welded to the gun chamber.

Philip then had fresh workers in Chalon and in October, 1377, ordered a 450-lb. gun, 58 cm. calibre, which was finished in January, 1378. In the period 1376–8, therefore, Philip had a formidable artillery. In 1390 a large copper gun was cast at Dijon, which burst in 1409 in the siege of Velloxon, outside Calais. Then bronze cannon were made. Froissart[116] refers to the use of guns outside Calais against the English. He mentions a 200-lb. shot; the largest mentioned above was 130 lb., so that this may have been for a still larger gun.

Christine of Pisa (1363–1431), daughter of the Venetian councillor and astrologer at the Court of Charles V, and educated in Paris,[184] wrote a well-informed *Livre des faicts d'armes et de chevalerie*, printed in 1488, in which she gives details of the large Burgundian and French cannon, which were given women's names.[185] Garite (Margaret) threw a 400–500-lb. stone ball, but "the best of all" (perhaps because it had a heavier charge) was Montfort, which threw a 300-lb. ball; these were Burgundian. French guns were all of iron except one copper gun for a 100-lb. shot, called "artique" (i.e. from Artois). They were used on the field in battles as well as in sieges. Christine says Greek fire was banned with excommunication by the Church, which apparently had no qualms about these enormous guns. In 1394 there is reference to a "gros canon" with "coing de fer" (breech strengthening ?) in Burgundian records. John the Fearless (1404–19) had more guns (bombardes) made for the siege of Velloxon (1409–10), outside Calais. These threw stone balls of 600, 320, etc., lb., two at least were of cast bronze, muzzle-loaders with chambers, and one of iron taking a charge of 25 lb. of powder. The wrought iron guns were often repaired on the field.[186]

The powder for the guns was brought ready-made from Paris and the guns fired an average of eight shots a day. The Burgundian name for a gun was, until 1400, "canon"; the Italian name "bombarde" was not used till 1409; the name "boite" was used in Flanders.[187]

The recoil of these large guns was formidable and elaborate structures were built to absorb it; the Dijon 1700-lb. copper bombard was controlled by eight men and an "engine." It was fired with a red-hot iron. Philip the Good, in the siege of Calais in 1436, had an army of 30,000, the Dutch fleet, iron and copper guns from Burgundy, Flanders, Zealand, Holland and Picardy, mangonels, a covered way, and perhaps Greek fire. The Dutch fleet soon made off, but the bombardment was pressed with vigour. At last the outer wall was demolished, when it was seen that the English had built another wall behind it. Sappers were then called in, since the guns were not getting on well enough, and they dropped the wall by a mine.[188]

A French bombard called "Bourgogne," made in 1436, was in two pieces screwed together which were transported separately; it threw a stone ball of 250–300 kg., 22 in. diameter. The upper part was of copper but went astray, so that it was replaced by one of forged iron.[189] The Burgundian "veuglaires" (see p. 121) were smaller than the large bombards and each had two or three separate chambers, ready charged with powder and a wooden plug. This permitted more rapid firing. They were thus breech-loaders, and one account says "6 gros veuglaires, à chascun deux chambres . . . et tous à mettre pierre par derrière." They were of cast bronze (alliage de cuivre) or more often forged iron. In 1364 veuglaires with three chambers were used on board ship.[190] Something is said later (p. 128) about the large Burgundian pieces; the iron (1421–30) and cast bronze (1474) mortars now in Basel (the bronze one made by Jehan of Malines) are Burgundian, as is perhaps the large iron gun, Dulle Grete (after 1430), in Ghent. Large iron bombards were made in 1446.[191]

The development of artillery by the Dukes of Burgundy is very impressive and an attempt was made by Rathgen to show that gunmakers and gunners from Germany took their knowledge to "all countries,"[192] including Burgundian territories, which were throughout under German influence (dauernd unter deutschem Einflüsse).[193] Artillery reached France through Flanders, which was linked with the Lower Rhine country to form a Germanised Kulturgebiet. German gunners went to Germanised Burgundy as they did to German Besançon, Basel and Mömpelgard. The Burgundians acquired Flanders by marriage, German settlers there adopted small German guns which they used on carts (Karren) against the English (who, it is true, had plenty of guns of their own), and thus spread the knowledge of artillery to France.[194] This account is mostly in the air, and when Rathgen quotes a French author[195] in support of it, he omits to mention that this authority lays more stress on influence from England than from Germany.

Charles the Bold, the last Valois Duke of Burgundy, was defeated in two battles by the Swiss in 1475, in the second of which, at Murten (Morat), he was killed. It is said that he lost 419 guns, some of which are still in Basle, and the fleeing Burgundians took to the trees, from which they were shot down like birds.

SOME NAMES OF GUNS

The name "artillery" is perhaps derived from the Latin ars, artis. In France and England it was first used for bows and arrows, as in Froissart, who distinguishes them from kanons (guns). It also sometimes meant a projectile. The meaning cannon first occurs in *Piers Plowman* (begun 1362, finally revised 1390), and all three meanings, machine, projectile and cannon, occur in poems of Chaucer of 1375–1400. Gun (gonne, pl. gonnes) in English at first meant a ballista; it may be a contraction of mangonel or come from "engine" or canna.[196] "Cannon" is derived from the Greek κάννη, Latin canna, a tube, through the Mediæval Latin cannonus, Italian cannone, a large tube. The English name "ordnance" for gun is derived by mistake from κανών, ordinance, instead of κάννη.[197]

The name "cannon" was more used in France, the common English name being "guns." Walshingham says: "et illi figere vel locare gunnas suas, quas Galli canones vocant."[198] Names used in France are canon (1339), tonnoire (1340–1), tonnoille (1346); the Burgundian guns (1362–1450) are qanons; in Holland the names dunrebussen (1348), donrebusse (1353) are used (from the German Donner-büchse). In 1345 the English name canon was used in the South of France, and in 1348 in the North; Edward III had landed in Flanders in 1340 after destroying the French fleet. The name bombarde is used in France in 1381.[199] The German name Büchsen sometimes appears as puchssen (Munich MS. 197) or Büchse, in Latin pixis (Kyeser, *Bellefortis*).[200]

One or two curious names for types of guns may be mentioned. The Latin cerbotana (derived from cerbus, bottle, and late Greek βοτάνη, gunpowder) was a cylindrical gun with a long tail-piece.[201] Ribaudequin or ribaude is a name often used by Froissart[202] for a number of small bombards mounted on a cart (the name may originally have meant a cart); other names are barricade, orgue, orgelgeschütz and totenorgel. Froissart says: "Iceulx ribaudequins sont trois ou quatre petits canons rangées de front sur haultes charettes," and they were used as quick-firing armaments. They were arranged in groups of twelve on each of the four sides of a square wooden log which could be turned when one set of twelve were fired, and there were three logs, carrying 144 ribaudequins in all, on each car.

The nomenclature used by recent German writers, e.g. Lotbüchse, Steinbüchse, Klotzbüchse, is very confusing.[203] Lotbüchse and Klotzbüchse had cylindrical barrels of uniform bore, Steinbüchse had a narrower powder barrel and a wider part for the stone ball. Klotzbüchsen fired iron balls (Klötze) and Lotbüchsen lead balls (Dutch lood = lead). Grabaldus[204] says:

> Nunc vero tormenta ænea, saxa rotunda immense ponderis, vi pulveris factitii et ignis emittentia cum sonitu, non ineleganter Bombardas appellamus. Tormentum generale vocabulum est omnium machinarum, saxa,

tela et id genus varia jaculantium, veluti est balista, bombarda, spingardæ, catapulta, scorpiones, arcubalistæ quibus spicula et veruta emittuntur.

The first clear description of a Steinbüchse is in an Italian work, the *Chronicon Tarvisinum* (1376) of Andrea Redusius de Quero, who was in Venetian service in 1427[205]:

bombardarum, quæ ante in Italia numquam visæ nec auditæ fuerant, quas Veneti mirabiliter fabricari fecerunt. Est enim bombarda instrumentum ferreum fortissimum cum trumba anteriore lata, in qua lapis rotundus ad formam trumbæ imponitur habens cannonem à parte posteriori secum conjugentem longum bis tanto quanto trumba, sed exiliorem, in quo imponitur pulvis niger artificiatus cum salm[=n]itris & sulphure, & ex carbonibus salicis per foramen cannonis prædicti versùs buccam. Et obtuso foramine illo cum concono uno ligneo intra calcato & lapide rotundo prædictæ buccæ imposito & assenato, ignis immittitur per foramen minus cannonis, & vi pulveris accensi magno cum impetu lapis emittitur.

The front part was the trumba, containing the stone ball, the middle part was the cannone, and the hinder part, containing the powder, was the bucca (buxa). The wood block over the powder (with a space empty of powder below it) was called a cheville or tampon in French, Klotz in German and coccone in Italian. The stone ball was put into the wider piece and fixed in place by wooden wedges, the clearance being packed with hay or (better) waxed cloth. The powder chamber was about three-fifths filled with powder, then one-fifth empty, then a wooden tamping hammered in to fill the remaining one-fifth. It was this tamping, thrown forward, which projected the ball. In some pictures the ball is shown partly projecting from the muzzle.[206] The German name "Böller" for a mortar is said[207] to be derived from πετροβόλος, stone-thrower, since it at first consisted of an upright metal pot with a flat stone laid across the mouth, but this seems to be a reminiscence of the legend of Berthold (p. 93) rather than a genuine etymology. Köhler[208] says true mortars were first used in 1430–60. The name is used by Sebastian Munster[209] for a "ball thrower."

The earliest cannon balls were of stone, which were used for a very long time since the early cannon were not strong enough to shoot heavy metal balls. The metal shot mentioned from 1326[100] must have been case-balls like those described by Doukas,[210] "as big as a hazel-nut." Salzman[211] says iron balls were not used for larger guns before the end of the fifteenth century; Biringuccio[212] speaks as if they were something new in his time, saying that the "fine and horrible invention" of iron cannon balls was first seen in Italy in 1495, when Charles VII of France used them in Naples. Small iron balls were made in Italy in the fourteenth century, but were rare elsewhere. Rathgen[213] says the first record of cast iron cannon balls is that of their use by the Duke of Burgundy in 1431; iron balls were made for use in Dijon in 1440. At Naumburg, wrought iron cannon balls were used in 1446 and cast iron balls in 1449. Lead shot, as has been seen above, is mentioned quite early. Gardner[125] says cast iron balls were very probably made at Buxted in Sussex much earlier than those made by Ulrich Behaim of Memminghem in 1388, and the usual date 1514 for them in England is much too late. Solid iron shot up to 65 lb. weight were found in the wreck of the ship *Mary Rose* sunk in 1545.

Hakenbüchse ("hook guns") of 1418 are first mentioned in the *Museribok* of Brunswick. They probably originated in Nürnberg and were used in the Hussite wars. Jacobs[214] thought the name implied that they were fired with red-hot iron hooks but Rathgen[215] assumed that it referred to metal hooks on the gun intended to carry ropes and take the recoil. They were quite small in 1449, the Hussite ones being of 19 mm. calibre weighing 22 lb.; they shot lead and stone balls and also arrows. Hunting (Jagd) pieces first appeared in Göttingen in 1425.[216] Couleuvrines or hand-guns appear in Burgundian records in 1420 and crapaudeau in Flanders and Burgundy (short breech-loaders? Vögler) in 1433.[217] Rathgen[218] says the Hussite guns of 1420 were like the Chinese gun of 1421 (ch. VI), which he thought was made in Europe; they were good. The total length was 420 mm. composed of a 280-mm. tube and a 140-mm. sheath (Tülle) for fixing on a stick. It is said[219] that John Ziska, leader of the Hussites in Bohemia (1419 f.), was the first European commander to make full use of artillery in the field and to see the value of mobile barricades of waggons (later effectively used in India by Bābur). A manuscript (Lat. 197, Hofbibliothek Munich)[220] of 1420 by a participant in the Hussite wars shows ships with paddles, diving apparatus, windmills, boring and grinding machinery, etc. In 1422 there were five large and forty-six other guns in use.

GUNS IN ITALY

After the earliest references for Florence in 1326[100] and Friuli in 1331,[101] there are many fourteenth and fifteenth century records for the use of guns in Italy. These include tromba marina, tuba marina and bombarda for Terni, north of Rome, in 1340[221]; iron guns and cannon balls (unum cannonem de ferro ad proiciendas pallas de ferro, tronum a sagittando palloctas, cannone de ferro ad tronum e pallo de ferro) at Lucca in 1341[222]; schioppo at Frassineto, on the Po, in 1346[223]; bronze schioppo 60 lb. weight and a magister sclopi in Turin in 1347[224]; a canon (tube?) and bombardam (complete gun?), ballotas of wrought iron 4 cm. diameter, etc., at Saluerolo, Modena, in 1350[225]; bombardam, iron pallottole (8 cm.) and fuoco (powder) at Perugia in 1351[226]; ballotis bombardarum ponere 33 liv. at Ravenna in 1358[227]; a 2000 lb. bombard at Pietrabona, besieged by Pisans in 1362[228]; bombardment of Pisa in 1364[229]; 500 bombards 220 mm. long to be fired from the hand, made in Perugia in 1364[230]; arrows and lances thrown from guns in 1364[231]; stones and lances thrown from guns by Pisans in 1371[232]; an inventory of 1371 for Formizane (Formigiona), near Modena, specifies schioppi for hand use[233];

bronze Steinbüchsen in Venice, and two spingards or bombards in Florence in 1376[234]; two tromba marina throwing 100–200-lb. stone balls made in Perugia in 1379[235]; a Bologna record for 1397 mentions unam bombardam pizolam, perhaps a kind of hand-mortar (pistol?)[236]; a cast iron bombarda a secchia (bucket gun, a name used in 1397) with wrought iron rings, made in Parma about 1400 and in the Turin Museum (with two similar ones from Montefeltro), total length 1020 mm., weight 635 kg., both barrel and front chamber conical in bore (100 to 140 mm., and 320 to 350 mm., respectively),[237] is perhaps the gun said by Rathgen[238] to be dated, apparently, 1405, but probably this was originally 1495, part of the 9 having rusted away; wrought iron guns are mentioned for Milan in 1427–33, and cast bronze guns are certain from 1418, but not larger than 50 cm. calibre[239]; cast iron guns, probably 50 cm. calibre, were made at Como in 1429 and 1433.[240]

Rathgen[241] refers repeatedly to Vatican Archives published (or edited) by K. H. Schäfer "durch deutsche Forschung." There were fire weapons in the Papal army in the siege of Terni in 1340, elsewhere in 1341, and several times in 1350. In 1340 there is mention of trumba marina and bombardam. The outcome is the "Beweis . . . das die Feuerwaffe ihren Anfang in Deutschland genommen hat." There is no trace of Italian gunmakers in France or Germany in 1376, and if the Castel Angelo in Rome was bombarded with bronze guns (cum pyxidibus æneis) in 1378,[242] the art of gun casting was brought from Germany to Italy in 1376. Muratori[243] suggested that one or two Italian names for arms might have been derived from German names. Hand-guns are later in Italy than in Germany; they are described as something new in 1430 in Lucca and unknown then in Florence.[244] Italy was isolated from France by land and sea and knowledge of artillery could not have gone from Italy to France as Köhler had suggested. There may have been small guns (up to 6 cm. calibre) in Italy in 1340–1, but all the Italian records may have been copied from some unknown German books, and this is the more likely since the powder charges given for Italian guns are the same as those used for German. Biringuccio, says Rathgen,[245] probably copied from German works, which alone could provide the information he gives; the information in the Munich MS. 600 (1380–90) and the manuscript of Hans Henntz of Nürnberg in the Weimar Bibliothek is in advance of Biringuccio or anything else Italian, and the mention in Biringuccio about German technique[246] was left out in the French translation. German gunmakers were in Siena in 1407.[247] These arguments are weak.

Welded iron guns are in Leonardo da Vinci's manuscripts.[248] The Genoese historian George Stella (d. c. 1420), in his Annals[249] refers to "instrumentorum pro jaciendis impetuosè lapidibus, quæ bombardas vulgus nominat," and to a "vas ferreum grossos emittens lapides, quod vulgaris Italorum lingua Bombardam asserit." Folengo (1491–1544) in his poem "Merlin Coccaius,"[250] spoke of "grossos cavallos doctus ad impressum rapidarum bombardarum." Details of Italian guns were given by Francesco di Giorgio Martini (c. 1465)[251] and by Biringuccio.[252]

GUNS IN GERMANY

Konrad von Megenburg (1309–74), who studied in Paris in 1329–37, and composed in Regensburg in 1349–51 a *Buch der Natur*,[253] mentions shooting from guns, apparently small hand-guns. He says: (*a*) sam ain geschôz, daz man auz püchsen scheuzet; (*b*) snell als ain geschoz, daz von ainen armprust vert oder auz ainer schozpüchsen, und toetent daz tier; but he does not mention gunpowder, *nitrum* being a transparent stone like mica: (*c*) Nitrum haizt spat. der stain ist weizlot und durchsichtich nâhent sam ain glas, und dar umb macht man in für die venster.

Rathgen has published some records in Frankfurt giving expenses for making guns. The earliest Frankfurt records available begin in 1348, when they refer to "um pyle zur büszen £3 2s.," an arrow for shooting. In 1349 "gezelt zu byssene 30s.," "steyne um eine bussen £6," "pulfir um formen zu den büszen und um büszenpyle und um andir arbeid £2½ 1s.," "coppir und zyn zu den fürbüszen £10 5s.," "geschütze 11 grossz." In 1350 "um salpetri £11 6s."[254]

Rathgen calculates the weight of the bronze gun as 34 lb. and (from reasonable relations to the weight) as 48 cm. long and 4 cm. bore, i.e. a small gun, not a cannon. He concludes[255] that in 1348–9: "in Deutschland, wenigstens in dessen südwestlichen Teilen, die Pulverwaffen etwas Altbekanntes gewesen ist." Guns must, he thinks, have been made in Frankfurt long before 1348, and if we have no earlier records they were probably burnt in wars, as were those in Strasbourg in 1870, otherwise we should have plenty of evidence, which we are justified in assuming in its absence.[256] Firearms (Pulverwaffen) were invented in Germany soon after 1320; Marinus Sanutus does not mention them in 1321 in his plan for a crusade (ch. I) but Konrad of Megenburg does, hence they appeared in 1320–31.[257] There is no evidence of an early development of artillery on the Lower Rhine, and Frankfurt in 1348 was far in advance of other parts of Germany, guns being first mentioned in Wesel in 1361.[258] Ingolstadt inventories mention four guns in 1330, and a Donnergeschütz is mentioned for Merseburg in 1334[259]; these are earlier than the Frankfurt records.

Frankfurt records speak of büszen, bly (lead) and salpeter in 1363 and of guns (some small) in 1366,[260] of an iron Steinbüchse (which burst) in 1377, and of a copper one which cost £102 and fired a 100-lb. stone ball in 1378.[261] In 1391 there were 101 Büchsen, some for lead shot (and hence quite small),[262] and in 1391 cast iron guns were, apparently, first made: "7 iserne buszen" were made by a foreign gunmaker Merckiln Gast (who may have been a German), who was paid "bussen uz isen gyessen." Cast iron is known in the Rhine region only about 1400, and Rathgen says the mention is crossed out in the manuscript "as something obvious" (actually the details he gives indicate that this part was *not* crossed out), and hence casting iron "by a German" (i.e. Merckiln, of unknown origin) is much earlier than has been supposed.[263] The mention once of brass (Messing) for a gun in 1381 probably refers to bronze.[264] The mediæval name "Metall" usually meant copper.[265]

Very large guns seem to have come into use in Germany after 1390, since a chronicle of Limburg (near Frankfurt) for 1393 says: "da gingen die grossen bossen an, die man numme gesehen enhatte uf ertrich von solchen grosse unde von solche swerde."[266] Very large Burgundian iron guns were made in 1375-7 by French and Majorcan gunsmiths long before this.[130,181] A large bronze gun was cast in Frankfurt in 1394 by a Nürnberg gunmaker. It, and six smaller guns, required 95 cwt. 12 lb. of bronze (with 12·6 per cent. of tin), the large piece weighing 70 cwt. 71 lb. This gun was used in the battle of Tannenberg in 1399.[267] The large gun, a muzzle-loader, was taken to Tannenberg in July, 1399, by a team of sixteen horses. It was 205 cm. long and 50 cm. calibre, firing a 350-lb. stone shot. It had a shield. The stone balls found on the site are much too large to have been used with this gun. The Archbishop of Mainz brought his own guns.[268]

Besides the large gun there were also "fustbusse," a name used first and last on this occasion (1399), perhaps meaning a gun with a fist-sized bore. These were either bronze or iron, perhaps wrought iron. Each was moved by four horses, perhaps on a carriage, and hence probably weighed about 16-20 cwt. or 1600-2000 lb. These were breech-loaders (Hinterlader) firing 10-lb. balls.[269]

Rathgen[270] thinks breech-loaders first appeared in 1398, a year before Tannenberg, and were, of course, a German invention; elsewhere[271] he says they were first used in Flanders in 1380, but these were probably invented in Strasbourg and taken to Flanders by a German Büchsenmeister of the Duke of Brabant. Breech-loaders (Hinterladegeschütze) or chamber guns (Kammerbüchse) are said by Rathgen to be mentioned for Frankfurt in 1412 and elsewhere in 1414, and in 1430 the name Vögler (fuggeler bussen) appears, which, on the basis of a remarkable etymological research, he concludes means a breech-loader.[272] There is no doubt whatever that breech-loading guns were used before the dates given by Rathgen for their appearance in Germany.

Köhler[273] refers to a Bologna inventory of 1397 as mentioning an iron wedge (bletum) for fixing the chamber against the barrel by driving it in between the back of the chamber and a wooden support, and says this is the first indication of breech-loaders (es ist dies die erste Andeutung, die wir vom Hinterlader haben).

In 1408 a large gun was cast at Marienburg for the Prussian Order under the direction of Johann von Christburg for use against the Poles in 1409. It was cast in two pieces from 231 cwt. of bronze containing some lead, so that it probably weighed 180 cwt., and Köhler[274] says the two pieces were screwed together (jedenfalls zusammengeschraubt worden sind).

There are several mentions of fugeler bussen in Frankfurt in 1444.[275] They were made both of pure copper and wrought iron in 1413; to allow of rapid firing several chambers were used with the same gun, and since it was very difficult to make a gas-tight joint they were quite small (this does not seem to follow, since quite large guns had the chambers tightened by wedges). They are called Hantbussen in 1439 and are perhaps the same as terrabüchsen (French tarabosse), which were small hand-guns.[276] Köhler[277] says the name Terrasbüchse was used in Brunswick and Prussia soon after 1410; the names voghe Büchsen, Vögler, Vögeler, voghelears, veuglaires, etc., are of Flemish origin (e.g. Malines, 1409–10); they were new in Lille in 1412, where they came from Douai for trial, and were sent from Lille to Tournay and Ypres in 1415. They appear in accounts of the Dukes of Burgundy of the first half of the fifteenth century.

Records of Naumburg an der Saale speak of Grecum ignem (gunpowder), telam (arrow) and instrumentum (gun?) in 1348; of pixta ad sagittandum and saltpetre in 1354; of pixta ereas (bronze guns) consumpimus in 1357; of heavier guns for arrows costing 8 gr. in 1380, and in 1396 of larger guns (Steinbüchse) (a grossin buchs cost only 4 gr.); large Steinbüchse for breaking walls, and czehnsz (tin) in 1448; and in 1449 payment "vor bly vor iszen und schosse darusz zu gissen," which suggests that cast iron balls were made.[278] Records of Trier[279] speak in 1373–4 of dy bussen (two guns) of 100 and 120 lb., gebort (bored), a clotz (ball) of forged iron covered with lead (blijs), arrows, polffer, etc.

Records of Cologne[280] are, apparently, available only from 1370, when they speak of herbis ad tonitrua (i.e. Donnerkraut, gunpowder); in 1371 of pulvis tonitruali and pro herbis tonitrualibus; in 1376 of dunreboissen, salpeter, sulphur, carbonibus, pixides, small guns kept in a chest, and lead balls (no arrows); in 1377 of large (9 cwt.) guns made by a French gunmaker (magistris gallicanis) from Arles (who is made into a German by Rathgen) and a Walloon.[281] Records of Augsburg from 1371 speak then of firearms and in 1377 of a grozzen Buch (whether iron or bronze is not said).[282]

A seventeenth-century chronicle of Erfurt by Hogel speaks of a gun (Steinbüchse) in 1377, which becomes 1362 in another chronicle of 1739.[283] In 1377, in Rothenburg, there are buhssen (of cast copper?) and bullfer; old stone balls of 70–83 kg., 40 cm. calibre, were found there; saltpetre was imported from Nürnberg.[284] At Nürnberg there were two iron guns (zwei eisnein puchsen) in 1356, bronze guns in 1357 and forged iron guns in 1378 and 1388, the last (called Chriemhilde) weighed 56 cwt. and shot a 500-lb. stone ball with a charge of 14 lb. of powder.[285]

Wiegleb[286] quoted a document naming "eyne Büchssen und eyn Schoss blyes, zwo Büchssen und dry schoss blyes . . . pro pulveribus . . . pro pulveribus ad pyxides . . . pro patellio ad pyxides," all in 1378, place not given. Cast copper donerbussen or dunerbussen are mentioned for 1388 at Dortmund,[287] and two cast iron guns (ghegatene ysern bussen), much more expensive than bronze guns, and shooting stone balls, for Wesel in 1398.[288] A cast iron Steinbüchse in Linz, which came from Carinthia, of uncertain date but probably late fourteenth century, is mentioned,[289] a 70-cwt. gun of 1402 at Göttingen,[290] an iron gun of 1365 and a large gun (Faule Mette) of 8750 kg. (?) of 1411 at Brunswick,[291] and a 150-cwt. Cologne gun (Unverzagt) of 1416.[292] Görlitz records mention a gun used in the Hussite wars (1419), arrows for a gun (1426), Orgelgeschützen, and cast bronze (7 per cent. tin) breech-loading guns.[293] Bocholt guns (1407–37)[294] and a bronze gun dredged up at Memel,

1450 at the earliest,[295] and fourteenth to fifteenth century Pest guns of bronze poor in tin,[295] are mentioned by Rathgen, who also deals inadequately[296] with Swiss guns.

Dutch guns were early; there were seventy-seven Lotbüchsen in Deventer in 1346 and some in Utrecht in 1355–9, dunrebussen and pixide in Deventer in 1348 and Utrecht in 1378, as well as in other parts of Holland.[297] A wrought iron bombard in the Rijksmuseum, Amsterdam, of 50 cm. calibre, has been dated 1377,[298] but is said to be later and to be a mortar.[299]

GUNS IN SPAIN AND PORTUGAL

The history of the use of artillery in Spain is related to that of the Arabs and this aspect is dealt with in chapter V. A single bombard was on an Aragonese ship in 1359 and in the attack on La Rochelle in 1371 the Spanish ships mounted guns.[300] The chronicles of King John II of Aragon say that in the siege of Setenil (1407) five bombards discharged only forty shots in one day.[301]

The guns (Steinbüchsen) in the Artillery Museum in Madrid, all of wrought iron, are mostly rather small. The oldest (1350–1400 ?) is 110 cm. long, 29 cm. calibre, with a chamber 9 cm. diameter in the upper and 3 cm. in the lower part. The others, of the last years of the fifteenth century, include breech-loaders. The iron work is very clever. A complete artillery was assembled only in 1522 under Charles V. Artillery appeared in Portugal in 1370; in 1384 it was used in the defence of Lisbon, and effectively in 1410 in the expedition against Ceuta.[302]

According to Tout,[303] the first great battle in which cannon played an important but not decisive part was at Aljubarrota in 1385, when the King of Portugal, with the assistance of English bowmen, defeated the Castilians and secured Portuguese independence; but even here, in the end, the English archers proved more valuable than the Spanish cannon.

In 1461 the Spanish rebels besieged Gerona with bombards which, with an obvious exaggeration, are said to have fired 5000 stone balls in a day. The French who came to the assistance of Isabella with artillery, beat off the rebels, who abandoned their guns.[304] In 1472 artillery and arquebuses were used by the Spanish and Portuguese at Zamora.[305] The Castilians used artillery and arquebuses in 1476 and the Portuguese abandoned their artillery.[306] In 1481 the Moors of Granada were using artillery, but were deficient in ordnance.[307] The Spanish army in that year was also poorly equipped and Isabella ordered large amounts of artillery and ammunition in Castile and Leon; small guns (lombards) were in use in 1482, cannon and stone balls were made at Huesca in 1483, and the Moors were then using artillery.[308] The Spanish guns from 1483 were also firing large incendiary balls composed of inflammable ingredients mixed with gunpowder.[309]

In 1487 the Spaniards laid in much larger supplies of artillery, inviting artisans from France, Italy and Germany, and importing large stocks of gunpowder from Portugal, Sicily and Flanders. The Spanish artillery assembled was such as was probably not possessed by any other European power.[310] The Spanish guns, more than twenty of which, used in the siege of Baza in 1487, are still in existence, were somewhat crude. The largest is 12 ft. long, consisting of iron bars 2 in. broad held together by iron rings and bolts. They were firmly attached to the carriages, so that they could shoot only in one direction (a point emphasised thirty years later by Machiavelli). They discharged marble balls 14 in. diameter, weighing 175 lb., or sometimes iron balls; the stone balls discharged in the siege of Balaguer at the beginning of the fifteenth century weighed 550 lb., but the guns were, apparently, not so good.[311]

The use of artillery in European countries other than those considered appears to have begun rather late and does not call for description here; it is said to have been introduced into Russia in 1389 and Sweden in 1400.[312] Karamsin suggested that Russia, cut off from Europe, obtained information about gunpowder from Germany through Novgorod; it was introduced into Russia in 1389, and cannon were used in the defence of Moscow and Galich but arms were not then used in the field. This section will be continued with a description of some particularly large guns or mortars, mostly of the fifteenth and later centuries.[313]

LARGE GUNS AND MORTARS

The most spectacular use of cannon in the fifteenth century was in the capture of Constantinople by the Turks in 1453,[314] accounts of which are given by the Byzantine historians Doukas and Chalkokondyles. The use of guns in the siege of Constantinople in 1422, which began on 10th June and was raised on 24th August, is not mentioned by Turkish authors, but is described by the contemporary Greek writer Joannes Kananos, who says the Turkish bombards (βουμπάρδαι) had very little effect; he also mentions the use of incendiaries and towers (beffroi) by Murad (Μουρὰτ Πεῖ, Murat-Bei) : καὶ ταῦτα μὲν περὶ τῶν βουμπάρδων τὴν ἀπραξίαν (et hæc quidem de bombardarum irritis et sine profectu ictibus). Chalkokondyles says : 'Αμουράτης μὲν οὖν ὡς ἐπέλασε, καὶ ἐπολιόρκει Βυζάντιον ἀπο θαλάττης εἰς θάλατταν, τηλεβόλοις τε ἔτυπτε τὸ τεῖχος καὶ ἐπειρᾶτο, οὐ μέντοι κατέβαλέ γε (Amurates cum venisset Byzantium urbemque obsedisset, mœnia tormentis quatiebat. tenabat quidem murum sternere, at cœpta parum succedebant). In the siege of Hexamilion on the Isthmus of Corinth in 1446, according to Wittek, the guns were cast on the spot, as they were by Muḥammad II in the siege of Constantinople in 1453.

Nikolaos Chalkokondyles (Chalcondyles) of Athens, writing about 1450, whose astonishing ignorance even for a Greek is commented upon by Gibbon,[315] makes the Germans the inventors of gunpowder and cannon, then diffused over the greatest part of the world[316] :

δοκεῖ μὲν οὖν ὁ τηλεβόλος οὐ πάνυ παλαιὸς εἶναι, ὥστε συνιέναι ἡμᾶς ἐπὶ νοῦν ἐληλυθέναι τοῖς παλαιοῖς τὸ τοιοῦτον. ὅθεν μέντοι ἀρχὴν ἐγένετο, καὶ τίνες ἀνθρώπων ἐς τὴν τοῦ τηλεβόλου ἀφίκοντο πεῖραν, οὐκ ἔχω διασημῆναι ἀσφαλῶς.

οἴονται μέντοι ἀπὸ Γερμανῶν γενέσθαι τε τούτους, καὶ τούτοις ἐπὶ νοῦν ἀφῖχθαι ταύτην τὴν μηχανέν. ἀλλ'οἱ μὲν τηλεβολίσκοι ἀπὸ Γερμανῶν καὶ ἐς τὴν ἄλλην κατὰ βραχὺ ἀφίκοντο οἰκουμένην.

Bombardam ego arbitror non esse inventum antiquum, ut quis credat a veteribus esse repertam. Unde autem originem sumpserint, aut qui mortales primi bombardas in usu habuerint, haud certo tradere possum. Quidam Germanos putant bombardis esse usos, iisdemque bombardarum inventum acceptum ferunt. Reliqui autem bombardarii ex Germania progressi in universum paulatim pervenere orbem.

Since he went as ambassador to the Turkish camp in the siege of Constantinople in 1446, Chalkokondyles probably derived this legend from the German or Hungarian gunners employed by the Sultan, to whom it was familiar. He explained that gunpowder is a mixture of powdered saltpetre, charcoal and sulphur; the first guns were iron, then the invention of casting them from copper and tin was discovered, since this mixture is stronger, and guns made from it throw balls further. The longer the gun the further the ball is thrown. The explosion arises either because there is no vacuum in nature or because fire gives efficacy, all the power is in fire. The powder is potential fire, and becoming fire it moves.

Doukas[317] reported that Ḥalil Bassa in 1452 made bronze (χαλκοῦς) guns throwing stones weighing 1500 lb. For the siege of 1453, Doukas says a Hungarian deserter from the Greeks set up a foundry at Adrianople and in three months cast an enormous bronze gun (ἐν τρισὶν οὖν μησὶ κατεσκευάσθη καὶ ἐχωνεύθη τέρας τι φοβερὸν καὶ ἐξαίσιον) which threw great stone balls with terrific noise and smoke. The bore of the gun was 12 palms and the ball weighed over 600 lb. A ball which struck a Venetian ship cut it in two and sank it. The Roman palm was 7·39 cm., so that 12 palms would be about 34·8 in. or nearly 3 ft.

Doukas goes on to say[318] that Greek fire was used as well as cannon. The bronze of the gun, he says, would become porous after two or three discharges, had not the artificer protected it from the fumes of nitre and sulphur by affusion of oil, which also cooled the metal: παρευθὺ κατέβρεχεν αὐτὴν ἐλαίῳ, καὶ σὺν τούτῳ ἐπληροῦντο τὰ ἔνδον αὐτῆς ἀερώδη βάθη, καὶ οὐκ ἐνήργει τὸ ψυχρὸν λεανθὲν ὑπὸ τῆς τοῦ ἐλαίου θερμότητος. A similar account is given in the old Italian translation of Doukas.[319]

Chalkokondyles[320] says the maker of the gun was a Dacian deserter from the Greeks named Urban ('Ορβανός, Δάξ). He[321] says the stone balls weighed 7 talents (875 lb.) and were shot a distance of 14 stadia (1½ miles). Two enormous guns were dragged to Constantinople, each requiring seventy oxen and up to 1000 men to move it. One gun burst on the first firing (as Doukas also says)[322] to the great annoyance of the Sultan, who suspected sabotage.[323]

An account of the casting of the great cannon is given by Kritoboulos,[324] who joined the Turks and wrote from their point of view about 1467 or later. He reports that guns were invented by the Germans or Kelts 150 years ago (i.e. in 1317) or a little earlier. He mentions the clay mould and core, strengthened with iron, wood, earth and stone. The walls of the cannon were a palm in thickness. Two furnaces built of large stones laid with cement, and lined with fire-bricks well smeared over with very rich and thoroughly kneaded clay, were erected near the moulds. One thousand five hundred talents of copper and tin were put in the furnaces, with logs of wood interspersed with coal and the holes in the stonework, except the tapping channels, were closed. A cordon of bellows operated for three days and nights till the metal was as fluid as water. The tapping channels were opened and the metal was run into the mould so as to rise above the top of the channel. The mass was then chiselled and polished.

In firing the gun, the chamber was filled with powder well rammed down, then a wooden plug or wad was hammered in. This plug had a depression for the stone ball. The ball struck the city walls, causing thousands of splinters of stone to fly into the air and killing the defenders. A single shot was sometimes sufficient to knock down half the height of the wall or one of the towers, for nothing could withstand the force of the impact.

The casting of bronze cannon in Constantinople in the seventeenth century is described by Evliyā Chelebī (Evliya Effendi).[325] Baron de Tott, the French son of a Hungarian, was called in by the Turks in 1769 to organise the defences of the Dardanelles against the possibility of Russian attacks. He describes the enormous old Turkish bronze cannon, which lay on the ground immovable, with heaps of stones at the back to take the recoil. The Turks were greatly afraid of firing them and implored de Tott not to attempt this, since the concussion would destroy the town. One gun was provided with stone balls weighing 1100 lb. and the charge of powder was 360 lb.[326]

Finlay[327] was informed by a Turkish captain that one of the stone balls was more than half a ton, and Hammer-Purgstall, says Finlay, had seen several stone balls 12 palms in circumference at Rhodes. A bombard of 11 palms circumference (78 cm. diameter) for 645 kg. stone balls was said to have been used by the Turks at Rhodes in 1480.[328]

Lefroy[324] gives an account of the eighteen guns remaining in the Dardanelles forts in 1868, one of which was presented by the Sultan to Queen Victoria. It was in the Museum of Artillery, Woolwich, but in 1929 was moved to the Tower.[329] It is of 25 in. calibre, 6 ft. 7 in. long, with a chamber diameter of 10 in. It was cast in 1464 by Munir 'Alī for the Sultan Muḥammad II. It is in two parts, screwed together. Lefroy gives a list of twenty-two extant guns, one of 1458, of large calibre (the largest 29·5 in.). In 1829 the Dardanelles cannon, the largest of 28 in. calibre and throwing a stone ball of 1578 lb., were kept loaded, each with 1 cwt. of gunpowder, and early in the nineteenth century English warships were severely damaged by stone balls from the great guns, the muzzles of which could be seen projecting from the forts. The Sultan Muḥammad II, in order to hit the decks of ships in 1453, used the guns as howitzers.

Doukas[330] describes the bronze gun used in 1439 as a tube like a cane with a receptacle filled with powder made from nitre, sulphur and willow charcoal, ground and mixed, with a bituminous odour. If a fiery spark is brought to the nostrils [of the gun] it inflames at once and the fiery spirit in contact with one ball expels it, and this ball in turn expels the others :

ἐξόπισθεν οὖν τῆς χαλκῆς καλάμου βοτάνης σκευασία ἐκ νίτρου τεάφης [θειάφι] καὶ καρβούνου ἰτέας πλήρης. ὀσμὴν οὖν ἀσπίθος ἢ τοῦ σπινθῆρος πυρός, εἰ πλησιάσειεν. τῇ ἀναμιγῇ ταύτῃ, αἴφνης ἐξάπτει, καὶ στενοχωρουμένου τοῦ πνεύματος ὑπὸ τῶν βολίδων, ἐξ ἀνάγκης ὠθεῖ τὰς βολίδας, καὶ ὠθουμένων ἡ πρὸς τὴν βοτάνην ἐγγὺς ὠθεῖ τὴν πρὸ αὐτῆς, ἡ δ'αὐτὴ τὴν ἔμπροσθεν.

expulsos post æneum autem tubum (fistulæ enim seu arundinis formam oblongam machina imitatur) receptaculum erat pulvere repletum, qui nitro sulphure et carbone saligno contritis ac commixtis conficitur. odorem bituminis. si naribus admoveatur ignis scintella refert, sic temperatus statim accenditur, ignisque spiritus in arto a globulis cohibitus eos necessario expellit, hac vi dum impelluntur, qui pulveri proximus est, contiguum propellit, hic antecedentem.

Doukas[331] also mentions lead balls the size of a Pontic (i.e. hazel) nut for use in this gun of 1439, a charge of five or ten being used for each round : ἔπεμπον γὰρ εἰς αὐτοὺς βολίδας μολιβδίνας, ὅσον καρύου Ποντικοῦ τὸ μέγεθος, ἀπὸ κατασκευῆς χαλκῆς ἐχούσης ἐντὸς τὰς βολίδας καθ' ὁρμαθὸν πέντε ἢ καὶ δέκα (in ipsos enim globulos plumbeos, nuci Ponticæ magnitudine pares, machina ænea, cui recta serie quini vel deni imponebantur expulsos).

He also describes oblong iron bombs filled with gunpowder compressed by a hammer : ἀλλ' ὅταν ἡ βολὶς τύχῃ σιδήρου ἢ ἄλλης τινὸς ὕλης ὁπλοποιίαν στενοχωρουμένης, τὸ σφυρῶδες εἰς γραμμὴν μετασχηματίζει, καὶ ὥσπερ ἧλος ὁ βόλος γίνεται, καὶ διέρχεται τοῖς τῶν ἐγκάτων ἐντέροις καθὰ ποταμὸς πύρινος (. . . in oblongam figuram versa clavo assimilatur; et interiora viscera ceu fluvius igneus pervadid).

The German historian Leunclavius (Johann Loewenklau) (1533–93), who knew Turkish and visited Constantinople, mentions the capture of Constantinople and the guns used only very briefly.[332] Leonard of Chios, Archbishop of Mitylene, who was an eye-witness of the capture of Constantinople and wrote a letter on it to Pope Nicholas V on 15th August, 1453, measured the stone ball of the second cannon (not the largest) and found it to be 12 palms[333] (Horribilem perinde bombardam, quanquam major alia, quæ compacta fuit, quam vix boum quinquaginta centum juga vehebant, ad partem illam muorum simplicium . . . figentes, lapide qui palmis undecim ex meis ambibat in gyro ex ea murum conterebant). He also says small arms (muskets, etc.) were used, as well as bows and arrows, towers and machines for throwing stones (sclopis, spingardis, zarbathanis, fundis, sagittis dies noctesque muros hominisque nostros vexabant nactabantque. Pulvis erat nitri modius, exiguus, tela modica : bombardæ si aderant, incommoditate loci primum hostes offendere). The account in Leonard of Chios is like that in an Italian chronicle of Zorzi Dolfin, in a St. Mark's (Venice) manuscript quoted by Thomas. He says the large bombard (la bombarda grande) was drawn by 150 pairs of oxen and its stone balls measured 11 palms (trazeua pietra uoltaua XI. palmi); small arms were used (ne manchua schiopetti, spingarde, zarabattano, funde, saggite . . .). The Greeks had some cannon, which could not reach the Turks, since they were short of saltpetre for powder (pocha poluere de salnitrio).

The sizes of some old cannon still in existence are given in the table below[334]:

	Length, m.	Bore, cm.	Weight, kg.	
Paris, 1404	3·65	39	4597	cast (Austrian)
Ghent, 1450–2	4·96	62	340	iron bars and rings
Edinburgh, before 1460 ..	4·0	50	147	do.
London, 1464	5·25	63·5	—	cast (Turkish)
Moscow, 1586	6·78	114·8	1955	cast
Vienna	2·56	88	—	—

The Ghent bombard, "Dulle Griete" (griete = great), had been dated 1382.[335] The cannon at Edinburgh Castle, "Mons Meg," made from iron bars, was formerly thought to have come from Mons[336]; it is described by Hewitt.[337] Mallet thought the Edinburgh bombard might have been made in Scotland, where tradition, according to Carman, says it was made about 1451; it is certainly older than 1489. It was used to fire a salute in 1682 and some of the iron rings at the breech were blown off, but the rest of the gun and the longitudinal bars of the breech were not appreciably disturbed, and the gun could probably still be fired.

The Moscow gun or mortar is very large.[338] The Vienna great bombard (Grosse Pumhart) of Steyr is of doubtful date[339]; it is a Steinbüchse. The barrel is externally cylindrical but the bore tapers from 790 to 882 mm. at the muzzle; the rear chamber, 1099 mm. long, has a bore of 180 mm. The thickness of the metal at the muzzle is 105 mm., the total external diameter there being 1·09 m. The thickness of the base is 213 mm. The chamber and barrel are not screwed together but fixed by pegs (Langschienen) and hoops (Reifen) on both pieces. The charge is estimated at 34 lb. or 26·288 kg. (Lippmann said 67 kg.) and the stone ball would weigh about 1000 lb. This Austrian gun is nearly as big as the bronze cannon of Muḥammad II in 1453. The Austrians have always been famous artillerists (a "Hungarian" made Muḥammad's cannon) and the Skoda factory in Pilsen (now Czech) made the 18-in. howitzers used in World War I; the present writer saw the top of a hill blown off by a shell from one of these in action. Some other fifteenth-century guns are[340]:

	Length, m.	Calibre, cm.	Weight, kg.
English, on Mt. St. Michel, before 1423 ..	3·64	48	150
" "	3·53	36	75
Burgundian at Basel, 1426–30	2·73	35	48
" "	2·27	23	12
Faule Magd, Dresden, c. 1450..	—	35	—
Catherine, Paris, 1404	3·65	61	—

Catherine, according to Carman, came from Rhodes, but was cast on the mainland, according to Feldhaus by Georg Endorfer for Austria. The Faule

Magd is probably from Mons and is said by Rathgen to be the earliest wrought iron muzzle-loading Steinbüchse. A Faule Grete used in 1413,[341] is commemorated in a song composed in Nürnberg in 1863 :

> Wie ein Bierfass war die Puderdose
> Und die Stärke des Metalls am Stosse
> Mass drei Achtel Ellen rheinisch gut
> Und ihr Korn war wie ein Zuckerhut.

It is presumably the bronze Donrebusse called "Faule Mette,"[342] cast (after a first unsuccessful attempt) at Brunswick in 1411. It weighed 160 cwt. and threw stone balls of nearly 2 cwt. It was melted down in 1728. In the 317 years of its life it fired only nine shots, all of which missed the target. These great guns, in fact, apart from their use against fortifications (which was not always successful, see p. 115), with their attendant masses of transport and technicians, were far from being as effective as might be supposed. The famous field artillery of Charles VIII of France, used in his invasion of Italy, fired continuously in the battle of Fornovo (1495) and killed ten men; in 1445 a rate of fire of two shots an hour was exceptionally good, and as late as 1563 German gunners had a motto : "Das Treffen is nicht allweg Kunst / Es liegt meistteils an Gottes gunst." As compared with guns, the Agincourt bowmen in 1415 shot off twelve arrows a minute and their formidable weapons pierced steel armour without difficulty; perhaps the use of bowmen in English armies until 1627 was not so absurd as it might now seem to be. The later history of artillery will not be discussed; it may be studied in the works quoted. English gunnery was studied by A. R. Hall.[343]

REFERENCES

1. Oscar Guttmann, *Monumenta Pulveris Pyrii, Reproductions of Ancient Pictures concerning the History of Gunpowder, with Explanatory Notes*, London, printed for the Author (270 copies), 1906, plates 5–7. (The text is useless.)

2. Hansjakob, *Der schwarze Berthold, der Erfinder des Schiesspulvers und der Feuerwaffen*, Freiburg im Breisgau, 1891, 91 pp. (the common reading in older German sources is "Schwartz," not "Schwarz"); summary in Romocki, *Geschichte der Explosivstoffe*, Berlin, 1895, i, 106–12; Lippmann, *Abhandlungen und Vorträge*, 1906, i, 186; Hime, *Origin of Artillery*, 1915, 124.

3. *Felicis malleoli . . . De Nobilitate et Rusticate Dialogus*, ed. Sebastian Brandt (BM IB.1784, s.l.e.a., catalogue dates 1490 ?); the passage is not in this first ed. but Hansjakob says it is in the second ed. s.l.e.a. (? Basel, ? 1495), ch. XXX, f. 117, also in the third ed., Basel, 1497; the second and third eds. were not available to me.

4. Hansjakob, ref. 2, 39f.

5. Ref. 2, 45; François de Belleforest (1530–83) wrote *La Cosmographie Universelle . . . auteur en partie Munster, mais beaucoup plus augmentée*, 2 vols., fº, Paris, 1575 (BN G.448–9); see ref. 24.

6. Ref. 2, 26, 43.

7. Aventinus, *Annalium Boiorum Libri VII*, bk. vii, ch. 21, § 20; Ingolstadt, 1554; fº; Leipzig, 1710, 763; J. G. Hoyer, *Geschichte der Kriegskunst*, Göttingen (vol. I, 1797; vol. II, i, 1799; vol. II, ii, 1800), 1797, I, Zusätze und Erläuterungen, 2.

8. Libri, *Histoire des Sciences Mathématiques en Italie*, 1838, ii, 225.

9. Lacabane, *Bibl. de l'École des Chartes*, 1844, i, 47.

10. Col. Ildéfonse Favé, in Napoleon III, *Études sur le Passé et l'Avenir de l'Artillerie*, 1862, iii, 87–8; Rathgen, ref. 16, 665. The Emperor Napoleon III began the book whilst a prisoner in Ham and was later assisted by Col. Favé, who wrote vols. iii ("à l'aide des notes de l'Empereur") and iv ("continué sur le plan de l'Empereur"). The book was extensively used by Köhler, ref. 12, and Rathgen. Napoleon III was greatly interested in artillery.

11. Max Jähns, *Handbuch einer Geschichte des Kriegswesens von der Urzeit bis zu der Renaissance*, Leipzig, 1880 (1288 pp.), 773.

12. G. Köhler, Generalmajor z. D., Ritter des eisernen Kreuzes I. u. II. Kl. und des rothen Adlerordens II. Kl. mit Eichenlaub. Mitglied des Gelehrten-Ausschusses vom germanischen National-Museum zu Nürnberg, *Die Entwickelung des Kriegswesens und der Kriegführung in der Ritterzeit von Mitte des 11. Jahrhunderts bis zu den Hussitenkriegen*, vols. I, II, IIIi, IIIii, Suppl., Breslau; only vol. IIIi, 1887, is of interest (pp. xlv, 528, with 6 lithogr. plates), III, i, 241.

13. Favé, ref. 10, 1863, iv, Pièces justificatives, no. 2, p. ix: la maniera de artilharia.

14. Ref. 12, III, i, 244.

15. W. Hassenstein, *Das Feuerwerkbuch von 1420, 600 Jahre Deutsche Pulverwaffen und Büchsenmeisterei*, Munich, 1941, 17, 47, 97; Jähns, ref. 11, 773, dates it 1432, and the date 1420 may be too early.

16. Bernhard Rathgen, *Das Geschütz im Mittelalter. Quellenkritische Untersuchungen von Bernhard Rathgen*, VDI Verlage GMBH, Berlin, 1928, 163. The book has portraits of General-leutnant Rathgen, Dr.phil.hon.causa, b. Copenhagen, 4th September, 1847, present at the siege of Paris, 1870–1, d. Marburg, 21st February, 1927, and of H. von Müller; *Quellen-verzeichnis*, ix ll., 718 pp., 14 plates. The book is confused, ill-digested, repetitious and tendentious, but represents much useful toil. It does not seem to be in any public library in Great Britain and I used the Berlin University Library copy.

17. Ref. 2, 47 f., 58, 68.

18. Hansjakob, ref. 2, 30, 32, 70, 74; cf. Wiegleb, Crell's *Annalen*, 1791, I, 206, 303.

19. Polydore Vergil, *De Inventoribus Rerum*, bk. ii, ch. 11; Amsterdam, 1671, 109. (For the battle of Fossa Clodia, see ref. 20.) The book is largely a compilation from classical authors. It first appeared in 1499, 4°, Venice (BM IA.23513), in three books, and after 1521 in eight, and went through a great number of eds. (BM has 1503, 1509, 1513, 1528, 1532, 1545, 1554, 1557, 1561, 1562, 1570, 1576, 1586, 1590, 1604, 1606 and 1671, and eds. of 1521, 1558, 1559 and 1560 are mentioned) and trs. See Ferguson, *Archæologia*, 1888, li, 107–41; *Hand List of Editions of Polydore Vergil's De Inventoribus Rerum*, ed. by Fulton from a MS. by Ferguson, New Haven, 1944 (in typescript) (BM Ac.2692, meg. (5.)); D. Hay, *Polydore Vergil, Renaissance Historian and Man of Letters*, 1952 (BM 10634.k.20). Ferguson (*loc. cit.*; cf. Thorndike, *History of Magic and Experimental Science*, New York, 1941, vi, 148) says the book was on the Index, since it refers with contempt to monks, but an expurgated ed. was printed in Rome, 1576, for use by Catholics.

20. Pancirollius, *Rerum Memorabilium et deperditarum libri duo*, bk. ii, ch. 18; tr. *The History of Many Memorable Things Lost which were in use among the Ancients, and an Account of many Excellent Things Found*, . . . 2 vols., 8°, London, 1715 (pagination cont.), 383 and Salmuth's note, 388, saying "whether he was a Monk of Friburg, Constantine Ancklitzen or Bertholdus Swartz (as some call him) a Monastick too, is not so very certain." He gives the story of the explosion in the mortar. Guido Panciroli, b. Reggio, 17th April, 1523, d. Padua, 16th May, 1598, or 1st June, 1599, was professor of law in Padua. His Italian MS., completed in Turin in the year of his death by order of Emmanuel Philibert, Duke of Saxony, was tr. into Latin by Heinrich Salmuth of Amberg (dates unknown) and published there, 8°, 1599, 1607, 1612; 4°, Frankfurt, 1622, 1629–31 (2 vols.), 1660; Leipzig, 1707; O. in *Nouvelle Biographie Générale*, 1862, xxxix, 126. The references to a battle between the Genoans and Venetians at Fossa Clodia "about 1380" is to that at Chioggia, a seaport on islands at the southern end of the Venetian lagoon, where in 1379–80 the Genoans were finally defeated by their Venetian rivals. See ref. 242.

21. Sebastian Franck, *Chronica, zeijt büch vnd geschicht*, f°, 1531, cxcvij: Anno 1380 ward das mördisch teüffelisch geschütz der Büchsen erfunden von eim Münch, vnd war dazümal der stet krieg.

22. Gasser, *Epitome Historiarum & Chronicorum Mundi*, 8°, Lyons, n.d. (after 1538), 224.

23. Hunter, *Archæologia*, 1847, xxxii, 385.

24. *Cosmographiæ Universalis*, bk. iii, ch. 174; f°, Basel, 1572, 602; the passage is not in the Basel, 1552, ed., 488, which has no chapter numbers, the section "De bombardarum apud Germanos inuentione" being much shorter and not mentioning Berthold; Köhler, ref. 12, III, i, 242–3, quotes from a German tr., 1556, 598 (not available to me) : Doch schreibt mir zu Doctor Achilles (Gasser) das anno Christi 1354 das buchsen in brauch gewesen seind an der see bei Denmark, vnd soll ir erster meister sein gewesen ein Alchimist vnd mönch, mit namen Berthold Schwarz.

25. [Achilles Gasser] *Chronica Der Weitberempten Keyserlichen Freyen . . . Statt Augspurg*, f°, Frankfurt, 1595, pt. ii, p. 108.

26. Feldhaus, ref. 55, 1914, 79, says the date is 1380 in a 1544 ed. of Munster and 1354 in a 1546 ed.; since it is not in the 1552 ed. (ref. 24) this statement requires checking (the 1544 and 1546 eds. were not available to me). Feldhaus also gives that date, 1354, from Gasser, ref. 25, instead of the correct 1353.

27. Cæsius, *De Mineralibus*, f°, Lyons, 1636, 388.

28. Gordon, *Opus Chronologicvm Annorvm Seriem*, f°, Paris, 1617, 547 : Ann. 1380. Nunc adinuenta maiora illa tormenta bellica, quatiendis & eueuendis vrbium mœnibus, explosione ferreorum globorum opportuna, apud Venetos, authore Bertholdo è Germania à quibusdam putatus est Monachus. que arte primi vsi sunt Veneti, in recupetenda hac anno Fossa Clodia He also gives the alternative date 1354.

29. Kircher, *Mundus Subterraneus*, f°, Amsterdam, 1665, ii, 467.

30. Becher, *Narrische Weisheit und weise Narrheit*, Frankfurt, 1682, 41 (no. 26).

31. Janssen, *Rerum Inventarum Onomasticon*, 8°, Amsterdam, 1684, 77; Lacabane, ref. 9, 47

32. H. Gram, "Om Bysse-Krud, Naar det er opfundet i Europa, Hvorlænge det har været i Brug i Danmark," in (a) *Skrifter som udi Det Københavnske Selskab af Lærdoms og Videnskabernes Elskere*, Copenhagen, 1745, i, 213–306; (b) Latin in *Scriptorum à Societate Hafniensi Bonis Artibus Promovendis Dedita . . . nunc autem in Latinum Sermonem conversum Interpretate*, Copenhagen, 1745, i, 211–306; (c) German, "Hans Gram über die Zeit der Erfindung des Pulvers in Europa, und das Alter desselben in Dännemark," in *Historische Abhandlungen der Königlichen Gesellschaft der Wissenschaften zu Kopenhagen, aus dem Dänischen übersetzt, und zum Theil mit Vermehrungen und Verbesserungen ihrer Verfasser . . .* von V. A. Heinze, Kiel, Dessau, and Leipzig, 1782, i, 1–160.

33. C. F. Temler, "Om den Tidspunkt da Bøsserkrud og Skydegevær ere opfundne i Europa," in (a) *Nye Samling af det Kongelige Danske Videnskabers Selskabs Skrifter*, Copenhagen, 1781, i, 1–54 (read November, 1778); (b) German tr., Christ. Friedr. Temler, "Von dem Zeitpunkt der Erfindung des Pulvers und Schiessgewehrs in Europa," in *Historische Abhandlungen*, ref. 32 (c), 1782, i, 161–242 (read 27th November–4th December, 1778). A report that cannon were used in Denmark in 1280 was shown by R. Watson, *Chemical Essays*, 1793, i, 333, to refer to some years after 1332.

34. Ref. 2, 22.

35. Ref. 32 (c), 1782, i, 33–46.

36. Jähns, ref. 11, 774; *id.*, *Geschichte der Kriegswissenschaft in Deutschland* (in *Geschichte der Wissenschaft in Deutschland*), Munich and Leipzig, 3 pts., 1889–91; 1889, i, 224.

37. Ref. 32 (c), 1782, i, 121 f., Appendix, 146 f.

38. Gram, "Verbesserung zur Geschichte des König's Waldemar," in ref. 32 (c), Kiel and Hamburg, 1786, ii, 1–444; Copenhagen, 1787, iii, 199, 321; for his account of King Waldemar IV of Denmark (1315–75), who introduced gunpowder into Denmark in 1372, see *id., ib.*, 1782, i, 151 f.

39. Ref. 32 (c), 1782, i, 6, 92.

40. Ref. 32 (c), 1782, i, 25–32.

41. Ref. 32 (c), 1782, i, 46, 67.

42. Morhof, *De Transmutatione Metallorum*, Hamburg, 1672, 132.

43. Ref. 32 (c), 1782, i, 147.

44. Doukas, *Historia Byzantina*, ed. Bekker, Bonn, 1834, 211, 596.

45. Ref. 32 (c), 1782, i, 43.

46. Ref. 32 (c), 1782, i, 81 f., 97 f., 124.

47. Ref. 32 (c), 1782, i, 91–9.

48. Petrarch, De Remediis Utriusque Fortunæ, bk. ii, ch. 99; see ref. 119.

49. Lalanne, Mém. div. Sav. Acad. des Inscriptions, II Sér., Paris, 1843, i, 348 f.

50. Ref. 32 (c), 1782, i, 167.

51. William Camden, Remaines Concerning Britaine, London, 1614, 240–1 (238 f., on Artillarie).

52. Robins, Traité de Mathématiques, tr. Dupuy, Grenoble, 1777, 94.

53. E.G. in Nouvelle Biographie Générale, 1864, xliii, 604, q. Jalofsky, De inventore pulveris pyrii et bombardæ, 4°, Jena, 1702; L. Larchey, Des Origines de l'Artillerie, 18°, Paris, 1862.

54. Köhler, ref. 12, III, i, 243, quotes this information from Ludovic Lalanne's Curiosités militaires (a work not available to me—it is not in the Bibliothèque Nationale catalogue), giving the name as "G. Crispi." The first important artist of that name was Giovanni Battista Crespi, c. 1557–1633.

55. Feldhaus, Z. angew. Chem., 1906, xix, 465; 1908, xxi, 639; Chem. Ztg., 1907, xxxi, 831; 1908, xxxii, 316; Die Technik der Vorzeit, Berlin, 1914, 78–80.

56. Ref. 12, III, i, 241–6.

57. Berthelot, Revue des deux Mondes, 1891, cvi, 786 (817).

58. Clephan, "The Ordnance of the Fourteenth and Fifteenth Centuries," in Archæological J., 1909, lxvi (sec. ser. xviii), 49–138 (a paper on sixteenth-century hand guns, ib., 1910, lxvii, 109, is not quoted here); the paper is fully illustrated but the text is not always trustworthy.

59. Ref. 16 (a), 3; (b), 118; (c), 703; (d), 163, 677; (e) 163.

60. Ref. 15, 97.

61. H. Diels, Antike Technik, Leipzig und Berlin, 1924, 112.

62. Z. angew. Chem., 1926, xxxix, 401.

63. Isis, 1929, xiii, 125.

64. Ref. 12, III, i, 225: Die bisherige Ansicht, dass Deutschland Anspruch hat, als die Wiege der Artillerie angesehen zu werden, lässt sich an der Hand der Urkunden, wie ich zeigen werde, nicht aufrecht erhalten. Rathgen, ref. 16, 667 f.

65. Rathgen, ref. 16, 701.

66. Rathgen, ref. 16, 665.

67. Rathgen, ref. 16, 136, 193, 566 f., 694.

68. P. A. Lenz, Nouvelles Archives, Philosophiques et Littéraires, Revue trimestrielle, 1840, ii, 589–609 (BM P.P.4475); Jähns, ref. 11, 1880, 774; Favé, ref. 10, iii, 71 (authenticity of date not certain); Hime, ref. 2, 1915, 120–1, 124, 127 (page ref. 559 to Lenz incorrect).

69. Mém. couronnées Acad. Roy. Belg., 1899, lviii, 116; Clephan, ref. 58, 145, and Rathgen, ref. 16, 703, also say the document is a forgery; Oman, ref. 139, 1924, ii, 213–4, who examined the Ghent records, says the entry is in a later hand in the margin, and dates it c. 1500.

70. Ref. 2, 120–1; G. Greenhill, The Kynoch Journal, 1906, vii, 138; O. Guttmann, ib., 1907, viii, 75; Greenhill, ib., 1907, viii, 78.

71. Nova Acta Eruditorum, 1769, 19: "Ambergæ, Palatinatus superioris, in officina armorum reperiatur tormentum militare, cui sit annus 1303, inscriptus" (in review of a book).

72. Feldhaus, ref. 55, 1914, 420; Rathgen, ref. 16, 703.

73. "Die Sage von der Erfindung des Schiesspulvers und der Deutsche Ursprung des abendländischen Geschützwesens," in Allgemeinen Militär-Zeitung, 1896, lxxi, 283 (no. 36).

74. G. Oppert, On the Weapons, Army Organisation, and Political Maxims of the Ancient Hindus, with Special Reference to Gunpowder and Firearms, Madras and London, 1880, 49; Guttmann, J. Soc. Chem. Ind., 1904, xxiii, 591 (15th June); Greenhill, The Kynoch Journal, 1906, vii, 138; Guttmann, ref. 1, fig. 69; Feldhaus, ref. 55, 1914, 409; Hime, ref. 2, 1915, 122, and frontispiece; Rathgen, ref. 16, 670–2, and plate 4; Sarton, ref. 122, plate fig. 14 (from Hime). The small details in the original are much clearer than in some of the reproductions.

75. M. R. James, *The Treatise of Walter de Milemete De Notabilitatibus, Sapientis, et Prudentia Regum, Reproduced from the Christ Church Manuscript. Together with a Selection of Pages from the Companion Manuscript of the Treatise De Secretis Secretorum Aristotelis preserved in the Library of the Earl of Leicester at Holkham Hall*, Roxburgh Club, Oxford, 1913 (CUL 879.bb.4.15); descr. xxxii, plate 140; Holkham Hall MS., plate 181; Rathgen, ref. 16, 670-2, and plate 4. Oman, ref. 139, 1924, ii, 212, 377, says the armour shown in the picture is "precisely suitable to the date," and went out of use soon after; P. Post, "Die frühste Geschützdarstellung von etwa 1330," in *Zeitschr. f. historische Waffen- und Kostümkunde*, Dresden, 1938, xv, 137–41.

76. Ref. 16, 123.

77. Rathgen, ref. 16, 440; on p. 703, however, he dates the touch-hole as a German invention of 1320–30.

78. Köhler, ref. 12, III, i, 249, 333; plate iii, fig. 11.

79. Rathgen, ref. 16, 670, 673–4.

80. Ref. 16, 124–5.

81. Köhler, ref. 12, III, i, 227.

82. Hime, ref. 2, 123, 129.

83. Köhler, ref. 12, III, i, 230.

84. Hassenstein, ref. 15, 85, 121, fig. 33.

85. H. T. Riley, *Memorials of London*, 1868, 205; Salzman, *English Industries of the Middle Ages*, Oxford, 1923, 156; Rathgen, ref. 16, 703.

86. Rathgen, ref. 16, 703; Hassenstein, ref. 15, 84.

87. Ref. 1, 27, figs. 73–6; Feldhaus, ref. 55, 1914, 411.

88. Ref. 16, 674, 705.

89. Froissart, *Chronicles*, bk. i, ch. 111; ed. J. A. Buchon, Paris, 1824, i, 310–11; Hime, ref. 2, 123, 131–2.

90. Archives of Lille, 16th June, 1349, q. by Kervyn de Lettenhove, *Œuvres de Froissart*, Brussels, 1867, iii, 499; "plonc" is the Walloon form of "plomb" (lead), Hime, ref. 2, 123.

91. Joseph Garnier, *L'Artillerie des ducs de Bourgogne, d'après les documents conservés aux archives de la Côte-d'Or*, Paris, 1895 (BN 8° Lk.² 4277, 298 pp.), q. by Hime, ref. 2, 123; the original (which Rathgen sometimes dates 1865 and sometimes 1895) was not available to me.

92. Reinaud and Favé, *Du Feu Grégeois*, 1845, 168; Berthelot, ref. 57, 818.

93. Hime, *Gunpowder and Ammunition*, 1904, 200; ref. 2, 123.

94. Ref. 16, 707.

95. E. de Dynter, *Chroniques des Ducs de Brabant*, Brussels, 1854, ii, 636; Rathgen, ref. 16, 683–4, says Tournay was one of the "freien Städte des deutschen Reiches."

96. Favé, ref. 10, iii, 81. Although Rathgen, ref. 16, 683–4, admits that this was earlier than the Frankfurt guns of 1348 (see p. 120), he does not think it served as a model for them.

97. J. F. Huguenin, *Les Chroniques de la Ville de Metz*, Metz, 1838, 44–5; accepted by Jähns, ref. 11, 775, and Hime, ref. 2, 127.

98. Favé, ref. 10, iii, 71; Rathgen, ref. 16, 703.

99. *Scriptores Rei Prussicarum*, ed. Töppen, Leipzig, 1861, i, 21–219; 1866, iii, 478–506.

100. Libri, ref. 8, 1838, ii, 73; 1841, iv, 487; text in Lacabane, ref. 9, 35 ("d'une authenticité irrecusable"), and Brackenbury, *Proc. Roy. Artillery Institution*, Woolwich, 1865, iv, 289 (text verified by Mas-Latric of the École des Chartes), accepted it as genuine. Favé, ref. 10, iii, 72, and Berthelot, *Forces des Matières Explosives*, 1883, i, 360, thought it was doubtful, since Libri was known to have falsified documents; Hime, ref. 2, 127–8 (authentic); Lippmann, ref. 2, i, 188 ("nicht glaubhaft"); *Chem. Ztg.*, 1928, lii, 2; *Isis*, 1928, xi, 429; *Beiträge zur Geschichte der Naturwissenschaften und der Technik*, 1953, ii, 83 ("die erste sichere urkundliche Bezeugung"); Davidsohn, *Geschichte von Florenz*, Berlin, 1912, III, 758; 1922, IV, i, 249; 1925, IV, ii, 21 (putting it beyond doubt); Rathgen, ref. 16, 674, 696, 703 f. ("ist im Datum gefälscht"; "wahrscheinlich gefälscht"), who recognised that it destroyed his thesis, belatedly returned to the suggestion that it was forged, which is not believed by any competent modern authority.

101. *Fragmenta Chronici Forojuliensis*, in Muratori, *Rerum Italicarum Scriptores*, Milan, 1738, xxiv, 1228–9; the account is in only one of the MSS. used. Ludovico-Antonio Muratori

(1672–1750) was one of the "dottori" or conservators of the Ambrosian Library at Milan, and from 1700 archivist of the Duke of Modena; accused of heresy, he boldly approached the Pope and was completely acquitted. His two main works are (i) *Rerum Italicarum Scriptores*, 25 vols., f°, Milan, 1723–51 (a new issue began to appear at Città di Castello, in 1900), and (ii) *Antiquitates Italicæ Medii Ævi*, 6 vols., f°, Milan, 1738–42, and Index, 1896 (with the famous essay XLIII, *ib.*, 1740, iii, 807 f. : "De literarum statu, neglectu et cultura in Italia post Barbaros in eam invectos usque ad annum Christi 1100"). The contents (together with those of Migne's *Patrologia*, the *Acta Sanctorum, Monumenta Germaniæ Historica*, etc.) are given in A. Potthast, *Bibliotheca Historica Medii Ævi. Wegweiser durch die Geschichte des europäischen Mittelalters bis* 1500, 2 ed., 2 vols., Berlin, 1896.

102. Ref. 2, i, 81.

103. Ref. 16, 29, 115 f., 670–1, 676, 687–8, 688, 694, 700, 703 f.; in one place (p. 676) he says: "Die erste sichere Nachricht über den Gebrauch der Pulverwaffe stammt aus dem Jahre 1331 [he rejects Florence, 1326] : Zwei deutsche Ritter, von Spilimberg und von Kreuzberg aus dem Friaul, bedienten sich, wenn auch erfolglos, bei der Berennung der Stadt Cividale der Büchse." In other places (pp. 687–8, 700) he says the passage in Muratori is interpolated.

104. K. H. Schaefer, *Deutsche Ritter und Edelknecht in Italien während des* 14 *Jahrhundert*, Paderborn, 1911.

105. Ref. 101, 1729, xv, 396; Rathgen, ref. 16, 685 f., 703, says the relevant passages in Muratori were excerpted in Angelucci, *Della Artigleria da fuocco*, Turin, 1862, a work which neither he nor I was able to see.

106. T. F. Tout, *English Historical Review*, 1911, xxvi, 666–702 (669, 688).

107. Lacabane, ref. 9, 28; Favé, ref. 10, iii, 73–4; Rathgen, ref. 16, 700, says Rouen was "deutsche Ursprungsland."

108. Brackenbury, ref. 100, 291; Hime, ref. 2, 129.

109. Tout, ref. 106, 669.

110. Ducange, *Glossarium Mediæ et Infimæ Latinitatis*, Paris, 1840, i, 718 (*s.v.* bombarda); Daunou, *Histoire Littéraire de France*, 1824, xvi, 109, and refs.; Temler, ref. 33 (*a*), 10; (*b*), 177; Lacabane, ref. 9, 36, and Köhler, ref. 12, III, i, 229, say the year is 1339 (O.S. 1338) and the place is not Puy-Guillaume but Puy-Guillem in Périgord, occupied by the English and besieged by the French in March and April, 1338 (O.S.) or 1339 (N.S.). Brackenbury, ref. 100, 287, says the document quoted by Ducange (an unusually trustworthy author) is not known, but he probably saw it in the Chambre des Comptes in Paris. Temler, however, says Ducange did not claim to have seen it.

111. Kervyn de Lettenhove, ref. 90, iii, 498; Tout, ref. 106, 670.

112. Lacabane, ref. 9, 28 f., 51; Favé, ref. 10, iii, 74–5.

113. Kervyn de Lettenhove, ref. 90, iii, 499.

114. Kervyn de Lettenhove, ref. 90, iii, 495–9, who says fire-arms were first used effectively in the siege of Tournay, 1340, and large cannon about 1360; Hime, ref. 2, 129.

115. Lacabane, ref. 9, 28 f.; the refs. to Froissart are given by Favé, ref. 10, iii, 71, 76, 79, 84, 91, 101, 102, 104; the "140 cannon" at Odruik in 1377 should, as Rathgen, ref. 16, 485, pointed out, be shots (jetoient les canons dont il avoit jusques à sept-vingt carreaux de deux cent pesants qui pertuisoient les murs).

116. Ref. 10, iii, 105; cf. Brackenbury, ref. 164.

117. Clephan, ref. 58, 54.

118. Ref. 90, iii, 499.

119. *De Remediis Utriusque Fortunæ Libri Duo*, bk. ii, ch. 99; Rotterdam, 1649, 275; Gram, ref. 32, dates the dialogue after 1354; Rathgen, ref. 16, 705, after 1366. See Brackenbury, ref. 100, 296; Favé, ref. 10, iii, 79; Jähns, ref. 36, 1889, i, 228; Clephan, ref. 58, 60 (wooden cannon); Biringuccio, *Pirotechnia*, bk. x, ch. 5, Venice, 1540, 159 *v.*; Jähns, ref. 11, 792; Feldhaus, ref. 25, 401 (three-barrel wooden gun of 1620); Carman, ref. 172a, 61 f. (leather), 64 f. (wood); Feldhaus, ref. 55, 1914, 402 (leather).

120. Lacabane, ref. 9, 28 f.; Brackenbury, ref. 100, 287; Favé, ref. 10, iii, 80; Köhler, ref. 12, III, i, 262; Rathgen, ref. 16, 705.

121. Ref. 10, iii, 76–107. For "canon" in French inventories of the fourteenth century, see also Napoleon, ref. 10, 1846, i, 43; De La Fons-Melicocq, *De l'Artillerie de la Ville de Lille aux XIVe, XVe et XVIe Siècles*, 1855 (only 100 copies printed), is quoted by Rathgen, ref. 16, 700, for a Lille gun of 1340. Rathgen, ref. 16, 535, also refers to small bronze muzzle-loading guns, mounted on trestles, cast in 1340 in Dijon by Martin de Cornuaille (Cornwall) for the attack on Calais in 1346.

122. F. Alvin, *Biographie Nationale de Belgique*, 1903, xvii, 430, q. by Sarton, *Introduction to the History of Science*, Baltimore, 1947, iii, 725 ("the first gun made in the Low Countries").

123. Elaborate details of the making of the great Caen gun of 1375 (grant canon de fer) are given from the original documents by Favé, ref. 10, 1863, IV, pièces justificatives, xviii–xxxvi; Rathgen, ref. 16, 223, calls it "the first Steinbüchse in France."

124. Laurent, *Aachener Stadtrechnungen*, Aachen, 1866, 182, q. by Romocki, ref. 2, i, 76; Rathgen, ref. 16, 7–8, 697, says it is possible, but not supported by any evidence, that the Aachen gun was German, imported from the Netherlands.

125. J. S. Gardner, *Archæologia*, 1898, lvi, 133–64.

126. Barbour, *Bruce*, bk. xvii, l. 250; bk. xix, ll. 394–9; Jähns, ref. 11, 776; Hime, ref. 2, 128; Oman, ref. 139, 1924, ii, 213, accepts the statements of Barbour as genuine and "crakys of war" were "certainly cannon."

127. Ref. 100, 290.

128. BM MS. Cotton Nero E II, pt. ii, f. 397; *Les Grandes Chroniques de France, selon que elles sont conservées en l'Église de Saint-Denis en France. Publiées par M. Paulin Paris*, Paris, 1837, v, 460; Clephan, ref. 58, 71, says there are many versions of the *Grandes Chroniques*, and since the passage is in the anonymous part of the MS. it cannot be relied upon; Lacabane, ref. 9, 44, accepted it without comment. Napoleon, ref. 10, 1846, i, 41 f., refers to BM Sloane MS. 2433 (fourteenth-century), tome ii, f. 115.

129. Kervyn de Lettenhove, ref. 90, 1873, I, ii, 153.

130. Kervyn de Lettenhove, ref. 90, 1868, v, 46.

131. Gibbon, *Decline and Fall of the Roman Empire*, ch. lxv; Everyman ed., vi, 345. Napoleon, ref. 10, 1846, i, 41, had emphasised that Froissart does refer to guns at Crecy.

132. Rigollot, *Mém. Soc. Antiquaires de Picardie*, 1840, iii, 133; Cayrol, *ib.*, 185; q. by Kervyn de Lettenhove, ref. 90, 1873, I, ii, 34 f.

133. Ref. 90, 1873, I, ii, 149.

134. Ref. 9.

135. L. Lalanne, ref. 49, 1843, 348 f.

136. *Istorie Pistolesi ovvero delle cose avvenute in Toscano*, 8°, Florence, 1733, 294 (BM 1440.k.1).

137. *Chroniche di Giovanni Villani*, bk. xii, chs. 67–8; Muratori, ref. 101, 1728, xiii, 947–9; *Bibliotheca Classica Italiana Secolo* XIV, Trieste, 1857, no. 21, pp. 483–4.

138. Muratori, *Dissertazioni sopra le Antichità Italiani*, Milan, 1751, i, 456.

139. C. W. C. Oman, *A History of the Art of War in the Middle Ages*, 1905, 611; 2 ed., 1924, ii, 138. Napoleon, ref. 10, 1846, i, 17, says the long bow shot ten to twelve arrows a minute as against two to three for a cross-bow.

140. Hunter, ref. 23, 379 f.

141. Ref. 33 (a), 16 f.; (b), 188 f.

142. Feldhaus, ref. 55, 1914, 413 ("Temler-Heinze . . . weist die Interpolation nach"); Rathgen, ref. 16, 705 ("Dieses Schulbeispiel der Unsterblichkeit mancher Lügen mag als Warnung . . . dienen"); Hassenstein, ref. 15, 83, 91, 98, 114 ("eine erdachte Mär").

143. Gram, ref. 32; Brackenbury, ref. 100, 297 f. (long discussion); Jähns, ref. 11, 84; Favé, ref. 10, 81 (three cannon throwing lead balls); Clephan, ref. 58, 70; Hime, 2, 132; Tout, ref. 106, 670 f., 688 f. ("it is uncritical to wave aside the evidence for cannon at Crecy"); Salzman, ref. 85, 156; Oman, ref. 139, 1924, ii, 142, was still doubtful and "no English chronicler mentions guns at Crecy," but he does not even mention the use of cannon at Aljubarrota (1385), *ib.*, ii, 190. He admits the use of "ribaulds" and "guns of the more ordinary size" before Calais in 1346, *ib.*, 218.

144. Ref. 16, 8, 678, 680–2, 694, 705.

145. Ref. 16, 694.

146. Ref. 16, 8.

147. Hunter, ref. 23, 379–87; Burtt, *Archæological J.*, 1862, xix, 68–75.

148. Burtt, ref. 147; Tout, ref. 106, 670 f., 688 f., 690.

149. Tout, ref. 106, 689; Salzman, ref. 85, 156.

150. Burtt, ref. 23, 71.

151. Ref. 106, 685.

152. Tout, ref. 106, 673–4, 681, 691; Salzman, ref. 85, 158 f.

153. Tout, ref. 106, 675 f., 691 f.

154. Jähns, ref. 11, 847.

155. Salzman, ref. 85, 158 f.

156. Rathgen, ref. 16, 707.

157. Tout, ref. 106, 676–7.

158. Tout, ref. 106, 680.

159. Tout, ref. 106, 684; Oman, ref. 139, 1924, ii, 228, quoting Tout, *ib.*, for their specification in English accounts of 1373–5, and also an Italian statement for their use in 1364.

160. Ducange, *Glossarium ad Scriptores Mediæ et Infimæ Latinitatis, s.v.* bronzina; Halle, 1772, i, 807.

161. Tout, ref. 106, 682; Salzman, ref. 85, 156 f.; *id., English Life in the Middle Ages,* Oxford, 1926, 207.

162. Clephan, ref. 58, 91.

163. Salzman, ref. 161, 1926, 214, plate p. 209, showing cannon and mortars in use in a siege in 1460; ref. 85, 156–63; for a Chinese bronze hand gun for mounting on a stick, of 1421, of a similar type, but with a firing pan, unknown on European pieces of the same period, see Feldhaus, ref. 55, 1914, 424; Carman, ref. 172a, 90 f.

164. Brackenbury, *Proc. Roy. Artillery Institution, Woolwich,* 1867, v, 1–38 (cannon in 1351–1400); the part on English guns is mostly from Hunter, ref. 23.

164a. *Chronique de Bertrand du Guesclin par Cuvelier,* publ. by E. Charrière, 2 vols., Paris, 1839, in *Collection de Documents inédites sur l'Histoire de France,* 1ᵉ Série, verses 11067–9, vol. i, 388.

165. Salzman, ref. 85, 163 f.; C. ffoulkes, *The Gun-founders of England,* Cambridge, 1937.

166. Hassenstein, ref. 15, 66, 118; Salzman, ref. 85, 163 f.

167. Ref. 33 (a), 1781, i, 13.

168. Tout, ref. 106, 683, 687, 698.

169. C. ffoulkes, ref. 165, 12.

170. Anon., *Archæologia,* 1792, x, 472; plate xxxvii (dated 1768).

171. Gardner, ref. 125; C. Dawson, *Sussex Archæol. Collect.,* 1903, xlvi, 1–62.

172. And was still by Rathgen, ref. 16, 576, although, *ib.,* 708, he mentions cast iron guns at Wesel in 1398.

172a. For Henry VIII's guns, see W. Y. Carman, *A History of Firearms,* 1955, 38 f.

173. Holinshead, *Chronicle,* 1808, iii, 832; he also, *ib.,* 1807, ii, 777, refers to guns on a French ship captured at Sluys in 1386.

174. John Stow, *The Annales of England,* 4°, 1601, 983; Rhys Jenkins, *Trans. Newcomen Soc.,* 1921, i, 16 (21), thought the English cast-iron industry was of French origin.

175. Tout, ref. 106, 682.

176. Köhler, ref. 12, III, i, 302.

177. Favé, ref. 10, iii, 109, plate 5; Köhler, ref. 12, III, i, 302; Hassenstein, ref. 15, 119, 145; Rathgen, ref. 16, 516, 522 (charge); ffoulkes, ref. 165, 8.

177a. Carman, ref. 172a, 31.

177b. Lefevre de Saint-Remy (1394–1468), in *Histoire de Charles VI. Roy de France. Escrite par les ordres & sur les Memoires & avec les avis de Guy de Monceaux, & de Philippes de Villette, Abbez de Sainct Denys, par vn Autheur contemporain Religieux de leur Abbeye . . . Traduite sur le Manuscrit de M. le President de Thou Par Mrᵉ I. Le Labovrevr,* 2 vols., f°, Paris, 1633, ii, Suppl., 90.

177c. Carman, ref. 172a, 33.

177d. [John Rous] *Pageant of the Birth, Life and Death of Richard Beauchamp, Earl of Warwick*, ed. by Viscount Dillon and W. J. St. John Hope, London, Longmans, 1914 (publ. at 21s.); on Rous, see S. Lee, *Dict. Nat. Biogr.*, 1909, xvii, 318–20 (dates him 1445–91); for the gun, see Carman, ref. 172a, 21 (fig. 3), 33.

177e. Favé, ref. 10, iii, 101; Carman, ref. 172a, 36.

178. Rathgen, ref. 16, 531; 1399 at Frankfurt, *ib.*, 17 f.

179. Köhler, ref. 12, III, i, 234.

180. Favé, ref. 10, iii, 97; ref. 123; Brackenbury, ref. 164, says the "morceaux de fer" given by Favé, ref. 10, iii, 99, and thought to be projectiles were "marteaux de fer," iron hammers.

181. Favé, ref. 10, iii, 101; *Mémoires pour servir à l'histoire de France et de Bourgogne, contenant un journal de Paris, 2e Partie, Mémoires pour l'histoire de Bourgogne. État des officiers et domestiques de Philippe, dit le Hardy, Duc de Bourgogne*, Paris, 1729, 64, q. by Brackenbury, ref. 164, 11–13.

182. Garnier, ref. 91; a rare book by Garnier, *L'artillerie de la commune de Dijon d'après les documents conservés dans les archives*, 1863, is mentioned by Rathgen, ref. 16, 555 f., 558, as showing that Dijon had guns from 1354; see also Favé, ref. 10, iii, 96 (gros canons getans pierres, in 1374); *ib.*, iv, pp. xviii–xxxvi (Caen, 1375); Napoleon, ref. 10, 1846, i, 60 f.; O. Cartellieri, *The Court of Burgundy*, 1929. The details of the battles of Granson and Morat in 1477, when Charles the Bold lost his artillery and his life, are so discrepant in the works of reference that I omit them. Napoleon, ref. 10, 1846, i, 77 f., says Charles the Bold lost 113 guns (two large) at Granson in March, 1476, and sixty-three guns at Morat in June, 1476, and was killed at the siege of Nanci in 1477. Charles's daughter married the Emperor Maximilian, and the bulk of the Burgundian possessions, including nearly all the Low Countries, passed to the Hapsburgs. In 1516, Maximilian's grandson, Charles V, inherited Spain and all the Spanish dominions, and in 1519, when he succeeded Maximilian as Emperor, he inherited the Hapsburg lands in Germany. In 1521 Charles retained the Spanish and Burgundian part and gave the rest, increased in 1526 by Bohemia and a claim to Hungary and Croatia, to his brother Ferdinand.

183. Rathgen, ref. 16, 484, 556.

184. Rathgen, ref. 16, 484–5, 707.

185. Napoleon, ref. 10, 1846, i, 44–5; 1851, ii, 61 (Garitte, Rose, Seneca, Maye (or Marye), Montfort, Artique); Favé, ref. 10, iii, 126–8; Rathgen, ref. 16, 488 f.; Rosenwald, in *Nouvelle Biographie Générale*, 1854, x, 443, says the work was "traduit en anglais et imprimé par ordre du roi Henri VII en 1489"; Klebs, *Incunabula Scientifica et Medica*, in *Osiris*, 1938 [recte 1937], iv, 103, 332, gives it as *Art de Chevalerie selon Végèce*, Paris, 1488, and *Fayttes of Arms*, Westminster, Caxton, 1489. It does not seem to have been reprinted. It is often quoted (from a Paris MS.) by Napoleon, ref. 10, 1846, i. See ref. 242.

186. Rathgen, ref. 16, 491 f.; details of the four largest guns; table summarising details of Burgundian cannon, 498–504.

187. Rathgen, ref. 16, 495; Köhler, ref. 12, III, i, 269 f.

188. Rathgen, ref. 16, 497.

189. Favé, ref. 10, 128–9. Napoleon, ref. 10, 1846, i, 49–51, says artillery first became important in France under Charles VII (1403–61). It was said to have been brought from Germany, but there is much fiction in the accounts; John Bureau, who is said to have brought it, had actually been in English service: etiam sub Anglorum servitio ac ditione tali officio incubuerat.

190. Favé, ref. 10, 130–2.

191. Rathgen, ref. 16, 509–13, 525, 710; a table of iron guns from 1376 (Burgundian, 11-in. shot) to 1890 (cast steel, 8-in. shot) is given in *ib.*, 522–3.

192. Ref. 16, 540: "Deutschland war das Land, in dem des Büchsenmeister trugen ihr können in alle Lande."

193. Ref. 16, 546.

194. Ref. 16, 678–81.

195. P. Vidal de la Blanche, *Tableau de la Géographie de la France*, in E. Lavisse, *Histoire de la France*, Paris, 1903, i, 57–83.

196. Jähns, ref. 11, 777; Hime, ref. 2, 4–7.

197. Jähns, ref. 11, 772, 847.

198. Brackenbury, ref. 164; Thomas Walsingham, *Historia Anglicana*, ed. H. T. Riley, Rolls Ser., 2 vols., 1863–4.

199. Rathgen, ref. 16, 530 f., 683.

200. Berthelot, *Ann. Chim.*, 1891, xxiv, 439; *J. des Savants*, 1900, 10.

201. Favé, ref. 10, iii, 114; Berthelot, ref. 200 (1891), 500, 506 (BN MS. 7239, *c.* A.D. 1450, falsely dated 1330 or 1340 in title). Promis, ref. 242 (*a*), 180, and Atlas, plate IV, has a cerbottana without a tail.

202. Froissart, *Œuvres*, ref. 90, iii, 498, and often; Favé, ref. 10, iii, 135; Köhler, ref. 12, III, i, 229, 279–80, 315; Hime, ref. 2, 179 f.; Rathgen, ref. 16, 89 (Nürnberg, 1449–62); the Burgundians are said to have had 2000 ribaudequins in 1411; Oman, ref. 139, 1924, ii, 222 ("incredible," but why?). Napoleon, ref. 10, 1846, i, 40, 173, says 148 small bombards on a car were used at Padua by an army from Carrara. For multiple-fire guns, see Carman, ref. 172*a*, 74 f.

203. Köhler, ref. 12, III, i, 256 f., 287 f., 313 f.; Lippmann, ref. 2, i, 161.

204. *Lexicon de partibus ædium*, bk. ii, ch. 10, p. 253, q. by Hoyer, ref. 7, 1797, i, 41.

205. In Muratori, ref. 101, 1730, xix, 754; Jähns, ref. 36, 1889, i, 236; Köhler, ref. 12, III, i, 228 (more primitive and precise than any German account), 266 f.; Rathgen, ref. 16, 687, 691 (Steinbüchsen in France in 1374, in Italy in 1376, and in Germany 1377, but the Italian account may have been copied from some unknown German work).

206. Hassenstein, ref. 15, 45, 100; Feldhaus, ref. 55, 1914, 411, fig. 272.

207. Lippmann, ref. 2, i, 158.

208. Ref. 12, III, i, 286. Promis, ref. 242 (*a*), 160–3, says Meyer was in error in saying that mortars were not used before 1480; they are named (mortari) by Martini, ref. 242 (*b*), 128, and Promis refers to two iron mortars of the fourteenth century, one 0·160 m. diam. in the bocca, 1·25 m. in the tromba, total length 0·600 m., and the other with corresponding dimensions of 0·350, 1·5 and 0·950. They must have been used before 1464, when they are mentioned by Martini, since they were not invented by him. Carman, ref. 172*a*, 55, says mortars were made before 1420, and the breeches of very large cannon were used as mortars.

209. Ref. 24, 1572, 602 : hoc tormentum non videtur multum differe à nostris uomiglobis, quos uulgò Böler uocant.

210. Ref. 44, 211.

211. Ref. 161, 1926, 157 f.

212. Ref. 119, bk. vi, ch. 3; 1540, 80; bk. vii, ch. 9; 1540, 117 *v*.

213. Rathgen, ref. 16, 532, 562, 665 (1450), 710 (Naumburg); Romocki, ref. 2, i, 187, refers to the mention in German works of *c.* 1450 of "ain ainiger büchsenschuss mit ysin oder plyin Klötzen" (with iron or lead balls?), which were expensive.

214. K. Jacobs, *Feuerwaffen am Niederrheine*, Bonn, 1910, q. by Rathgen, ref. 215, and not available to me.

215. Ref. 16, 62–7, 78 (Hussite guns of 1420 weighed 2·9 kg.).

216. Köhler, ref. 12, III, i, 329 f.; Rathgen, ref. 16, 91.

217. Rathgen, ref. 16, 530 f., 549, 710; for hand guns at Higuerula (Spanish) in 1431, see Oman, ref. 139, 1924, ii, 180.

218. Ref. 16, 78 f.; for the Chinese gun, see ref. 163.

219. Hoyer, ref. 7, 1797, i, 104; H. A. L. Fisher, *A History of Europe*, 1949, i, 357; Oman, ref. 139, 1924, ii, 361 f., who says "real cannon," not ribaudequins, were mounted on the carts, "we may almost call them field-guns," and *ib.*, 229, that the Hussites were the first to use hand-guns, fired by a match and aimed from the shoulder, on a large scale. In the siege of Karlstein (1422) the Hussites threw in putrid matter from ballistæ, which the defenders covered with white arsenic; Napoleon, ref. 10, 1851, ii, 78–9. Some records say that corpses were so disposed of by defenders of towns.

220. Jähns, ref. 11, 1889, i, 275; Berthelot, ref. 200, 1891, 433; Feldhaus, ref. 55, 1914, 23; Rathgen, ref. 16, 709.

221. Rathgen, ref. 16, 686.

222. Köhler, ref. 12, III, i, 225–6; Rathgen, ref. 16, 686, 690, 705.

223. Rathgen, ref. 16, 686; sclopetum was the later Latin name for a pistol; Oman, ref. 139, 1924, ii, 228.

224. Köhler, ref. 12, III, i, 226; Rathgen, ref. 16, 686, 705.

225. Rathgen, ref. 16, 686; Köhler, ref. 12, III, i, 228, quoted a canonem a bombardis for 1341.

226. Köhler, ref. 12, III, i, 226; Rathgen, ref. 16, 686, 705 (mit Schaft).

227. Muratori, q. (in full) by Brackenbury, ref. 164, 26 f.; Favé, ref. 10, iii, 92; Rathgen, ref. 16, 686, 706.

228. Muratori, ref. 101, 1729, xv, 1037 (written after 1376); Köhler, ref. 12, III, i, 227; Rathgen, ref. 16, 686, 706 (1363).

229. Muratori, ref. 228, and ib., p. 134, for date (written after 1409); Rathgen, ref. 16, 686–8.

230. Köhler, ref. 12, III, i, 227; Rathgen, ref. 16, 706; Oman, ref. 139, 1924, ii, 228 (emphasising that these, and scioppi for Modena in 1364, were hand-guns).

231. Muratori, ref. 101, 1729, xv, 1042; Köhler, ref. 12, III, i, 227.

232. Muratori, ref. 101, 1729, xv, 182; Köhler, ref. 12, III, i, 227.

233. Köhler, ref. 12, III, i, 227; Rathgen, ref. 16, 706.

234. Rathgen, ref. 16, 707.

235. Köhler, ref. 12, III, i, 228; Rathgen, ref. 16, 707.

236. Köhler, ref. 12, III, i, 281.

237. Köhler, ref. 12, III, i, 287 f.; conical bores were intended to accommodate balls of different sizes; they were introduced in the early fifteenth century but were found to be unsuccessful; ib., 301, 309.

238. Rathgen, ref. 16, 572, 699.

239. Rathgen, ref. 16, 574–7.

240. Rathgen, ref. 16, 29.

241. Rathgen, ref. 16, 685 f., 700; B. Rathgen and K. H. Schaefer, "Feuer- und Fern-waffen beim papstlichen Heere im 14 Jahrhundert," in Zeitschr. f. historische Waffenkunde, Dresden, 1915–17, vii, 1–15 (referring to "Schaefer, Vatikanische Quellen," a work I have been unable to trace).

242. C. Promis, (a) Dell' arte dell' ingegnera e dell' artigliere in Italia sua origine sino al principio del XVI secolo, 4°, Turin, 1841 (BM 718.k.9), 123 f. and (b) Atlas (Atlante), Turin, 1841 (BM 1265.i.7); (c) Tratto di Architettura Civile e Militari di Francesco di Giorgio Martini, ed. Promis, Turin, 1841 (BM 1265.f.14). The work (c) contains two parts, each described as extracted from : Trattato di Architettura civile e militare di Francesco di Giorgio Martini, architetto del seculo XV, ora per la prima volta pubblicato per cura del Cavaliero Cesare Saluzzo, con dissertazioni e note, per servir alla storia dell'arte militaire Italiana, 2 vols., 4°, Turin, 1841, with atlas of 38 plates. This work was not available to me, but the Atlas seems to be the same as that named under (a) above. The work (c) is in two parts, the first a life of Martini and the second deals with the Saluzzo manuscript and gives extracts from it. The work (a) deals with Christina of Pisa (p. 17), Taccola (p. 23), Santini (p. 25), Alberti (p. 29), Valturio (p. 34), Leonardo da Vinci (pp. 44–52), Giambattista della Valle (p. 60), Biringuccio (pp. 63–6), Tartaglia (p. 69), etc. Pages 123–30 contain "Osservazioni preliminari sopra l'antichità e la nomenclatura delle artiglerie," with many references to books and MSS., and it then deals with the plates in the Atlas. Mines are dealt with in pp. 329 f. Some material from this book is included in other chapters. Muratori, ref. 101, Antiquitates, 1739, ii, 514, said that cannon were first used in Italy in 1378 in the war between the Genoans and the Venetians at Fossa Clodia.

243. Antiquitates Italicæ Medii Ævi, 1739, ii, 513–18.

244. Köhler, ref. 12, III, i, 335.

245. Ref. 16, 115 f., 136, 693.

246. Presumably in ref. 119, bk. vi, ch. 3, 1540, 78 v., where Biringuccio says (q. Cornazzano, probably his Opera bellissima dell'arte militare, Venice, 1493; Klebs, ref. 185, 112) that it is not known who invented the gun but it is believed that it came from Germany, where it was discovered by chance at least 300 years ago (i.e. about 1240). The French tr. of

Biringuccio by Jacques Vincent (1556) omitted many other parts of the work. C. S. Smith and M. T. Gnudi, *The Pirotechnia of Vanoccio Biringuccio*, New York, 1942, xxi, say the parts on gun founding were tr. by Rieffel, *Traité de la fabrication des bouches à feu de bronze au XVIᵉ siècle en Italie*, Paris, 1856.

247. Rathgen, ref. 16, 29.

248. Rathgen, ref. 16, 229.

249. Muratori, ref. 101, 1730, xvii, 1224.

250. *Le Opere Maccheronichi de Merlin Cocai*, Mantua, 1882, i, 117.

251. *Trattato di Architettura Civile e Militari di Francesco di Giorgio Martini*. Promis, ref. 242 (c), Turin, 1841, ii, 131; Romocki, ref. 2, i, 246.

252. Biringuccio, ref. 119, bk. v, ch. 3; 1540, 78 v.

253. Konrad of Megenburg, *Das Buch der Natur*, ed. Pfeiffer, 1861; (a), 91; (b), 274; (c), 453; Rathgen, ref. 16, 9.

254. Rathgen, ref. 16, 5 f.

255. Rathgen, ref. 16, 10.

256. Rathgen, ref. 16, 677.

257. Rathgen, ref. 16, 9–10; Sanutus was mentioned in ch. I. Since Konrad of Megenburg wrote as late as 1350, it is hard to see how Rathgen gets his date 1320–31 from him; he was a young student then.

258. Rathgen, ref. 16, 680–1.

259. Rathgen, ref. 16, 703.

260. Rathgen, ref. 16, 11–12.

261. Rathgen, ref. 16, 17 f.

262. Rathgen, ref. 16, 24 f. (he says, p. 23, that Büchse could also mean a box for gunpowder).

263. Rathgen, ref. 16, 27–8, 71; but see ref. 171 for early English cast iron.

264. Rathgen, ref. 16, 71.

265. Köhler, ref. 12, III, i, 263.

266. Köhler, ref. 12, III, i, 302.

267. Rathgen, ref. 16, 33–7, 708; the text is very confused but I have done my best to sort it out; on p. 705 it is said that only light guns were made before 1450.

268. Rathgen, ref. 16, 45, 48 f., 53, 698; Sarton, *Introduction to the History of Science*, 1947, iii, 1555, quoting information given him by Granscay, speaks of a "brass cannon fully charged with powder and lead," found on the site in excavations in 1849, with very large stone "cannon balls," arrows and cross-bow bolts, other weapons, and armour, adding that the castle of Tannenberg was destroyed in 1399 by Ruprecht von der Pfalz and his allies. Rathgen, 72, says the Tannenberg remains may be as late as 1450. There were two famous battles of Tannenberg, that in 1410 when the Prussians were defeated by the Poles and Lithuanians, and that in 1914 when the Russians were defeated by the Germans near Hohenstein in a battle conventionally called "Tannenberg" to obliterate the bitter memory of 1410.

269. Rathgen, ref. 16, 37 f., 53, 708.

270. Ref. 16, 58 f. (wrought-iron).

271. Ref. 16, 181.

272. Ref. 16, 54 f., 538 f., 709.

273. Ref. 12, III, i, 282.

274. Ref. 12, III, i, 295, 304; Rathgen, ref. 16, 417; for the description by Leonardo da Vinci, see Feldhaus, ref. 55 (1914), 400.

275. Rathgen, ref. 16, 59, 61.

276. Rathgen, ref. 16, 54 f., 57 f., 61, 68.

277. Ref. 12, III, i, 306–7, 316 f.; he says, 279, hand guns (Handbüchsen) first appeared in Germany in 1388.

278. Rathgen, ref. 16, 151–67, 705.

279. Rathgen, ref. 16, 168–85.

280. Rathgen, ref. 16, 280.

281. Rathgen, ref. 16, 190 f.

282. Rathgen, ref. 16, 205, 706, and (mixed up with Nürnberg records) 233 f.

283. Rathgen, ref. 16, 197, 706 (1371), 707 (1375).

284. Rathgen, ref. 16, 199.

285. Rathgen, ref. 16, 233, 237, 706; Köhler, ref. 12, III, i, 275 f.

286. *Geschichte des Wachsthums und der Erfindung in der Chemie in der ältesten und mittlern Zeit,* Berlin und Stettin, 1792, 139.

287. Rathgen, ref. 16, 262.

288. Feldhaus, ref. 55, 1914, 234, q. Jacobs, ref. 214, 83; Rathgen, ref. 16, 708.

289. Köhler, ref. 12, III, i, 251.

290. Rathgen, ref. 16, 269.

291. Rathgen, ref. 16, 272–4, 705; gunpowder, 1354–88.

292. Rathgen, ref. 16, 308.

293. Rathgen, ref. 16, 322–48; the last use of arrows in guns is otherwise dated 1352 (p. 705), 1373 (p. 706) and 1382 (p. 707). Rathgen did not live to revise the proofs of his book.

294. Rathgen, ref. 16, 448.

295. Rathgen, ref. 16, 74.

296. Rathgen, ref. 16, 477, 706.

297. Rathgen, ref. 16, 378–9, 680, 705.

298. Feldhaus, ref. 55, 1914, 415, fig. 273.

299. Rathgen, ref. 16, 675, 707.

300. Hoyer, ref. 7, 1797, i, 48; Köhler, ref. 12, III, i, 224; Rathgen, ref. 16, 706.

301. W. H. Prescott, *History of the Reign of Ferdinand and Isabella,* London, 1849, i, 430.

302. Rathgen, ref. 16, 566–73, 708, who says, p. 694, that the Spanish artillery was a century behind the German, never reached the standard of that in other lands, and, since Spain and Portugal were rather backward, Köhler was wrong in assuming a Moorish origin of artillery.

303. Ref. 106, 676.

304. Prescott, ref. 301, i, 133, 135.

305. Prescott, ref. 301, i, 229, 231.

306. Prescott, ref. 301, i, 229–31.

307. Prescott, ref. 301, i, 382, 387, 431.

308. Prescott, ref. 301, i, 389, 391–2, 404, 412.

309. Prescott, ref. 301, i, 430.

310. Prescott, ref. 301, i, 428; cf. ref. 302.

311. Prescott, ref. 301, i, 429.

312. Berthelot, *Science et Philosophie,* 1905, 131; N. M. Karamsin, *Histoire de l'Empire de Russe,* 11 vols., Paris, 1819–26; 1820, v, 143–4, 473, 485.

313. Some large guns (some only reconstructed from descriptions, e.g. those of Frankfurt) are shown in plates 5–7 in Rathgen, ref. 16. They include Frankfurt Steinbüchsen of 1349 (17 kg.), 1377 (636 kg.) and 1399 (3270 kg.), all bronze (plate 5); the Brunswick "Mette" (1411; 9000 kg., bronze); the Dresden "Faule Magd" (*c.* 1410; 1300 kg., wrought iron); a Burgundian Steinbüchse at Basel (*c.* 1430; 2000 kg., wrought iron) (plate 6); The Vienna Steinbüchse (*c.* 1425; 10,000 kg., made from wrought iron rings and bars); a Burgundian Steinbüchse at Basel (*c.* 1450; 2000 kg., iron) (plate 7); cast iron guns from Landskron, nr. Neuenahr (*c.* 1400; 115 kg., in Berlin) and Pössneck, Thuringia (*c.* 1400, in Nürnberg); Vögler, Göttingen (*c.* 1400); a stone ball, 47 cm. diam., 110 kg., at Tannenberg, 1399 (plate 8); hand guns (Tannenberg, bronze, 1·2 kg.), before 1399; Berlin Zeughaus (8·5 kg., *c.* 1350), and others (plate 9); old reconstructions of wrought iron chamber guns from Metz (*c.* 1420; in the Berlin Zeughaus) and from Danzig (*c.* 1450; 10·5 cm.) (plate 10); for illustrations of old guns and mortars, see Carman, ref. 172a, *passim;* Feldhaus, ref. 55, 1914, 399, 401–30, 437, 441–7, 493; Hassenstein, ref. 15, *passim,* especially, 142 f.; and Demmin, ref. 334.

314. For a connected account of the siege of Constantinople in 1453, see Gibbon, ref. 131, ch. lxviii; Everyman ed., vi, 417–61; Finlay, *History of Greece,* Oxford, 1877, iii, 500–24, for the guns, 506, where Urban is called a Wallachian; for a Turkish account of Saʿd al-Dîn,

G. de Tassy, *J. Asiatique*, 1826, viii, 340; J. Kananos, *De Bello Constantinopolitane. Anni . . . Christi* 1422, in Migne, *Patrologia Græca*, 1866, clvi, 65; Chalkokondyles, *De Rebus Turcicis*, v; ed. Bekker, Bonn, 1843, 231; P. Wittek, in D. Ayalon, *Gunpowder and Firearms in the Mamluk Kingdom*, 1956, 142; G. Schlumberger, *La siège, la prise, et le sac de Constantinople par les Turcs en 1453*, 3 ed., Paris, 1914, and L. Brehier, *Vie et Morte de Byzance*, Paris, 1947, add nothing to Gibbon's account which is of any interest to us; Ayalon, 4, 7, 140 f., says the first reliable report of the use of firearms by the Ottomans is in the siege of Antalia in 1425, but Kananos was a contemporary witness; Finlay, 490, says cannon were first used by the Ottoman Turks in a siege of Constantinople in 1422, but they were inferior and produced little effect.

315. Ref. 131, ch. lxvi; Everyman ed., vi, 356.

316. Chalkokondyles, *De Rebus Turcicis*, v; ed. Bekker, Bonn, 1843, 231.

317. Doukas, *Historia Byzantina*, xxxv; ref. 44, 246–7; for the Hungarian nationality, see Gibbon, ref. 131, p. 426. Reinaud and Favé, *Du Feu Grégéois*, 1845, 213 f., at first thought that the use of gunpowder originated in Hungary, near the mouth of the Danube, but they later gave up this idea; Jähns, ref. 36, 1889, i, 224 f.

318. Ref. 317, 273.

319. Ref. 317, 465.

320. Ref. 316, viii, p. 385.

321. Ref. 316, 451. Promis, ref. 242 (*a*), 140, 144, says the Turkish guns threw balls made from Italian marble, weighing 689 kg.

322. Ref. 317, xxxviii, p. 272.

323. Chalkokondyles, ref. 316, 382, 385–9.

324. Kritoboulos, *De Rebus Gestis Mechemetis*, in C. Müller, *Fragmenta Historicorum Græcorum*, Didot, Paris, 1870, v, 40–161 (76–9); Gen. J. H. Lefroy, F.R.S., *Archæological J.*, 1868, xxv, 261 (tr. from Paris MS.); Clephan, *ib.*, 1911, lxviii, 107; picture in C. ffoulkes, *Arms and the Tower*, 1939, 180. A description of the cannon and some analyses of the metal from which it was made are given by Abel, *British Assoc. Report*, 1868 (1869), II, 34. The bronze is not uniform in composition, the percentage of tin varying from 4·71 to 9·75 in different parts. Abel thought the gun had been cast from a number of smaller pieces melted together, but this is doubtful.

325. Evliya, *Narrative of Travels in Europe, Asia and Africa*, tr. Hammer, London, 1846, I, ii, 55 f.; Carra de Vaux, *Penseurs de l'Islam*, 1921, i, 251.

326. *Memoirs of the Baron de Tott on the Turks and the Tartars*, tr. from the French, 2 vols., London, 1785, ii, 40 f., 71 f.; for a later description see Baron von Moltke, *The Russians in Bulgaria and Rumelia in 1828 and 1829*, tr. by Lady Duff-Gordon, London, 1854, 428 (28 in. bore gun).

327. Ref. 314, iii, 506.

328. Köhler, ref. 12, III, i, 291.

329. C. ffoulkes, ref. 165, 15; ref. 324, 180. Three similar cannon from Rhodes of 1420, 1482 and 1500 (?) were presented by the Sultan to the Germanisches Museum, Nürnberg: Rathgen, *Ostasiatische Zeitschrift*, 1925, ii, 204.

330. Ref. 44, 211.

331. Ref. 44, 211; the note by Bulliaud, p. 596, says : τεάφης per τ scribitur in Cod. MS., sed in dictionariis Græcobarbaris θειάφι sulphur. verba porro illa, ὀσμὴν οὖν ἀσπίνθος τῆς σπινθῆρος πυρός, εἰ πλησιάσειεν, reddidi ut in textu leguntur. et pro ἀσπίθῳ ἄσφαλτον voluisse dicere probabile est, cum ἀσπίδος Græcobarbara vox idem significet ac σπινθήρ. deinde quis odor scintillæ illius pulveris accensi, nisi sulphureus, cui bituminis odor proximus est. Hime, ref. 2, 3, mentions that the Latin word nochus, a hazelnut, was used in Germany for bullets, particularly projectiles discharged from machines, although Konrad Kyeser, in his *Bellifortis*, used it for a smoke-ball.

332. Joannes Leunclavius, *De Rebus Turcicis*, in Migne, *Patrologia Græca*, Paris, 1866, clix, 201/375, 209/390.

333. Leonardus Chiensis, *De Expugnatione Constantinopolis*, in Migne, ref. 332, 1866, clix, 923–44 (927–9); Gibbon, ref. 131, ch. lxviii; Everyman ed., vi, 420, 426; [G. M.] Thomas, "Die Eroberung Constantinopels im Jahre 1453 aus einer venetianischen Chronik," in

THE LEGEND OF BLACK BERTHOLD

Sitzungsber. königl. bayer. Akad. Wiss., Philos.-philol. Classe, Munich, 1868, pt. ii, pp.1–41; cf. *id., ib.*, 1864, pt. ii, pp. 67–80.

334. From Feldhaus, ref. 55, 1914, 407; Favé, ref. 10, iii, 168 f., plate 10 gives rather different figures; the Vienna one is from Köhler, ref. 12, III, i, 289, plate IV, 3, and Clephan, ref. 324, 94. See also R. Mallet, *Trans. Roy. Irish Acad.*, 1856, xxiii, 313 f., for good descriptions and illustrations. The Mont St. Michel guns mentioned above, ref. 177, are shown in a plate in Rathgen, ref. 329, 196 f. For some interesting illustrations of guns and firearms of all types, see A. Demmin, *Die Kriegswaffen in ihrer historischen Entwicklung*, 2 ed., Leipzig, 1886, 661 f. See also the descriptions and plates in Napoleon, ref. 10, 1846, i (with the Pièces justificatives, 357–83, covering material dealt with in refs. 121 f.) and 1851, ii.

335. Feldhaus, *Lexikon der Erfindungen*, Heidelberg, 1904, 16; Hime, ref. 2, 126; this old date was still given by Rathgen, ref. 16, 517, 707.

336. Rathgen, ref. 16, 510 f., 518, says it is probably from Mons and "cast" before 1460; but, *ib.*, 522, that it is older than Dulle Griete and the two, together with the Burgundian iron cannon at Basel, are earlier than 1430; they are all cannon rather than mortars or Steinbüchsen, although some later guns, e.g. Vogel Greif, 1529, in the Paris Artillery Museum, have chambers like Steinbüchsen, *ib.*, 524. Mallet, ref. 332, 321; Carman, ref. 172a, 27, 45.

337. Hewitt, *Archæological J.*, 1853, x, 25; picture in ffoulkes, ref. 324, 174.

338. Rathgen, ref. 16, 524, gives its weight as 39,000 kg. instead of 1955 given by Feldhaus, and *ib.*, 529, mentions a 35-in. diameter iron mortar of date 1858, in Woolwich Arsenal. A 26-in. ball said to have been shot by the English against the French at Louviers in 1431 is doubtful; Köhler, ref. 12, III, i, 292.

339. Rathgen, ref. 16, 230 f.; Lippmann, ref. 2, i, 160, dated it *c.* 1350; Hewitt says it was reported to have been captured by the Turks and recaptured in 1529 by the Austrians. Carman, ref. 172a, 55, calls it a mortar.

340. Rathgen, ref. 16, 230, 233 (Faule Magd), 514–5; Favé, ref. 10, iii, 128 f.; Köhler, ref. 12, III, i, 292 f. (who says a 519 cwt. gun cast at Nürnberg in 1445 is a myth); Feldhaus, *Die Technik der Antike und des Mittelalters*, 1931, 360 (Catherine); Carman, ref. 172a, 30 (Catherine).

341. Rathgen, ref. 16, 459, 468, 709.

342. Köhler, ref. 12, III, i, 292, 296–8, for other guns, *ib.*, 299 f.; French guns in Favé, ref. 10, iii, 170, plate 11, figs. 1, 2, are similar, in pieces screwed together, and Köhler, 301, thinks they may have had conical bores.

343. *Ballistics in the Seventeenth Century*, Cambridge, 1952; H. J. Webb, *Isis*, 1954, xlv, 10–21; see also ffoulkes, refs. 165 and 324. References to many older works dealing with artillery are given in Sir H. Elliot, *The History of India as told by its own Historians*, ed. by J. Dowson, 1875, vi, 455–82. For English guns in the Tower and Woolwich Arsenal, see Carman, ref. 172a, 32–3; for seventeenth-century English guns, *ib.*, 43 f.

Chapter IV

MISCELLANEOUS TREATISES ON MILITARY ARTS

THE present section deals with some treatises on military arts mostly written in the fifteenth century and with miscellaneous topics. It may be regarded as a supplement to the sections on the discovery of gunpowder and the invention of guns. Some of the material is still in manuscripts, descriptions of which have been given, some has been published.

The oldest illustrated manuscript of this type (Essenwein dated it 1390–1400, Jähns 1350) is Codex german. 600 of the Munich Staats-Bibliothek (paper, 22 ll., folio), a copy of which is in Germanische Museum (Nürnberg) MS. 25661.[1]* This consists mostly of crude coloured illustrations of guns, making, testing and purifying saltpetre, a recipe for gunpowder (4 lb. salniter, 1 lb. sulphur, 1 lb. charcoal, 1 oz. salpetri, 1 oz. sal ammoniac and $\frac{1}{12}$ of camphor; the salpetri may be what the *Feuerwerkbuch*, ϵ (see p. 155), calls sal practica) by grinding the components in a mortar; how to make bad gunpowder good, purifying saltpetre by melting and adding sulphur, loading the gun with a stone ball, firing a gun, testing a new gun by firing it without a ball. Sulphur is tested by holding a piece in the hand, when it should crackle (as it does). There are pictures of hand-guns and a multiple-gun (four guns on a flat board; Jähns says there is an actual gun of this type with five barrels, early fifteenth century, in an Italian museum). It is noteworthy that no large cannon are shown.

Berthelot[2] described some manuscript treatises on military arts with numerous reproductions of illustrations (redrawn and retouched):

α Munich Staat-Bibliothek, Codex Latinus Monacensis (CLM) 197, fifteenth century (see ch. II for the *Liber Ignium* in this MS.).
β Paris, Bibliothèque Nationale (BN) Latin 11015 (A.D. 1335).
γ Paris BN Latin 7239 (fifteenth century).
δ Göttingen Univ. 63 phil., written 1395–1405, of Kyeser's *Bellifortis*.

α *Munich CLM 197.*

This is a composite work; the dates 1427, 1438 (on a bombard) and 1441 occur in it. Parts of it are copied into Paris 7239 (*c.* 1450) and also into a St. Mark's Venice manuscript (1449), and one in Vienna. Part I (ff. 1–48) is by a German military engineer and is in German; it describes and illustrates guns (puchssen) and two guns pointing opposite ways on a carriage; a diving suit and helmet with a breathing tube on floats (crudely shown, without tube, in Kyeser's *Bellifortis*, see δ, p. 146), devices for lifting cannon, an armoured car called "der Hussenwagen" (as used by Ziska in the Hussite wars, see p. 118), and what seems to be a box of cannon balls. Part II is in Latin and is the notebook of an Italian military engineer. It contains (f. 13) recipes for gunpowder and the text of Marcus Græcus (ff. 74[77] *r.*–75[78] *r.*; the

* For references see p. 177.

144

manuscript has been renumbered, the present numbers being those in square brackets) ending : Explicit liber ignium a Marcho Greco compositus; recipes for waters, candles, aqua ardens, aqua inextinguibilis, etc. It includes (see p. 52) a preparation of alcohol, and mentions balm, camphor, sulphur, native sulphur, filtered olive oil, naval pitch, Greek pitch, turpentine, peghola (a kind of pitch), dry varnish, filtered honey, boiled wine, alcohol (aqua ardens), lard, whale oil, fats of various animals and serpents, and gunpowder, as incendiary materials or components, which are thrown by hand or by machines.

There are pictures of chimneys, flat bellows worked by an overshot water-wheel blowing a fire, very like those in Agricola's *De Re Metallica* (1556), all kinds of mechanical and hydraulic apparatus, and a profusion of figures of guns, some mounted on boats, one throwing a large pike (as thick as the bore of the gun) with two weights attached by chains, and others throwing large incendiary arrows. Guns made in detachable pieces which can be fitted together (the old British army "mountain guns") are shown. There are methods for attacking ships and figures of a rotating mitrailleuse different from the ribaudequins or clustered cannon of other treatises. A recipe for gunpowder gives 16 oz. saltpetre (salnitri), 4 sulphur and 3 willow charcoal. All kinds of more or less useless additions (arsenic, camphor, spirit of wine, vinegar, etc.) to the powder are specified. The powder is also directed to be wetted with alcohol and dried in the sun, which is not really a preparation of corned gunpowder, described in another work, the *Feuerwerkbuch* (see ϵ, p. 154). There is also a picture of a man, the microcosm, in a circle. Thorndike,[3] who has recently described the manuscript, says Marianus Jacobus (Taccola, see p. 172) is once named in it, but the German part of the manuscript has probably no connection with him.

After a recipe for gunpowder containing arsenic sulphide, the writer adds in Italian : "Diav(o)lo aj(u)taci te; amen (may the Devil help you)."[4] It probably made the powder more brisant, since arsenic is included among a list of things in Paris 7239 (moisture, camphor, arsenic, alcohol (aqua vitis) and mercury) which when mixed with powder cause guns to burst.[5] The Munich manuscript has the earliest illustration of a dam with a movable door, and there are illustrations of ships with paddle wheels, divers in dress, etc.[6]

γ   *Paris BN 7239 (fifteenth century).*

This manuscript (very clearly written) shows large siphons for inundating enemy country or passing water over a mountain (*sic*), flat bellows used as water-pumps, a gun on a carriage with a shield in front, mortars shooting incendiary bombs nearly vertically, one having the powder chamber and touch-hole horizontal and the barrel vertical; a bombard with tail (cerbotane), a mounted man with a small gun and burning match (like rope), and various mechanical apparatus. It concludes with a theory of the structure of the earth, which is partly covered with water by reason of its weight, the other part rising above because of the air in its pores and the central fire; this is reminiscent of Leonardo da Vinci.[7] Thorndike[8] says BN 7239 and St. Mark's

VIII.40 are by Marianus Jacobus Taccola, not Paolo Santini. The title is *De Machinis Libri Decem.* It is in the Munich manuscript used by Berthelot, and in a manuscript belonging to Count Wilczek, which in 1924 was in the castle of Kreutzenstein in Austria, and had been described in that year by Laborde. But, Thorndike says: "I am inclined to question whether the text in BN 7239 is not more the work of Santini than of Taccola." Jacopo Mariano Taccola, also called Marianus Jacobus, Marianus Taccola, or "Archimede," of Siena, fl. *c.* 1438–49, made many annotated technical drawings contained in several manuscripts (Munich CLM 197 (1438); St. Mark's Venice Lat. XIX.5 (1449); the so-called "Constantinople manuscript," Paris, BN 7239, fifteenth century, copy in a Vienna manuscript, 23172, Count Wilczek collection). The St. Mark's manuscript treatise, *De Machinis Libri Decem,* has pictures of incendiary ships.[9] It contains material from Valturio. Incendiary mixtures contain sulphur, naval pitch, tow, turpentine, alena ex oleo uncta, but the word pulvere seems to mean dust and not gunpowder. The mixtures were contained in tubs (like Joinville's, ch. I), which could be thrown from ballista or lowered by cranes on to other ships. There are also pictures of incendiary contrivances for land warfare, including armoured incendiary cars. Some of the pictures in BN 7239 are found in a *De Rebus Bellicis* of the later fourth century.[10] Thorndike, who (as usual) criticises Berthelot mildly, gives a summary of the figures in BN 7239 and Munich 197, with comments on the mechanical devices. He says BN 7239 is in an Italian hand of the fifteenth century; the date "1330 vel 1340" in the manuscript itself is "a century too early." BN 7239 and CLM 197 differ in several respects. The MS. BN 7239 is supposed to be earlier than the capture of Constantinople (1453) and to have come from the Seraglio ("dit du Serail"). It was purchased by the King of France from the Turkish Ambassador in 1687–8. The work was previously ascribed to an otherwise unknown Paolo Santini Ducensis (from Duccio) or Lucensis (from Lucca).

Thorndike doubts the frequent assertion that Taccola borrowed from Valturio's *De Re Militaria* (see p. 164). Taccola dedicated his work to the Emperor Sigismund on his way to be crowned in January, 1433. Valturio did not die till 1482; his work was printed in 1472, there is a manuscript of 1463 (BN 7236). "It would . . . seem more likely that Valturio made use of Taccola's designs, or that they drew from common sources." Valturio's manuscript has guns, cannon, eight guns on a revolving wheel, and "a spiked ball or bomb whence flames issue." Another book of Taccola's of 1433 on engines and edifices (Florence MS. FN Palatine 766) is mostly illustrations.

δ   Kyeser's *Bellifortis.*

A very interesting work is the *Bellifortis* of Conrad Kyeser of Eystadt (b. 25th August, 1366; d. *c.* 1405), a pompous military engineer. It contains hundreds of illustrations, some copied from Valturio and Vegetius. The Göttingen manuscript, written in 1395–1405, is a copy of an unknown original; it consists of 140 parchment leaves and 243 written pages, and has a portrait of Kyeser. A later manuscript (*c.* 1410) in the library of the Prince of

Fürstenberg, Donaueschingen, has additional material.[11] There are several other manuscripts.[12] It is written in bad Latin verse and prose and divided into ten books, on war chariots, siege engines, hydraulic machinery, incendiary weapons, the peaceful use of fire, and tools. It mentions a ship with paddles for sailing against the current (perhaps by an axle connected with the paddles winding up a chain),[13] two kinds of diving suits, with short tubes not reaching to the surface of the water and apparently with some kind of air-reservoirs, scaling-ladders, devices (spikes in covered pits, etc.) for defence and an early type of portable firearm on a stick (Stangenbüchse).[14]

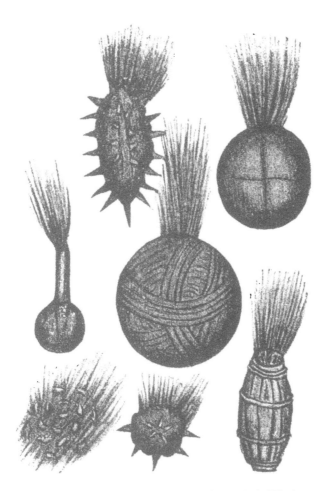

Fig. 1. Bombs, one bursting, shown in Kyeser's *Bellifortis*.

The first part has a section on astrology, the signs of the zodiac and the six planets (Venus is omitted) being depicted as horsemen of appropriate colours and carrying symbolic banners. The work shows war chariots, some with multiple-barrel guns (Froissart's ribeaudeaux, ribaudequins, German Totenorgel). There are incendiary arrows and a rocket (an illustration apparently from an Arabic manuscript since the man has Arabic dress). The rocket is a cylindrical bag, tied at the ends and is shown without a stick, although the text says it should have a stick fastened to the body, which will serve to steer it (ei quasi remus). A rocket with a heavy charge must have a hollow (German "Seele") in the charge so that this can burn rapidly. Kyeser was aware of this, since he says the charge must be bored (in extremitate perforabis cum sobula vel cum stylo). He also knew that the propulsion was caused by the rush of gas, since he says the powder container must be gas-tight (quod aer non possit exire). What Favé[15] calls rockets (fusées) as used at Amiens in 1418 for throwing fire and Greek fire seem to have been mortars; the fire was made from 1½ oz. of white varnish, 10 lb. of "soufre en brique," 1 oz. of amber, 1 lb. of black pitch and 1 oz. of camphor. It was kept in earthenware jars (terrines) or in bags of sheepskin (etuis à fusées de feu gregéois faits de peau de mouton).

The guns have a cylindrical rear part for the powder (sometimes with a solid extension) with a touch-hole, and a wider front portion, sometimes cylindrical, sometimes bell-mouthed, for the stone shot, which is held by wooden wedges and packed round with straw. Sometimes the gun has a large ring cast on it through which a rope is passed for moving the piece. As compared with the Munich and Paris manuscripts written about fifty years later, there are few pictures of guns and firearms; a small piece carried by a soldier is shown, also a horse with an incendiary on its back, which appears in Arabic manuscripts (ch. V).

An illustration shows six small hand-guns fixed round a hexagonal block of wood and all pointing forward; another illustration shows six small guns (with tail-extensions, cerbotane) fixed by metal straps to a flat board with an extension fitted with a small wheel. These are evidently intended to revolve and the text says:

> Contus ille magnus pyxidum sex stat revolvendus
> Emissa prima redit altera demum secuta
> Decipiuntur hostes, post primam non timent ullam

for the first figure, and for the second:

> Hæc rota movetur per circumferentiam istam
> Pyxis nam post pyxidem statim mittit lapidem
> Hostis sic decipitur per hoc atque fallitur.

These are primitive revolvers (the name is in the text) and apparently fired stone balls, so that the guns were probably small Steinbüchse rather than ribaudequins.

The text of the *Liber Ignium* of Marcus Græcus (anonymous) is included,[16] and there is a recipe for alcohol (aqua aurea), made by distilling red wine with orpiment, arsenic and sulphur, or quicklime, or with calcined tartar and then quicklime (which would give a very strong alcohol). There are recipes for magic candles and lamps, and phosphorescent materials (including glow-worms) reminiscent of Marcus Græcus. Greek fire, however, now includes powdered dried human blood. The cracker or rocket case for gunpowder is of parchment (as in Roger Bacon) bound with string. Various recipes for gunpowder include sublimed arsenic, realgar, sublimed mercury, crude mercury and metallic mercury (mercurium vivum). Mercury as an addition to gunpowder is useless,[17] although it long continued to be specified.[18] In one place the mercury is called servus fugitivus (the Arabic name) and it was perhaps added because of its supposed volatile nature. Amber is also specified. One illustration[19] shows a bomb bursting into fragments, and it is called a dracon (ad dracones istos; evolat ignis laceratus frangens id quod carpit: ita damna plura sequuntur). The powder for it (pulveres . . . ad dracones) contained small amounts of orpiment and amber, and was moistened with strong brandy. This bomb is more fully described in the *Feuerwerkbuch* of 1420 (see ε, p. 157). A bomb (bombis) is mentioned in Flemish archives of 1300.[20]

Sebastian Munster[21] says: Hoc tormentum non videtur multum differe à nostris uomiglobis, quos uulgò Böler uocant, nisi quòd incendiarium, id est, globus, puluere bombardico extorquetur . . . postquam bombardæ usus cœpit, machinæ omnium impetuossimæ quæ à Germanis paulò supra 100. annos fuit inuenta. This weapon is not very different from our shot-thrower, commonly called a mortar, except that incendiary which is a globe stuffed with gunpowder, which began to be used after bombards, the most violent of all machines, which was invented by the Germans not much above a century ago (Munster wrote about 1540).

In Kyeser the quicklime-sulphur mixture of Julius Africanus (ch. I), which inflames with water, is improved by adding saltpetre and succo ceparum and making into balls. The addition of quicklime to gunpowder to keep it dry is mentioned. Flying dragons towed by a string and containing burning oil or oil mixed with gunpowder (as in Chinese accounts) are described and illustrated; the recipes include the words "Npcfndp" and "ut scis." The device is called a "tygace volante."[22]

There is a description of a cask of poisoned wine left for consumption by enemy troops.[23] A poisonous smoke is generated by throwing powdered "magnet" on a fire; in Hippolytos (c. A.D. 230) the sensation of an earthquake is said to be produced by burning the ordure of a weasel with the stone magnet on an open fire.[24] Kyeser also gives directions for making fragrant smokes, also given later by Kircher[25]; these smokes may well have served to conceal the smell of arsenic with which they were poisoned by mixing orpiment with the ingredients, since poisonous arsenical smokes are described in the *Feuerwerkbuch* (ε, p. 158).

A curious "mine detector" is the use of thick smoke generated by burning resin under an inverted tub which is then lifted off; the smoke spreads over the surface of the ground and if it rises in any place, the earth there has recently been disturbed (nam ubi se movit fumus vel exerit idem ibi fode: certe invenies quæque terrata).[26]

A rather confused description of a nitre bed is given (which Berthelot said is the earliest known), wood-ashes being added as well as lime, and Kyeser also proposed to work in large pots[27]:

Ollis magnis paribus simile poteris operare
Imponendo stratum, super stratumque limphando
Conditis prædictis, præparatisque condimentis
Hæc locabis humo vel cellario adæquato
Infra quindenam collige, sed pluries funde
Hæc ars nitri salis, ab exule stat repertata.

What Berthelot and Romocki thought was a turpentine blow-lamp (æolipile) is really a Püstrich, a device for blowing a fire by steam emitted through a small orifice from a vessel containing water and laid on a fire. If surreptitiously filled with turpentine it would produce a disastrous fire in a castle or house to be attacked. The Püstrich was in the form of a human head, the steam being emitted through the mouth. An old one found in a castle at Sondershausen was thought to be an old Slav idol, Perkunas.[28]

Hollow bones filled with explosive and fitted with fuses were to be strewn on the floors of dining-rooms (in the prevailing fashion), but were to be used only against Turks. Illustrations of bombs (some, according to Hime,[29] hand-grenades) are given, and there is a picture of a tree blown up by gun-powder inside the trunk.

A recipe for nitric acid or aqua regia[30] (aqua corrosiva, aqua Martis) "according to the books of the goldsmiths," specifies distilling Roman vitriol with saltpetre, or with sal ammoniac (sal Armoniacum) mixed with saltpetre (permisce Assionis; assio is the old Arabic name for saltpetre) and camphor, in a cucurbit with an alembic joined to it with philosophical lute, over a gentle fire (in cucurbita cum alembico lutato luto sapientiæ et destilla cum lento igne). The "first water" (prima aqua), which dissolves iron with bubbles (hence aqua Martis), is used as a cautery, for removing writing, and for dissolving (devouring) all metals. When mixed with an equal weight of alcohol (aqua vitæ), and a candle brought near, it inflames (eadem aqua cum pondere simili aquæ vitæ admiscatur generat ignem in continenti ita quod candela possit accendi). Common water distilled over good quicklime seven times is supposed to become very acute (erit acutissima), so that if sal ammoniac is added it cannot be kept in a copper vessel.

The mixture of all kinds of recipes is like that in Marcus Græcus and in later "books of secrets." Romocki emphasised that Kyeser was in Italy and probably got many of his ideas there; many parts of his work are reminiscent of Leonardo da Vinci, who was born fifty years after Kyeser's death and probably used the same Italian sources.

Many recipes are the same or almost the same, and they are copied without order and in most cases apparently without understanding of their meaning. They were taken from collections like those used by Albertus Magnus, probably several small complete treatises being incorporated unchanged. (Romocki's suggestion that they were all taken from a large single book is improbable.) Kyeser probably parted with good money to Jews for the privilege of copying these often unintelligible texts, in the hope that at some future time they might be interpreted by some expert and would come in useful. There is a striking resemblance of many recipes with those given by Albertus Magnus (see pp. 81–6), and they almost certainly came ultimately from Arabic works, probably translated in Spain. One or two specimens will suffice for comparison with those in Albertus Magnus.

1. Accipe pannum aut stuppam et intingue in petroleum id est oleum extractum de lateribus, cui circumpone cera aut sepum, candelam formabis ad libitum, post aqua perfunde et accende et non extinguetur nec a vento nec a pluvia.

This is a recipe for a torch. Petroleum is not "oil of bricks" (see p. 156), and there is no point in pouring water over the torch before lighting it. The writer of the recipe did not understand it.

2. Quum vis accendere lampadem et videre quod sit homo niger astans et in eius manu virga ita ut fugiant homines ædis ex eo. Accipe ranam viridem et decolla eam super pannos exequiarum virides et madefac ipsam cum sambucino et impone in ipsam licmen et illumina cum eo in lampade viridi nam tu videbis nigrum stantem inter cuius manus erit lampas et in manu eius virga et est mirabile multam.

3. . . . Accipe ergo piscem qui dicitur elrahed in basare et est piscis Æthiopiæ et nominatur delphinus, tere ergo ipsum, deinde accipe pannum funeris et sperge super ipsum aliquot zemar, deinde funde super illud ex illa pinguedine liquefacta, deinde accende ipsum in lampade viridi et erit illud quod dixi tibi.

4. Accipe sulphur pallidum et commisce cum oleo viridi et pone in lampade in quo sit licmen de panno nigro, glauco vel flaveo, quæ dum accensa fuerit in domo videbunt se astantes pallidos fore, quod si apposueris de cerebro asini aut de lacrimis eius, videbis homines habere capita asinina.

5. Ut scintillæ exeant de aqua. Recipe ovum et evacua subtili foramine, post imple testam illam cum pulveribus sulphuris vivi et calcis vivæ, foramen cum cera obstruendo, proiciendo ad aquam ex exsiliet magna flamma. Recipe vinum ut scis imponendo sal et loca in carbones vivos, adde desursum candelam et ardebit, sed vinum ardens cum mercurio impositum clarius lucebit et fortius.

6. Ignis græcus sic fit. Recipe nitri .iij. lotones salfanii .viij. lotones sulphuris lotones pulverisa simul et adde .ij. lotones petrolei. Item salis nitri .ij. lotones sulphuris .j. lotonem carbonum tiliæ .j. lotonem pulverisa postea recipe calcem vivam .j. lotonem et dispone sacculum ut scis ad telum ponendo calcem vivam in imo sacculi.

The first recipe here is for gunpowder which, since it also contains the old component, petroleum, is still called "Greek fire." A similar composition is in a manuscript of 1338[31] (Munich Staats Bibliothek 4350, which is mostly in Latin and of theological content), which Jähns calls "das älteste abendländische Pulverrezept" (he has forgotten Roger Bacon's nearly a century earlier), in which another old component, varnish (firnis glaz, glassy resin), is added with a drop of mercury:

> Item dass ist daz pvlver damit man auss der physs schiest. Da soll man nemen zv zway tail lindains oder sælwaidens kols vnd zway tail sal petre. vnd denn daz fvnftail sol man avch nemen svlfvr vivi oder rechten swefel. vnd derzv sol man avch nemen firnis glaz den zehnden tail. vnd daz sol man alz derr machen. vnd sol ez inder ain ander stozzen zv ainem pvlver in ainem morser. vnd wenn man ez in die pvhss tvt . . . zv dem pvlver giezzen ainen tropphen kecksilvers.

The second recipe in (6) is for a gunpowder containing some quicklime to keep it dry (the saltpetre used being impure). The addition of quicklime to old gunpowder to make it usable is definitely specified in another place[32]:

> Pulveres si pyxidum cum viva calce commisces
> Albea præclara, poteris sagittare recenter.

7. Fortissimi pulveres pyxidum. Recipe arsenici sublimati, mercurii sublimati, mercurii crudi, reubarbi ana .j. lotonem realgaris etiam .j. lotonem, mercurii vivi adponendo .ij. guttas simul misce, superadde .j. libram salis nitri, fac pulveres et admisce octavam partem tiliæ et quartem partem sulphuris.

The useless addition of rhubarb to the equally useless mercury and mercury sublimate (mercuric chloride) is a curiosity.

ε   The *Feuerwerkbuch*.

An important work contained in many manuscripts, most of which also contain the *Liber Ignium* and other works, calls itself the *Feuerwerkbuch* (in diesem buch das da haisset das fürwerckbuch). Some manuscripts in which it is found are:

A.   Laurentianus (Biblioteca Medicea Laurenziana, Florence), MS. LV 4, *c.* 1437.[33]

B.   Vienna Nationalbibliothek MS. 3062, *c.* 1437.

C.   Berlin General Stab 117, now *m.* 4, *c.* 1450.[34]

D.   Dresden C.62.

E.   Paris Dépôt général de la guerre 276, *Bellicorum instrumentorum*, 1420.[35]

Manuscripts B and E were mentioned by the Benedictine Bernard de Montfaucon[36] (1655–1741). The fifteenth-century manuscripts are discussed by Jähns.[37] The *Feuerwerkbuch* was formerly attributed to Abraham von Memmingen (1417),[38] who was master-gunner to Duke Friedrich IV of Austria, or to Conrad von Schongau,[39] but the actual author is unknown.[40] It has been dated *c.* 1425,[41] or *c.* 1422,[42] or even 1400,[43] but the latest editor, Hassenstein,[40] dates it 1420. It is contained in numerous manuscripts,[44] the text varying somewhat. Hassenstein's is in no sense a critical text, since it merely reproduces a late printed form of the work. His date, 1420, is only conjectural and his text probably contains later interpolations, so that conclusions drawn from it, that this or that was known to German gunners in

1420, are unwarranted.   The repetitions in it show that it was pieced together from several sources, no doubt of very different dates.

Ginsburger[45] referred to a manuscript of the *Feuerwerkbuch* written in Hebrew script, and claimed the invention of gunpowder for a Jew, Typsiles, in Augsburg in 1353 (which is a century too late).   What Hassenstein, a fervent Nazi, would have said of the last is perhaps best not imagined.   The *Feuerwerkbuch* was printed as an appendix to *Flavii Vegetii Renati vier Büchern*

Hye nachuolget vonn Büchsen geschoß / Puluer / fewerwerck / wie man sich darmit auß ainer Statt / Feste / oder Schloß / so von Feynden belägeret wer / erretten / Auch sich der Feind darmit erwören möchte.

Fig. 2.   Beginning of the *Feuerwerkbuch* in Vegetius, *Vier Büchern von der Ritterschaft*, Augsburg, 1529.

von der Ritterschaft, in Augsburg in 1529,[46] from which it was reproduced in facsimile, with a transcription in modern German and many reproductions of illustrations in other manuscripts, by Hassenstein.[40]

Le livre de secret de l'art de l'artillerie et canonnerie, from the first half of the fifteenth century, regarded by Favé[47] as the oldest treatise on artillery, contains recipes for purifying saltpetre and making gunpowder and Greek fire, which appear to be taken from Marcus Græcus, etc. According to Jähns[48] the work, taken by Reinaud and Favé from a fifteenth-century (*c.* 1430) Paris MS. BN 4653, is largely a translation of the German *Feuerwerkbuch*. It was printed in 1561.[49]

The *Feuerwerkbuch*, after a short introduction, begins with a dialogue between a master-gunner and an apprentice, and then continues with the descriptive text in five parts:

(1) Descriptions of the materials of gunpowder:
    (*a*) Saltpetre.
    (*b*) Sulphur.
    (*c*) Charcoal.
(2) On ordinary gunpowder.
(3) On special powders.
(4) On special guns and methods of shooting.
(5) On different incendiary compositions, etc. (from Marcus Græcus).

An analysis of the contents was given by Romocki.[50] It contains the first description of making "lump powder" (Knollenpulver): the powder is stamped with good wine vinegar with a wooden pestle, and the stiff paste is rolled into balls (zu rechten knollen) which are dried in the sun or in a warm room. Another way of making it is to melt sulphur and then stir in powdered saltpetre and charcoal. In the nineteenth century, as Romocki showed, powder was heated until it baked together, or was melted to a hard mass in shells.[51] The process is dangerous, but the powder is more resistant to atmospheric moisture; it was not uncommon in the eighteenth to early nineteenth century for large stocks of powder to become useless from moisture (partly because of impurities in the saltpetre).[52] Although it has been claimed[53] that this was "corned gunpowder," Rathgen[54] emphasised that the product is not corned powder but "lump powder" (Knollenpulver), and says that true corned powder was first used in Nürnberg in 1449–50. It was too strong for use with the earlier cannon.[55]

A kind of tracer bullet for use at night was made by coating the stone cannon ball with a melted mixture of resin and "inslatt" (tallow), then dipping it into gunpowder; the ball leaves the gun burning.[56] Various kinds of coloured gunpowders are described: white (saltpetre, sulphur, "Felber-baumholz," sal ammoniac and camphor, or wyszekolen, perhaps starch); red (saltpetre, sulphur and sandaly = sanders-wood); blue (saltpetre, korn-plumen (cornflowers) and sawdust); yellow (saltpetre, sulphur and spicanardi (a yellow flower? or Indian nard?)).[57]

Fig. 3. Testing gunpowder by burning a sample and afterwards examining the residue. (Vienna Hofmuseum MS., 15 cent.)

In a series of twelve questions and answers for a master and pupil,[58] a theory of the explosion of gunpowder is given: "Has saltpetre or sulphur the force to drive the stone? I answer, both, since when powder is kindled in the gun (entzündt wird in der büchse) the sulphur is hot (hytzig) and the saltpetre cold, and heat cannot suffer (leyden) cold, nor cold heat; they are two opposite things. Although the one cannot suffer the other, yet one without the other is useless." The function of the charcoal is to preserve the fire (hegt aber das füer). There is mention of a gas (tunst) evolved in the explosion of powder in an experiment in which a pound of good powder is exploded in a closed wine cask, and the gas bursts the cask[59]: "das bulffer ze hant verbrunnen vnd bricht der tunst das vass." The explosive force of gunpowder was later ascribed to the evolution of gas by van Helmont[60]: "Constat enim sale petræ . . . sulfure & carbona, quia juncta, si accendantur, vas non est in natura, quod occlusum, non dissiliat, propter Gas. . . . Ergo illa apposita se mutuo in Gas convertunt, per destructionem."

A large number of recipes for purifying saltpetre by crystallisation contain nothing new (alkali is not used); they were obviously collected from various sources. It is said that saltpetre from Venice (imported from the East) is often adulterated, as is found by dipping the hands in it, when they become wet (from deliquescent calcium salts). Saltpetre is said to be collected from walls wetted with saltpetre mother-liquor or urine, or solutions of tartar and common salt. The processes of purification, e.g. with sharp lye, vinegar, etc., are not good, since common salt (an impurity which should be removed) is sometimes directed to be added to the solution to be crystallised. The writer was, however, aware that common salt is soluble in cold water and can be washed out of saltpetre by this means.[61] The powder was tested by burning it on a board and examining the residue. Powder is said to keep better when mixed with barrasz (? borax, or sugar) and vinegar.[62]

A recommended addition to gunpowder is a "salpractica," which is pronounced "salportica" or "salpertica," which quickens and strengthens all powder. It is made by taking saltpetre (salbetter) recrystallised three times, covering it with brandy, adding 4 lot (1 lot $= \frac{1}{32}$ lb.) of sal ammoniac and 1 lot of camphor, boiling down to a quarter, then pouring into a porous earthenware pot which is kept in a cellar. The efflorescence forming outside the pot is scraped off with a hare's foot and is salpractica. If 1 lot is added to 30 lb. of powder it improves it greatly, but 1 lb. to 30 lb. is better.[63] It was, of course, useless. Tin powder is also directed to be added to the gunpowder.[62]

The supposed "improvement" of sulphur by grinding it with mercury and then "pouring" it into brandy is useless. Apparently dark-coloured sulphur was thought to be the best. Even charcoal was thought to be improved by treating it with ink (atriment). The compositions for powder are 4, 5 or (best) 6 saltpetre, 2 sulphur and 1 charcoal (or 4 : 1 : 1 in one case).

There is a very obscure recipe for making "good oil of sulphur" (güt schwebel öl).[64] Powdered sulphur is mixed with salpracticum (saltpetre treated with brandy, sal ammoniac and camphor) and distilled vinegar.

The mass is dried and put into a cucurbit and alembic, then put at first over a slow fire in an ash-bath, then the fire is increased "till it begins to drop" and until no vapour is seen to come from it.

> . . . wie man gut schwebel öl machen soll . . . nimm schwebel wie vil du wildt vnnd stoss in gar wol vnnd thu darzu Salpracticum auch wol gestossen, vnd wol gemischt vnder einander, vnd schütt darzu acetum bene distillatum, vnd lass es wol sieden in einem verdektan hafen biss es wol trucken werd, vnd thu es in einen kukurbit vnd alembic [alent in printed text] darauf, vnd lutier [leuter in printed text] das gar wol, vnd setz das auff ein öfelin vnd äschen, vnd mach ein gutt fewr darzu, biss es anfahe tropffen, vnd mach darnach ein gross fewr daz du keinen dunst mehr sehest davon gan.

Romocki[65] said this oil of sulphur was "nothing else than sulphuric acid," which was produced in the eighteenth century by Ward by burning sulphur and saltpetre under a bell. Romocki refers to the descriptions in the works attributed to Basil Valentine[66] (really composed about 1590 by Thölde) and in Libavius,[67] only the first of which resembles the description in the *Feuerwerkbuch*.

There is also a recipe for a water to be used instead of powder (mit wasser onn pulver)[68]:

> Take saltpetre and distil that to water and sulphur to oil and sal ammoniac also to water, and add oleum benedictum according to the weight you will hear of . . . six parts of saltpetre water [nitric acid], two parts of sulphur water, three parts of sal ammoniac water, two parts of oleum benedictum. Load the gun tight with blocks (klotzen) and stones, then pour in the water to the tenth part, behind the tamping (klotz), light it as quickly as you can. See, the gun is very powerful. With an ordinary gun you can shoot with this water 3000 instead of 2500 paces. But it is very expensive (ist aber gar köstlich).

On the assumption that oleum benedictum was a light coal-tar fraction (Teeröl, Benzol), Romocki and Hassenstein concluded that the result was "ein gefährlicher Nitro-Explosivstoff ganz moderner Natur." Even if the oleum *was* benzene, the resulting nitrobenzene would not be an explosive and the gun could not be fired. It was probably only an imaginary experiment.

The printed text does not say how oleum benedictum is made, but Romocki[69] quotes a recipe from the Berlin manuscript (Zeughaus 1, A.D. 1454); old olive oil (allt pawmöll) is taken and bits of broken stone heated to redness are dropped into it. The mass is powdered and put into a glazed retort coated with clay. The retort is gradually heated until, when it is hot enough, a red oil is seen to come over (bis dv siechst rott öll fliessen). This red oil must be kept away from fire since when it inflames it cannot be extinguished. This is "oil of bricks," as Libavius[70] says: "ole. benedictum dicitur oleum destillatum ex lateribus imbutis oleo oliuarum, &c." Its composition is doubtful but it would certainly not be a tar-oil.   A poem of the early fifteenth century quoted by Romocki[71] also refers to "shooting with water" and gives the same description, as does Reinhardt von Solm in his *Kriegsbuch* (1559) also quoted by Romocki.   The translation in the *Livre du canonnerie et artifice de feu*[72] is in this case quite incorrect, since the "waters" are given as mere solutions of saltpetre, etc.; otherwise the translation is good.[73]

Romocki[74] refers to the later well-known experiments of Borrichius and Rouelle on the inflaming of mixtures of nitric acid and turpentine, but these do not seem to have much bearing on the *Feuerwerkbuch*.   The oil of bricks (oleum laterinum), also called oil of the philosophers or oleum benedictinum, is mentioned in Arabic treatises (ch. V), and by Roger Bacon and Albertus Magnus (ch. II).   It was a red fetid empyreumatic oil distilled from pieces of bricks soaked in olive, nut, hemp, or linseed oil, and was long described in textbooks of chemistry.[75]   Romocki's suggestion that it was a tar-oil containing benzene or aromatic hydrocarbons was adopted by Lippman,[76] Rathgen[77] and Hassenstein.[78]   Poisson[79] had correctly identified it.   (Benzene was discovered in a liquid separating from oil-gas by Faraday in 1825.)

A kind of shell (Kyeser's dracon, see p. 149) was made by cramming gunpowder made into a paste with brandy into a hollow iron ball, plastering over the outside with molten sulphur, and covering this with saltpetre and sulphur and fustian.   Then a hole is made through to the hole in the shell, and through the wooden plug in the gun (durch den klotzen), which communicate with each other, so that when the gun is fired the flame starts in the hole in the shell: the powder in the shell begins to burn slowly, so that the shot can be fired before it goes off on the target.[80]   Since the composition could not very well be filled through a small hole in the shell, this was probably in two hemispherical pieces, afterwards fastened together, as illustrated in a work of 1472 by Valturio (see p. 165).   A "Furkugel," a wooden ball filled with powder and scrap iron; a "Sprinckloit," a hollow stone ball with a sulphured wick (so beweyset er wase crafft er kan) and a fuse (Federkiel), are bombs.[81]

A paraphrase of the *Feuerwerkbuch* (Germanische Museum, Nürnberg MS. 1481a)[82] has a description of a hand-grenade, the shell being fitted with a fuse and thrown.   Senfftenberg, commander of artillery in Danzig in 1550–1600, in a manuscript written some years later than 1566, describes and illustrates in detail bombs with fuses.[83]   A longer version of the *Feuerwerkbuch* in a Berlin manuscript [Kgl. Zeughaus 2] contains several interpolations (which seem to be largely incorporated into Hassenstein's text).   In this it is said that gunpowder will explode under water, three centuries before the Amercan David Bushnell, in a submarine, torpedoed the English ship *Eagle* in 1776.[84]   It says a ball of gunpowder and oil, coated with sulphur, burns in water (so prynnen flamen vnter dem wasser), and powder in a wooden box explodes under water (zerpricht vnd zerslecht waz vmb sy ist, es sey ym wasser oder auswendig).

The *Feuerwerkbuch* says nothing of Roman candles or rockets, and in some paraphrases of it in which they are mentioned they are for use as fireworks rather than weapons. It does refer to a straight gun-tube with alternate charges of powder and iron or lead plugs (klötzen) bored so that the flame can pass from the top, where the powder is kindled, through the borings filled with powder (ain klotzbuchs mit vil klötzen).[85] For defending thick walls against attack the *Feuerwerkbuch* describes making a hole through into the cavity excavated by the enemy and shooting through it balls the size of an apple composed of a mixture of 30 lb. of resin, 30 lb. saltpetre, 10 lb. sulphur, 6 lb. charcoal and 10 lb. arsenic, when a thick poisonous smoke is formed (gewinnt also grossen tunst vnd rauch, denn Arsenicum mit dem schwebel ist ain grosses gift vnd brinnent also sere).[86] Mouldy cheese water (containing sulphuretted hydrogen) in bottles was also used in attacks underground.[87] Large mines or masses of the then very expensive gunpowder appear only in somewhat later treatises (some of which are contained in manuscripts of the *Feuerwerkbuch*).[88]

The mediæval method of measuring time by prayers (Paternosters) or the recitation of creeds, very common in Europe, especially in Catholic countries, was used by gunners in timing their fuses.[89] The *Feuerwerkbuch* contains a formula for blessing a gun, ending in the magic words "Eler Elphat Sebastian non sit Emanuel benedicte," and gunners in Germany were expected to be highly religious men, making the sign of the cross over the cannon ball when fixing it in the gun.[90] The Italian potters in the sixteenth century also lighted their kilns "always in the blessed name of God," or with "prayers offered to God with all the heart," or "in the name of Jesus Christ,"[91] and they were probably no better men than the German gunners.

η *Miscellaneous Manuscript Technical Treatises.*

Besides the manuscripts mentioned above, there are many others dealing with military matters, a few of which have been published. A very brief survey of some of these will be given; some of them have not yet been examined in detail.

I. Giovanni di San Gimignano, or Giovanni de Goro, Joannes Gorinus de Sancto Geminiano (d. 1323), an Italian Dominican, mentions Greek fire in bk. ix of his *Summa de exemplis et rerum similitudinibus*, completed not later than 1313 and first published (Deventer, 1477) under the pseudonym "Helvicus natione Teutonicus."[92]

II. MS. Bodleian Digby 67 (fifteenth century) contains part of a letter from Ferrarius, a Spanish monk, addressed to Anselm, which is regarded by Clephan[93] as in the same class as Roger Bacon and Marcus Græcus. This letter "suo amico Anselmo Ferrarius" is also contained in BM Sloane 2579 (fifteenth century; Sloane 1754, fourteenth century, has the prologue only). Mrs. Singer[94] identifies him with Adam Ferrarius, d. 1383, or Johannes Ferrarius, 1386–1446 (who would perhaps be too late). The Bodleian (Digby 67, f. 33 *v.*) and British Museum (Sloane 2579, f. 85 *r.*) manuscripts give a recipe for an inextinguishable fire which contains sulphur, colophony, amber (glassa), turpentine, etc., which are to be distilled in an alembic, put on cotton (bombax), and thrown from a ballista. The British Museum manuscript only (Sloane 2579, ff. 85 *v.*–86 *r.*) contains a recipe for gunpowder which is made from 1 lb. of sulphur vivum, 2 lb. of charcoal of willow or lime (carbonum salicis sive tilie), finely ground on a marble or porphyry, and put into a short and thick case for "thunder" or a long and thin case for "flying fire." This seems to be the same account as that in Roger Bacon and Marcus Græcus.

III. MS. Ambraser 49, *c.* 1405, of the Kaiserhaus (now National Bibliothek), Vienna, *Allerley Kriegsrüstung*, twenty-eight parchment and fifteen paper leaves, folio, has rough drawings of war chariots, a diver with helmet, swimming apparatus and a gun.[95] The Ambraser manuscripts were taken from Ambrass Castle to the Vienna library in 1665.

IV. MS. fol. 328, *Wunderbuch*, in the Grossherzogliche Bibliothek, Weimar, begun *c.* 1430, with pictures of court life made *c.* 1520; the work of several hands, 329 parchment leaves, quarto, incorporates Kyeser's *Bellifortis* and has pictures of a corkscrew borer, a flying dragon, a püstrich, etc.[96]

V. MS. Ambraser 5135 (old Ambraser 52), Kunstsammlung, Kaiserhaus, Vienna, *c.* 1435, anonymous, *Streydt-Buch von Pixen, Kriegsrüstung, Sturmzeuch und Feuerwerckh*, partly in verse, illustrated. It says the master gunner (Büchsenmeister) should be able to read and write and should not be drunken; he should know the chemical operations, have a knowledge of old siege engines and new guns, and be able to build fortifications. In particular, he should be able to shoot accurately. The purification of saltpetre is described. The illustrations show "Rollschuss" (a beginning of ricochet shot), "hail shot" (Hagelschuss, a kind of grape shot), hedgehog shot (a kind of shrapnel), a subterranean powder mine, and hand-grenades. There are the "twelve questions," found in the *Feuerwerkbuch*.[97]

VI. Martin Mercz (Mertz, Merz), a gunmaker of the Palatinate, composed in 1471 an illustrated *Kunst aus Büchsen zu Schiessen* (Hauslab Library, Vienna; later Prince of Lichtenstein Library MS.; also in Munich MS. of 1475), which deals mostly with "harquebusery." Mercz died in 1501 at Amberg, and his tombstone in the church has his portrait in low relief, showing him wearing an eye-shield (he lost his right eye, probably in the course of his work). This is said to be the earliest monument of an artillerist.[98]

VII. The famous *Mittelalterliche Hausbuch*, compiled about 1480 by an anonymous arquebus maker in South Germany (Heidelberg or Speyer?), in the library of the princes of Waldburg-Wolfegg-Waldsee in Wolfegg Castle, Würtemberg. The artist is the so-called Master of the Amsterdam Cabinet, and the work, a kind of pictorial encyclopædia, is an important source for mediæval life and technology. The page and plate references below are to the 1912 edition.[99] The *Hausbuch* has some words written in Hebrew letters, so that the author may have been a Jew. It describes the treatment and assaying of ores with fluxes and antimony (anthiorum) (pp. 27–9), and illustrates very accurately (plates 38–9) various kinds of metallurgical furnaces, mines (plates 32, 37, 40), stamps worked with water-wheels (plates 43–4) and double bellows (plates 41–2). It describes the manufacture of alum from alumstone and its crystallisation and purification; saltpetre is made in pits containing quicklime, dry straw and earth stratified, wetted daily with urine for three weeks, then extracted in four copper boilers up to 12 ft. diameter with hot water, and crystallised (p. 29).

Gunpowder is described somewhat as in the *Feuerwerkbuch*, but equal parts of saltpetre, sulphur and charcoal are used, with or without camphor and amber; incendiary objects and a mixture of gunpowder, realgar, orpiment and Hüttenrauch (arsenic oxide) for making poisoned smoke in saps, are described (p. 41). A powder mill (?), guns and ballista with carriages, instruments, a waggon barricade and an army on the march are illustrated. One heavy gun is shown underslung below the axle; a gun mounting permitting a change of elevation (shown in Favé) is, according to Jähns, a considerable advance over earlier ones. Soap (including scented) made from tallow and causticised potash solution, artificial pearls and aqua fortis (akua firtis) are mentioned (pp. 20-1). Brandy (aqua vitæ) is distilled "per alempicum" and used for adding to gunpowder and making tinctures; a test of strength is that it burns off the finger without causing pain; vinegar is made from poor wine or beer (pp. 19-25, 40-1). There are sections on dyeing, cleaning and preserving from moths (pp. 25-6). The large number of animal, vegetable and mineral substances specified in recipes include sugar (zucker, cucker) and sugar candy (cucker candit).

VIII. Ulreuch Bassnitzer, born in Landshut, composed in 1489 an illustrated manuscript (Heidelberg University Palat. Germ. MS. 130) on machines, *Der Gezewg mit seiner Zugehorunge* (Gezewg = Gezeug, MHG geziug), including weapons.[100]

IX. Philip Monch, born in the Palatinate, composed in 1496 an illustrated *Buch der Stryt vn(d) Buchsse(n)* (Heidelberg University Palat. Germ. MS. 126, with Monch's portrait) on guns and harquebuses, including a gun borer worked by horse-power. He seems to have been an ironsmith.[101]

X. Ludwig von Eybe zum Hartenstein composed in 1500 at Amberg an elaborate *Kriegsbuch* (Erlangen University MS. 1390) based on Kyeser,[102] and containing descriptions of waggon parks used in the Hussite wars. Most of the above manuscripts, and a number in the British Museum Sloane collection,[103] await detailed study.

FONTANA

Giovanni da Fontana (1395?–1455?), a Venetian engineer, M.A. Padua, doctor of arts 1418, refers to "the horrid machine called a bombarda," and marvels that so much force is generated by such a very little powder (ex quibus est orrida machina quā(m) bō(m)bardā(m) appellamus at diruē(n)dā(m) o(mn)ēm fortē(m) duritiē(m) etiam marmoreā(m) turrē(m) non minus impietatis q̄(uam) ingenii fuisse existimo qui primo adinuenerit . . . vnde tā(n)tā(m) vim habeat a pusillo pulvere nō(n) admirari).[104] Thorndike[105] showed that the book containing this statement, printed as a work of Pompilius Azalus, is probably really by Fontana.

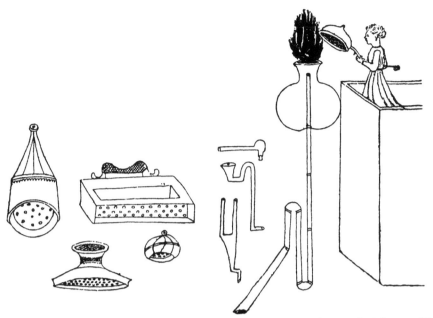

Fig. 4. Apparatus depicted by Fontana. *Left*, forms of clepsydras and siphons. *Right*, raising water by suction; a fire is kindled in the globe, then the lid put on and, as the heated air cools, the water rises.

A long analysis of an illustrated manuscript by Fontana in Munich (Stadt Bibliothek Cod. iconogr. 242), written in secret script with a later decipherment in Latin, is given by Romocki.[106] It is entitled *Bellicorum instrumentorum liber cum figuris et fictivis literis conscriptus*. Birkenmajer[107] says the manuscript is later than 1420 (as suggested by Romocki). The Munich manuscript speaks of Arabian, Persian and Egyptian weapons and musical instruments; it mentions guns only three times, but suggests the use of rockets in propelling torpedos on the surface of water. It also illustrates a magic lantern (without lenses), automata, organs, a letter-lock and dredges.[108]

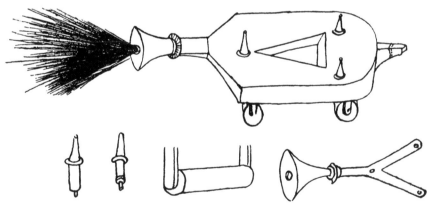

Fig. 5. Jet-propelled car (Fontana).

The many applications of rockets are noteworthy.   One illustration shows what appear to be bombs shot from an immense bombard mounted in the entrance of a castle against ships on a river.   Artificial birds, hares and fish filled with incendiaries are shown as flying, rolling and swimming rockets, and there is a clear illustration of jet propulsion of a car on wheels by a rocket, and rocket torpedos (aquaticum igniferum, quia ignem per aquas defert, fic sicud illud quod per terram et quod per aërem defert ignem).   Fontana was within reach of the jet-propelled kite.   A very few years ago these illustrations and suggestions would have produced a languid smile of amusement on the self-satisfied face of the up-to-date scientist and technician, but now the smile would be Fontana's.

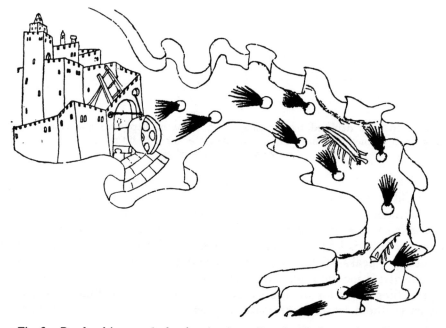

Fig. 6.   Bombard in a castle for shooting incendiary bombs into a river (Fontana).

A *Metrologum de pisce, cane, et volucre* by Fontana (Bologna MS. 2705)[109] deals with clocks, mechanical devices, a hot-air balloon, a diving bell and experiments on the amounts of gunpowder needed to shoot rockets of different weights.   It describes how to give a rocket the appearance of a flying dragon or a demon emitting fire from the mouth and leaving a stench; to make a torpedo or magic candle which rises and sinks in water as stronger and weaker powders burn in succession; also a diabolical figure filled with chemicals which burn under water and emit rays of fire.   Fontana notes that any animal with lungs needs air just as a candle does to burn, but not for the same reason, although many think so; perhaps he is anticipating Leonardo da Vinci here.

A *Speculum almukesi compositio*, ascribed in a manuscript to Roger Bacon and dealing with burning mirrors, may be by Fontana, and a *Protheus*, perhaps also by Fontana, deals with hydrostatics and devices on the lines of Heron of Alexandria, also (in a section on hydraulic organs) on the natures and specific gravities of metals in relation to musical consonance.[110]

A large work by Fontana, *Tractatus de trigono balistario abbreviatus*, in an Oxford manuscript, Bodleian Canonicus Misc. 47,[111] did not seem to me to contain anything of present interest, but I was not able to examine it in detail. It was written in 1440, and is part of a collection purchased in 1817 from the Abbot M. C. Canonicus of Venice.

FRANCESCO DI GIORGIO MARTINI

Francesco di Giorgio Martini, born in Siena in September, 1439, an architect, sculptor and painter, studied in Rome, and then worked in his native town; he wrote about 1465 a treatise on civil and military architecture, printed from a Turin manuscript in 1841 by Saluzzo and Promis,[112] which gives recipes for powders for guns of different sizes:

	Saltpetre.	Sulphur.	Charcoal.
Bombards and mortars throwing 200 lb. and over	7	4	3
Smaller bombards and mortars, and spingardes ..	4	2	1
Passevolants, basilisks, cerbottanes, arquebuses ..	8	3	2
Escopettes	14	3	2

For the smaller arms the saltpetre must be well refined, the sulphur pale yellow and the charcoal freshly made. In making large quantities of powder, the three substances must be kept separately and mixed at the time the powder is wanted, since nitre and sulphur change and neutralise one another (l'uno la virtù dell'altro corrode a impedisce) with time. It is better to store wood than charcoal, since charcoal absorbs much moisture, which spoils the powder. A secret for keeping powder for a long time is to wet it to a paste with clear and very strong vinegar (aceto fortissimo e chiaro), the "loaves" of 4 or 8 lb. being dried in the shade or even in the sun or in a stove.

PEDER MANSSON

Peder Månsson, a Swedish monk (1460–1534) who was in Rome in 1508–24, in his *Stridhs-Konsth*[113] describes the preparation of gunpowder and incendiary mixtures. Gunpowder is "byssopwlffer," saltpetre is "salnitrum." The recipes include camphor, terpentin, olio de petra, linolyo, brenth win, peteroleum, tucia alexandrina, brennesten, calamita, oslæktan kalk and lutum sapientie. Recipes are given for powder for large guns (pwlwer til stora byssor; saltpetre 2, charcoal 1, sulphur 1), for culverins (pwlwer til slangor; saltpetre 3, charcoal 1, sulphur 1), and for hand-guns (handbyssor; saltpetre 5, charcoal 1, sulphur 1); for incendiary balls (ffyrballa), a composition for burning a town (brenna en stadh), flying or wild fire (göra eld wtan eldföre), and automatic fire (göra en sten som tændher eld) of quicklime, tutia, saltpetre, sulphur, camphor, calamita, etc., kept in a box closed with luta sapientiæ. The recipes were probably copied from Italian manuscripts.

VALTURIO

Roberto Valturio was born in Rimini, an old Roman town on the Adriatic coast of Italy, about 1413, and died there after 1482. He was in 1446 Apostolic Secretary in Rome, afterwards councillor to Sigismondo Malatesta in Rimini. Besides being an accomplished architect and inventor, he was well acquainted with classical literature and made a special study of ancient engines of war. His principal work, *De Re Militari Libri XII*, which exists in many beautiful manuscripts in Italy, France and Germany, was written before 1463, since a letter written before 1463 from him to the Sultan Muḥammad II, the capturer of Constantinople in 1453, and accompanying a manuscript copy of his book, is given by the French historian Baluze[114] (1630–1718), the librarian of Colbert, for whom he collected manuscripts.

The book was first printed at Verona in 1472 (excessively rare), a second edition (revised by Paolo Ramusio) in 1483, and an Italian translation at Verona in 1483; an Italian verse translation by Cornazzano was published in Venice in 1493, and a French translation by Meigret in 1554. The illustrations in the first editions are superior to those in the later ones; they are full-page woodcuts printed separately and inserted in the book. They were reproduced in the German translation of the *Epitome Rei Militaris* of Vegetius (fl. A.D. 383–450), Augsburg, 1475, which was the first technical book to appear in German, as the 1472 Latin edition of Valturio was the first technical book in Italy and the third Italian illustrated book.[115]

Valturio shows a peculiar bombard consisting of two parts of nearly the same length at right-angles, and also a picture of a shell or a bomb, apparently in two hemispherical parts fastened together. A later edition of the book[116] gives the figures reversed from right to left. The right-angled bombard ("elbow-gun") is pictured in an edition of Vegetius,[117] where it is shown being fired at the bottom of the upright piece by a gunner protected by a strong wall of wooden planks. In another edition of the same year,[118] the gun is shown straight and not right-angled. A fearsome-looking contrivance (machina Arabica) in the shape of a great dragon, with a gun between its teeth, throwing an arrow, is depicted by Valturio,[119] but it is purely imaginary.

There is no doubt that Valturio's shells were intended to burst, since an interpolation in the Berlin MS. Germ. qu. 1018 (before A.D. 1450) says: "dann so zerspringt vnd zerschlecht dye chugel alls umb vnd hylfft darfür keynerlay sach" (then the sphere bursts and strikes all about it and nothing can prevent this happening).[120] This manuscript also says that if the shell is to be thrown by hand, a slower powder (eins tregen puluers) must be used, and the fuse (vederchiel) made longer. Earthenware shells can also be used for grenades (iron ones, it says, were usually for hand-throwing). Hime[121] gives reasons for supposing that these shells would not, in fact, burst explosively, and were probably only an inventor's idea and never actually tried. Valturio's was an incendiary, not a shell. Hime[122] says: "The step from Valturio's shell to common shell may seem to us now to have been a short and an easy one, yet it took nearly a century to make it; the obstacle that barred the way being neither the envelope nor the bursting charge, but the fuze.   It is impossible to say exactly when, where, or by whom explosive shell were first employed."

Small hollow bronze shells in one piece were probably first described by Baptista della Valle.[123]   These are said to be cast by the *cire perdue* process from bronze containing 25 per cent. of tin (rame parte tre, et stagno parte una), which would be very brittle.   To slow down the explosion, a mixture of resin and ground cannon powder was interposed between the touch-hole and the charge of small-arm (stronger) powder (bona poluere de schiopetto). The fact that these were of bronze (like Valturio's) suggests that they were incendiaries only.

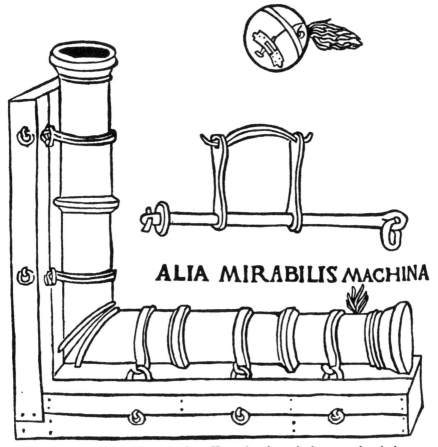

ALIA MIRABILIS MACHINA

Fig. 7.   Devices shown by Valturio.   *Top*, a bomb made from two hemispheres.
Below, an Elbow-gun.

John Stow[124] (c. 1525–1605) says that in 1543 Peter Bawde, a Frenchman and gunfounder, and Peter van Collen, another alien, made in England cast iron mortars of 11–19 in. calibre, with cast iron shells.   The shell was stuffed with "fireworke or wildefire," with a match (fuse) for setting it on fire "for to break in small pieces," any one of which pieces would be lethal.

A collection of recipes was offered for sale to Queen Elizabeth I by Ralph Rabbards in 1574.[125] It contained accounts of the distillation of herbs and flowers by decensory, simple and compound waters, the purification of saltpetre, Greek fire, oils for fireworks, flying fire, incendiary weapons, shot, fiery chariots, mines of firework for land and sea, small artillery, a musket with divers strange and forcible shot, a boat going without sails, and "hollowe tronckes" (infernal machines ?).[126]

Explosive shells were apparently first used effectively at the sieges of Bergen-op-Zoom and Wachtendonck in 1588, and it is said they were made by an Italian deserter to the Dutch, who devoted himself to the art, hitherto unknown, of making hollow balls of iron or stone, which when filled with a certain composition and ignited burst into innumerable fragments like grape stones (in innumeros quasi acinos dissilirent).[127] According to Hime[127] the first methodical and successful shell-fire from guns was by the English at Gibraltar in 1779–83, time fuses, essential for success, being used then. Shrapnel, named after its inventor, Lt. Henry Shrapnel, R.A., was invented in 1784 but was first used in 1803, the German claim for Zimmermann (1573) being without foundation.

TARTAGLIA

Niccolo Tartaglia, born at Brescia about 1500, died in Venice in 1559,[128] ascribed the invention of gunpowder to Archimedes.[129] He gives several compositions of gunpowder[130] for all kinds of guns :

						Large.	Muskets.
Saltpetre ..	..	1	10	12	100	4	48
Sulphur ..	..	1	3	3	10	1	7
Charcoal ..	..	1	3	2	36	1	8

He refers to different kinds of charcoals and to the addition of mercury, camphor, sal ammoniac and alcohol (aqua vita). The work contains plates showing guns and mortars, and there is a picture of a simple gauge for measuring the elevation of a gun. Cyprian Lucar, in a translation of the works of Tartaglia, published in 1588 but not available to me, added a chapter on the "properties, office and duty of a gunner," dealing with the refining of saltpetre and sulphur, reviving spoiled gunpowder, and making fireworks for civil and military purposes.

UFANO

Diego Ufano, a Spanish captain of artillery who was in Antwerp early in the seventeenth century,[131] describes the Spanish artillery used in the Low Countries.[132] He also describes and figures[133] an explosive bullet. It is a hollow ball filled with gunpowder, having a hole into which is inserted a tubular fuse, which is perforated; it is not said whether it is of wood or metal. The bomb or shell was not previously used in the war in the Low Countries, but had been tried, not without danger to the spectators ("a quoy toutesfois on ne doibt adiouter foy"), since when fired at a wall the pieces flew back on the people who fired it, and "on la veu les espreuves bien dangereuse." Ufano also describes pots with narrow necks, filled with incendiary and bullets, which were thrown by cords and were (apparently) fitted with wicks to ignite the contents.   He describes a grenade containing the bomb in a spherical case with a cord, an incendiary arrow thrown from a cross-bow, and a "trompe à feu" (called a "bombe" in the French translation) consisting of a cylinder into which the composition is pressed by a wooden stick.   The incendiary mixture is a Greek fire composed of equal weights of gunpowder and refined saltpetre, and half the weight each of purified sulphur, resin, Greek pitch, grains of varnish and sal ammoniac, with a small quantity of camphor.   These are finely powdered separately and thrown a little at a time into a sufficient quantity of linseed oil, nut oil and Spanish alqitran (pitch), well heated. The whole is allowed to cool.   It is filled into "les grandes et les petits balles à main, de pots de feu et autres telles inuentions pour l'endommager par terre et par eau"; also "les trompes à feu et tous les artifices qu'on peut employer à repousser un assaut ou à incendier sur terre et sur mer"—the words of Marcus Græcus.[134]   He also describes rockets (fusées volantes) in great detail, and a petard for blowing open an armoured door.[135]   The Venetians made much use of incendiaries in the fourteenth to fifteenth century.[136]

FURTENBACH

Joseph Furtenbach (d. 1667), a mathematician of Leutkirchen, was Rath- und Bau-Herr in Ulm, and wrote a number of works on civil, military and naval architecture, as well as one on gunpowder and artillery:

(a) Halinitro-Pyrbolia. / Schreibung Eine newen Büchsenmeisterey / nemlichen : Gründlicher Bericht wie die Salpeter / Schwefel / Kohlen / vnnd das Pulfer zu præpariren / zu probieren / auch langwirzig gut zu behalten : Das Fewerwerck zu Kurtzweil vnd Ernst zu laboriren. . . . Alles aufz eygener Experientza, 4⁰, Ulm, 1627, two title-pages, 107 pp. (BM 534.m.19).

(b) J. Furtenbach dess Aeltern / Mannhaffter Künst-spiegel . . . f⁰ in 6's, Augsburg, 1663 (BM 1265.f.25), with a section, p. 209 f., on Büchsen- meisterey, casting shot, etc.

(c) Architectura Civilis : Das ist Eigentlich Beschreibung wie mā nach besten form / vnd gerechter Regel / Fürs Erste : Palläst . . . Spitaler / Lazureten . . . alles aufs vilfaltiger Erfahrung, 4⁰, Ulm, 1628 (BM 1265.f.26(1)).

(d) Architectura Navalis. Das ist : Von dem Schiff Bebäw, 4⁰, Ulm, 1629 (BM 1265.f.26(2)).

(e) Architectura Martialis : Das ist / Aussführliches Bedencken / vber das / zu dem / Geschütz vnd Waffen gehörige Bebäw, 4⁰, Ulm, 1630 (BM 1265.f.26(3), with other works).

(*f*) Architectvra Recreationis Das ist Von Allerhand Nutzlich: vnd Erfrewlichen Civilischen Gebäuen, 4⁰, Augsburg, 1640 (BM 1265.f.24(2)).

In the *Halinitro-Pyrbolia*, Furtenbach describes rather briefly the purification and preparation of saltpetre, sulphur and charcoal, the manufacture of gunpowder, the testing of the force of the powder by an arrangement in which a lid blown off a box moves in vertical guides with swivelled projecting arms, adapted from an English one described by Bourne in 1578.[138] It also gives descriptions of fireworks, rockets, guns, mortars, a battery of guns on a raft, etc., with many plates.

The *Architectura Navalis* (p. 107 f.) describes the use of rockets (Ragetten) with iron heads and containing lead balls in the powder charge, fired from tubes (holen Teuchel) in the defence of ships from pirates. It also describes a bomb (Pettardo) dropped from a movable yard-arm on a ship below.

SIEMIENOWITZ

Casimir Siemienowitz, formerly Lieutenant-General of Ordnance to the King of Poland,[139] describes rockets[140] and bombs,[141] including bombs floating on water, those emitting perfumed smokes, and grenades.[142] He mentions "German poison gas,"[143] describes the preparation of fulminating gold[144] and flowers of benzoin (florum Belzoi) by sublimation.[145] He describes the construction of a flying dragon[146] and an infernal machine.[147]

KIRCHER

Athanasius Kircher[148] describes the composition of an incendiary from sulphur dissolved in oil of turpentine; an automatic fire composed of sulphur, quicklime and sal ammoniac; the preparation of gunpowder, fulminating gold, fuses, fire-resisting compositions; the compositions of powders for rockets and coloured fires (including iron powder and antimony); a metal globe filled with powder burning under water; some compositions giving a perfumed smoke; incendiary mixtures (from Porta and Cardan); a flying dragon (with a picture like that in the Kyeser manuscript) propelled by rockets, flying birds, a large Catherine wheel turned by Roman candles on the periphery, and an organ worked by Roman candles inside the pipes.

INFERNAL MACHINES

Before leaving the subject, a few words will be said on gun locks, infernal machines, explosive mines and rockets, with a few other miscellaneous observations.

The earliest cannon were fired by a red-hot iron or slow match (smouldering rope) applied to gunpowder in the touch-hole, and the second method long continued in use for small arms such as muskets, the match being later fastened to a cock for security in the hand when shooting. Some time about 1500, Leonardo da Vinci made a drawing of a flint-lock with a spring, and about

1510–20 the contrivance was made into a successful "flint-lock" in which a spark was produced by pyrites striking on steel, flint and steel being used only later. The place and date of the invention are not certainly known, Nürnberg in 1517 being one suggestion, but it was certainly used for cavalry pistols in the Schmalkaldic war in 1547 in Germany.[149] The invention at once opened the possibility of firing charges at a distance by means of a string operating a lock, or a lock operated by clockwork, and the second contrivance gave birth to the time-bomb and the infernal machine.

This use is mentioned in a work written by Samuel Zimmermann of Augsburg in 1573, dealing mostly with fireworks.[150] These include balls of a mixture of saltpetre, mercury and sulphur, which "resolve themselves into smoke," and various kinds of fireworks for amusement, including set-pieces— some of which, however, the "Feuerwerks-Künstler" saw only in dreams, after he had "lange genug darinnen speculirt und imaginirt." These included fiery fish and flying dragons. On waking, he set to work and made contrivances operated by hidden springs (verborgene Feder) and a hidden string (verborgene Schnur). At the request of the gunner, the pyrotechnist reveals that the device is a spring mechanism in which stone and steel come together, so that fire at all times but in a concealed manner comes to light, but not until there is a strong motion between the two. After this the fire-locks on hand-guns were discovered, by means of which a rapid fire is obtained :

> ein Federwerck, wo Stein und Stahl zusammenkombt, in welchen beyden das Fewer allezeit und alle Augenblick verborgener Weise sich eröffnet und sichbar wirdt, vnd doch verborgen lieget . . . vnd mag sich nicht eröffnen noch sehen lassen, so lange biss das zwischen ihnen beyden eine starcke Bewegung geschieht. . . . Nach diesem sind auch die Feuerschlösser erfunden worden an die Handtbüchsen, damit man gar schnelligsten Feuer geben vnd lossschiessen kann.

Apart from its use in fireworks and guns, the device can be applied to such pleasing purposes as to make a stool or chair with such a spring fire-work inside it, so that anyone sitting on it will be shot or else terribly burnt. Similarly, one can lay in the street what looks like a purse of gold, upon taking up which anyone is at once shot :

> ein Stuhl oder Sessel zu machen ist mit dergleichen inliegenden Feder-feuerwerck, der darin setzt alsbald erschossen, oder sonsten greulich verbrant würde. Dessgleichen möchte man wohl vf freyer Strass ein Seckel vol Geldes legen, welcher deshalbigen auffheb, alssbalt davon erschossen blieb.

Apart from such booby-traps, the pyrotechnist had also learnt something from the "Alchimisticis," viz. to make a fire-work which will be kindled by water, powdered unslaked lime being mixed with tin powder and a device made from the mixture which, by rain or water, is kindled and begins to burn :

> Daruon Weiss ich dir wohl zu sagen denn da ich selbst in Alchimia gearbeydet, sahe ich viel dergleichen gegenwürff. So ist auch nicht weniger, das fewerwerk zu machen seindt die vom Wasser angezündet

werden, als denn der pulverisirte vngeleschte Kalkstein, so er auch mit guten Zintpulver vermischt und ein Zeug darauss gemachet wird, dieser Zeug sich von Regen und Wasser anzündet und anhebet zu brinnen.

This is the old "automatic fire" of Julius Africanus (ch. I) in a modern form "made in Germany." A striking clock is adapted to fire off a gun by pressing its trigger; a glass globe filled with gunpowder is exploded by a mixture of aqua fortis and oil of vitriol or aqua fortis and iron filings composed inside it (a modern device); and "the Radios oder Wiederschein" of sunlight on mirrors or lenses can be concentrated on gunpowder, "dadurch also ein ganzes Blockhauss, Schiff auff dem Meer und Seen angezündet vnd verbrennt möchte werden."

The work has been supposed to contain the first mention of shrapnel: "Hagelgeschret, das sich über hundert Schritt von Stück aufthut." Jähns says shrapnel is "eine deutsche Erfindung aus dem 16. Jhdt.," and this "angeblich englische Erfindung" is "deutschnational." The true shrapnel shell, in which the charge is only just enough to fracture the envelope of the shell and allow the bullets contained in it to travel forward with the velocity and direction of the fractured case, was really invented by Lt. Henry Shrapnel in the siege of Gibraltar in 1779–83, in which red-hot shot were also used. Shrapnel shells were first used effectively at Surinam in 1804. The English historians of artillery reject the claim made by Jähns, who mentions another manuscript work by Zimmermann: *Bezaar, Wider alle stich, schuss vnd straich, voller grosser geheimnussen, genannt Pyromachia*, which, after a theological introduction, deals with protection from fire, the fire in crystals, whether witches can cause lightning, the ordeal by fire, etc., i.e. a truly "deutschnational" production reminiscent of Paracelsus.

A large collection of infernal devices for use in warfare is found in a sixteenth-century work, *Von allerlei Kriegsgewehr und Geschütz*, by the Austrian Wulff von Senfftenberg.[151] The artillery includes large mortars (19 in. calibre) with explosive shells, and clustered guns (Orgelgeschütz) on waggons which can be fired by a string contacted by an enemy approaching by night. A speciality was some kind of packet, box, or chest, filled with gunpowder and supposed to contain valuables, which exploded on opening. Explosive letters were also sent. The first recorded use of an infernal machine (a chest), Romocki says, is in the siege of Pskow (1581) by Polish, German, Hungarian and Scottish troops under Stephan Báthory, King of Poland. The Russian defender, Ivan Petrovich Shujski, was sent the chest by a freed Russian prisoner. It was made by Johann Ostromecki. Shujski did not open it, but a few of his companions were killed by doing so. Von Senfftenberg also gives recipes for poisons, poisoned (arsenical) smoke, etc., to be used, however, only against Turks or other infidels, and the explosive letters are to be carried by Jews.

Although Machievelli was an innocent child in comparison with Senfftenberg, it should be noticed that the infernal machines were operated by their victims, and a learned Jesuit, Mariana (*De Rege et Regis Institutione*

Fig. 8. Infernal machine with time-fuse operated by clockwork (Senfftenberg).

Libri III, Toledo, 1599), was later to rule that, although the elimination of a "tyrant" by putting poison in his food was sinful, since all men must eat, yet a poisoned nail in the seat of his chair was "licet," since he was not under any necessity to sit on it and did so of his own free choice.

MINES

"For 'tis the sport to have the enginer
Hoist with his own petar; and't shall go hard
But I will delve one yard below their mines,
And blow them at the moon."

(Shakespeare, *Hamlet*, act iii, scene 4).

Mining and counter-mining probably spread from the East; perhaps they were used at Jericho (Joshua vi. 20); from the Persians they passed to the Romans and Byzantines, and a knowledge of them reached Europe with the Crusaders. Herodotos[152] reports that in the siege of Barqa (North Africa) by the Persians, mines were driven to the walls. These were detected in the city by laying a bronze shield on the ground, when it sounded over a mined place. The Barcæans then countermined in these places, and killed the Persian miners.

Josephus[153] described mining under a building in Jerusalem and putting in wooden props. These were set on fire and burnt out, when the ground collapsed and the tower fell. This method continued in use for centuries. In some cases hand-to-hand conflicts in saps, sometimes with incendiary weapons like those described by Anna Comnena (ch. I), took place.

Interesting remains of mines and counter-mines executed by professional engineer troops were found on the site of Dura-Europos in Syria, which was besieged by the Persians in A.D. 265.[154] The Persian mine was timbered, including the roof (which was unusual, but was used in the Laurium mines). The foundations of a tower were extensively cut away and the weight was supported by timbers which were set on fire to drop the tower. The Roman soldiers, who were Oriental auxiliaries, countermined and there was an underground battle. Many bodies, arms and a skeleton in Persian armour were found; some bodies were not completely decomposed. Some pitch in a jar and some yellow crystals believed to be sulphur, were found. These were probably used to set fire to straw and faggots to burn the mine props; extensive remains of combustion were found.

Some accounts of *explosive* mines are based on misunderstanding, as when Thwrocz (John Thwrocz or Thurocz, b. 1420) in his Hungarian chronicle says that in the siege of Belgrade in 1439 the defenders destroyed a Turkish mine by exploding a counter-mine[155]:

Zowanus . . . aliam de ipsis castri corpore progredientem fossam, consimiliter, fossæ Cæsaris [the Turkish mine] obuiam, celeri cum labore effici, ac saletro bombardarumque pulueribus, ac alliis rebus, ardere, flammasque repentinas ac fumum vehementem subdito edere valentibus, impleri, orificiaque fossæ eiusdem intrinseca, artificiose fortiterque recludi, et tantummodo pro succendendis rebus et polueribus prænotatis, quoddam foramen irreclusum dimitti, hostemque occultas parentem insidias, mira spectatione, indesinenter, plurimos per custodes, subaudire fecit. Cumque hostilis fossa intra mœnia castri, ad fossam per castrenses egestam, deducta est, et rumor effodientium illam per custodes extitit auditus: mox ignis adhibitus, omnes prædictas res compositas succendit, et vehementi flamma fumoque omne vivens, homines scilicet et pecora, quæ intra eandem fossam erant, subita morte necauit.

This was taken to mean a gunpowder mine.[156] Yet Callimachus Experiens, a Polish secretary (d. 1486), in his chronicle[157] speaks only of incendiaries; the Turkish cannon had breached the walls and the besiegers had filled the moat with brushwood; they then called for surrender. The besieged, however, used the armistice to impregnate (imbuere) the brushwood with pitch, oil, sulphur and incendiary powder (pice oleo sulfure incendiarioque puluere). When the Turks attacked, the mass was set on fire and many of them perished.

The destructive effect of the explosion of a mass of gunpowder could have been made familiar by such events as the blowing up of the town hall (Rathaus) of Lübeck in 1358, which was used as a gunpowder store, if this story is true, which is very doubtful.[158]

According to Romocki[159] the first clearly recorded case of the use of an explosive mine is in Italy in 1403. Buonaccorso Pitti of Florence in his chronicle[160] says that in that year Florence and Pisa were at war. A Pisan refugee told the Florentine balia of a badly walled-up and forgotten passage through the wall. The balia sent the engineer Domenico di Matteo to inspect it, and he proposed to put a large charge of powder into the passage and explode it to breach the wall:

> . . . come metterebbe cierta quantità di polvere da Bombarda nel voto di quella porta per quelle buche, e che poi le darebbe il fuoco, e che sanza dotta inuno momento la forza di quel fuoco gitterebbe quelle mura di mattoni per terra dentro, e di fuori.

Mariano of Siena, called "il Taccola" and "il Archimede," was born in Siena on 4th February, 1381, the son of a vintner (vinajolo), executed statuary and architecture for that town, and died in 1458.[161] He wrote a *De Machinis Libri X*, extant in manuscript as Cod. lat. 197 of the Munich Stadt Bibliothek, a copy of which is in the St. Mark's Library, Venice, in which town he is said to have written it in 1449.[162] This gives a picture of mines in a hill under a building shown falling down, two barrels of gunpowder, and flames issuing from doorways leading to the mines and says:

> ibi faciant cavernam latam ad modum furni; in eam immittant tres aut quattuor vigites sursum apertos plenos pulvere bombardæ; inde ab ipsis vigitibus ad portam cavernæ ducitur funiculus sulphuratus, qui, obturata porta cavernæ lapidibus et arena et calce, accendatur. Sic ignis pervenit ad vigites et concitata flamma, arx in medio posita comburitur.

The Paris manuscript BN Latin 7239 has the picture from Taccola and practically the same text, except that it ends: "Tunc ignis transit per funiculum usque ad pulverem caratelli, illico elevatur flamma, ruit tota roca," showing that "comburitur" in the Munich manuscript means "is exploded."[163]

A manuscript written by Francesco di Giorgio Martini, found at Pisa in 1831 and now in Turin, containing a translation of Marcus Græcus in the Sienese dialect, has a drawing similar to Taccola's and a description of the caves, powder barrels and sulphured fuse like his, but ending:

> E dope el fuocho dato, assai lontano è da fuggire, perchè grande ruina à da stimare. E quando questo si possi exercitare, non sarà fortezza alchuna che resistare possa.

There had been a ruinous explosion in 1459 in Ragusa, when a large store of gunpowder in the cellars of the palace became ignited and caused a great devastation. Martini was aware of the danger of a blow-back through the mine entrance, which occurred in such cases. In another Turin manuscript by Martini there is a picture of a zig-zag passage to a mine, and in a Siena manuscript of 1470–80 there is a description of such a passage, in digging which a compass was used.[164]

In a battle between Florentines and Genoese in 1487, an attempt was made to blow up the fortress of Sarzanello by a new kind of mine said to have been laid by Francisco Giorgio of Siena, i.e. Martini,[165] and since the mine exploded in the wrong place it may have had a zig-zag gallery; in any case it could not have been laid by Martini, who was then in Siena and Urbino.[166]

Fig. 9. Gunpowder mine with zigzag passage containing the fuse (Biringuccio, 1540).

Biringuccio[167] gives a picture of a mine with a zig-zag passage, which he says was invented by Pedro Navarro in 1503, when the Castello dell'Uovo was blown up. Romocki[168] thinks Martini may have been concerned in this, since he was alive in 1506. In 1495 Castel Nuovo in Naples was taken by the Spaniards, who used a mine with wooden props and gunpowder, the exact construction of which is not clear,[169] and Romocki[170] again thinks Martini was concerned, and a confusion of events led Biringuccio to attribute the first-mentioned event (1503) to a Spaniard, Pedro Navarro. Gibbon,[171] who attributes the invention of gunpowder mines to George of Siena (i.e. Martini), also mentions an account of the explosion of a mine in Navarre in 1512[172] as dealing with : "L'invention d'entrouvir la terre avec de la poudre ensevelie dans ses entrailles, et de faire voler en l'air avec un fracas terrible les bâtimens les plus massifs," and "estoit récente."

Watson[173] reports that gunpowder was, according to one account, introduced for blasting rocks in German mines in 1627, being introduced from Hungary, and according to another account it was used in 1613, being invented by Martin Freygold at Freiberg. Watson was told by some old miners that it was a tradition in Staffordshire that blasting was introduced by German miners sent over by Prince Rupert between 1636 and 1645. It was first used in Somersetshire about 1684. In 1665 Sir Robert Moray reported on M. du Son's process of splitting rocks in mines by cartridges of gunpowder inserted in borings made by an iron tool steeled at the end and driven in by a hammer.[174] Paper cartridges are said to have been used in Augsburg in 1377.[175] Linen or paper bags containing gunpowder were used for charging cannon in 1560 and cartridges with powder and ball combined for small arms in 1590.[176]

It has been mentioned above that the "lump powder" referred to in the *Feuerwerkbuch* was not corned gunpowder. Hime says this was in use to a limited extent for grenades and small arms in England long before 1560, but it was too strong for use in early cannon and was expensive; glazing of gunpowder was mentioned in 1684.[177]

ROCKETS

Incendiary rockets were used by the Chinese (ch. VI). The name rochette or rochet was early in use in Italy and they were used there at least as early as 1380 at Chioggia in the war between the Genoans and Venetians.[178] They were little used in the West, and even in 1783–4 there were none in Constantinople, those presented by Tipu Sultan of Mysore being greatly admired.[179]

According to Elliot,[180] rockets were largely used in India from Akbar's time; the ironwork of one weighed 30 lb. Shells and rockets are spoken of in 1599, but apparently were European, and they were used earlier. Rockets were used in India in 1783–4, and much larger explosive rockets caused such damage to British troops at Seringapatam in 1799 that Col. Congreve perfected English war rockets, which proved very effective at Copenhagen in 1807.[181] The rocket is now the most modern military weapon.

LEONARDO DA VINCI

It was not my intention to say anything of Leonardo da Vinci (1452–1519), since he has been so often written about, but a few words on his military work may perhaps be added. Recent historians such as Duhem and Thorndike have laboured to show that he had many Aristotelian ideas and that some things he said which were thought to be original can be found, more or less, in writings of scholastics such as Nicolas of Cusa, Albertus Magnus, Buridan, Blasius of Parma, and even such a modest figure as Themo the Jew. The comparison of Leonardo, a man full of originality and inventiveness, with such otherwise estimable compilers, calls to mind a statement of his own[182] :

> Those who are inventors and interpreters between nature and man as compared with the reciters and trumpeters of the works of others, are to be considered simply as is an object in front of a mirror in comparison with its image when seen in the mirror, the one being something in itself, the other nothing.

Leonardo had a good and probably practical knowledge of artillery. He studied the compositions of bronzes for casting guns.[183] His anticipations of the submarine, the aeroplane, the military tank and the steam gun are well known. He gives the composition of Greek fire as made from willow charcoal, saltpetre, spirit of wine (aqua vitæ), sulphur, pitch, frankincense and camphor, boiled together and spread over Ethiopian wool.[184] Asphyxiating sulphur bombs[185] are mentioned; "German" poison gas, consisting of the fumes of burnt feathers, sulphur and realgar (arsenic sulphide), which lasts seven or eight hours, and an improved poison gas (fumo mortale) from arsenic, sulphur and realgar, are blown by smith's bellows.[186] A composition from the venom of toads, slaver of a mad dog, and tarantula spiders, and a poisonous dust for throwing on ships, containing verdigris or chalk mixed with poison, are prescribed; the dispenser should wear a mask of fine cloth.[187] Among the collection of drawings in the famous *Codice Atlantico* in Milan[188] are representations of the steam cannon, apparatus for boring cannon (apparently with spiral rifling), apparatus for drying saltpetre and subliming sulphur for making gunpowder, an edge-roller mill and an ordinary mill with flat stones for mixing the powder, a revolving multiple-barrelled gun (up to twelve guns), hand-guns, large and small cannon with apparatus for training on the target, arrows (bolts) for shooting from guns, mortars for throwing bombs, rockets, stone and lead (not iron) balls, etc.

BIRINGUCCIO

The remarkable book, *Pirotechnia*, of Vanoccio Biringuccio, born in Siena in 1480, who travelled widely in Italy, visited Germany in earlier life, and died in 1539, has often been referred to in other places. It was greatly superior to the German "firework books" and contains the first detailed information on the casting of bronze cannon and bells. It appeared in 1540 in Venice :

> De la Pirotechnia. Libri I.X. Dove ampiamente si trata non solo di ogni sorte & diuersita di Miniere, ma anchora quanto si ricerca intorno à la prattica di quelle cose di quel che si appartiene a l'arte de la fusione ouer gitto de metalli come d'ogni altra cosa simile à questa. Composti per il. S. Vanoccio Biringuccio Sennese. Con Priuilegio Apostolico & de la Cesarea Maesta & del Illustris. Senato Veneto. MDXL, la. 8°, viii ll., 168 folios.

It was reprinted in Italian in 1550, 1558–9 (twice) and 1678, and in a French translation in 1556. The first German translation appeared in 1928 and the first English in 1942.[189] Gun casting is described in Books 6 and 7; Book 7 describes the boring of guns and the casting of iron balls ("a horrible invention"); Book 10 the preparation of gunpowder and fireworks, and a description of gunpowder mines and of counter-mining. Most of the items of interest to us are mentioned elsewhere.

JEAN APPIER (HANZELET)

Jean Appier, who took the name of Hanzelet, was a printer, engraver and technician of Lorraine. He was born on 15th November, 1596 (ten days too late!) in Harancourt, a village near Nancy, and died in Nancy in 1647. He was educated by his father, who was one of the engineers employed by the Duke Charles III to direct the new fortifications of Nancy. A skilled engraver on copper, producing portraits, Jean Appier himself prepared the excellent illustrations for two books on pyrotechny, both of which he printed at Pont-à-Mousson.

The first book[190] contains 101 figures. He was associated in its production with François Thybourel, a physician, mathematician and poet, of the University of Pont-à-Mousson, who contributed a poem in the preliminaries of the book, containing the epitaph of Berthold Schwartz:

> Cy gist Berthold le noir, le plus abominable
> D' entre les inhumains,
> Qui, par son art, a rendu misérable
> Le reste des humains.

Having so absolved his conscience, Thybourel felt able to put his name to a book describing all kinds of military equipment and engines, not omitting the explosive basket of eggs, and chests apparently containing treasure which when opened exploded and blew the inquisitive examiner to pieces.

Book iv (pp. 1–112) deals with fire-balls burning under water (p. 44), poisonous smoke-balls (p. 53), fire-lances (p. 58) and incendiary arrows (p. 72). The last do not contain mercury in the composition, as do those in Appier's second book. The explosive egg-basket is described and pictured (p. 103). Alexander's fire contains oil of bricks and oil of philosophers, to be used for burning enemy countries, the recipe being taken from "Ruscel" (p. 108; Ruscelli or Rosello, who wrote a famous book of recipes under the name Alessio of Piedmont, although his authorship has been questioned). The automatic fire of Julius Africanus (who is not named) appears (p. 109).

The second book by Appier[191] is based on the first and contains 136 figures. These are mostly counter-proofs of those in the first book and are inferior to them. Appier was accused of appropriating material from Boillot (see ch. VII), but his book seems to be largely original; he admits that his account of cannon (p. 1 f.) is from Ufano; it is followed by one on mortars (p. 48) and one on petards for attaching to doors (p. 102 f.; see p. 167). The section on fireworks (pp. 161–264) is of great interest. It describes incendiary arrows from which the flame is projected *forwards* (p. 162), fire-lances (p. 164), and a recipe for a poisonous incendiary mixture containing mercury to be put into a bag attached to the point of an arrow which, after lighting the composition, is shot from a cross-bow. Fire-darts for shooting from an arbalest (p. 166) and grenades (p. 172) are described. Infernal machines include an explosive basket of eggs (p. 189), an explosive chest on a truck (p. 193) and an explosive cask of wine (p. 199). Multiple-barrel guns (des Orgues) (p. 208), double-effect grenades throwing out small cannon which in turn fire shots (p. 218), rockets (fusées) and compositions for them (pp. 233–91), including iron-headed military rockets (p. 239), an awl or borer for boring the middle of the rocket charge, a 6–7 lb. rocket with a grenade in the head for frightening men and horses in cavalry troops, rockets throwing out rip-rap crackers, and fire-lances (masses ou trompes à feu) with handles (p. 250), all appear.

The section on fireworks for amusement is quite short (pp. 252–64). It includes fire-balls which burn under water (p. 259). A composition which inflames with water or brine (saline) contains quicklime, "tuthie non preparée, salspestre en roche, pierre d'aymant," 1 lb. each, "soulfre vif" and camphor, 2 parts each. The mixture is put into a crucible covered with another crucible and is heated in a lime-furnace until it is converted into a stone, "laquelle estant humectee d'eau ou de saline, s'allume facilement" (p. 227). This is a variation of the old automatic fire (ch. I).

FREZIER

Amédée-François Frezier was born in Chambery in 1682 and died in Brest in 1773. He came of an English or Scottish family which migrated to France at the end of the sixteenth century and changed its name from Frazer.[192] He was an all-round military expert and fortified San Domingo and other places. His book[193] describes coloured fires, Roman candles, set-pieces, fireworks burning under water, etc., and is particularly detailed on rockets, including those of repeated flight (plusieurs vols), i.e. the prototype of the modern stratosphere rockets, and also military rockets containing devices which caused them to rotate in flight. The last, which reminds us of the rifled shell, may yet reappear as a modern invention.

REFERENCES

1. Description in Jähns, ref. 6, i, 229 f.; reproductions of illustrations in Essenwein, *Quellen zur Geschichte der Feuerwaffen*, Leipzig, 1877 (not available to me), and some in O. Guttmann, *Monumenta Pulveris Pyrii*, London, 1906.

2. Berthelot, (*a*) *Ann. Chim.*, 1891, xxiv, 433–521; (*b*) *ib.*, 1900, xix, 289–420; (*c*) *J. des Savants*, 1900, 1, 85, 171; (*d*) *Mém. Acad. Sciences*, 1906, xlix, 1–377 (266), and issued separately as *Archéologie et Histoire des Sciences*, 1906.

3. Thorndike, *Archives d'Histoire des Sciences*, 1955, viii, 12. The name is also given as Taccolo.

4. Berthelot, ref. 2 (*a*), 480; *Rev. Deux Mondes*, 1891, cvi, 786 (815).

5. Berthelot, ref. 2 (*a*), 500.

6. M. Jähns, *Geschichte der Kriegswissenschaft in Deutschland* (in *Geschichte der Wissenschaften in Deutschland*, Munich and Leipzig, 1889–91, xxi, pts. i–iii), 1889, i, 278; Favé, in Napoleon,

178   A HISTORY OF GREEK FIRE AND GUNPOWDER

Études sur le Passé et l'Avenir de l'Artillerie, 1862, iii, 43, 112; Berthelot, ref. 2 (*a*); Feldhaus, *Technik der Vorzeit*, Berlin, 1914, 687; Laborde, ref. 8.

7. Berthelot, ref. 2 (*a*), 498–521; the correct date of BN 7239 was given by Jähns, ref. 6, 1889, i, 280; for figures of guns from it, Napoleon, ref. 6, 1846, i, 39 and plate II; 1851, ii, 130; C. Promis, *Dell'arte dell'ingegnere e dell'artigliere in Italia dalla sua origine sino al principio del XVI secolo*, 4º, Turin, 1841, 25; see ref. 164. Romocki, ref. 12, i, 245, said the earlier date attributed to BN 7239 was based on a note at the beginning of the MS. written by Villoison: "Tractatus Pauli Sanctini Ducensis de re militari et machinis bellicis eleganter ibi depictis, scriptus sub eo tempore quo primum in usu fuit pulvis tormentarius, hoc est circa 1330 vel 1340."

8. Ref. 3, 7–26; the contents of Taccola's work in the Munich and St. Mark's MSS. had been noted by Jähns, ref. 6, 1889, i, 278–9, before Thorndike. See also Horowitz, *Geschichtsblätter der Technik*, 1922, ix, 38–40; A. de Laborde, "Un manuscrit de Marianus Taccola" (i.e. the Wilczek MS.), in *Mélanges offerts à M. Gustave Schlumberger*, Paris, 1924, 494–505.

9. Favé, ref. 6, iii, 43, 112.

10. S. Reinach, *Revue Archéologique*, 1922, xvi, 205 (from BN MS. Latin 9661; mentions Berthelot); Thorndike, ref. 3, 12, who adds Oxford Bodl. MS. Canon Misc. 378.

11. Feldhaus, ref. 6, 23.

12. Jähns, ref. 6, 1889, i, 250 f.; Berthelot, ref. 2 (*a*), 505; (*b*), 289 f.; (*c*), 85, 71; S. J. von Romocki, *Geschichte der Explosivstoffe*, 1895, i, frontis., 133–78. A related MS. is perhaps that of Augustinus Dachssberg of Munich, dated 1443, which also shows only six planets, as in the *Bellifortis*; Jähns, 260.

13. Romocki, ref. 12, i, 138.

14. Romocki, ref. 12, i, 162, merely says: "Auf S. 104b findet sich noch, offenbar nur ad vocem 'pyxis,' die Abbildung eines Schützen, welcher ein Handrohr abfeuert."

15. Ref. 6, iii, 125.

16. Romocki, ref. 12, i, 149–151; cf. *ib.*, 140 f.

17. Romocki, ref. 12, i, 157.

18. E.g. by Francis Bacon, *Novum Organum*, in *Works*, ed. Spedding and Ellis, 1857, i, 302–3; iv, 188: sed maxime cerintur hoc in argento vivo, quod non male dicitur aqua mineralis . . . quod etiam admixtum pulveri pyrio, ejus vires multiplicare dicitur. The effect is ascribed to its vapour. Napoleon, ref. 6, 1846, i, p. xi, says it was used to the end of the seventeenth century.

19. Romocki, ref. 12, i, 169, fig. 29.

20. Kervyn de Lettenhove, *Œuvres de Froissart*, Brussels, iii, 498.

21. Munster, *Cosmographiæ*, Basel, 1572, 602.

22. Romocki, ref. 12, i, 161, fig. 22; on flying dragons, see F. Denk, *Sitzungsber. Phys. Med. Soc. Erlangen*, 1940, lxxi, 353–68 (Vienna, Nat. Bibl. MS. 3064, *c.* 1450), who says it was a kind of box-kite, not a hot-air balloon; see Kircher, ref. 25, 479.

23. Romocki, ref. 12, i, 140; Berthelot, ref. 2 (*b*), 383.

24. Hippolytos, *Refutationis Omnium Hæresium*, bk. iv, ch. 39; ed. Duncker and Schneidewin, Göttingen, 1859, 106; tr. in *Ante-Nicene Christian Library*, 1868, vi, 106.

25. Athanasius Kircher, S.J., *Mundus Subterraneus*, fº, Amsterdam, 1665, ii, 476.

26. Romocki, ref. 12, i, 174, who says it is *not* for the detection of buried explosives.

27. Full text in Romocki, ref. 12, i, 163.

28. J. Toll, *Epistolæ Itinerariæ*, Amsterdam, 1714, 34, and plate; Berthelot, *J. des Savants*, 1899, 276; 1900, 90; *id.*, *Science et Morale*, 1897, 486; Romocki, ref. 12, i, 146; Feldhaus, ref. 6, 26, 844. In a fifteenth century manuscript (Montpellier Ecole de Médicine 277; *Catalogue Gén. des Manuscrits des Bibliothèques Publiques des Départements*, Paris, 1849, i, 808) alcohol (which is called Greek fire; ignis inextinguibilis quem dicunt ignem grecum), the preparation of which is described, is to be filled into a Püstrich (ymaginem).

29. Hime, *Origin of Artillery*, 1915, 141; Romocki, ref. 12, i, 169, figs. 25, 27, 30.

30. Romocki, ref. 12, i, 171.

31. Jähns, ref. 6, i, 228; this part is almost certainly a later addition in the "gemischten Handschrift."

32. Romocki, ref. 12, i, 167–8.

33. Jähns, ref. 6, 1889, i, 5 f., 261–3, 393–408 (good summary).

34. Jähns, ref. 6, 1889, i, 545.

35. Jähns, ref. 6, 1889, i, 408, 633.

36. B. Montfaucon, *Bibliotheca Bibliothecarum Manuscriptorum Nova*, Paris, 1739, i, 565 (B = De Pyrotechnia tam seria et militare quam jocose); ii, 762 (E = Pyrotechnia sive de arte metallica, no. 6638).

37. Ref. 6, 1889, i, 382 f., 393–408.

38. Jähns, ref. 6, 1889, i, 392 f.; *id.*, *Handbuch einer Geschichte des Kriegswesens von der Urzeit bis zu der Renaissance*, Leipzig, 1880, 773, 804; Romocki, ref. 12, i, 178 f.; Feldhaus, ref. 6, 23.

39. Jähns, ref. 38, 1880, 804; Hime, ref. 29, 154.

40. Wilhelm Hassenstein, *Das Feuerwerkbuch von 1420. 600 Jahre Deutsche Pulverwaffen und Büchsenmeisterei. Neudruck des Erstdruckes aus dem Jahre 1529 mit Übertragung ins Hochdeutsche und Erläuterungen*, . . . *mit 98 Bildern*, Munich, Verlag der Deutschen Technik GMBH., 1941 (187 pp.), 79.

41. Jähns, ref. 6, 1889, i, 401.

42. Thorndike, *A History of Magic and Experimental Science*, New York, 1941, v, 34.

43. B. Rathgen, *Das Geschütz im Mittelalter. Quellenkritische Untersuchungen*, Berlin, 1928, 109.

44. List in Hassenstein, ref. 40, 85 f.

45. M. Ginsburger, "Les Juifs et l'art militaire au Moyen-Âge," in *Revue des études juives*, 1929, lxxxviii, 156–66.

46. *Flauij Vegetii Renati vier bücher der Ritterschaft* . . . *Mit einem zusatz / von Büchsen geschoss / Puluer / Fewrwerck / Auff ain newes gemeeret vnnd gebessert*, D.M.XXIX; colophon: Gedruckt zü Augspurg durch heinrich Stainer Im M.D.XXIX Jar (BM 8562.g.3(3)), fº in 6's unpaged, 200 pp., the *Feuerwerkbuch* being pp. 180–200; Hye nachuolget vonn Büchsen geschoss / Puluer / Fewerwerck / wie man sich damit auf ainer Statt / Feste / oder Schloss / so von Feynden belägeret wer / erretten / Auch sich der Feind darmit erwözen möchte; other eds. of Vegetius, *Fl. Vegeti Renati* . . . *de re militari libri quatuor*, Paris, Wechelum, 1533, fº in 6's (BM 433.e.20), 1535 (279 pp.). A text of the *Feuerwerkbuch* (which he dated 1425) was printed from a MS. of 1445 by J. G. Hoyer, *Geschichte der Kriegskunst*, Göttingen, 1799–1801, II, ii, 1108–39 (shooting "with water," 1134); G. Köhler, *Die Entwickelung des Kriegswesens und der Kriegführung*, Breslau, 1887, III, i, 267.

47. Ref. 6, iii, 138–161; Reinaud and Favé, *Histoire de l'Artillerie Iʳᵉ Partie. Du Feu Grégeois des feux à guerre et des origines de la poudre à canon d'après des textes nouveau*, Paris, 1845, 133 f.; *id.*, *J. Asiatique*, 1849, xiv, 283.

48. Ref. 6, 1889, i, 269, 408; Hassenstein, ref. 40, 91–2.

49. *Livre du canonnerie et artifice de feu*, Appendix: *Petit traicté contenant plusieurs artifices du feu, très-utile pour l'estat de canonnerie, recueilly d'un vieil livre escrit à la main et nouvellement mis en lumière*, Paris, 1561.

50. Ref. 12, i, 179–230.

51. Romocki, ref. 12, i, 186; Hassenstein, ref. 40, 62.

52. R. Watson, *Chemical Essays*, 1793, ii, 10; Hime, ref. 29, 152.

53. Jähns, ref. 38, 1880, 804 (but in ref. 6, 1881, i, 401, he says it is not corned powder); Hime, ref. 29, 154; Hassenstein, ref. 40, 61, 117; Köhler, ref. 46, III, i, 336 (Knollenpulver is still described in 1472).

54. Rathgen, ref. 43, 77, 109–36 (see Jähns, ref. 53).

55. Hime, *Gunpowder and Ammunition*, 1904, 183.

56. Hassenstein, ref. 40, 30, 67.

57. Hassenstein, ref. 40, 32, 70; Rathgen, ref. 54, 109; Romocki, ref. 12, i, 206, refers to Alexander of Tralles, bk. ii, ch. 5, in *Medicæ Artis Principes*, Paris, 1567, 173, who merely gives spicæ nardi without explanation.

58. Hassenstein, ref. 40, 16, 43, 98.

59. Jähns, ref. 6, i, 401; Romocki, ref. 12, i, 181.

60. Van Helmont, *Ortus Medicinæ*, Amsterdam, 1648, 107; 1652, 87.

61. Hassenstein, ref. 40, 21 f., 55 f.: lime, alum, vitriol, Spanish green (verdigris), etc., are specified, but not wood ashes.

62. Rathgen, ref. 54, 109 f.

63. Hassenstein, ref. 40, 30, 67.

64. Hassenstein, ref. 40, 33, 73, 127; corrections from MS. in Romocki, ref. 12, i, 203.

65. Ref. 12, i, 204–5.

66. *Triumph-Wagen des Antimonii*, 1604; in *Chymische Schriften*, Hamburg, 1740, i, 429.

67. *Alchemia*, Frankfurt, 1597, 343; 2 ed., *Alchymia*, 1606, 159–60.

68. Hassenstein, ref. 40, 34, 74, 127; another text in Romocki, ref. 12, i, 207; this item is in Ambraser MS. 67 (*c.* 1410): Jähns, ref. 6, i, 389 ("wie man mit Wasser schüsst, wie man oleum benedictum macht").

69. Ref. 12, i, 142.

70. Ref. 67, bk. II, tract. ii, ch. 17; 1606, 146.

71. Ref. 12, i, 212–4.

72. Ref. 49.

73. Romocki, ref. 12, i, 216.

74. Ref. 12, i, 217 f.

75. Lemery, *Cours de Chymie*, Paris, 1756, 417; Macquer, *Dictionnaire de Chymie*, Paris, 1778, ii, 433, 456 (huile fétide empyreumatique). I have been unable to find any more recent statements of its composition.

76. *Abhandlungen und Vorträge*, Leipzig, 1906, i, 144; *Entstehung und Ausbreitung der Alchemie*, Berlin, 1919, i, 480.

77. Ref. 43, 127.

78. Ref. 40, 40.

79. *Revue Scientifique*, 1891, xlvii, 457.

80. Hassenstein, ref. 40, 167.

81. Rathgen, ref. 43, 112–3.

82. Jähns, ref. 6, i, 390–2 (early fifteenth century, *c.* 1425); Romocki, ref. 12, i, 189; Rathgen, ref. 43, 710, dates it 1435. It contains a version of Marcus Græcus (also in Nürnberg MS. 3227a).

83. Favé, ref. 6, iii, 265, 274, plate 46.

84. Jähns, ref. 6, 1889, i, 542, 631; Romocki, ref. 12, i, 189, 197–9; Feldhaus, ref. 6, 1122. See ref. 151.

85. Hassenstein, ref. 40, 34, 74; Romocki, ref. 12, i, 200, says this is the same as the "Espignol" used by the Prussians in the war against Denmark in 1864.

86. Hassenstein, ref. 40, 31, 70; he omits the arsenic, which Romocki, ref. 12, i, 200–1, says was done in some MSS.

87. Rathgen, ref. 43, 109 f.

88. Romocki, ref. 12, i, 201.

89. Hime, ref. 29, 210.

90. Hassenstein, ref. 40, 95, 116.

91. Piccolpasso, *Li tre libri dell' arte del vasaio*. *The Three Books of the Potter's Art*, ed. and tr. by Rackham and Van der Put, London, 1934, 40, 53, 67, 69.

92. Quétif and Echard, *Scriptores Ordinis Prædicatorum*, Paris, 1719, i, 528; Klebs, *Incunabula Scientifica et Medica*, in *Osiris*, 1938, iv, 188 (later eds. 1485, 1497, 1499).

93. Clephan, *Archæological J.*, 1909, lxvi, 146.

94. Mrs. D. W. Singer, *Alchemical Manuscripts in Great Britain and Ireland*, Brussels, 1930, ii, 634, 712 (no. 1066); 1931, iii, 935. See Thorndike, ref. 42, 1923, ii, 758; 1934, iii, 149; who quotes the Bodleian MS. Digby 164 (early fifteenth century), f. 17, for *Extracta de tractatu fratis Ferrarii super arte alkymie. Dirigit epistolam suam Papæ et primus ponit artis impedimenta*, and says the same MS., ff. 8–12 *v.*, contains Roger Bacon's *Epistola de Secretis operibus*; W. Y. Carman, *A History of Firearms from Earliest Times to 1914*, 1955, 160, incorrectly says the gunpowder recipe is in the Bodleian MS. Digby 67; Mrs. D. W. Singer, *Ambix*, 1959, vii, 25.

95. Jähns, ref. 6, 1889, i, 258; Feldhaus, ref. 6, 24 (many more MSS. are described by Jähns than are mentioned here).

96. Jähns, ref. 6, 1889, i, 274; Feldhaus, ref. 6, 22, 847; Köhler, ref. 46, III, i, 333 (said to have belonged to Scanderbeg, the Albanian chief, c. 1403–67).

97. Jähns, ref. 6, 1889, i, 382–7 (for the similar MS., Ambraser 67, ib., 387); Feldhaus, ref. 6, 22; Strunz, Die Vergangenheit der Naturforschung, Jena, 1913, 104; A. Köhler, Mitteilungen zur Geschichte der Medizin und der Naturwissenschaften, 1908, vii, 525–8; Feldhaus, Rühmsblätter der Technik, Leipzig, 1910, 107; a list of German Büchsenmeister and their seals is given by Rathgen, ref. 43, 137–56.

98. Jähns, ref. 6, 1889, i, 409; Feldhaus, ref. 6, 703 (some of these private German and Austrian libraries have since been dispersed).

99. Das Mittelalterliche Hausbuch, ed. H. T. Bossert and W. F. Storck, Leipzig, 1912; full summary by Lippmann, Beiträge zur Geschichte der Naturwissenschaften und der Technik, Berlin, 1923, i, 200–10 (which I have checked throughout with the original and found to be correct); Jähns, ref. 6, 1889, i, 269; Feldhaus, ref. 6, 516, 1061–2, fig. 710. There were two earlier eds., Leipzig, 1866, and Frankfurt, 1887 (ed. A. von Essenwein); Jähns, ref. 6, 1889, i, 269, who says it is of Swabian origin; Favé, ref. 6, iii, plate 15, fig. 3.

100. Jähns, ref. 6, 1889, i, 412; Feldhaus, ref. 6, 82.

101. Jähns, ref. 6, 1889, i, 276; Feldhaus, ref. 6, 717; Jähns also mentions a Rust-vnd feuerwerck-buych, Frankfurt Stadtbibliothek MS. 40, with good illustrations.

102. Jähns, ref. 6, 1889, i, 272, 309; Feldhaus, ref. 6, 270.

103. See Thorndike, ref. 42, London, 1923, ii, 751–808 (many of the authors appear to have been Spanish Jews and the original sources were probably Arabic).

104. Liber Pompilii Azali Placentini de omnibvs rebvs natvralibvs qvæ continentvr in mvndo videlicet. Cælestibvs et Terrestribvs necnon Mathematicis. Et de Angelis Motoribvs quæ (sic) Cælorvm, sm. f° in 6's, Venice, Octavian Scot, 1544 (BM 536.l.7), f. 111 v.; f. 15 r. refers, under Mars, to chirurgicos, Alchimistos, coquos, meliuolos, ætatemq; virilem.

105. Ref. 42, 1934, iv, 150; 1941, v, 4.

106. Ref. 12, i, 231–40; Jähns, ref. 6, 1889, i, 276; Feldhaus, ref. 6, 334; id., Technik der Antike und des Mittelalters, 1931, 347.

107. A. Birkenmajer, Isis, 1932, xvii, 34–53.

108. Feldhaus, ref. 6, 48–9 (automata), 65–6 (dredge), 823 (magic lantern), 969 (letter-lock).

109. Thorndike, ref. 42, 1934, iv, 174–5.

110. Thorndike, ref. 42, 1934, iv, 178–80.

111. Thorndike, Isis, 1930, xiv, 221; 1931, xv, 31 (42).

112. C. Promis, Trattato di Architettura Civile e Militare di Francesco di Giorgio Martini, Turin, 1841 (BM 1265.f.14), ii, 128 f.; Favé, ref. 6, iii, 198; Jähns, ref. 6, 1889, i, 282; Romocki, ref. 12, i, 245–9.

113. Peder Månssons Strids-Konst och Strids-Lag, in Samlingar utgifne af Svenska Fornskrift-Sällskapet, Stockholm, 1845, iii; Peder Månssons Skrifter, in ib., 1913–15, cxliii–cxlviii, 177–85. The MS. is also of interest for the history of glass. There is no translation known to me.

114. Étienne Baluze, Miscellaniorvm, 8°, Paris, 1683, iv, 524–7; Miscellanea, f°, Lucca, 1762, iii, 113.

115. W. M. Ivins, Junr., Bull. Metropolitan Museum of Art, New York, 1926, xxi, 267.

116. En Tibi Lector Robertvm Valtvrivm . . . de Re Militari Libris XII . . . Paris, Wechelum, 1532 (BM 534.m.2.(2)), 267; for a summary of the German version, see Jähns, ref. 6, i, 264, for the work in general, 358–62.

117. Fl. Vegetii Renati . . . de re Militari libri quatuor, f°, Paris, 1535, 125. Carman, ref. 94, 30, says the L-shaped gun was called in Italian a bombardo cubito.

118. Fl. Vegetii Renati . . . de re Militari libri quatuor, f°, Paris, 1535, 128.

119. De Re Militari, 1532, 239; Favé, ref. 6, iii, plate 4.

120. Romocki, ref. 12, i, 191.

121. Hime, ref. 29, 192–4.

122. Ref. 29, 195, 202. Jähns, ref. 6, i, 268, says the "bomb" is too fragile to be thrown from a gun and since the powder is so tightly compressed it would not explode, so that it is an incendiary: "kein Sprenggeschoss, sondern bloss ein Brandkugel."

123. *Vallo Libro continente appertinenti à Capitanii retenere & fortificare vna Citta*, sm. 8º, Vineggia, 1543, BM 534.c.2.(1), unpaged, Capitoli tre de artificii, di fuoco nuouamente aggionti (pref.); ch. 1, Per fare Trombe de fvoco; ch. 2, Per fare balle de bronzo da trazere in vn battaglioni de fanti; plate of bombs; Romocki, ref. 12, i, 194–5. The British Museum has eds. of della Valle's book (all 8º) of Venice, 1524, 1529; Vineggia, 1535, 1543, 1550, 1558; French tr., *Vallo. Du faict de la guerre et art militaire*, Lyons, [1554]; Romocki gives Venice, 1521, presumably the first ed.

124. Stow, *The Annales of England*, 4º, 1601, 983 (1 ed., 1580, as *Chronicles*).

125. J. O. Halliwell [-Phillips], *A Collection of Letters illustrative of the Progress of Science in England*, 1841, 7–12.

126. See Romocki, ref. 12, i, 275, fig. 65; Tout, *English Histor. Review*, 1911, xxvi, 685, said a "trunk" was a wooden support for a cannon, sometimes on wheels.

127. Everhart van Reyd, *Belgarum Aliarumque Gentium*, tr. D. Voss, fº, Leyden, 1633, bk. viii, p. 182 (Bergen-op-Zoom), as quoted in the text; Famianus Strada, *De Bello Belgico*, fº, Rome, 1640, ii, 448; *Histoire de la Guerre de Flandre, de Famianus Strada*, traduite par P. Du-Reyer, fº, Antwerp, 1705, iii, 503–4 (une nouvelle espece de balles qu'on appelle Bombe; for a petard, *ib.*, 490)—he says they were thrown from large mortars, and were hollow iron or stone bullets filled with explosive and with a fuse (amorce) and used at Wachtendonck in 1588; they were invented "par un artisan de Venlo," although others said they were invented by an Italian who burnt himself with them; Hime, ref. 29, 197; Jähns, ref. 6, 1889, i, 351, 641 (Zimmermann).

128. Jöcher, *Gelehrten-Lexicon*, Leipzig, 1751, iv, 1013 (who says de Thou gave 1577 for the date of his death); Merlieux, *Nouvelle Biographie Générale*, 1864, xliv, 887.

129. *Nuova Scienza, cioè Invenzione nuovamente trovata, utile per ciascuno, speculativo, matematico, bombardiero*, ed. altri, 4º, Venice, 1537 and later eds.; *Quesiti et Inventioni diverso*, 4º, Venice, 1550 and later eds., of which I have used that of 1554 (CUL O.28.14); 1554, 38 *v*.

130. Tartaglia, *Quesiti*, ref. 129, 1554, 39–40; for Tartaglia's description of artillery, see Napoleon, ref. 6, 1846, i, 164 (26 sizes).

131. Jöcher, ref. 128, 1751, iv, 1560.

132. *Tratado de la Artilleria y uso del practicado*, 4º, Antwerp, 1613; I have seen only the translation: *Artillerie. / C'est a dire / Vraye Instrv-/ction de l'Artil-/lerie et de Tovtes / ses Appartenan-/ces . . . Le tout recueilly de l'experience es guerres du Pays-bas & publié en langue Espagnolle. Par Diego Vfano Capitaine de l'artillerie au Chasteau d'Anuers. Mais maintenant traduit en langue Francoise, & orné de belles & necessaires figures. Par Iean Theodore de Bry, Bourgeois d'Oppenheim*, 4º, Frankfurt, Engolf Emmel, 1614, iv ll., 164 pp., ii ll. (plates not paginated but included in signatures); shows Tartaglia's quadrant altimeter, and a diver attaching a sunken gun to a screw in a boat for raising it; BM C.47.i.12.(1); Favé, ref. 6, iii, 300 f., plates 48–50.

133. Ref. 132, 1614, 141–3.

134. Favé, ref. 6, iii, 310–11.

135. Favé, ref. 6, iii, 313, plate 50.

136. Favé, ref. 6, iii, 352.

137. According to Guttmann, ref. 2, plate 6, a portrait of Berthold Schwarz is in item *f*, but it is not in the copy I examined.

138. Hime, ref. 29, 163.

139. *Ars Magnæ Artilleriæ pars prima, Studi & operâ Casimiri Siemienowitz Equites Lithuani, Olim Artilleriæ Regni Poloniæ Propræfecti*, la. 4º, Amsterdam, 1650 (BM 64.f.1). Although described as "pars prima," as far as I know no further parts were printed. The book, illustrated with many excellent plates, does not deal with cannon, which were perhaps reserved for another volume. *Volkommene Geschütz-, Feuer-werck- und Büchsenmeisterey-Kunst; hiebevor in Lateinischer Spraach beschrieben . . . Anitzo in die Hochdeutsche Spraach übersetzet von T. L. Beeren. Mit Kupffern und einem gantzen neuen Theil vermehret durch D. Elrich*, 2 pts., fº, Frankfurt-am-Main, 1676 (BM 717.*l*.9); this contains a second part by Daniel Elrich dealing with instructions

for making gunpowder, purifying saltpetre and sulphur, etc., and making fireworks, and has 27 extra plates. See ch. VII, ref. 81.

140. Ref. 139, bk. iii, De Rochetis, pp. 92–120.

141. Ref. 139, bk. iv, De Globis, pp. 120–48.

142. Ref. 139, 149 f.

143. Ref. 139, 107.

144. Ref. 139, 84.

145. Ref. 139, 85.

146. Ref. 139, 240.

147. Ref. 139, 273.

148. Kircher, ref. 25, ii, 467–80 : bk. xi, sect. 5, pars. 4, Ars Pyrabolica, quam & Pyrotechniam vocant; q. "Siemienovius Artigleria magna," Freitag, Oswald Croll, Scaliger, Paracelsus, Quercetanus, Tartaglia, Prechtel, Cardan, Porta, etc. The section on "Rochetas, quas Itali *Raggi* Germani *Rakettas* vocant" is from Siemienowitz; they are really Roman candles. For MSS. of Bodo von Liebhard (seventeenth to eighteenth century) on gunpowder, etc., see Clauss, *Archiv f. d. Geschichte der Naturwissenschaften und der Technik*, 1911, iii, 269. See Jerome Cardan, *De Subtilitate*, bk. ii, 8°, Basel, 1560, 92, 107, 109; *id.*, *De Rerum Varietate*, bk. x, ch. 49 (De ignis artificiis), 8°, Basel, 1557, 659; Baptista Porta, *Magiæ Naturalis Libri Viginti*, bk. xii (Portentosos ignium exitialium machinationes molitur), Leyden, 1651, 461–88; Robert Flud (also called Fludd), *Utriusque Cosmi Maioris scilicet et Minoris Metaphysica, Physica atque Technia Historia*, f°, Oppenheim, 1618, Tract II, Part VI, bk. iv, ch. 7, p. 422 (incendiary balls thrown by an arcubalista).

149. J. Beckmann, *History of Inventions*, 1846, ii, 533; Romocki, ref. 12, i, 254; Feldhaus, ref. 6, 441. Carman, ref. 117, 131, says a German flintlock pistol dated 1423 is very doubtful; they were used in England in the time of Henry VIII (1491–1547).

150. *Dialogus oder Gespräch zweier Personen, nämlich eines Büchsenmeisters mit einem Feuerwerks-Künstler, von der wahren Kunst und rechten Gebrauch des Büchsengeschosses und Feuerwerks*; Jähns, ref. 6, 1889, i, 640 (Zümermann); Romocki, i, 257, giving MSS. dated 1574, 1575, 1577, and three undated, all in Germany. Chain-shot (two balls connected by a chain) was used by the Italians at Bicocca in 1522; Napoleon, ref. 6, 1846, i, 182.

151. Favé, ref. 6, iii, 265 f.; Napoleon, *ib.*, 1846, i, 166, etc., from MS. in Dépôt général de la Guerre, on artillery; the following account is based on an anonymous MS. in the Herzoglich Anhaltinischen Behörden-Bibliothek, Dessau, assumed to be by von Senfftenberg, as given by Romocki, ref. 12, i, 263 f.; see also Feldhaus, ref. 6, 528.

152. Herodotos, iv, 200; Aineias, *Tactica*, 37.

153. Josephus, *De bell. Jud.*, II, xvii, 8.

154. R. du Mesnil du Buisson, in *The Excavations at Dura-Europos. Conducted by Yale University and the French Academy of Inscriptions and Letters, Sixth Season Report*, ed. Rostovtzeff, Bellinger, Hopkins and Welles, New Haven, 1935, 188–205 and plate XVIII. The article includes a brief history of military mining.

155. J. Thwrocz, *Chronica Hungarorum*, in Johannes George Schwandtner, *Scriptores Rervm Hungaricarvm Veteres . . . ex MSS. Codicibvs, et Rarissimis Editionibvs, Bibliothecæ Augustæ Vindobonensis . . . cvra, et stvdio Ioannis Georgii Schwandtneri*, 3 vols., 4°, Vienna, 1746–6–8; 1746, i, 247; repunctuated text in Romocki, ref. 12, i, 242.

156. Gibbon, *Decline and Fall of the Roman Empire*, ch. lxv, Everyman ed., vi, 346, refers to the mention of Turkish cannon in the siege of Constantinople in 1423 by Chalkokondyles, *De Rebus Turcici*, bk. v, ed. Bekker, Bonn, 1843, 231; and before Belgrade in 1439 by Doukas, *Historia Byzantina*, ch. 30, ed. Bekker, Bonn, 1834, 211; in ch. lxviii, Everyman ed., vi, 435, Gibbon says gunpowder mines were first used at Sarzanello in 1487.

157. *De Rebus Vladislai*, in Schwandtner, ref. 155, i, 468.

158. *Chronicon Sclavicum quod vulgo dicitur parochi Susebenis*, ed. E. A. Th. Laspeyres, Lübeck, 1865, 132–3 (van vosumnisse der dede det bussen krude makeden; the note says Detmar dated the event 1358); *Chronik des Franciscaner Lesemeisters Detmar*, ed. F. H. Grautoff, 2 vols., Hamburg, 1829–30, i, 281 (de materie untfenk des vures, alse swevel unde ander tuch, dat

to des Stades behof was; date 1358). Köhler, ref. 46, III, i, 237, says it was an incendiary, but Romocki, ref. 12, i, 242, says it was an explosion. Both give the date 1360.

159. Ref. 12, i, 243.

160. *Cronica di Buonaccorso Pitti Con Annotazioni ed All'Ilustriss. e Clariss. Sig. Senatore Raimondino Pitti*, ed. Manni, 4º, Florence, 1720, 75. (Pitti fl. *c.* 1400).

161. Milanesi, *Documenti per la Storia dell'Arte Senese*, Siena, 1854, ii, 284–6; see ref. 8.

162. Romocki, ref. 12, i, 243.

163. Romocki, ref. 12, i, 245.

164. Romocki, ref. 12, i, 247–8; C. Promis, *Dell'Arte dell'Ingegnere e dell'Artigliere in Italia dalla sua origine sino al principio del XVI secolo*, Turin, 1841, 329 f.; and *Atlas*, Turin, 1841, 5, and plates XXXVI–XXXVIII.

165. Tiraboschi, *Storia della Letteratura Italiana*, Florence, 1807, VI, i, 426.

166. Romocki, ref. 12, i, 249.

167. *Pirotechnia*, Venice, 1540, 175 *v.*; Cardan, *De Subtilitate*, bk. ii, 8º, Basel, 1560, 109, refers to Franciscus Georgius Senensis for "cuniculorum igneorum, quos minas vocant," and gives a picture of a zigzag mine passage.

168. Ref. 12, i, 251.

169. Silvestro Guarino, *Diario*, in [A. A. Pelliccia] *Raccolta di Varie Croniche, Diarj, ed altri Opuscoli cosi Italiani, come Latini appartenenti alla Storia del Regno di Napoli*, 5 vols., 4º, Naples, 1780–82; i, 223 : . . . fatte chiu tagliate nella fabrica e fosso pontana con travi, et in quilli travi ce fo posto fuoco, e con fassine, polvere e [= di] bombarde, in modo che tutta cascao insieme.

170. Ref. 12, i, 252.

171. Ref. 156, ch. lxvii, Everyman ed., vi, 435.

172. *Histoire de la Ligue Faite A Cambrai . . . Contre La Republique de Venise*, Paris, 1709, iii, 93–7, q. Paul Jovius.

173. Ref. 52, 1793, i, 341 f.

174. Sir R. Moray, *Phil. Trans.*, 1665, i, 82 (from a letter from Du Son).

175. Ref. 43, 215; this is doubtful.

176. Hime, ref. 29, 151; Jähns, ref. 6, 1889, i, 664, says paper cartridges *without balls* were used after 1550.

177. Mieth, *Artilleriæ Recentior Praxis*, Frankfurt, 1684, pt. ii, ch. 55; q. by Hime, ref. 29, 154.

178. *Danduli Chronicon*, in Muratori, *Rerum Italicarum Scriptores*, Milan, 1728, xii, 448 (igne imissio cum rochetis); 1729, xv, 769.

179. Hime, ref. 29, 145.

180. Sir H. M. Elliot, *The History of India as told by its own Historians*, ed. J. Dowson, 1875, vi, 470.

181. Hime, ref. 29, 147–8.

182. E. MacCurdy, *Leonardo da Vinci's Note-books, arranged and rendered into English*, 2 vols., 1938, i, 61.

183. MacCurdy, ref. 182, ii, 405; J. P. Richter, *The Literary Works of Leonardo da Vinci*, 2 vols., London, 1883, ii, 24 (I am aware of the fact that a revised and enlarged edition was published in 1939).

184. Richter, ref. 183, ii, 280; MacCurdy, ref. 182, ii, 186, 194, 207, 217; both incorrect.

185. MacCurdy, ref. 182, ii, 198.

186. MacCurdy, ref. 182, ii, 201, 210, 217.

187. MacCurdy, ref. 182, ii, 210, 217–9.

188. *Il Codice Atlantico di Leonardo da Vinci nella Biblioteca Ambrosiana di Milano*, 6 enormous vols., Milan, 1894–1904.

189. *The Pirotechnia of Vannoccio* (sic) *Biringuccio translated from the Italian with an Introduction and Notes*, by Cyril Stanley Smith and Martha Teach Gnudi, New York, 1942. The notes are especially valuable. There is a summary of the *Pirotechnia* in Jähns, ref. 6, 1889, i, 591–6, who mentions Latin trs., Paris, 1572, and Cologne, 1658, which Smith and Gnudi were unable to trace.

190.   *Recueil de plusieurs machines Militaires, et feux Artificiels pour la Guerre & Recreation. Avec l'Alphabete de Tritthemius* . . . *De la diligence de Iean Appier dit Hanzelet Calcographe et de Francois Thybovrelm*[e] [*sic*], Chirurg[n], sm. 4°, Au Pont-a-Movsson Par Charles Marchant Imprimeur de L A 1620 (BM 8828.bbb.11). The "books" 1 to 5 have separate title-pages and pagination; the work on secret writing by Trithemius and another work (with separate title-page) are unpaged.   On Hanzelet, see J. Lamoureux, *Nouvelle Biographie Générale*, 1858, xxiii, 315–7 (bibl.).

191.   *La Pyrotechnie de Hanzelet ou sont representez les plus rares & plus appreuuez secrets des machines & des feux artificiels propres pour assieger battre surprendre & deffendre toutes places*, Au Pont a Mousson par I & Gaspard Bernard, 4°, 1630, 264 pp. (BM 534.i.4.(3)).   The dedication is signed Iean Appier Henzelet. See T. L. Davis and J. R. Ware, *J. Chem. Education*, 1947, xxiv, 522.   Some illustrations are reproduced by Romocki, ref. 12, i, 327, fig. 80 (from Berlin MS. germ.qu.169) of petard; 353, fig. 88 (from Antoine de Ville, *Traité de Fortification*, Paris, 1614) of petard; and by Feldhaus, ref. 6, 529, fig. 360, 1175 (from Thybourel, 1620) of explosive egg-basket.

192.   P. Levot, *Nouvelle Biographie Générale*, 1857, xviii, 859–65.

193.   *Traité des Feux d'Artifice Pour le Spectacle, Par le sieur Frezier*, 12°, Paris, 1706 (engr. title-page dated 1707; title-page dated 1706) (CUL 7360.d.1), preface, 394 pp., contents; 1715 (engr. title-page dated 1707) (BM 55.a.3); 2nd ed., 8°, Paris, 1747 (BM 8715.cc.18), liv pp., 1 blank, 496 pp., 13 plates.

Chapter V

GUNPOWDER AND FIREARMS IN MUSLIM LANDS

THE Arabs were a Semitic race known to the Jews, Persians, Greeks ('Αραβία) and Romans (Arabi) for centuries before the time of Muḥammad the Prophet (A.D. *c*.570–632). Another Greek name for Arabic tribes was Saracens (Σαρακηνοί); in the time of Prokopios (*c*. A.D. 550) the Saracens came from a large region reaching from Egypt to the Euphrates. Belisarius (*c*. A.D. 505–65) described the Arab troops under his command as incapable of building fortifications. The name Saracen became familiar in the West in the Crusades, but it was then often applied to Turks. Arabic-speaking peoples appear in our pages several times, and a very brief sketch of their history is necessary in order to appreciate their significance in the origin and distribution of a knowledge of gunpowder and firearms. The story is somewhat complicated.

Immediately after the death of Muḥammad in A.D. 632 the Arabs began a career of conquest which over-ran with surprising rapidity vast areas of the civilised world. They were brave and formidable fighters, and for the conquered peoples they had a simple formula: Islām (the religion of Muḥammad), tribute or death. On the whole they ruled wisely and humanely, they were reasonably tolerant and at first exacted only a moderate tribute, and they restored settled government in regions under their power. The whole complex of races dominated by the Arabs and their Islamic successors is conveniently called Muslim, although many were not Arabs or believers in Islām.

In 635 the capture of Damascus and the battle of Qādisīya gave the Arabs the control of 'Irāq. In 637 the Persian capital al-Madā' (Ctesiphon, near the later Baghdād) was abandoned to them, and the conquest of Persia was completed in 642. Mesopotamia was in turn reduced, and in 638 the camp-towns of Baṣra and Kūfa were founded. In December, 639, the Arabs entered Egypt from the desert on the east and worked their way to the remains of the ancient capital of Memphis, near which they founded the camp of Fusṭāṭ, which in the tenth century was to become the city of Cairo. From Memphis the Arabs proceeded to the siege of the fortress called Babylon (near Cairo), which capitulated in April, 641, and they then turned their attention to the great city of Alexandria. This was fully garrisoned with seasoned troops, had practically impregnable walls to the landward side, and free access to the sea. The Arabs had no experience of, and no equipment for, siege warfare, but for some unknown reason Alexandria capitulated without resistance in November, 641.

It is more than likely that the native population of Alexandria played a part in its surrender; from the times of the Ptolemies they had suffered a system of taxation which was adjusted so as to leave only a bare subsistence, and if an Egyptian wished to fish in the river he not only paid a tax for this but was accompanied to his pitch by tax officials who relieved him of the appropriate due for every fish he caught. The Greek rulers were hated; the over-taxation in other places was later to shift the best elements of the Byzantine population into lands under Muslim rule. In turn, the Muslims later made the same mistake, which inevitably led to their decline. History has no meaning or appeal to governments.

From Egypt the flood of Arab conquest moved slowly over North Africa, where the camp-city of Qairawān was founded. It reached Spain, Toledo being occupied in 711. Crete and Sicily followed, Palermo remaining Muslim from 827 until the Norman conquest in 1091. The Muslims were now in direct contact with the South Italian and Spanish cultures. Further east, the way was blocked by Constantinople, which never fell to the Arabs; its conquest in 1453 was by non-Semitic Turks.

Muḥammad left no son. After the first two khalifs Abū Bakr and 'Umar, the khalifate was assumed by 'Alī, the husband of Muḥammad's daughter Fāṭima, but 'Alī's son Ḥasan abdicated in favour of Mu'āwiya, the governor of Syria (whom we met in ch. I). Mu'āwiya founded the Umayyad khalifate, ruling at Damascus from A.D. 661. A revolt under 'Alī's son, which led to the Shī'ite schism, was suppressed, but opposition to the Umayyads was organised under Shī'ite cover by the descendants of the Prophet's uncle.

The Shī'a party held that the khalifate was created for the descendants of Muḥammad, his cousin and son-in-law 'Alī and his lineal descendants. In 750 the Umayyads were defeated by a Persian army from Khurāsān, and the 'Abbāsid khalifate was founded, which was under strong Persian influence. The seat of the 'Abbāsid khalifate was later the new city of Baghdād, built by the khalif al-Manṣūr in 762–6. In 750 a surviving Umayyad prince fled to Spain and established independent rule, but the title of khalif was not adopted by the Umayyads in Spain until 929. The 'Abbāsid khalifate made Baghdād a great centre of culture, industry, trade and learning. In the tenth century the effective ruler in Baghdād was a Persian Shī'ite of the Buwayhids, who divided among themselves most of Persia and 'Irāq.

Under the 'Abbāsid khalifate the old Arab simplicity of life disappeared, and the cultures of the regions under its control were absorbed. The 'Abbāsids were of Persian origin, and Persia in its turn had assimilated Babylonian elements. Through Syria, Persia and Egypt a knowledge of Greek and Hellenistic science and pseudo-science reached the Muslims, who had little interest in old Greek literature as such. Among the gifts they received was alchemy, which came to them from Alexandria, partly by way of Syria.

Alongside the 'Abbāsid khalifate in Baghdād and the Umayyads in Spain, an independent Shī'ite Fāṭimid khalifate was founded near Tunis, but ruled in Egypt (Cairo) and Syria (910–1171). This was ended by the conquest of Egypt by Saladin (Ṣalāḥ al-Dīn), a Kurd, in 1171, and his successors were the Ayyūb sultans of Egypt and Syria.

The Turks emerged from Central Asia, and the Saljūq Turks under Tughril Bey conquered Khurāsān in 1040. Under Alp Arslān (1063–72) and Malikshāh (1072–92) they ruled over a great empire, including Persia,

Mesopotamia and Syria. In 1071 the Byzantine defeat at Manzikart in Armenia, near Lake Van, opened the way into Asia Minor, and Turkish rule spread over Anatolia (Rūm). One branch of the Saljūqs was established at Izniq (Niqiya, Nicæa), from whence they were ejected by the Crusaders and moved to Qūniya (Konieh, Iconium). They were defeated by the Mongols in 1243 and became subject to the Mongol Ilkhān dynasty of Persia. The third Mongol invasion under Hūlāgū sacked Baghdād in 1256, and the last 'Abbāsid khalif, Mustaʿsim (1242–56), was executed. A fictitious 'Abbāsid khalifate installed by the Mamlūk sultans of Egypt lasted until the 'Uthmān (Ottoman) conquest of 1517.

The Ottoman ('Uthmān, Osmanli) Turks, who take their name from Osman Bey (d. 1326), were drawn into the Byzantine complex; they crossed the Dardanelles in 1345 at the invitation of John Cantacuzenus during a civil war in Constantinople, and expanded towards Anqara. Murād I (the Amourat of the Byzantine historians) conquered most of the Balkans and Macedonia; his successor Bāyazīd I (Yildirim, "the thunderbolt") made widespread conquests which led to the organised opposition of the Crusades. Bāyazīd defeated the Crusaders at Nikopolis in 1396, but in 1402 he was captured by Tīmūr (Tamerlane) the Tartar, at Anqara.

Murād II (1421–51) besieged Constantinople without success in 1422 and his army was defeated in 1442 by the Hungarians under John Hunyadi, who in turn was defeated by the Turks in 1448. Muhammad II (1451–1512) captured Constantinople in 1453, as we saw in detail in chapter I, but he failed to take Rhodes from the Crusaders in 1480.

We must now retrace our steps and pick up another line of Turkish history. Turkish slaves who originally formed the bodyguard of the 'Abbāsid khalif revolted and set up independent states, one of which, Ghaznī in eastern Afghānistān, was founded in 977 by Sabuktigīn, whose son Mahmūd took Peshāwar in 1001. Mahmūd did not settle in India but raided it annually in the cooler season. In his second raid, in 1008, a confederation of Indian rulers was defeated.

Mahmūd built a college at Ghaznī and the Persian poet Firdawsī and scholar al-Bīrūnī were at his court; al-Bīrūnī composed an account of India in which he mentions Indian alchemy. The Muslim sultanate of Delhi, which was recognised by the 'Abbāsid khalif in 1226, lasted with varying fortunes, the most famous ruler being Fīrūz Shāh (1351–88).

The Mogul (Mongol) empire in India began with Bābur, who was a Barlas Turk on his father's side but a Mongol on his mother's; he detested the name of Mongol. He captured Kābul in 1504 and in 1526 at Pānīpat, in a great battle in which he used artillery, he defeated the native princes and became Sultan, ruling in Agra. The further fortunes of the Mogul empire in India will occupy us later.

The Ayyūb dynasty in Egypt ended in 1249 and in 1250 Shajar al-Durr, the widow of the last Ayyūb sultan, established the military rule of the Mamlūks (mamālīk, plural of mamlūk, slave). Their sultans were elected from Turkish and Circassian slaves captured as boys (many originally Christian) and specially trained as soldiers. The Mamlūks fought the Crusaders, their most famous sultan being Baybars al-Bunduqdāri (1260–77), a Turk (Marco Polo's "Bendocquedar"), who fought both Crusaders and Mongols. The Mamlūk rule was ended by the Ottoman Turks in 1517, and the Mamlūk troops became Janissaries. A Mamlūk line of rulers was re-established early in the eighteenth century by Circassians; Napoleon defeated Mamlūk troops in the battle of the Pyramids in 1798, and the line ended with Dā'ūd Pasha in 1831.

The Muslim culture in Spain is often called Moorish, since most of its representatives came from Barbary. In 1036 the Umayyads were ousted by the Almoravids, whose rule gave way in 1130 to the Almohades, and the history of Spanish Islām is a dreary story of murder and treachery. In 1085 Toledo was taken by Alfonso VI of Castile and became Christian. The last strongholds of the Muslims in Spain were Cordova and Granada, and Muslim rule in Spain ended with the capture of Granada in 1492 in the reign of Ferdinand and Isabella.

It was in Spain that the Muslim civilisation came into effective contact with Europe. The khalifs usually encouraged learning. They built magnificent cities, provided with schools and great libraries, and European scholars visited Toledo, Cordova and Granada to seek the treasures of learning to be found there, including translations of works of Aristotle unknown in the West. Translations of works in Arabic were made and among these was the first book on alchemy known in Europe, translated by Robert of Chester in 1144. The much older Greek works on alchemy, from Syriac translations of which the Arabs had derived their first knowledge of the subject, were unknown in Europe until the sixteenth century. The activities of Michael Scot have been mentioned in chapter II.

MUSLIM INCENDIARIES AND GUNPOWDER

The earliest reference to the use by the Arabs of ballistas (manjanīqs) for throwing stones is the tradition that they were made in the third century A.D. by the King of Ḥira, and since this is a locality in the north-west province of 'Irāq, the machines were probably adopted from the Persians, who in turn got them from the Greeks.[1] From the same source the Arabs probably acquired incendiary weapons. When Muḥammad the Prophet besieged Ṭā'if in A.D. 630, the defenders threw down red-hot clay balls,[2] and Elmacin (Al-Makīn, 1205–73; called Ibn al-'Amīd by Arabs) reported that in A.D. 690 Mecca was attacked with catapults which threw burning balls of pitch at the Caaba, burning the roof to cinders: arctissime Meccam, & Caabam catapultis impetebat, ita ut dirueret eam: quin & pilas ex pice & igne in eam jaciebat, ut comburentur tecta ejus & in cinerem redigerentur.

Incendiary arrows were used in the first Muslim invasion of India in A.D. 712, when the hawdaj of the elephant of Dahir, King of Alor in Scinde, was set on fire.[2] There is abundant evidence of the use of incendiaries by the Arabs from A.D. 780 to 950, and also by Muslims in the Crusades; e.g.

mixtures of pitch, wax, oil and sulphur in 1097 and 1147; naphtha shells at Acre in 1189–91; and tubs of incendiaries thrown from ballistas in 1229.[3] There is no clear evidence that the Muslims used firearms or gunpowder as an explosive throughout the whole period of the Crusades (1097–1291).[4]

A supposed use of gunpowder by the Arabs in Spain in the thirteenth century is based on the dating of an Arabic manuscript in the Escorial by Michael Casiri, a Maronite Christian who was born in Tripoli in 1710. He was trained in Rome and became a priest in 1734. After travelling with Assemani and teaching in Rome, he went to Madrid in 1748, where he became director of the Escorial Library. He began his book (ref. 6) in 1750 and died in Madrid in 1791. The Arabic manuscripts in the Escorial were captured in a ship from North Africa in 1671; a large collection (about 24,000) of Arabic manuscripts at Granada was burnt by the order of Cardinal Ximenes.[5]

The author of the Escorial manuscript, called by Casiri[6] Shebah ben Fadhl and dated 1249, is now supposed to be Shibah ibn Faḍlallāh al-'Umarī (d. 1349),[7] and the work may be even later than 1349.[8] Casiri translated the Arabic text (which he reproduces) as follows: Serpunt, susurrantque scorpiones circumligati ac pulvere nitrato incensi, unde explosi fulgurant ac incendunt. Jam videre erat Manganum excussum veluti nubem per aëra extendi ac tonitrûs instar horrendum edere fragorem ignemque undequaque vomens, omnia dirumpere, incendere, in cineres redigere.

Pulvis nitratus at that time meant gunpowder and Hallam[9] said he would be glad to know if this was a fair translation of the Arabic word. The Arabic word is bārūd (the third word from the left in the first line of the text reproduced by Romocki[10]), and this could mean either saltpetre or gunpowder. Lalanne[11] (who dated the manuscript 1248) took it to mean gunpowder, and the "scorpion" a cask like that described by Joinville (p. 25), filled with burning gunpowder. "Scorpions," however, were engines for throwing stones or other objects by means of slings,[12] and a "cask of burning gunpowder" would be a precarious weapon. Quatremère[8] translated the Arabic: on se défendit à l'aide de scorpions de poudre bien ficelés, et on y met le feu. Ce feu, partout où on le lance, brûle. Romocki and Hime both say that Casiri's translation is incorrect, but they do not give a better one, and they omit to draw attention to an important note in which Casiri says:

> Ex dictis sanè liquet Scriptorum nostrum de globis ferreis ope ignis artificialis excussis loqui; vocabulis enim *Naphtha* & *Barud* ubique utitur, ex quibus per ea tempora pulvis tormentarius conficiebatur. Nomine autem *Barud* Persæ, Turcæ & Arabes olim *Nitrum*, hodie *Pulverem nitratum* intelligunt; *Naphtha* verò bituminis genus sulphure mistum significat. (Casiri's spelling "naphtha," from νάφθα, instead of the commonly used "naphta" or "naptha," is noteworthy.)

That is, bārūd among the Persians, Turks and Arabs formerly meant saltpetre but came to mean gunpowder, and bārūd and naphtha were both used at the time the manuscript was written to mean gunpowder, although naphtha is properly a kind of bitumen mixed with sulphur. This statement has recently appeared as a new discovery.

There is no reason why bārūd in 1349 should not mean gunpowder (saltpetre itself would not burn or explode). The object thrown from a ballista could very well have been a gunpowder bomb. Hallam,[9] in discussing the passage, said that he had, on the whole, no doubt that bārūd meant gunpowder, which was known to Roger Bacon, and cannon or mortars were invented early in the fourteenth century. This is perfectly correct and Romocki and Hime could well have mentioned it. This passage in Casiri has been dealt with first because it was supposed to refer to the date 1249, whereas it is probably a century later. Some references to the supposed early use of artillery will now be considered.

Ibn Khaldūn (1332–1406) says[12] that the Sultan Abū Yūsuf in the siege of Sijilmāsa in 1274 used:

> machines de siège, telles que catapultes, balistes et l'engin à feu [hindām al-naft] qui lance du gravier de fer. Cette mitraille est chassée hors de l'âme [hindām] de la pièce par le moyen de la poudre enflammée [bārūd] dont la propriété singulière opère des effets qui rivalisent avec la puissance du Créateur.

This apparatus, discharging fragments of iron by means of a burning powder, reminds one of the paper or bamboo huo ch'iang (tuyau de feu) said to have been used in China in 1232 and 1259 (see ch. VI).

The assumption that cannon were used in 1331 depends on a document quoted by the Spanish historian Zurita (1512–80), archivist to Philip II, who spent twenty years in writing his great work[14] and has a reputation for accuracy. He says that when the Moorish King of Granada besieged Alicante in 1331 he used a new machine which caused great terror. It threw iron balls with fire:

> tambien se publicava, que el Rey de Granada, con todo su poder, por mar y por tierra vernia sobre Alicante; y puso en aquel tiempo grande terror una nueva invencion de combate, que entre las otras machinas que el Rey de Granada tenia para combatir los muros llevava pelotas de hierro, que se lançavan con fuego.

The correct translation of "con fuego" is "[along] with fire," although Köhler,[15] who (after Hoyer) drew attention to the passage, thought it could mean "by means of fire." The same passage is quoted by Andres[16] (1740–1817), who speaks of "moltes pilotes de fer *per* gitarlas llunys *ab* foch (many iron balls to be thrown with or by means of fire)." It is seldom that a preposition introduces such uncertainty!

An event in the siege of Baza, variously dated 1323, 1324 and 1325 by different authors, was reported by Casiri.[6] Lalanne[17] accepted Casiri's translation from Ibn al-Khaṭīb (1313–74) for events of 1312 and 1323 (Casiri gives the Arabic text): Ille . . . castra movens, multo milite hostium urbem *Baza* obsedit, ubi machinam illam maximam Naphtha & globo instructam, admodo igne, in munitam arcem cum strepitu explosit; he discharged [explosit] with much noise a great machine provided with naphtha and a ball. Romocki[18] and Hime[19] say there is no word corresponding with "explosit" in the original, and that Casiri (who had no intention of deceiving his readers, since he gave the Arabic text) had inserted "&" between "naphtha" and "ball". Hime gives as the correct translation : By means of a great machine provided with naphtha [made up in] hot [burning] balls, he struck the arch of an inaccessible tower. Ballistas and incendiary balls had been provided by 'Abd al-Raḥmān II in A.D. 844 for his army in Spain, and gunpowder was then unknown. Hence, say Romocki and Hime, the same sort of apparatus was used in 1323-5.

What seems to be a better translation was given by Casiri's successor at the Escorial, Jose Antonio Conde (1765–1820), who also drew upon unpublished Arabic manuscripts for his book.[20] Quatremère[8] says Conde translated the original Arabic texts he used "more or less correctly"; Romocki[6] that his book was in general very useful (im übrigen sehr verdienstvoll); but Hime[21] (who devotes nearly a page to an attack on Conde) quotes Renan to the effect that Conde "ne possédait guère que les premiers éléments de l'arabe," and adds that "suspicion hangs over every page of this book." The chief source of complaint seems to be that Conde does not quote his sources, so that the kind of author who pillages the works of his predecessors, yet supplies only the references they give without troubling to look at them, is compelled in this case to quote his actual source when he appropriates material from Conde's book.

Conde says of the siege of Baza and that of Martos which followed it[22] : Asi que en la luna de Regeb del año 724 (A.D. 1325) fué a cercar la ciudad de Baza que habitan tomado los Christianos; a campó y fortificó su real; combatió la ciudad de dia y noche con máquinas é ingenios que lanzaban globos de fuego con grandes truenos, todo semejantes á los rayos de las tempestades, y hacian gran estrago en los muros y torres de la ciudad.

Al año siguiente de 725 (A.D. 1326) fué el rey con poderosa hueste y bien provisto de máquinas é ingenios á cercar la ciudad de Martos; la combatió desde el dia 10 de Regeb con incesante fuego de las máquinas de truenos y se apoderó por fuerza de la fortaleza.

Allouche[23] says the text of the account of the siege of Huescar Baza in 1324 (sic) given by Casiri is incomplete. He gives a fuller text from a manuscript of Lisān al-dīn Ibn al-Khaṭīb of Granada (1313–74).[24] Allouche assumes (with Casiri) that nafṭ meant gunpowder, that "a machine worked by naphtha" (a cannon) and not "a machine throwing naphtha" was concerned (since it is said to have demolished walls), and that the account is "the first witness (témoinage) of the efficacious use of cannon," as Quatremère[25] had supposed, Reinaud and Favé's[26] conclusion that only naphtha was used as an incendiary being less probable.

The additional text quoted by Allouche does not seem to me to justify the conclusion that cannon were used, and I agree with Rathgen[27] that the truenos used at Baza and Martos were some sort of bomb, perhaps containing gunpowder, thrown by a ballista, and not by a gun. I do not believe that cannon were used in Spain in 1324–5, and the first mention of them otherwise is in Florence in 1326 (see p. 101).

The Arabs in Spain do not appear to have had guns in the attack on Tarifa in 1340. Conde says[28] machines throwing large iron balls and naphtha were used : . . . y fueron delante de Tarifa y acamparon alli en 3 del siguiente mes, y principiaron á combatirla con máquinas é ingenios de truenos que lanzaban balas de hierro grandes con *nafta*, causando gran destruccion en sus bien torreados muros. It is important to realise that the destructive effects were caused by the large stone or iron balls thrown by ballistas, and the incendiary effects by naphtha thrown in the same way, and not to assume that the balls were thrown from cannon by means of gunpowder, which was also called naft. Ballistas were still used when cannon were known (p. 127).

Pedro Mexia (1496–1552) said[29] that the Moors did not use gunpowder until they were besieged in Algeciras in 1343, and that Peter, Bishop of Leon, in his chronicle of Alfonso XI (1312–50) reported that in a naval battle between the Moorish kings of Tunis and Seville, the ships of the king of Tunis carried iron bombards which projected fire. Mexia begins with a short history of gunpowder and artillery, taken from Biondo Flavis (Flavius Blondus, 1388–1463) and Raphael Maffei (Raphael Volaterrano, 1451–1522), saying that gunpowder and artillery were invented by a German (i.e. Berthold Schwartz) and first used by the Venetians against the Genoese in 1338 (see p. 93). He then proceeds to describe the siege of Algeciras in 1343 and the naval battle mentioned above :

Todo esto era liuiano, à todo esto vence en crueldad la inuencion de la puluora y artilleria : la qual dizen que hizo y ymaginó vn hombre natural de Alemaña, cuyo nombre no se sabe, ni merecio que del quedasse memoria. Los primeros que della vsaron, segun dize Blondo y Raphael Volaterano, fueron los Venecianos contra los Ginoueses, en el año del Señor de mil y tres y trezientos y ochenta años. Aunque à mi vermas antigua cosa deue ser esta inuencion, porque en la coronica del Rey don Alfonso el onzeno de Castilla, que gano las Algeziras, se escriue, que teniendo el cercada el Algezira en el año del Señor de mil y trezientos y quarenta y tres años, los Moros cercadós tirauan desde la ciudad ciertos truenos con tiros de hierro, lo qual es quarenta años antes de lo que dize Blondo. Y aun mucho tiempo antes desto en la coronica del Rey don Alfonso, que ganó à Toledo, escriue don Pedro Obrispo de Leon, que en vna batalla de mer, que vuo entre l'armada del Rey de Tunes y del Rey de Seuilla Moros, à quien fauorecia el Rey don Alfonso, los nauiros del Rey de Tunez trayan ciertos tiros de hierro o bombardos, con que tirauan muchos truenos de fuego, lo qual si assi es, deuio de ser artilleria, aunque no en la perfecion de agora, y ha esto mas de quatrocientos años.

Casiri[6] quotes from a Spanish chronicle of King Alfonso XI for the siege of Algeciras (1342–4) and translates into Latin :

Y los Moros de la ciudad lançaban muchos truenos contra la huesta en que lançaban pellas de fierro grandes tamañas como mançanas muy grandes, y lançabanlas tan lexos de la ciudad, que passavan allende della hueste algunas dellas, è algunas dellas ferian en la hueste.

Multa Mauros ab oppido in exercitum displosisse [not "explosisse," as Favé says] tonitrua, quibus ferreas pilas malis Matianis prægrandibus pares emittebant; idque tam longè ut aliæ obsidentium copiarum stationem præterirent, aliæ ipsas offenderent copias.

Although this leaves it doubtful whether "iron balls thrown by fire" or "iron balls thrown together with fire" is meant, Conde's[27] translation makes it probable that the second meaning, i.e. incendiary balls, is the one intended:

Levantoron los christianos grandes máquinas y torres de madera para combatir la ciudad, y los muslimes las destruian con piedras que tiraban desde sus muros, y con ardientes balas de hierro que lanzaban con tronante *nafta* que las derribaba y hacia gran daño en los del campo.

Juan de Mariana (1536–1623) reported[30] that gunpowder and iron bullets were used at Algeciras; the Latin text and the old Spanish translation read:

[Algeziræ] excussa tormentis tela lapidesque à nostris. Barbari ex urbe contrà ferreos globos cum igne et fragore, pulvereque tormentario eiaculabantur, haud levi sæpe nostrorum damno. Huius tormenti mentionem, nunc primum in historia factam inuenio.

En especial tirauan muchas balas de hierro, con tiros de puluora, que con grande estampido, y no poco daño de los contrarios, las lançauan en los reales. Esta es la primera vez que doste genero de tiros de poluera hallo hecha mencion en las historias.

The English translation by Stevens reads: The besieged did great harm among the Christians with iron bullets they shot. This is the first time we find any mention of gunpowder and ball in our histories.

Mariana says the earls of Derby and Salisbury were in the battle of Algeciras, and Bishop Watson[31] suggested that they brought back a knowledge of firearms to England. Prescott[32] later believed that the Spaniards derived their knowledge of artillery from the Arabs in Granada, who were early acquainted with gunpowder. Although Gardner[33] speaks of the use of cannon by the sultans of Fez and Granada in 1340, he gives no authority for the statement, which may be Conde's.[28]

Casiri,[34] under the date 1382, quotes: quinque Zabræ & Sagitiæ oppidi portum subires farina, melle, butyro &c, quo tonitrus emittibatur pulvero onustæ; entraron en la ciudad cinco Zabras y Sætias cargadas de farina y de miel, y de manuteca, y de polvoracon que lançaban del trueno.

It has been seen that Casiri (1770) suggested that the Arabic word naft (as well as bārūd) could mean gunpowder. Quatremère[35] in his translation (1836) of Rashīd al-dīn added two long notes on gunpowder and firearms, in which he quotes the Arabic texts for the technical terms, giving many more original references than do more recent authors. In the first note he says naft was originally a bituminous oil which, with other ingredients, was the basis of Greek fire, but "it appears that the same word was later sometimes (quelquefois) used to designate gunpowder (la poudre)," as several quoted texts show. In the second note, when he came to discuss various kinds of firearms, Quatremère pointed out that the modern names of these originally meant bows and ballistas for throwing stones. These names were applied to firearms after the invention of gunpowder. Quatremère said that, although the name naft was sometimes used for gunpowder, he had come to the conclusion that when mentioned in relation with exercises of Janissaries it meant naphtha. He quotes several Arabic texts in support of this view (which I believe is correct), and showed that naphtha continued to be used after the invention of gunpowder. In some cases it was said to burn under water and it was used in pyrotechnic displays, but contemporary travellers who had tried to find out what was used were told by the artisans that it was a profound secret, to reveal which was punishable by death—a reminiscence of the old story of Greek fire (see p. 21).

Köhler[36] and Mercier[3] also supposed that naft could mean gunpowder, Köhler saying that, taken in conjunction with the mention of guns in Florence in 1326 (see p. 101), the Arabic accounts suggest that the Arabs introduced firearms into Spain, from where they passed to Italy, going from there to France, and finally Germany. Mercier thought that "it would not be too adventurous to consider that the balls were thrown by gunpowder."

In his second note, Quatremère[35] says jarkh originally meant an arbalest for throwing stones or naphtha, and zanbūrak a kind of bow, although both names were later used for firearms. Quoting many (mostly late) Arabic authors, he concluded that the name midfaʿ meant a cannon, and the name mukhula meant a culverin. The name bunduq originally meant a hazel nut, then a lead bullet, and finally a musket. In an account by Ibn Iyās, a culverin is said to be made by a person called Domingo, presumably a Spaniard.

From the Arabs in Spain we now turn to the Mamlūks of Egypt and Syria. Ibn Khaldūn (1332–1406), whose Kitāb al-ʿibar was not completed until after 1377 (when cannon were well known in Europe),[37] reported that in December, 1366, the amir Yalbughā al-Naṣiri used near the citadel in Cairo an offensive weapon called a ṣawāʿiq al-naft against his opponents. Al-ʿAynī (b. 1360) uses the name makāḥil al-naft; al-Maqrīzī (d. 1442) uses simply naft; they, and Yūsuf ibn Taghrībirdī of Cairo (1411–69), who usually copied earlier works carefully, all refer to this incident.[38]

Al-Qalqashandī (1355–1418), who spent his life in the Mamlūk chancellery, reports that in 1376 he saw in Alexandria a midfaʿ made of copper and lead, hooped with iron bands, which threw a great bullet (bunduq) of red-hot (muḥmāt) iron from the Hippodrome as far as the chain closing the port beyond the Bāb al-Baḥr, a great distance.[39]

Ayalon,[38] who supposes that naft means gunpowder, makes these accounts speak of the "first use of artillery in Cairo" in 1366, adding that in 1389–90, during the fierce battles for the throne, artillery figures prominently in the sieges of Cairo and Damascus. Thereafter its use increased until it became commonplace. Wiet[39] speaks of zanbūrak (light artillery), of artillery in Constantinople of bronze with iron hoops, and a large gun ("folly") in Cairo in 1500 for use against the Ottoman Turks. He thinks the last Mamlūks had large cannon, but their citadels in Syria were without artillery. Ayalon supposes that midfaʿ al-naft̤, mukhulat al-naft̤, s̤awāʿiq al-naft̤, s̤awārikh al-naft̤, ālāt al-naft̤ and hindām al-naft̤ were all names used for firearms of various kinds, and the last four were short-lived names for cannon.[40] In quoting al-Qalqashandī, however, he omits the words "of red-hot iron." Possibly the "copper and lead" in al-Qalqashandī's account may mean "copper and tin," i.e. bronze, as in the artillery mentioned by Wiet, but this is conjectural.

Ibn Khaldūn says the hindām al-naft̤ threw iron pebbles from a magazine in front of a fire kindled by means of bārūd, and bārūd here may mean gunpowder.[41] Al-Qalqashandī says makāḥil al-bārūd are al-madāfīʿ shooting big arrows which almost pierce a stone, and iron balls weighing from 10 to over 100 Egyptian rut̤l (11½–112 lb. troy), and the qawārīr al-naft̤ are qudūr (literally, jugs) in which naft̤ is put for throwing at fortresses for the purpose of burning them.[42] The last are naphtha bottles (naphtha grenades); the troops throwing them are called zarrāqūn naffat̤ūn, and another name for them is naft̤īya. Ayalon[43] supposes (which I think is wrong) that the last means midfaʿ troops using a firearm (midfaʿ) charged with gunpowder (naft̤), although he agrees that Orientalists do not make this rather pedantic distinction. What Ayalon calls "guns" are, I think, probably ballista or trebuchets; he admits[44] that the "noise like thunder" which they are said to make is a common Oriental exaggeration for such weapons, which we shall also encounter in the Chinese descriptions (ch. VI).

Ballistas (manjāniq) were used in 1389, 1400, 1404, 1412, 1415, 1433, 1434, 1442, and even in 1500–16, when cannon were known, as Ayalon says, and in the cases when they threw naphtha they were most likely the ballista described by the Crusaders (see ch. I) and not cannon, as Ayalon supposed. That makāḥil al-bārūd meant a ballista or trebuchet throwing stones and incendiaries, rather than a gun, seems probable from a description by Ibn Fad̤lallāh al-ʿUmarī (d. 1349) in his al-Taʿrīf, who says it is something which throws both fire (nār) and solid projectiles (bandāniq).[45]

Ayalon proposes what he thinks is a new hypothesis that from about 1360 naft̤ was the common Mamlūk name for a firearm, naft̤ first and then bārūd being names for gunpowder, bārūd originally meaning saltpetre. He says[46] (I think incorrectly) that the Mamlūks used naphtha only once, and without success, in the period of the Crusades, at Wādī al-Khāzindar in 1299, because they had found its effects to be capricious and dangerous. Hence he assumes that all accounts speaking of the use of naft̤ must be understood as referring to firearms. I believe that the material collected in this chapter throws doubt on Ayalon's hypothesis, and the repeated use of naphtha in the Crusades (see ch. I) seems to me to be established beyond all doubt. Ayalon points out that an obsolete Ethiopic name for a gun (probably derived from Egypt) is näft̤, and that Maghribi cannon of the period 1557–74 have the inscription nuft, but these do not seem very strong arguments in favour of Ayalon's hypothesis, as he admits.

Ayalon[47] supposed that gunpowder was first used as an incendiary only, replacing naphtha, and hence was called naft, and then, since saltpetre (bārūd) is more important than its other two constituents, sulphur (kibrīt) and charcoal (fahm), it took the name, bārūd, of saltpetre. If knowledge of gunpowder reached the Arabs from China (see ch. VI) this is just possible, but not if it came from Europe, where gunpowder was not used as an incendiary.

Mercier[3] says that the naffāṭa is a copper instrument for throwing naphtha (naft); that the qarūra ("wine bottle") was a glass jar for containing naphtha (there is one in the Invalides); that Amari's suggestion[48] that the Muslims first learnt the use of Greek fire from the Crusaders is incorrect—they were using naphtha at least as early as A.D. 780 (see p. 189); and that the idea of Brooks[49] (and apparently of Ayalon) that Greek fire was not used in attack but only for defending towns had been corrected by Canard.[3] I agree with all this.

MUSLIM FIRE BOOKS

The Muslim works dealing with the use of incendiaries in war are sometimes called furūsīya (horsemanship) treatises. Some of the most important were published in extracts and translations by Reinaud and Favé and Quatremère, and Ayalon examined a number of them in French and English libraries without finding any important information on firearms not already given by these authors. A list of the contents of the most important furūsīya manuscripts in the Istanbul libraries, many of them written in the early and late Circassian periods,[50] appears to show that these also are of a similar character. A brief summary of a few of these works, in so far as they are of interest to us, will be given, some points which seem to have been overlooked by previous writers being emphasised.

I.   The work by Murdā ibn 'Alī (twelfth century), composed for Saladin (d. A.D. 1193),[51] contains a section on burning mirrors compiled from an extant work by 'Uṭārid ibn Muḥammad al-Ḥāsib, and several anonymous recipes for incendiaries (nufūṭ) which are of the same character as those in the other works described below. (Cahen is in error in saying that the latter do not mention the quicklime and sulphur "automatic fire"). The early part, on swords, gives a fanciful recipe for making steel, taken, it is said, from Jābir ibn Hayyān's *Book of the Seventy*. The section on naphtha (pp. 145–8) has recipes like those in the other works; it does not mention saltpetre or gunpowder.

An incendiary oil is made by distilling oil with quicklime in an alembic. Several of the usual recipes for Greek fire (which is called naphtha) are given. A "naphtha thrown with arrows" (not "for throwing arrows") contains equal parts of sulphur, resin, colophonium, tow or straw, and wheat-chaff, all ground and mixed. It is wrapped in a fine linen cloth and covered with mulberry bark. The "heart" of the iron arrow is filled with it, it is kindled at the extremity, and the arrow is thrown. Another mixture for the same purpose contains equal parts of sulphur, lacquer ("laque"), colophonium, yellow amber and crumbled dry horse-dung; it is enveloped in wool fluff (bourre de laine) and lime.

A mixture for throwing in an empty egg-shell contains sandarac, "water of sulphur, that is, naphtha," and colophonium. An "active naphtha" for throwing from mangonels contains 10 of tar, 3 of resin, $1\frac{1}{2}$ each of sandarac and lacquer, 3 of pure sulphur separated from all earth, 5 of dolphin fat, and 5 of "dissolved" fat of goat's kidney. These are melted together, and when used a tenth of "mineral sulphur called naphtha," which is like old oil, is added, the mixture being heated in an earthenware pot (marmite) till it inflames, when it is thrown. A similar mixture floats on water and is excellent for burning ships. (The name "mineral sulphur" for petroleum is noteworthy.)

Oil of bricks (see p. 156) is mentioned, also an "ingenious" (automatic?) fire made from quicklime, sulphur, oil and cypress resin. A statue covered with quicklime and oils inflames when sprinkled with water. Cantharides or "shining worms" are squeezed into black oil, and the bile of the sea-tortoise, etc., added to make a composition which shines at night. These recipes are among the older components of the *Liber Ignium* (see Ch. II), and since they are found here in an Arabic work composed in the twelfth century, what was said concerning the *Liber Ignium* is supported.

II. The Leyden Arabic manuscripts 92 and 499, one dated at the end as a copy of an original of 1225, are of the same work, entitled "Treatise on stratagems, wars, the capture of towns, and the defence of passes, according to the instructions of Alexander son of Philip." This follows the Arabic custom of attributing all the discoveries of war and peace to Alexander the Great, or his "vizier" Aristotle.[52] It quotes no authorities except Aristotle, and parts of it could be derived from older sources.

Some recipes translated by Reinaud and Favé[53] deal with the preparation and use of various kinds of naphtha: white, red, Persian, of Qulzum (Suez) etc. Also various tars of Maghrib, Barqa and Syria; galbanum (which is said to be a liquid pitch); various oils: linseed, coconut, of Palestine, Maghrib, etc.; resins: pine, turpentine, pistachio, mastic, colophony, etc.; and fats of dogfish, pond and river dogs (?), domestic dogs, bears, wolves and cows. Also specified are yellow and white of egg; milk of the dôm-palm; yellow, white and black sulphur; arsenic and alum. The Arabic text (which is given) abounds in technical words not found in dictionaries and the translation is often unintelligible.[54] The original was perhaps intentionally obscure,[55] and some "cover-names" may be used.

The white naphtha must have been distilled petrol, since it is said to be tested by dipping the hand (Quatremère says a cloth) into it, and putting it over a fire, when "the naphtha at once takes fire." Black naphtha is tested by kindling it and seeing if the flame sets fire to a reed. To make white naphtha from black naphtha, this is[56]: (a) boiled with water and decanted, digested with oil of Palestine and boiled with olibanum, then, after cooling, mixed with white of egg and a little alum; or (b) beaten with yolk and white of eggs and calcined egg-shells, heated to boiling, cooled and filtered two or three times through a mixture of sand, sulphur and charcoal, when it becomes limpid and clear; or (c) distilled in a cucurbit and alembic. It has already been mentioned (see p. 4) that the "white naphtha" mentioned by early Greek authors may have been obtained by processes like (a) and (b). This is obviously old material copied from earlier sources. Reinaud translated: "you distil it by the cucurbit and by alembic whilst it is moist"; Quatremère[57]: "you distil it with a cucurbit and an alembic and you receive it in a liquid," but "in a liquid form," would make better sense. Reinaud's "huile graisse" is replaced by "huile chaude" by Quatremère, but "incendiary oil" would be more suitable.

The distilled naphtha was of very low boiling-point, since in a "process of Baṣra" the naphtha is distilled and then heated (perhaps redistilled?) in a vessel by burning under it reeds "one by one," i.e. by a very gentle heat. A mixture of resin, fat, black pitch and naphtha is said to burn on water; "if you wish the flame to be very pure, add sulphur and powdered colophony."[58] Gentle heating in horse-dung is prescribed and distillation was well-known.[59]

To "draw fire from cold water," mix powdered quicklime, soda (naṭrūn) and sulphur in a pot, then add water "in the sun" (cf. Julius Africanus, p. 9), or sprinkle it on water, when fire comes out. "If you add naphtha it inflames at once. This is a curious process."[60] Fire which burns on water is made from oil "or other light matter." Distilled water of camphor mixed with oil of balm inflames everything.[61]

A fire-resisting paint is made from equal parts of powdered pots, red clay, scales of "telline" (?) and salt, ground together and then wetted with a solution of gum in strong vinegar and water of egg.[62] Protection from fire by compositions containing talc (in one Arabic dictionary ṭulq is said to be a drug pressed from a plant) was well known to the Arabs.[63] (Dr. Holmyard suggests that ṭulq from a plant may be ṭabāshīr, a siliceous concretion in the bamboo.) A method of "striking the enemy with syringes" (the Byzantine σιφῶνα) specifies hollow reeds cut section by section. A fire-lance is tipped with tow soaked in naphtha and sprinkled with sulphur.[64]

Black sulphur (crude petroleum) is distilled in an alembic until it is moist (? a liquid distils), when an admirable oil comes over which can be used in all operations.[65] In a description[66] of the burning of a castle of stone and mortar, a liquid is made from powdered yellow marcasite (cf. Julius Africanus, p. 9), urine of a child, sulphur, "blue naphtha," vinegar and old sour wine. Jars filled with the mixture are thrown when the wind is towards the enemy. The other troops advance "with fire and naphtha," and the exhalation of the mixture then inflames and the castle crumbles with "a noise of thunder and a frightful whistling." If any débris is left, it is destroyed by naphtha troops using a "prepared water" and naphtha, the naphtha taking fire and burning with a thick black smoke, which is lethal by its terrible and suffocating stench and fire (cf. Marcus Græcus). This is typical Oriental exaggeration.

Alternatively, naphtha is thrown on a great brushwood fire, and when the defenders smell "the odour of this water" they perish. Even an iron door is broken in pieces by burning naphtha. Al-Dahabī in his *Ta'rīkh al-Islām* describes the naphtha troops in Baghdād as throwing glass bottles of naphtha in exercise "so that the field was filled with fire."[67]

Although MS. 92 has the title "Treatise on strategems and wars, warlike instruments, the siege of fortresses, the way of striking the sword and throwing arrows, also the manufacture of bārūd," there is no mention of saltpetre in either manuscript and the word was probably added by a later copyist,[68] unless this part, at the end of the manuscript, has been lost. Saltpetre was apparently unknown in 1225, but it is mentioned by Marcus Græcus in 1250.[69]

Alexander the Great is associated in Eastern legends with the invention of incendiary weapons, probably as a result of misreading the works of ψ-Kallisthenes and Philostratos (see p. 209). Firdawsī (d. A.D. 1020) says Alexander's philosophers devised an iron horse with an iron rider filled with naphtha, which was on wheels like a carriage. When it came in contact with elephants, the whole exploded. Alexander had a thousand of these machines made. This may be the "rolling horse" mentioned in a Chinese account of India. "Prester John" had bronze figures on horse-back, filled with fire, with a man behind who threw something within them, which immediately produced an immense smoke. Qazwīnī (d. 1283) says Alexander invented infernal machines of a bull's hide stuffed with pitch, sulphur, quick-lime and arsenic, as a trap for a dragon which infested an island of the Indian Sea; the dragon swallowed two of them, which kindled a fire in his stomach.[70]

III. Al-Ḥasan al-Rammāḥ (the lancer) Najm al-dīn al-Aḥdab (the hunchback) flourished probably in Syria, and died in A.D. 1294–5 in his thirties, so that his writing can hardly be before 1280. His work *Kitāb al-furūsīya wa al-munāṣab al-ḥarbīya* ("Treatise on horsemanship and stratagems of war") exists in two Paris Arabic manuscripts, BN ancien fonds 2825[71] (old 1128) and fons Asselin 643. It is carefully written (but sometimes without diacritical points) and with coloured figures, and was probably provided by the government for confidential use. It frequently cites Muḥammad ibn al-Shayẓamī and Ibrāhīm ibn Sallām, who are otherwise unknown. The introduction says that the book contains "all that is necessary for the masters, men of war, gallants, and artificers, in fact of military operations, the different ways of using the lance, the mace, and the arrow; ways of mixing materials, constructing machines, communication of fire, etc.; naval combats, and other things no less curious," all for the advancement of Islam. Pyrotechnics, however, play the most important rôle.

The manuscript says : "The second part treats of machines of fire to be used for amusement or for a useful purpose, machines of fire required in war on land or sea, for the defence of fortresses, in sieges, when a place is to be set on fire, in saps when doors covered with iron are to be burnt, when pots are to be thrown by mangonels, pots with narrow necks, clubs (or hammers), fire-lances, instruments for distillation, the proportions [i.e. recipes], smokes, flying fire or rockets (fusées volantes), flowers [i.e. fireworks], lance-heads, cups, birds, and moons." The "birds" may be expendable carriers or fireworks; the "moons" are a kind of Bengal fire. The stone-throwing engines described are the same as those used in Europe in the twelfth and thirteenth centuries.

An anonymous work in MS. 2825 cites the same two authors as Ḥasan al-Rammāḥ, and another (different) al-Rammāḥ. It professes to reveal all the secrets and describes the purification of saltpetre (not given in the Leyden manuscript, II above) by a process of simple crystallisation[72]:

Take white, clean, bright (or natural) saltpetre ad. lib. and two new (earthen) jars. Put the saltpetre into one of them and add some water. Put the jar on a gentle fire until it gets warm. [Skim off] the scum that rises [and] throw it away. Stir up the fire until the liquid becomes quite clear. Then pour it into the other jar in such a way that no scum remains attached to it. Place this jar on a low fire until the contents begin to coagulate. Then take it off the fire and beat gently.

Ḥasan al-Rammāḥ describes the purification of saltpetre (bārūd) by treatment of the solution with wood ashes (which would precipitate deliquescent calcium and magnesium salts) and crystallisation[73]:

Take dry willow wood, burn it, and plunge it into water according to the recipe for its incineration. Take three parts by weight of the saltpetre and the third part of the wood-ash which has been carefully pulverised, and put the mixture into a jar—if made of brass so much the better. Add water and apply heat, until the ashes and saltpetre no longer adhere together. Beware of sparks.

The passage was translated differently by Quatremère,[74] who thought (incorrectly) that it described the preparation of gunpowder (bārūd) from saltpetre and charcoal. But there is no mention of sulphur, and if "burnt willow wood" meant charcoal, this would be used as a clarifying agent, but it probably meant wood-ashes. It will be remembered that a claim has been made (p. 73) that the purification of saltpetre with wood-ashes was known to Roger Bacon,[75] but this is based on an arbitrary manipulation of a text, and the first clear account of the process known to me is that of Ḥasan al-Rammāḥ. It may be, as Romocki suggested,[76] that it was adopted because of the age-old use of ashes or lye (as well as soda) as detergents (see ch. VII).

It has been said[77] that there was a saltpetre (shūraj) industry in Baṣra in A.D. 869, but the word translated "saltpetre" could mean "alkali or soda,"[78] although it is the Persian (and Indian vernacular) name for saltpetre.

Ḥasan al-Rammāḥ gives a long account of the use of artificial fire in war by Alexander the Great, who is said to have set Tyre on fire by incendiary arrows. He mentions magic appliances like those in Marcus Græcus and other European works: magic lamps which made men look black, coloured flying birds, the appearance of skeletons, etc., and the spectators look as if they were covered with blood. There are also recipes for fire-proofing wooden engines with a mixture of talc, clay, white of egg, gum and "salamander skins" (asbestos?). These mixtures were also used to fire-proof the clothes and hat of a rider, who was then covered with naphtha and sulphur and set on fire to ride among the enemy cavalry. These devices nearly all appear in European works, e.g. of Albertus Magnus (ch. II), but the last is novel.

There is no mention of the use of gunpowder as a propellant. There are descriptions of rockets, Roman candles throwing "chick-peas" or incendiary balls composed of saltpetre, sulphur, charcoal, resin, linseed oil, etc., and sometimes metal filings, but not of guns or firearms of any kind.[79]

The most notable feature of Ḥasan al-Rammāḥ's work is the extensive use it makes of Chinese material, although he does not use the name "snow of China (thalj al-Ṣīn)" for saltpetre (see ch. VII). The fireworks are both static and projected, and are called jasmine flower, mūraq (?) flower,[80] flower of China, moonlight, rays of the sun, garlands of gold, white or green nenuphar, yellow tongue, wheel, wheel of China, stars, etc.; and there are also yellow, green, white, red and blue smokes (as in the Chinese *Wu Pei Chih*, ch. VI). The mixtures contain more saltpetre than modern compositions and sometimes contain Chinese red arsenic (realgar), iron filings, Chinese iron, bronze filings, white lead, "incense stone" (?), camphor, sal ammoniac, coral, mastic, gall nuts, etc. The collection of recipes was probably taken from various sources at different times in the author's family and handed down, in the same way as those in the *Liber Ignium* of Marcus Græcus (ch. II). Some recipes are described as "tested."

In the selection of recipes in the following table the numbers give the proportions of saltpetre, sulphur and charcoal *in this order** (an ingredient omitted is denoted by –) followed (when these are specified) by the other ingredients :

Flying fire $10:1:3$.

Flower of China $10:1\frac{7}{8}:2$ or $10:2:3\frac{1}{4}$ and Chinese iron 10.

Rocket compositions $10:1\frac{1}{2}:3$, $10:2\frac{1}{8}:2\frac{1}{4}$, $10:1\frac{1}{4}:2\frac{1}{4}$, $10:1\frac{1}{2}:2\frac{1}{2}$.

Rays of the sun $10:1\frac{1}{8}:2\frac{1}{4}$, $10:1\frac{1}{8}:2$.

Jasmine flower $10:2:3$, iron filings 5 or 6.

Mūraq flower $10:\frac{3}{4}:4$, and steel $3\frac{1}{2}$.

Chickpeas $10:2\frac{1}{2}:\frac{1}{2}$, incense stone $\frac{1}{4}$; $10:1\frac{3}{4}:1\frac{7}{8}$, Chinese iron 2.

Moon flower $10:3:\frac{1}{2}$, incense stone $\frac{1}{2}$; $10:2\frac{1}{8}$ –, arsenic $\frac{7}{8}$; $10:2\frac{1}{2}$ –, arsenic $1\frac{3}{4}$; $10:2\frac{1}{2}$ –, arsenic $2\frac{1}{4}$, white lead $\frac{1}{2}$.

Stars $10:3\frac{1}{2}$ –, arsenic $3\frac{1}{2}$, mastic $1\frac{1}{2}$; $10:2\frac{1}{2}$ –, white mastic $\frac{5}{8}$, coarse iron filings 7; $10:3$ –, camphor grains 3.

Garlands of gold leaves $10:1:4$, filings of needles $\frac{1}{2}$, bronze filings $\frac{1}{2}$; $10:1:2$, Andarani (rock) salt $\frac{1}{2}$, coral $\frac{1}{2}$.

Yellow tongue $10:2$ –, $10:2:2\frac{5}{8}$, Andarānī salt 2, sugar 2, incense stone 2.

---

* The order given, which is adopted throughout this book, is more or less standard among authorities on gunpowder.

Concealed flying fire (volant caché) $10 : \frac{1}{2} : 3$, gall nuts $\frac{1}{4}$; mix the charcoal
and saltpetre, the sulphur and gall nuts, then mix the two and beat strongly.
"Liquid preparations" $10 : 3 : \frac{3}{4}$; $10 : 2$ and "half the composition for a fuse."
Lacquer is specified (as in Amiot's Chinese recipes, ch. VI) but is not called
Chinese; also camphor as in Chinese recipes. Verdigris is used to colour the
outside of flames green, the colour being deepened by adding sal ammoniac
and mercury sublimate (which is correct). Indigo is also used, as in China
(ch. VI). A soporific smoke composition contains 10 saltpetre, $4\frac{1}{2}$ sulphur,
18 arsenic and 3 narcotic opium.[81]

Out of a total of seventy-one compositions, in sixty-one saltpetre is 10 parts,
sulphur varies from $\frac{1}{4}$ to 3, three times 4, once 11, and once 32, but mostly
between 1 and $1\frac{1}{8}$ and $1\frac{7}{8}$. Charcoal appears in only fifty-two compositions,
usually in small amounts, mostly below 1, varying from $\frac{1}{2}$ to 3, the highest
(exceptionally) 4. The compositions for coloured fires, "chickpeas" (projected balls), rockets, flying fire, fuses, etc., all contain saltpetre.[82] The MS.
BN 2825 gives the composition for "Chinese arrows" as saltpetre 10, sulphur
$1\frac{1}{8}$ or $1\frac{1}{4}$, charcoal $2\frac{5}{8}$, all powdered, filled into the cartridge, and provided
with a cotton fuse.[83]

Ḥasan al-Rammāḥ describes various kinds of incendiary arrows and
lances,[84] and he describes and illustrates what has been supposed to be a
torpedo.[85] This is called "the egg which moves itself and burns," and the
illustration and text suggest at least that it was intended to move on the
surface of water. Two sheet iron pans were fastened together and made
tight by felt; the flattened pear-shaped vessel was filled with "naphtha,
metal filings, and good mixtures" (probably containing saltpetre), and the
apparatus was provided with two rods (as a rudder?) and propelled by a large
rocket. Such an apparatus is not described in Chinese works.

The fire-pots were of earthenware, glass, paper, tree bark, leather, or
metal, filled with mixtures containing saltpetre, and covered outside with
tar, wax, naphtha, sulphur, etc. They were thrown or carried on lances.
The metal ones may have functioned as bombs, the others as incendiaries.
Instead of a sulphured wick, as in the corresponding Chinese "destroying
fires," the vessel is provided with what seems to be a fuse (ikrīkh) filled with
incendiary, sometimes two or three fuses were used to one pot (qidr).[86]
The Chinese thirteenth-century fuse was a sulphured cord (ch. VI); that of
Ḥasan al-Rammāḥ was a cord made from cotton and palm leaves soaked in
naphtha and dried[87] (unless "naphtha" here means gunpowder). Ḥasan
al-Rammāḥ's descriptions of igniters are obscure.[88] He had two kinds, which
he calls the "rose" (warda) and the ikrīkh. The second word means a
duct, channel, or tube, but it was also used for the composition filling it.
The two igniters were used together on the same incendiary bomb and the
rose was perhaps lighted first: "quand tu voudras attaquer ton adversaire,
met le feu à la rose; tu mets le feu aux roses et tu lances la marmite." Although
it is usually supposed that the ikrīkh was a fuse, Mercier,[89] in a long discussion,
concluded that it was an incendiary torch, "or rather, a reserve of fire."

The question as to whether Ḥasan al-Rammāḥ was acquainted with the explosive, as distinguished from the incendiary or pyrotechnic, use of gunpowder was discussed by Favé,[90] who said:

> Plusieurs des compositions donnée par Hassan Alrammah pouvait aussi faire explosion, mais on n'a pas trouvé dans son livre la mention du pétard; les passages où il en parle sont peut-être restés inintelligibles, ou bien il ne le décrit pas parce que le pétard était très-connu et sans emploi à la guerre. Quoi qu'il en soit, il est certain que les Arabes maniant ainsi de véritables poudres dans des proportions variées, devaient éprouver des explosions imprévues. . . . Les explosions étaient plus à craindre encore au temps d'Hassan Alrammah qu'au temps de Marchus [Marcus Græcus], parce que la préparation du salpêtre avait été perfectionnée [i.e. its purification with wood-ashes].

Although Hime[91] said Ḥasan al-Rammāḥ "knew nothing of explosives," it is perhaps more accurate to say that he says nothing definite on the subject. Reinaud[92] suggested that, in addition to their contact with the Crusaders, the Muslims encountered the Mongols who had penetrated into Persia, 'Irāq, Syria and Asia Minor, and had reached Egypt, and "from this concourse of nations a new military art was born." The next treatise we shall consider introduces the use of gunpowder as a propellant in firearms.

IV. A St. Petersburg Arabic manuscript,[93] copied at the end of the fifteenth century for a Mamlūk Sultan of Egypt, which belonged to Count de Rzevuski, is entitled "Collection combining the various branches of the art," and is ornamented with paintings. Reinaud and Favé suggested A.D. 1300–50 for its original date of composition, since it cites Ḥasan al-Rammāḥ and mentions Ghāzān, Mongol khān of Persia, who died in 1304. The work is very methodical, but the sections on incendiaries are less detailed than those of Ḥasan al-Rammāḥ. It contains some Greek words. Reinaud and Favé attributed it to Shams al-dīn Muḥammad, who died at Damascus in 1350. This is a fairly common name, but this one is the only Arabic writer of the time mentioned by Ḥājjī Khalīfa. Romocki regarded the author as unknown. The manuscript uses the name midfa' for an instrument for projecting arrows or bullets, and Reinaud and Favé[94] regarded it as a gun. It says[95]:

> Description de la drogue à introduire dans les madfaa, avec sa proportion: baroud, dix; charbon, deux drachmes; soufre, un drachme et demi. Tu le réduiras en poudre fine, et tu rempliras un tiers du madfaa; tu n'en mettras pas davantage, de peur qu'il ne crève. Pour cela, tu feras faire par le tourneur, un madfaa de bois, qui sera pour la grandeur en rapport avec sa bouche; tu y poussons la drogue avec force; tu y ajouteras, soit le bondoc, soit la flèche, et tu mettras le feu à l'amorce. La mesure du mafaa sera en rapport avec le trou; s'il était plus profond que l'embouchure n'est large, ce serait un défaut; gare aux tireurs; fais bien attention.

Dr. E. J. Holmyard has kindly checked Reinaud's translation for me and finds no fault with it, except that the word translated "amorce" (literally percussion cap) more nearly means "priming." The "drug" (dawā') is a fine powder of 10 parts of saltpetre (bārūd), 2 of charcoal (faḥm) and $1\frac{1}{2}$ of sulphur (kibrīt) and is a $10 : 1\frac{1}{2} : 2$ gunpowder. The midfa' is apparently a wooden tube.

Guttman[93] threw doubt on the date (first half of the fourteenth century) of the manuscript proposed by Reinaud and Favé, since Vassaf in 1313 knew nothing of gunpowder or firearms (see p. 250). Contemporary travellers report[96] that in 1471 Ūzūn Ḥasan (Ussun Cassano) of Persia asked the Venetians for gunpowder and artillery. These were sent, but the Persians, who had previously won a battle, were defeated on the Euphrates by the superior artillery of the Turks in 1473–74. Ismā'īl Ṣafawī (the Ṣūfī), the Persian, was also defeated by superior Turkish artillery in September, 1514, at Chāldirān on the Euphrates. From about 1550 Turkish artillery was supreme; about 1598 it was cast by an Englishman, Sherley. Guttmann wrongly says the Persians in 1514 could not make artillery; an Italian traveller in 1507 reports two large bronze mortars for attacking castles; one of five spans bore at the muzzle was cast in one piece when he was there.

Fig. 10.   Incendiary arrow (left); bomb (?), pedestal of incendiary cartridges (see Fig. 11) and a ball for throwing from a short midfa' (centre); a midfa' with ball on a carrying stick, and a bomb, gunpowder container, or naphtha jar (qarūra) (right).

The Arabic bundūq originally meant hazel-nut, then a clay pellet, then a lead bullet shot from some kind of bow, and finally a firearm.[97] In the present case it probably meant a lead bullet. Quatremère[98] says the name midfa', which originally meant an instrument of war, was first used in 1383 in Egypt with the meaning "cannon," but was more often used in the fifteenth century for a small gun now carried on a camel and called a zanbūrak in Persian. Romocki thought the "arrow" named was really a ball, since an arrow could not have been used in a gun only a diameter long (as he interpreted the passage given in French above). A second midfa' described as having a piercing and boring action was, says Romocki, not a firearm at all but a pointed iron spike in a wooden tube (he gives the Arabic text). The arrow, however, is clearly shown in the figures as inside the tube; and short dumpy cylinders on the end of more or less long carrying sticks are also shown with a ball projecting from the mouth (Fig. 10). A very difficult passage is translated by Reinaud and Favé as follows[99]:

> Chapter of the lance which you point towards the enemy and there issues from it an arrow that enters his breast. Tu prendra une lance que tu creuseras dans toute sa longeur à une étendue de quatre doigts près; tu foreras cette lance avec une forte tarière, et tu ménageras un madfaa; tu disposeras aussi un pousse-flèche (un madfaa de flèches) en rapport avec la largeur de l'ouverture. Le madfaa sera en fer. Ensuite tu perceras sur le côté de la lance un petit trou; tu perceras également un trou dans le madfaa; puis tu prendras un fil de soie brute que tu attacheras au trou du madfaa; tu le feras entrer par le trou qui est sur le côté de la lance. Tu te procureras pour cette lance une pointe évidée à sa partie antérieure, de manière que, lorsque tu tireras, le madfaa pousse fortement la flèche par la force de l'impulsion que tu auras communiquée; le madfaa marchera avec le fil; mais le fil retiendra le madfaa de manière à l'empêcher de sortir de la lance avec la flèche. Quand tu monteras à cheval, ainsi armé, tu auras soin de te munir d'un troussequin, afin que la flèche ne sorte pas de la lance.

Dr. E. J. Holmyard has kindly verified the accuracy of Reinaud's translation for me. He remarks that midfa' means something for repelling or thrusting and, in consequence of the use of the silk thread, which could not restrain a charge of gunpowder, he asks whether the propulsive agent was a spring. The first recipe, for gunpowder "for the midfa'," seems to me to imply that gunpowder was used.

The arrangement described consists of a wooden tube closed at the bottom and containing gunpowder. Over this powder is an iron tube, also closed at the bottom, fitting inside the first tube and containing an arrow. The inner iron tube is tied to the outer wooden tube by a thread of raw silk, passing through holes in the two tubes. When the powder is fired, the inner tube starts forward but is checked by the thread, the arrow inside it proceeding in its course by its momentum, and leaving the open mouth of the outer tube. As Favé says: "Si les choses se passaient réellement ainsi, le projectile ne pouvait avoir qu'une très-petite vitesse, sans quoi le fil qui retenait le madfaa à flèches aurait été rompu."

A silk thread seems, in fact, to be a very unsuitable thing to use, and it may be asked why, if the propulsive force of gunpowder was known, the arrow or the ball was not simply put in front of the powder charge, with a suitable tamping interposed, and a single metal tube used? This would have been a gun. Favé said:

> ... le poudre à canon lançant à faible charge de petites balles ou de petites flèches dans des armes de petit calibre et sans solidité, était encore plus un épouvantail ou un jouet qu'une force redoutable; mais l'art devait passer par ces essais impuissants avant d'entreprendre de lancer des projectiles capables de renverser tous les obstacles.

Reinaud and Favé[52] thought the propulsive apparatus was invented by Arabs in Syria or Egypt. The St. Petersburg manuscript describes pots filled with incendiary mixtures thrown by ballistas or trebuchets, but not cannon, so that small arms are here in use before cannon, which is unusual. It mentions an "arrow of China" and describes in detail, with illustrations, a mounted naphtha thrower attended by two men on foot dressed in black and carrying sprinkling clubs (massues à asperger). The Mongoloid features of some of the men illustrated are noteworthy. The sight and sound of this outfit would, it is said, put a large force into flight. The MS. III (BN 2825) also describes terrifying outfits on horseback, the man and horse being protected from fire by felt and compositions of vinegar, fish-glue, talc, brick-dust, etc., and accompanied by men provided with incendiary clubs and madāfi' (the plural of midfa').[100] Such anti-incendiary mixtures are described by Marcus Græcus (see p. 53).

V. There are seven recipes for gunpowder in an Arabic chemical work written in Syriac characters which Berthelot[101] dated tenth to eleventh century, although this part may be later (Lippmann put it as late as 1300). The proportions for various purposes are similar to those in Ḥasan al-Rammāḥ: saltpetre : sulphur : charcoal $= 10 : 1\frac{1}{2} : 2\frac{1}{2}$, $10 : 1 : 1$, $10 : 2 : 2$, $11 : 3 : 3$; and for fireworks $10 : 1\frac{1}{2} : 3$ and $10 : 3 : \frac{1}{2}$. The "Frankish grains" also mentioned, not an incendiary but a mixture of alum, verdigris, vinegar, etc., may indicate a Western origin.

Fig. 11. Left : foot soldier carrying a sprinkling club and in the left hand some undefined object, with fireproofed clothing to which are attached firework cartridges. Centre : mounted man with fireproofed clothing and helmet and carrying a spear, to all of which are attached firework cartridges; the horse also has a fireproofed covering with firework cartridges attached. Right : foot soldier with naphtha flask and short midfa' with ball; clothed in the same way as the man on the left.

MAMLŪK ARTILLERY AND FIREARMS

Some interesting details on the use of artillery by the Mamlūks were assembled by Ayalon. He thinks[102] it was almost entirely used in siege warfare, its use in the field being opposed by the very conservative traditions and the difficulty of adapting it to cavalry tactics.   In siege warfare, cannon could replace or supplement the old stone-throwing machines (manjanīqs) without much interference with the old organisation of troops, but in the field no similar weapon had preceded it, and its introduction would have called for radical changes.   This thesis, although reasonable, is somewhat at variance with Ayalon's idea[43] that the troops which we have assumed were armed with naphtha bottles (an old and tried weapon) were in fact armed with muskets.

The Sultan al-Ghawrī (1500–16) seems to have introduced the manufacture of artillery on a large scale.   He established an arsenal for casting cannon at Cairo, and Ibn Iyās says that fifteen, seventy, seventy-four and seventy-five guns were cast on different occasions.   These were partly used in defending Alexandria and the Egyptian coast against Ottoman attack, partly at sea against the Portuguese in the Red Sea and the Indian Ocean, and partly in coastal fortresses defending the Red Sea coast.   In the battles against the Portuguese the Mamlūks even found it necessary to obtain artillery from the Ottoman Turks.

The traditional conservatism of the Mamlūks is brought out in some passages from Muslim authors quoted by Ayalon,[103] in which it is said that the Ottomans used cannon and arquebuses unfairly; they were unknightly weapons and called for no bravery; even women could beat brave troops if they used firearms: "God curse the man who invented them, and God curse the man who fires on Muslims with them."   The Ottomans were reproached with having assembled armies containing "Christians, Greeks, and others," and having "brought with you this contrivance artfully devised by the Christians of Europe when they were incapable of meeting the Muslim armies on the battlefield."   This is the age-long cry of the cavalry against the gunner, who but rarely involves himself in the unpleasantness of counter-battery tactics.

The arquebus, hand-gun, or portable firearm, was apparently first used in the time of the Mamlūk al-Ashraf Qāytbāy, in late December, 1489, or early January, 1490.   After an interval of some years it is mentioned again in January, 1497, and from then it occurs frequently.   Arquebuses and their ammunition are called al-bunduq al-raṣāṣ.   The meaning of bunduq has already been discussed (p. 205), raṣāṣ is lead; bunduqīya is the common name for a hand-gun.   The arquebus was a small but deadly weapon; the Ottomans caused heavy losses among the Mamlūks with them.   In 1514 an Ottoman army containing 12,000 soldiers equipped with arquebuses completely routed a much larger force of Mamlūks.   Here again the Mamlūk conservatism prevented the early use of the new weapon, since it was not adapted for cavalry, but Qāytbāy in 1489–90 ordered troops to learn the use of al-bunduq al-raṣāṣ and they drilled under his personal inspection. They were inferior in pay and rank to cavalry.   Later, the hand-gun was adapted for use by cavalry, and became more respected.

There are references to the use of firearms by the Ottomans in 1378 (Ragusa), 1389 (at Kossovo), 1395–1402 (first siege of Constantinople, see ch. III) and 1396 (Nikopolis), but Wittek[104] says there is no clear evidence of their use before 1400; they seem without much doubt to have used cannon in defending Adalia in 1424, i.e. in siege warfare, and in the field in 1488 (if not in 1473), and muskets in 1465.

GUNPOWDER AND FIREARMS IN INDIA

Ktesias of Knidos (*c.* 398 B.C.) mentions[105] an oil from a huge Indian worm living in the Indus which set everything on fire. (Ktesias was a Greek who was a physician in the Persian court and his information came from Persian sources.) The captured animal was hung up in the sun for a month and the oil which dripped from it was collected in earthenware pots. It was sent to the king in sealed jars. These were thrown into besieged towns and set fire to everything when they broke; neither man nor animal nor anything could withstand this combustible, and only mud or sweepings could extinguish the flames. This story is repeated by two authors of the second to third centuries A.D.

Ailian,[106] who quotes Ktesias, said the oil was used by the King of India; it burnt up everything—men, animals and arms, and no battering ram or military engine could resist it. Philostratos,[107] in his *Life of Apollonios of Tyana*, says the white worm was found in the river Hyphasis in the Panjāb; the oil made by melting it down could be kept only in glass vessels and when once set on fire was inextinguishable by all ordinary means. It was used by the king in war to burn walls and capture cities.[a] Philostratos also says that when Apollonios asked the Indians why Alexander the Great refrained from attacking the Oxydrakes ('Οξυδράκαι), a nation of the Panjāb living between the rivers Hyphasis and Ganges, he was told that these holy men overthrew their enemies, Herakles of Egypt and Dionysos, with storms, lightning and flaming thunderbolts from above, which fell on their armour[b] (πρηστῆρες αὐτοὺς ἀπεώσαντο καὶ βρονταὶ κάτω στρεφόμεναι καὶ ἐμπίπτουσαι τοῖς ὅπλοις).

Francis Bacon (d. 1626) had apparently read Philostratos on India and had got some information about China before Voss (see ch. VI), since he says :

"For certain it is, that ordnance (tormenta ænea) was known in the city of the Oxidrakes in India; and was that which the Macedonians called thunder and lightning, and magic. And it is well known that the use of ordnance (pulveris pyrii et tormentorum igneorum) hath been in China above two thousand years."

Von Bohlen,[109] who discussed the use of fireworks and gunpowder in India, quoting Ktesias (whose "oil" was probably naphtha), Halhed, and others, was unable to reach a definite conclusion, but suggested an early use of gunpowder in China and perhaps India.

Many recipes for incendiary mixtures, toxic smokes and military devices are contained in the *Arthaśāstra*, which is a work attributed to Kauṭilya (*c.* 300 B.C.) but in its present form contains interpolations at least as late as the first to the fifth centuries A.D.[110] "Nitre" (yavakshāra) and a "salt extracted from fertile soil" are mentioned (p. 110)* as well as "explosives" (agniyoga, agnisamyogas) (pp. 57, 154, 468), the meaning of agniyoga being "inflammable powder (p. 468). No weapons using gunpowder are mentioned, so that it is doubtful if it was known. The equipment of a fort, besides agnisamyogas, includes machines (yantra) and "such weapons as can destroy a hundred persons at once" (pp. 57, 468). "Fiery spies" (tīkshṇa) (p. 21, and often) were engaged on all kinds of work, not necessarily incendiary. At the time the work was compiled there was intercourse with China, since it mentions Chinese silk (pp. 90, 137, 246; chīnapaṭṭa). Considerable knowledge of metallurgy and technical arts is shown (pp. 91 f.).

Fire-pots (agnidhāna, p. 424), fire thrown over a sleeping person (p. 455) and "explosive fire" (p. 451) are mentioned. The walls of a chamber were smeared with poisonous or "explosive" substances, or with lac which can be set on fire (p. 455). A mass of "sea-foam" mixed with burning oil and placed under water exhibited a spontaneous outbreak of fire (p. 457). Showers of firebrands (ulkā) were accompanied by a noise of thunder from the sky (p. 458). Poisonous smokes, some of them scented, are specified, the compositions for making which (including arsenic sulphides, or green vitriol) are given in detail (pp. 475–8); they are mostly fabricated from powders of all kinds of plants (some poisonous), insects, animals and reptiles.

Sending volumes of smoke from the mouth (pp. 458, 461), blowing through tubes or hollow reeds the fire contained in a few pots (p. 461), a body burning with magical fire at night (p. 460, perhaps a phosphorescent preparation from glow-worms), a body besmeared with burning oil (tejantaila), and an image of a god covered with a layer of mica (talc?) besmeared with burning oil (p. 461), are some tricks described. A commentary on the work says agniyoga is "a terrific combustible prepared from the powder or oil (chārṇa) of various kinds of snakes, lizards (kṛkalāsa) and worms" (p. 154).

Vegetable oils (linseed, nimba seed, sesamum seed, mustard seed, etc., and castor oil) (pp. 112, 136, 481) and perhaps petroleum (pp. 457, 461), as well as bitumen (p. 92), are specified. Inflammable powders are attached to birds, cats, monkeys, etc., and set on fire, for the purpose of burning thatch and forts (p. 468). Small fire balls are made from a mixture of sarala (Pinus longifolia), devadāru (deodar), pūtitṛṇa (stinking grass), guggulu (bdellium), śrīveshṭaka (turpentine), juice of sajja (Vatica rubusta) and lākshā (lac), combined with dungs of an ass, camel, sheep and goat. An inflammable powder to be hurled against an enemy is a mixture of the powder of priyāla (Chironjia Sapida), charcoal of avalguja (oanyza, serratula, anthelmintica), madhūcchiṣṭa (wax) and the dung of a horse, ass, camel and cow. Another inflammable powder is the powder of all the metals (sarvaloha), as red as fire; or the mixture of the powder of kumbhī (gmelia arborea), sīsa (lead), trapu (tin, or zinc?), mixed with powdered charcoal made from the flowers of pāribhadraka (deodar), palāśa (butea frondosa) and hair, with oil, wax and turpentine.

* The page references are to Shamsastry's translation (1923); see ref. 110, p. 231.

These and many similar recipes are all for incendiaries and not gunpowder or explosive mixtures, since no salts or sulphur appear in them. They are combined with recipes for all kinds of poisons, in some cases accompanied by specified exhortations or prayers to gods to expedite their use. It is said that powder of fire-fly (khadyota) mixed with oil of mustard seed emits light at night. Many other "wonderful and delusive contrivances" are described (pp. 475–92), and it will be seen from the above that the collection of recipes is very similar to those in Albertus Magnus, Marcus Græcus, and the various European and Arabic firework books already discussed, and in the next chapter similar Chinese material will be met with.

GUNPOWDER IN INDIA

The legend that gunpowder was known to the ancient Hindus goes back to two authors, both quoted by Gmelin[111] in 1797, viz. Halhed[112] and Craufurd.[113] Halhed, who used a Persian translation of a Sanskrit digest of laws prepared by some pandits, thought that agni astra mentioned in them meant "firearms" or "fire-arrow discharged from bamboo," and śataghnī, literally "hundred-killer," was a cannon. Craufurd thought the old Hindus used gunpowder, but whether they did so before the Europeans was doubtful. Knowledge of gunpowder may have reached China from India.

The agni astra ("agnee aster," "agny aster") was a fire-weapon, the śataghnī ("shet agnee," "shet agny"), literally "hundred-killer," had been translated "cannon," but its meaning is doubtful. Craufurd thought it might be a hole bored in a rock and charged with gunpowder and a ball. This is quite possible, since in 1771 the Royal Artillery, at the suggestion of Lt. Healy, bored wine-glass-shaped cavities in the rock at Gibraltar and used them as mortars for throwing masses of small stones, the gunpowder charge being tamped with a wood block and fired by a central fuse in a copper tube. The powder took five minutes to burn before successfully throwing the stones. The device was used several times. The Russians about 1750 are said to have used mortars scooped out of ice and throwing cannon balls of ice.[114] Craufurd emphasised that the Indian rockets (see p. 226), fireworks and fire-balls in his time were very perfect weapons.

The "hundred-killer" is probably some fabulous weapon, perhaps an iron mace, since in the Raghuvaṃsa it is said[115] "the demon laid his iron-headed śataghnī upon Rāma, as Kuvera laid his famous club upon Jamrāj." Rāy[116] found that Halhed's quotation is from the seventh book of the code of laws of Manu (300 B.C.? or A.D. 150?), in which it is forbidden to use "any blade made red-hot by fire or tipped with burning materials," i.e. incendiary lances or arrows, such as are mentioned in other Indian works. Halhed translated, incorrectly, "cannon or guns or any other kind of firearms." Sinclair[117] supposed that, even if the whole work quoted by Halhed was not of the sixteenth century, the parts in question were interpolated, but it is really the translation which is faulty.

Oppert,[118] however, maintained that the passage in the Code of Manu referred to firearms, and supported this by quoting two "ancient Sanskrit manuscripts," neither of which he attempts to date, viz. the *Śukranīti* and the *Nītiprakāśikā* (which he first discovered). The *Nītiprakāśikā* is alleged to be an extract from a larger work devoted to the *Nītiśastra* ascribed to Vaiśampāyana, and it gives a full account of the *Dhanurveda*; it mentions the Huns (Hūṇas), and contains a detailed description of the various arms and war implements of the Hindus. The *Śukranīti* is ascribed to Uśanas or Śukrācārya. Although Vaiśampāyana has been called the author of the *Yajurveda*, it is evident that both works in their present form have no claim to be regarded as "ancient." Oppert also refers to the *Agnipurāṇa* (see ref. 117).

The *Nītiprakāśikā* mentions the nālikā or "musket" (p. 14),* the vajra or thunderbolt (p. 16), and the śataghnī or "hundred-killer." The allusions to the nālikā and to smoke-balls, which a later commentator says were made of gunpowder, shows (according to Oppert) that gunpowder was known to Vaiśampāyana.

The *Śukranīti* is ascribed to a purely mythical Uśanas or Śukra, mentioned in the *Mahābhārata* as the ruler of the planet Venus (Śukra), and is said by native authors to contain the principles of all the sciences (p. 34). It contains 2200 ślokas or double verses, said to be a residue of a purely mythical work of Brahma which contained 10,000,000 ślokas. The *Śukranīti* has four sections, and a fifth as an appendix which contains the description of firearms and a full account of the manufacture of gunpowder (agni-cūrṇa, "fire powder") (p. 42). The work is said to be older than any corresponding Chinese work (p. 45). The size of the work of the divine Brahma would rival that of a Chinese encyclopædia.

The *Śukranīti* describes small guns, five spans long with a perpendicular and horizontal hole at the breech, powder and ball being compressed by a ramrod, and also large guns drawn on cars. The gun is usually of iron but sometimes (in the *Nītiprakāśikā*) of stone. Balls are of iron, lead, or "any other material"; some big ones have small ones inside (pp. 65–6, 105 f.). The name nālikā for a gun, derived from nāla, a reed, is said by Oppert to be well known but not given in this sense in Sanskrit dictionaries (pp. 66, 129); he says bamboo guns were still used in Burma.

The *Bhāratacampū* of Anantabhaṭṭa (*c.* A.D. 1580) speaks of "heaps of leaden balls which emerge quickly from the gun lighted by a wick" as like hailstones descending from black clouds illuminated by lightning (which reminds us of Chinese and Japanese descriptions). An earlier poem (twelfth century), the *Naiṣadha* ascribed to a Brahman Śrīharṣa (not the King of Kaśmīra of the same name), is said to speak of two "guns" (nālikā) throwing balls (pp. 67–8). But these, says Oppert, are obviously all later interpolations or interpretations not earlier than the seventeenth century (cf. p. 80).

An anonymous reviewer of Oppert's work[119] refers to Beckmann[120] for the view that the Indians knew of gunpowder in the earliest periods, and the

---

* The page references in the text are to Oppert's book (1880); see ref. 118, p. 232.

Arabs learnt of it from them. It was brought by the Arabs "from Africa to the Europeans, who however improved the preparation of it, and found out different ways of employing it in war, as well as small arms and cannon."

Oppert (p. 63) had found it "peculiar that powder should not have been described in Sanskrit works" in the earliest periods, but this was perhaps because it was so familiar that it was not worth mentioning. Although the Arabs are supposed to have learnt of gunpowder from the Indians, the latter must themselves have forgotten their discovery completely, since they made no use of it. Some supposed representations of firearms in stone carvings are, as Oppert (pp. 76 f.) admits, fairly modern or of doubtful meaning.

There is a recipe in the *Śukranīti* for gunpowder (pp. 62–3, 106 f.) in which the proportions of saltpetre, sulphur and charcoal are given as 5 : 1 : 1, or for a gun 4 : 1 : 1 or 6 : 1 : 1. As translated by Rāy[121] (who gives the Sanskrit text) this reads :

Take 5 palas of saltpetre, 1 pala of sulphur, and 1 pala of charcoal prepared from the wood of Caloptris gigantea and Euphorbia neriifolia by charring by smoke circulating through it. Powder and mix them intimately; macerate in the juice of the plants named and of garlic and afterwards dry the mixture in the sun and pulverise it to the fineness of sugar. If the fire powder is to be used for a gun, 6 or 4 palas of saltpetre are to be taken, the proportion of charcoal and sulphur being as before. By varying the proportions of the ingredients, viz. charcoal, sulphur, saltpetre, realgar, orpiment, litharge, asafœtida, iron powder, camphor, lac [*sic*; lacquer in the Chinese *Wu Pei Chih*, see ch. VI], indigo, and the resin of Shorea robusta, different kinds of fires are devised by pyrotechnists, giving forth flashes of starlight.

Rāy thought the passage was interpolated some time after the introduction of Indian warfare during the Muslim period, and Hime,[122] who remarked that since the best (4 : 1 : 1) mixture for cannon was not known in Europe until the sixteenth to seventeenth centuries, concluded that the interpolation, which is "the handiwork of some charlatan," was added in that period.

Rājendralāla Mitra[123] says the *Nitiśara* of Śukrācārya (the *Nītiprakāśikā*, says Rāy) is of uncertain date, and the text is inaccurate; "the work is said to be an old one, and quotations from it, or references to it, occur in ancient books; but the last chapter is apparently spurious, as it describes guns as they existed a hundred years ago." The work is legendary: it makes Brahma constitute an army of 2,187,000,000 foot, 21,870,000 horse, 218,700 elephants and 21,870 chariots,[124] and no one is likely to accept the conclusions drawn from it by Oppert that "gunpowder and firearms were known in India in the most ancient times . . . that the knowledge of making gunpowder was never forgotten in India . . . [but] was kept everywhere a deep secret . . . [and] the existence of guns and cannons in India seems . . . to be satisfactorily proved from evidence supplied by some of the earliest Indian writings." Oppert nowhere gave any satisfactory critical examination of the date of the material which he used and his conclusions must be rejected. At the same time, the recipe quoted above, with its additions of plant materials, realgar, orpiment, litharge, etc., is remarkably like those in the Chinese *Wu*

Ching Tsung Yao, which has been dated A.D. 1044 (see ch. VI). The Chinese texts are usually fairly precisely dated, whilst Indian works are often not. This difficulty must not be allowed to impair the interest or value of Indian works, but they must also be examined from the point of view of their scientific and technical contents with due care and with a suitably critical attitude. I feel that Oppert's treatment does not satisfy this requirement. Another point is that Indian chemical and technical treatises are mostly composed in a dry and very concise style, without emphasis on things we consider important. This can mislead a reader who is accustomed to the more flamboyant treatment found in some of the Chinese works. Whereas the Chinese works usually begin with a history of the subject which is recognisably mythical, in Indian the old material, some very old, is not well separated from modern additions.

Oppert (pp. 60 f.)* says saltpetre was then (1880) gathered in large masses in Bengal whenever it effloresces on the soil, more particularly after the rainy season. In the *Śukranīti* it is called suvarcilavaṇa, "well-shining salt." The *Dhanvantarinighaṇṭu* describes it as a tonic, as a sonchal salt; it is also called tilakam (black), kṛṣṇalavaṇam, and kālalavaṇam; it is light, shiny, very hot in digestion, and acid. It is good for indigestion, acute stomach-ache, and constipation, and is a common medical prescription. Sulphur is found in India but is mostly imported from the East; some Indian gunpowder is made without it. Various kinds of charcoal for gunpowder are specified in the *Śukranīti*. Besides the three usual constituents, realgar, orpiment, graphite, vermilion, magnetic oxide of iron, camphor, lac, indigo and pine-gum are also mentioned (pp. 63, 107). These are also mentioned as constituents of incendiaries in the *Arthaśāstra* and appear in Chinese accounts (see ch. VI).

The *Rājalakṣmīnārāyaṇahṛdaya*, part of the *Atharvaṇarahasya*, speaks of "the fire prepared by the combination of charcoal, sulphur and other material," and is dated by Oppert in "a very remote period" (p. 64)—a sufficiently vague statement. The Sanskrit name for gunpowder, it is said, is agnicūrṇa, "fire powder," occasionally shortened to cūrṇa, "powder." The Dravidian languages all have the same name for "drug" and "gunpowder": Tamil marundu, Telugu mandu, Kanarese maddu, Malayālam marunnu; sometimes tupāki, "gun," is prefixed, in Malayālam veḍi, "explosion." The word marundu may, it is said, be derived from the Sanskrit past-participle mardita, "pounded." The Dravidian equivalent of cūrṇa is śuṇṇambu (Tamil) or śunnamu (Telugu), meaning "chalk" (pp. 64–5). These are probably all quite late names when used in this sense.

Pliny[125] mentions caves in Asia which produced an efflorescence of aphronitrum (aphronitrum tradidit in Asia colligi, in speluncis mollibus distillans specus eos colligas vocant, dein siccant sole). Hoefer thought this was saltpetre. The meaning of aphronitrum will be discussed later. There is a plentiful supply of saltpetre in the Ganges Valley, but this does not ensure that it was used in an early period for making incendiary compositions or

*The page references are to the work quoted in ref. 118.

gunpowder. There is no name in Classical Sanskrit for saltpetre, shoraka in late Sanskrit is from the vernacular shūraj, which is a Persian word. In Classical Sanskrit sauvarchala means natural salts corresponding with aphronitrum in Latin,[127] and it is perhaps most correct to say that the evidence that either meant saltpetre is slender.

Elliot[128] says "the earth of Gangetic India is richly impregnated with [saltpetre] in a natural state" which may be extracted "by lixiviation and crystallisation without the aid of fire," and sulphur "is abundant in the northwest of India"; yet if the Hindus had invented gunpowder they must have forgotten it at the time of the Muslim invasion, when the only inflammable projectiles known to them were composed chiefly, if not entirely, of bituminous substances varying from naphtha, the most liquid, to asphaltum, the most solid; and even these were very rarely brought into action. "There is nothing in the testimony of either native or foreign witnesses sufficiently positive to lead to the conclusion that, in modern times at least, the knowledge of firearms was indigenous in India, and antecedent to their use in Europe."

Hime, perhaps too definitely, concluded that "early Indian gunpowder is a fiction." If this is so, then early Indian firearms are also imaginary, and both Hime[129] and Rathgen[130] concluded that there were no large guns in India even as late as 1498, only small matchlocks imported from Europe.

The ruins of some ancient Kashmīr temples have been thought to show the effects of gunpowder, and when Sikandar (d. 1416), who subverted the Hindu religion in Kashmīr, ordered all places of worship to be razed to the ground, a temple, it is said, emitted from its foundations volumes of fire and smoke. General Cunningham[131] supposed that the temples must have been blown up with gunpowder, their overthrow being too complete to have been the effect of earthquakes, the stones being all completely separated (a known effect of a gunpowder demolition), and Sikandar may have learnt of gunpowder from Tīmūr, who was then invading India. In the battle between Tīmūr and the Ottoman Bajazet (Bāyazīd) in 1402, Greek fire was used by both sides, but not cannon; the Turks had metal cannon at Constantinople in 1422 (see ch. III), and a knowledge of gunpowder may have reached Kashmīr in the reign of Sikandar.[132] Elliot[133] thought that the damage to the temples was perhaps due to a petroleum well (petroleum occurs in many places in India), but Hime[134] agreed with General Maclagan[135] that the ruins were due to earth-movements and old age.

In describing the capture (A.D. 712) of Alor in India, Mīr Ma'sūm Bhakkarī in his *History of Scinde* and Ḥaydar Rāzī in his *General History* mention the use by the Muslims of ātish bāzī or fire-throwing machines "which the Arabs had seen in use with the Greeks and Persians." The Arabs "took vessels filled with fire (ḥuqqahā-y-i-ātish bāzī) to dart fire at the seats on the elephants," which (as they are notably afraid of fire) ran off in a panic.[136] These would be naphtha pots or siphons which, if they reached the seat on an elephant, threw fire for a fair distance. Firishta (d. c. A.D. 1611), in an account of a battle fought by Maḥmūd Ghaznī near Peshāwar in A.D. 1008, says that an elephant on which the Hindu commander rode "becoming unruly from the effects of the naphtha balls and the flights of arrows, turned and fled," and the soldiers (as was the military custom) followed their commander, thus losing the battle.[137]

Gibbon,[138] who had read Dow's translation of Firishta which mentions "the report of a gun," said: "as I am slow in believing this premature (A.D. 1008) use of artillery, I must desire to scrutinise first the text and then the authority of Ferishta," and, as in many other cases when he refused to be over-awed by Sanskrit or Chinese specialists, his scepticism is fully justified. Briggs, a later translator of Firishta and a military expert, said[139]:

> Bābar was the first invader who introduced great guns into Upper India, in 1526, so that the words tope and toofung have been, probably, introduced by ignorant transcribers of the modern copies of this work, which are in general very faulty throughout.

The passage, said Briggs, is different in different manuscripts; in the India House manuscript, copied in 1648, tope (gun) and toofung [tufang] (musket) do appear, also in two copies used by Wilken, but no Persian or Arabic history speaks of gunpowder before the fourteenth century, and in some manuscripts "tope is replaced by nupth (naphtha) and toofung by khudung (khadang, arrow)." It has been mentioned (p. 194) that some modern writers think naft may also mean gunpowder, and could probably make out a case for the use of guns in India in A.D. 1008. Dow, said Elliot,[140] "boldly translated the word as guns," and Elliot adds:

> The *Tārīkh-i Yamīnī*, the *Jāmi'u-t Tawārīkh* of Rashīdu-d dīn, the *Tārīkh-i Guzida* of Abū'l Fidā, the *Tabakat-i Nasiri*, the *Rauzatu-s Safā*, the *Tārīkh-i Alfi* and the *Tabakāt-i Akbarī*, though almost all of them mention this important engagement, in A.D. 1008 . . . and mention the capture of thirty elephants, yet none of them speak of either naft or tope.

We may safely dismiss the use of gunpowder and guns in India in A.D.1008.

Maḥmūd Ghaznī, who made six invasions of India in A.D. 1000–1026, is said to have built a fleet at Multān in 1026, with twenty archers on each of his 1400 boats, which engaged the 4000 vessels of the Jats, and by ramming them and by the use of naphtha sank or burned their craft. As Lane Poole[141] said: "whatever really happened, we may be sure that there never were 5000 boats on the upper Indus, and that mountain tribes [the Jats] do not usually fight naval battles." Elliot[142] quotes several authors as showing that in the country round Peshāwar and also near Multān there are sources of petroleum, and Hime[143] concluded that Firishta actually used the words "naphtha" and "arrow"; Hime supposed that the guns of Upper India later entered through Afghanistan, whilst those of Western India came by sea on ships. Briggs, in his translation of Firishta, said[144] (the spelling is Briggs's):

> If any reliance is to be placed on Moola Daud Bidury, the author of the *Tohfutu-oos-Saluteen*, guns were used at this time [1368] by the Hindoos, and in a subsequent passage it is remarked that the Mahomedans used them for the first time during the next campaign. But I am disposed to

doubt the validity of both these statements. . . . Ferishta . . . also observes, that Toorks and Europeans, skilled in gunnery, worked the artillery. That guns were in common use before the arrival of the Portuguese in India, in 1498, seems certain, from the mention made of them by Faria-e-Souza.[145] . . . All these circumstances, however, do not lead to the conclusion that the Hindoos had guns before they were introduced from the West by the Mahomedans, who adopted their use from Europe.

Dowson,[146] pointed out that the word 'araba in Firishta, which in modern India means a field gun, originally meant a cart, and only later a cart for a gun. (Wheeled gun carriages were first used generally in Europe only during the reign of Louis XI of France, 1461–83.)[147] In Bābur's time, gun carts were chained together to form a sort of rampart.[146]

A passage in the *Kanauj-Khand* of the Hindu poet Chānd, c. A.D. 1200, referring to cannon "which made a loud report when they were fired off and the noise which issued from the ball was heard at a distance of 10 kos," and another which speaks of "fire machines," are later interpolations and are not found in manuscripts.[148]

'Abd al-Razzāq, who was sent on an embassy to India by Shāh Rukh in 1441 and was at Bījānagar during the great feast of Mahanāwī, describes naphtha-throwers mounted on elephants. Although he goes on to say: "One cannot, without entering into great detail, mention all the various kinds of pyrotechnics and squibs (mūshksāz) and various other amusements which were exhibited," he says nothing of guns, and the name ātish bāzī means pyrotechnic displays as well as artillery.[149] Ma Huan, an interpreter in the Chinese expedition under Chêng Ho sent out by the Emperor Yung Lo in 1405, reports in his book *Ying Yai Shêng Lan* (A General Account of the Shores of the Ocean)[150] that in the kingdom of Bengal he saw tree-bark paper, granulated and white sugar, candied fruits, ardent spirits sold in the market places (but no tea), and also "guns, knives and scissors." He does not mention fireworks.

Sharaf al-dīn 'Alī Yazdī (d. A.D. 1454) in his Persian history of Tīmūr (Tamerlane) says[151] the Sultan Maḥmud opposed him near Delhi with: "jetteurs de pots à feu et de poix enflammée ainsi que les fusées volantes [rockets] pointées de fer, qui donnent plusieurs coups de suite dans le lieu où elles tombent." Gibbon says "Voltaire's strange suspicion that some cannon, inscribed with strange characters, must have been sent by Tīmūr to Delhi, is refuted by the universal silence of contemporaries."

The Muslims were certainly familiar with the use of naphtha and Greek fire at that time, and they were acquainted with the distillation of petroleum, which they probably learnt in Syria (see ch. I). Its use in sea-battles was probably a Byzantine invention (see ch. I). The material collected in preceding pages lends no support to the assertion of Mercier[152] that: "il apparaît d'une façon évident que ce sont les Arabes qui ont la maîtrise incontestable et même l'exclusivité de l'utilisation du naphthe, pur ou en mélange, dans les feux à guerre." This was a Greek speciality.

Ziā al-dīn Barnī (the main source of Firishta) in his *Tārīkh-i Fīrūz Shāhī* says[153] that in an abortive attack on Rantambhor in 1290 the Sultan Jalā al-dīn ordered maghrībīhā (Western) machines (i.e. manjanīqs or stone-throwers) to be erected. Amīr Khusrū (d. 1325)[154] says that the Hindus, besieged in this fortress by Sultan Alā al-dīn in 1300, had:

> collected fire in each bastion. Every day the fire of those infernals fell on the light of the Muslims, and as there were no means of extinguishing it they filled bags with earth and prepared entrenchments. . . . The Royal Westerns shot large earthen balls against the infidel fort. The victorious army . . . made vigorous attacks, rushing like salamanders through the fire which surrounded them. The stones which were shot from the catapults and ballistas, within and without the fort, encountered each other half-way, and emitted lightning.

This is rather like the Indian story in the *Mahābhārata* of shooting from the mysterious fiery weapon brahmāstra, when "the fury of the two fiery darts acting against each other, overspread the heavens and earth, and waxed strong like the burning rays of the sun."[155] C'est magnifique, mais ce n'est pas la guerre.

Barnī says[156] that in an attack on Arangal in 1309 "the Westerns (ballistas) were played on both sides and many were wounded"; the mud wall of the fort, says Khusrū,[157] was so strong that a ball from a Western rebounded from it "like a nut which children play with," yet ultimately the walls "were pounded by the stones discharged at them." Elliot[158] remarked that Khusrū "is full of illustrations, and leaves no manner of doubt that nothing like gunpowder was known to him," although other authors have attempted to make out that cannon balls were meant.

The *Autobiography* of Tīmūr says that in 1398–9 the Hindus besieged by Tīmūr in Bhatnīr "cast down in showers arrows and stones and fireworks [incendiaries] upon the heads of the assailants," who "treated them as mere rubbish."[159] The same work says[160] the elephants in the army of the Sultan Maḥmūd which Tīmūr defeated at Delhi in 1399 carried throwers of grenades (ra'd-andāzān), fireworks (ātish bāzī) and rockets (takhsh-andāzān), which Tīmūr's troops attacked with arrows and swords. The *Zafar-Nāma* of Sharaf al-dīn 'Alī Yazdī (d. 1446), in its account of this battle says[161] rocket-men (takhsh-afgan) and grenade-throwers marched by the side of the elephants, which were covered with armour and carried cages in which crossbowmen and discus throwers were concealed, and they were attacked with arrows and swords. It is not said that Tīmūr used rockets.

In 1482, Mahmūd Shāh I of Gujarāt is said by Firishta to have fitted out a fleet carrying gunners and musketeers against the pirates of Bulsar, and two years later used cannon to breach the walls of Champanīr and to fire "shells" (ḥuqqa) at the Rājā's palace.[162] But in 1511–12, according to Arabic manuscripts in the Bibliothèque Nationale, a Shāh of Gujarāt ("Modhaffershah") sent to Kansuh, King of Egypt, asking him for arms and cannon to enable the Gujarātis to defend themselves against the Franks (the Portuguese), "for the peoples of India had not then possessed artillery of any kind."[163] Faria y Souza[164] reported a battle between the King of Bisnagar and the Khān of Visiapur in which the former captured 400 large cannon and many small ones, but these were apparently Portuguese.

THE CANNON OF THE MUGHAL EMPERORS

The Mughal (the Arabic spelling of Mongol) Emperors of India ruled from 1526, nominally until 1858, although the last great Emperor was Aurangzeb, who died in 1707. They began with Bābur. His descendants introduced a strong Rājput strain by marriage with Hindu princesses. Later, Mughal came to mean any fair-skinned man from Central Asia or Afghanistan.[165] Bābur's memoirs are regarded as truthful and reliable; they "rank but little below the Commentaries of Cæsar."[166]

The *Tūzak-i Bāburī* (Autobiography of Bābur) written in Turkish and translated into Persian by 'Abd al-Rahīm in 1590, describes[167] an attack on Chanderī in 1527–8, when "the Pagans exerted themselves to the utmost, hurling down stones and throwing inflammable substances on their heads," but the troops of Bābur persevered and entered the fort. In the action, Ustād 'Alī Kūlī, Bābur's gunner, fired some shots; before the battle he had fired a large ball from a cannon, which burst and killed eight men.[168] In 1528–9 the Hindus of Bengal were assisting Bābur, who ironically says they were "famous for their skill in artillery," since when they fire their batteries "they do not direct their fire against a particular point, but discharge at random." At Lucknow the Hindus succeeded in setting fire to some hay which the Mughals had collected in a fort, "by fireworks, turpentine and other combustibles thrown on it," when the fort became as hot as an oven and had to be evacuated.[170]

Bābur was well supplied with cannon and small arms (matchlocks), which he used with effect in the decisive battle of Pānīpat on 21st April, 1526, in which Ibrāhīm, Sultan of Delhi, was killed and his army of hired troops was routed.[171] Particulars of Bābur's guns are given by Ḥaydar Mirzā, a military leader in his armies.[172] In his train he had 700 carriages (gardūn), each drawn by four bullocks and carrying a swivel-gun (zarb-zan), which discharged a ball (kalula) of 500 mitqāls in weight (3·5 lb. if a mitqāl is $1^3/_7$ drachms), used against horsemen; and twenty-one carriages, each drawn by eight pairs of bullocks, for which stone balls were of no use, cast brass balls weighing 5000 mitqāls (32 lb.) being used. They would hit anything at a range of a parasang (about 4 miles). Davies[172] says that 'araba, of which 700 were used, means a cart, not a gun carriage, and that Bābur in his autobiography speaks of having only two guns and says the battle was decided by bowmen.

Bābur uses the name firingihā for guns, which has been supposed to mean "Franks," i.e. guns of European origin although made in India.[173] It will be seen later (ch. VI) that this interpretation is doubtful. Pavet de Courteille, the French translator of Bābur, had rendered 'araba as "chariot" or at most "gun carriage," but there is little doubt that Bābur had guns on the occasion when these are mentioned.[174]

Bābur says Ustād 'Alī Kūlī, his master-gunner, was able to discharge his gun eight times in one day, and for three days after, sixteen times a day, "remarkably well." Another gun, larger than this, burst the first time it was fired.[175] Bābur spoke of guns being *cast* in his capital, but a century later welded iron guns are said to have been made at Dacca.[176] When Akbar took the fort of Jūnāgarh in Kathiāwar in 1591, he found a gun which had been captured bearing the name of the Turkish Sultan Sulayman the Great,[177] and the same Sultan had fortified Surat with mortars to keep out the Portuguese.

Bābur was succeeded in 1530 by his son Humāyūn, in whose reign, in 1548, an enormous howitzer, still in existence,[164] was cast by Muḥammad Rūmi (probably a Turk).[178] Humāyūn's son Akbar (1556–1605), who was born in exile and was unable to read or write to the end of his life, was a great patron of literature, the books in his library being read to him. Akbar was very interested in artillery. The *Ā'īn-i Akbarī*, the third part of the *Akbarnāma* which was written by his minister Abu'-l-Faḍl 'Allāmī (Abu'-l-Fazl), says[179] that Akbar designed a gun which could be taken to pieces for transport, a machine which cleaned sixteen gun-barrels at once, and a combination of seventeen guns which could be fired by a single match. The work does not deal with the manufacture of gunpowder or the casting of cannon, but describes the manufacture of iron matchlock barrels from welded rolled sheet, or rods pierced when red-hot with an iron pin, two, three or four such pieces making up one gun, the longest being 2 yards. The guns, it is said, were fired without a match, by a slight movement of the cock. Akbar invented an alloy called kaulpatr of 9 parts of copper and 1 part of tin (modern gun-metal) and was very interested in alchemy. He named his guns and often fired them himself. He also introduced a universal religion, and was altogether a man of advanced ideas.[180]

A mine in a sap (sabat) made by Akbar's troops in the siege of Chitor in 1567 was accidentally exploded when, according to a native historian in the *Tārīkh-i Alfi* (compiled by scholars under Akbar's orders), bodies were hurled "miles away."[181] One of Bābur's guns is shown in a British Museum manuscript (Or. 3716) as a long nearly cylindrical tube thicker at the breech end and tapering towards the muzzle, which has a collar; and another manuscript (South Kensington No. 74) shows one of Akbar's guns (Akbar Nāma) as evidently very heavy, practically cylindrical with a flange on the muzzle and a boss on the breech end, and it was mounted on a car.[182] Some very large Indian cannon will be described later; some of them are too large to be really useful and would be dangerous to fire, so that it has been suggested[183] that they were made by native workers with the intention of surpassing the size of European cannon.

Aurangzeb used rockets in attacking the strong fort of Bitar in 1657. One of them fell accidentally into a hole made by the defenders and filled with gunpowder, grenades, etc., which they intended to blow up when the besiegers entered; the rocket set fire to these, producing so much loss and confusion among the garrison that the place was carried after a short struggle.[184] Gamelli Curari in 1695 says that in the time of the Emperor Aurangzeb (1659–1707) most of the soldiers had bows and arrows and the swords were so poor that they imported English ones. All the artillery was directed by Portuguese, English, Dutch, German or French gunners.[185]

Besides cannon and small firearms, bows and arrows were always used in warfare by Akbar and his successors. When Akbar's army captured a great fortress they found it stocked with more than 1300 pieces of artillery (zarb-zan) with balls in weight from nearly 2 mans to a sīr or half a sīr, also many mortars (huqqa-dān). But there were also manjanīqs throwing stones of 1000 or 2000 mans, and on every bastion large iron cauldrons in each of which 20 or 30 mans of oil could be boiled and poured down on the assailants.[186] One of the guns was so large that when it was fired the gun and part of the tower in which it was mounted fell down.[187]

Gode[118] quotes the *Ākāśabhairavantra* (*c.* 1550) as advising the king to worship thirty-two weapons, including nālikā (gun). In the Sanskrit poem *Rāṣṭraudhavaṁśamahākāvya* of Rudrakavi, at the court of the Bāgalaṇ king in 1596, a red-hot iron cannon ball (āyasagolakaiḥ) with which the gun was charged (nālikāvinihata) contained within it sharp arrows (śitaśaraiḥ) and stones or gravel (uplaiḥ). Nīlakaṇṭha Caturdhara (later seventeenth century) described guns as machines (yantraṇi) made of iron (lohamayāni) capable of throwing by the force of gunpowder (āgneyauṣadhabalena) balls of lead (sīsa) and bell-metal (kāṁsya) and stones (dṛṣadgola). The historical poem *Kaṇṭhīravanarasarājavijayam* (1648) mentions guns (pirangi) carried on carts together with thousands of bags of gunpowder in the army of Raṇadullakhān against the King of the Karnatak and his feudatories. The poem *Śambhurā-jacarita* by Harikavi (Bhānubhaṭṭa) (1685) mentions a series of cannon balls (sugolakatai) issuing from the mouths of cannon (nālikāvaktrataḥ . . . udgatā) and looking like rows of suns. Some technical names are to be found in the Sanskrit *Yāvanaparipāṭī-anukrama* by Dalapatirāya (*c.* 1764): yantraśālā is an arsenal, gulikāṅgāra cūrṇa is gunpowder (older agnicūrṇa), portable guns to be carried on camel's back are uṣṭranālikā, on horse-back hayanālikā, and on the back of an elephant gajanālikā. Gode adds some references to iron guns of 1554–5 and 1589 (?), bronze guns of 1663 and 1664 made at Burhānapur; 100 brass guns and 12,000 (*sic*) muskets at Gujarāt; guns fired at marriage parties, and supposed references to firearms in the *Mahābhārata* (which contains material dating from 200 B.C. to A.D. 200 with later additions).

THE PORTUGUESE IN INDIA

From 1412 the Portuguese had been searching for a sea route to India by way of the southern extremity of Africa. Bartholomew Diaz in his outward voyage in 1486 had passed the Cape but, either because he was too far out at sea or because it was obscured by fog, he did not see it. He discovered it on his return voyage and called it the Cape of Storms (Cabo dos Tormentos), but when Diaz reported it on his return to Portugal in 1487, King John called it the Cape of Good Hope (Cabo de boa Esperança), a name which it has ever since retained. The sea route to India was now open.

King Manuel of Portugal appointed Vasco da Gama to command an expedition of four ships to go further east with the hope of reaching India, and this left Lisbon in 1497. It took up an Indian pilot at Malindi, near Mombasa, sailed across the Indian Ocean, and on 23rd May, 1498, landed at Calicut in South India. The Zamorin ("emperor" or chief Raja) of Calicut, at the instigation of jealous Arab merchants in the town, became hostile, and Vasco da Gama had finally to fight his way out of the harbour.

A squadron of ships under Cabral was sent out by King Manuel of Portugal; it went out of its way and reached the previously unknown coast of Brazil, but finally half the fleet reached Calicut. The expedition left forty Portuguese in Calicut, who after Cabral's departure were murdered. Manuel in 1502 sent out twenty ships commanded by Vasco da Gama, which founded the African colonies of Sofala and Mozambique; Vasco da Gama bombarded Calicut and destroyed the Zamorin's fleet, returning with booty to Lisbon in 1503. Goa was captured by Albuquerque in 1510.

An account of Vasco da Gama's first visit to Calicut is contained in an anonymous *Roteiro da Viagem* (Sailing Route) probably written by a sailor on one of the ships. This document, discovered in the public library of Oporto and published in 1838, says[188] that when Vasco da Gama was led to the Zamorin's palace with sounds of music someone in the crowd fired a musket, apparently the only one in the town: o qual vinha pera jr com o capitam e trazia mujtos tambores e anafis e charamellas e hũa espingarda a qual hia tirando amte nos. Since Calicut was full of Arab traders when Vasco da Gama arrived in 1498, it is reasonable to suppose that the musket carried by the Indian had come from them.

Fernaõ Lopes de Castanheda, the natural son of the first Portuguese auditor (ouvidor) of Goa, travelled in India from 1528, and was later appointed conservator of the archives in the University of Coimbra, where he died in 1559. His account of the incident says[189]: & vinha por mandado del rey pera ho acompanhar ate ho paço, & leuaua consigo muytos Naires, & diante muytos trombetas & anafis que yão tangendo, & assi hũ Naire que leuaua hũa espingarda com que tiraua de quando em quãdo; which the translation in Kerr renders: The procession again set out, preceded by many trumpets and sacbuts sounding all the way; and one of the Nayres carried a caliver, which he fired off at intervals.

The Nayres (or Naires) was the name given to nobles in the service of the Zamorin, and one of them had a caliver or musket. Castanheda also says[190] that two Italian lapidaries from Milan deserted from Vasco da Gama and entered the service of the Zamorin of Calicut. They instructed the native smiths in making artillery like the Portuguese guns.

An anonymous account by "a Florentine nobleman" of Vasco da Gama's landing, which was printed by Giovanni Battista Ramusio (1485–1557), says[191] none of the very numerous Arab ships had arms or artillery (& nõ portano arme ne artegliaria). An Indian pilot who accompanied the expedition on its return to Lisbon in 1499 told this Florentine nobleman that foreign ships had landed in Calicut eighty years before (i.e. in 1419). These ships carried bombards (bombarde) which were much shorter than the modern ones (ma piu curte di quelle che si usan' al presente). Twenty or twenty-five of these ships returned every two or three years. These ships may have been Chinese.

Faria y Souza (1590–1649) speaks[192] of a Gujarāt ship in 1500 firing a cloud of arrows and some cannon balls (una nube de flechas sobre nuestra gente y algunas balas) at Portuguese ships; of the Indians at Calicut using fire-ships in 1502; and of the Zamorin's fleet in 1503 carrying 382 guns. He reports that the Moors of Sumatra, Malacca and the Moluccas "were well disciplined and much better stored with artillery than we that attacked them" (in 1506). Elliot[193] said: "There is certain testimony to the use of cannon in Gujerāt before the arrival of the Portuguese; which is easily accounted for by the constant communication at that time with the Turks of Egypt and Arabia."

Cabral's ships were fired on by some pieces of ordnance on shore in Calicut, but without much effect.[194] In 1502 the Zamorin of Calicut is reported[195] to have had a large number of guns of all sizes, of cast and wrought iron, one of which shot iron balls weighing 1 lb. The Caymal of an island subject to the Raja of Cochin (south of Calicut) had deserted to the Zamorin of Calicut with an army of 3000 nayres, including 700 archers and forty men armed with matchlocks, and he had a battery of five cannon which fired cast iron balls. He had several paraws (ships with oars) provided with ordnance supplied by the Zamorin, which he used to defend the harbour against the Portuguese.[196]

Thome Lopez,[197] who was on one of the Portuguese ships in 1502, reports that a large Indian fleet had only a few inferior guns and relied for defence on bows and arrows and stones, but in 1503 all the Indian ships at Calicut had guns (tutte portauano bombarde). Castanheda[198] says many guns opposed the Portuguese landing in 1504 and they had been made under the supervision of the two Italian lapidaries; a Moor had also devised fire-ships, and floating castles reinforced with iron and carrying cannon and men armed with muskets.

Ludovicho Barthema of Bologna,[199] who went to Calicut from Java in 1504, found that the Zamorin had 450 to 500 guns, large and small, made by the two Milanese, Pietro Antonio and Giovanni Maria, who had lodgings in a mosque. They showed him drawings and a model for a large bronze bombard weighing 150 cantaras (about 10 tons), and four large iron bombards they had made. He describes an Indian attack on the fort at Cananor in 1507 in which 140 guns, large and small (artigliaria infra grosse & minute), including twenty-four large cannon, were used, also "artificial fire" (perhaps rockets). In 1508 the Portuguese were defeated in a naval battle lasting two days, the Portuguese captain being killed by two cannon balls which struck him in succession.[200]

Odoardo Barbosa[201] of Lisbon said the Indian ships had many guns (piene di artegliaria), some heavy, when Albuquerque attacked Hormuz and Goa in 1510; there were very heavy cannon at Dīū, a small island harbour of Combaja (Kathiāwar), which became Portuguese in 1534, and was defended against Gujarāti attacks in 1538 and 1546; one of these guns is now in Lisbon. He says that large ships from Goa opposing the Portuguese had guns of copper (di rame) and iron. In 1511 Albuquerque attacked Malacca, the Sultan of which assembled muskets (schioppi) and heavy artillery (artiglieria grossa), one piece of which is now in Lisbon.

Barbosa found at Java (Giaua maggiore) great activity in building ships, and says the Javanese were expert gun casters (gran maestri di gittar artegliaria). They had abundance of light arms (spingarde, schioppi) and Greek fire (or fireworks?) (fuochi artificiate) as well as guns, and the large ships were very heavily armed. Faria y Souza[192] reports that the Portuguese in 1511 were opposed in Malacca by a people using cannon, mining the streets with gunpowder and using floats of wild-fire (fuego) at sea, and that Muḥammad, the King of Java, had 8000 guns.

Cæsar Frederick, a Venetian merchant who travelled in 1563–81, found that the Moorish Niẓām of Chaul in Gujarāt had many very large guns, some in pieces (screwed together?), which threw very large stone balls; some of these balls were sent to Portugal as curiosities.[202] Captain Downton records the use of wild fire (Greek fire) by a Portuguese ship against an English ship in 1594, causing surprise and damage; the English ship used its large cannon elevated as howitzers.[203]

Portuguese guns, according to Rathgen,[204] were rather late (1370) and their use first became important at Ceuta in 1410. Vasco da Gama's three small ships carried guns which the accounts call berços; one of this type in the Lisbon Museum is a fairly small breech-loader, 1·63 m. long, weighing 102 kg., and in Spain the versos were quite small guns. Vasco da Gama also had larger guns, called "camels." Rathgen says the Portuguese entering India found that firearms were known in restricted areas but they were much inferior to those then used in Europe. The guns in India probably came from Europe, and from India they went to China and other further eastern lands. Any other view is impossible (mit voller Beweiskraft widerlegt). Since a contemporary document[205] written in Portuguese with a name ending in "Alemā" was probably written by a German (Alemanna) and mentions German gunners (bombardarii) in Portugal, the Portuguese guns also came from Germany, and it is Germany's "Ehrenverdienst . . . die den Weltenlauf bestimmende Pulverwaffe geschaffen . . . zu haben." The author's name, however (like the language of the manuscript), appears to be Portuguese, and the early accounts of the making of guns and cannon in Portuguese India all agree that the makers were Italians and natives trained by them.

Indian steel, including wootz, occurs in graves of the seventh to sixth centuries B.C. Early Indian ironworkers were outstandingly skilful; the famous Delhi pillar, made in A.D. 415, weighing over 6 tons, is lasting evidence of this. The famous Damascus steel was really made from Indian metal. About the same age as the Delhi pillar is a copper statue of Buddha from Sultanganj, which is 7½ ft. high, cast in sections probably by the cire-perdue process; bronze casting was well developed in South India from the tenth century, large figures being made for temples and ceremonial processions.[206] There is no need to go outside India for technical skill in iron or bronze work. Large guns, both of iron and bronze, were made in India.[207]

LARGE GUNS MADE IN INDIA

The very large iron gun, Raja Gopal, in Tanjore,[208] probably sixteenth century, is 7·445 m. (about 25 ft.) long, 1·140 m. (over 3 ft.) diameter, and weighs 40,000 kg. (nearly 40 tons). Although the gun is cylindrical, the powder chamber is narrower than the barrel containing the shot (the Stein-büchse type). It is similar to, but some two to three times as heavy as, the Dulle Griete at Ghent (p. 128), and is made from iron bars and shrunk-on rings, the workmanship being better than that of the contemporary European gun. There is a legend that it was fired only once; it was so big that no gunner would venture to fire it in the normal way, so that a train of gunpowder 2 miles long was laid to the touch-hole, and when the gun went off the shock damaged the health of people living near it.

The large wrought-iron gun of Murshidabad (near Calcutta) is breech-loading. The chamber is not screwed to the barrel or chase, but is provided with rings for lashing it on. The chase is 12 ft. 6 in. long, calibre 18½ in., the chamber is 4 ft. 3 in. long. It is ornamented with Oriental characters and was probably made in India rather than in Europe and afterwards ornamented in India. The date is unknown but the gun is of the same type as European guns of the end of the fourteenth century. It had been thrown into a river, and slowly emerged from 5 ft. of water on account of the growth of a pipal tree underneath it (which a Buddhist would regard as an omen).

Large iron guns mentioned by Rāy are: (i) one at Dacca, which has now fallen into the river, 30 tons with iron cannon balls weighing 465 lb.; (ii) one at Murshidabad, 17 ft. 6 in. long, 5 ft. 3 in. circumference, made in 1637; (iii) some at Bījāpur, one 21 ft. 7 in. long, 4 ft. 4 in. diameter at the breech, calibre 1 ft. 7 in., bore length 18 ft. 7 in., weight 47 tons, with trunnions; another 30 ft. 8 in. long, 1 ft. diameter bore; (iv) one at Gulbarga, Hyderabad, with a double row of iron rings, ten on each side, for transport, and a projection at the muzzle for sighting (like the English gun mentioned on p. 112). Those undated are probably of the sixteenth and seventeenth centuries.

In an old fort at Raichore, south-west Hyderabad, with stones carrying the dates 1563 to 1619, there was an old cannon made from twelve iron bars 1¼ in. square, surrounded by three wrought-iron coils, the inner one 1¼ in. thick and the middle and outer coils 1¾ in. The outer coil extends only 2 ft. 7 in. from the part where the breech, just below the touch-hole, had been blown away. The length of the remaining part is 20 ft. 4 in., the external circumference near the muzzle 3 ft., at the trunnions 3 ft. 3 in., and at the breech 4 ft. 2 in.   It has four pairs of eyes for slinging it by means of ropes. A projection just in front of the trunnions fits on a wrought-iron cross bar 4 ft. 6 in. long which passes through the two arms of a Y-shaped support built into the masonry base.   Another broken gun of similar construction is in a bastion in a wall of the fort.[209]

There are also some Indian bronze guns of excellent make and some are very large.   The earliest are those captured at Calicut (1504, 11,250 kg.) and Dīū (1533, 19,494 kg.), now in Lisbon.[210]   The Bījāpur bronze gun (Malik-i-Maidan) was cast in 1551 by a Rūmī (either a Turk or a European, perhaps under Jesuit supervision).   It is 14 ft. (4·270 m.) long, 28 in. (71 cm.) bore and 13½ ft. (4·118 m.) circumference, weighing about 34½ tons (35,000 kg.).

Another great bronze gun ("basilisk" in European nomenclature) was cast in 1553.   The Dhūl Danī, cast in Agra in 1628, is 20 ft. 4 in. (6·2 m.) long, 2 ft. 6 in. (51 cm.) calibre, and weighs nearly 27 tons.   The Gohlke of Bhurtport, captured in 1826 and taken to Woolwich, was cast in 1677; it is 16 ft. 4 in. long, calibre 8 in., diameter 3 ft. 3 in. at the base ring, and 2 ft. at the muzzle.   A gun partly of bronze and partly of iron, captured in Malacca, is in Lisbon.

Mallet mentions an Arakan (Burma) bronze gun 30 ft. long and 10 in. calibre, 2½ ft. across the muzzle; the date is unknown but it is very old in appearance.   These very long guns were called serpentines in Europe. The largest bronze gun in the world must have been the Zafar Bakht ("giver of victory"), cast in Delhi in 1627 and weighing 52½ tons (50,000 kg.).[211]   There is, apparently, no information as to its location if it still exists.   It would be much heavier than the Moscow gun (39 tons).   A bronze gun cast in the Admenugger Arsenal in Agra in 1820 was 14 ft. long and 21 in. calibre.[212]

Edward Terry, who visited India in 1616, says[213] the Mogul's troops had all kinds of firearms and artillery.   The infantry were armed with match-locks.   The elephants carried 6 ft. long heavy guns mounted on wooden frames strapped to the elephant, and the gun, which fired an iron ball as big as a tennis-ball, could be trained and fired by a mounted gunner.   The gun-powder was of excellent quality.

A large gun of gold and silver from Candia (in Crete), perhaps Turkish, was in an arsenal in Verona; silver (or at least silver-plated) guns are reported from India, and the Gaekwar of Baroda is said to have silver cannon and a field battery of solid gold guns.[214]

The Indian war rockets described and depicted by Craufurd,[113] were formidable weapons before such rockets were used in Europe.   They had bam-boo rods, a rocket-body lashed to the rod, and iron points.   They were directed at the target and fired by lighting the fuse, but the trajectory was rather erratic.   The use of mines and counter-mines with explosive charges of gunpowder is mentioned for the times of Akbar and Jahāngīr.[215]   The use of a gunpowder mine, as "something entirely new in the Deccan," to blow down the wall of a fort, Belgām, in 1472, is reported by the *Burhānu-i-ma'āthir* and Firishta.[216]

FIREWORKS IN INDIA

Some information about the use of fireworks in India is available. Gode[216] quotes from Sanskrit manuscripts on fireworks by Gajapati Pratāparudradeva of Orissa (1497–1539), the reputed author of some works on dharmaśāstra such as the *Sarasvatīvilāsa*. His work on fireworks, *Kautukacintāmaṇi*, is quoted from two manuscripts, one dated 1778 but a copy of one dated 1670. The work is in verse. Materials specified are saltpetre, sulphur, charcoal of bamboo, pine, willow, birch-bark, of a special plant, etc.; steel powder, powder of roasted iron, green exudation from copper, yellow orpiment, five kinds of salt (natron, saltpetre and borax, and two others not translated), lodestone, mercury, cinnabar, ochre, tin or lead powder; also a special wood, a wick, a hollow bamboo, pulp of castor-oil plant seeds, paste of food (rice, etc.), and cow's urine.

Guns and displays of fireworks are mentioned in the large *Ākāśabhairava-Kalpa*, a Sanskrit work later than 1400, in which a gun (nālikā) is one of thirty-two weapons to be worshipped by the king, also firework displays (vinodas) and a rocket (bāṇa), fired off at the end. The Marāthī poem *Rukmiṇī-Svayaṁvara* of Saint Ekanātha (1570) mentions rockets, fireworks producing garlands of flowers, a flow of fiery sparks, a moonlight effect, and a hissing noise. Some are operated on the ground, one is held in the hand. Another Mahārāṣtra author, Saint Rāmadāsa (1608–82) also mentions displays of specified kinds of fireworks, cannon, muskets and small gunpowder mortars for producing thundering sounds. Although bāṇa has been called a rocket, Bernier, who was in India in 1656–68, calls it a grenade attached to a stick, and speaks of frightening elephants with cherkys, which are probably crackers, now called carkī in Bengal and carkhī in the United Provinces (where there are nine varieties); the name may come from the Sanskrit cakra. The Sanskrit baṇa means an arrow; it seems to have been used for a rocket after 1400; in 1676 a dictionary of non-Sanskrit terms by Raghunātha Paṇḍita said it meant a tube filled with gunpowder.

Tavernier, in 1676, saw Chinese fireworks in India and Java. By the end of the eighteenth century firework displays were popular in Rājputāna and were used in Royal entertainments on a lavish scale. An English display at Bibipur near Lucknow in 1790 took an artist Karār (Karrar?) six months to prepare and was very lavish: it included coloured fires, large set-pieces, rockets (which sent out firework fish), fire-flowers whirling in the sky, etc. This was apparently the first European display in India. An Indian booklet on fireworks by Lakṣmaṇa Pāmaji Khopkar (Bombay, 1886), includes antimony among the constituents, and it was not used in China in the eighteenth century (see ch. VI).

REFERENCES

1. Hime, *The Origin of Artillery*, 1915, 64.
2. *The History of India as told by its own Historians. The Muhammadan Period. The Post-humous Papers of the late Sir H. M. Elliot, K.C.B.*, edited and continued by Professor John Dawson,

M.R.A.S., *Staff College, Sandhurst*, 1867, i, 170 (from the *Shāh-Nāma* of Firdawsī, completed in A.D. 1010); Elmacin, *Historia Saracenica*, Arabic text and Latin tr. by Erhenius, Leyden, 1625, 61. Casiri, ref. 6, ii, 8, translated: manganis & mortariis ope naphthæ & ignis in Cabam jactis, illius tecta diruit, combussit ac in cinerem redegit. The first Muslim use of naphtha given by Quatremère, ref. 35, 132–7, was in the invasion of Sind, when bottles of it, which they had seen used by Greeks and Persians, were thrown at elephants. See ref. 136.

3. Joseph Toussaint Reinaud and Ildephonse Favé, "Du feu grégeois, des feux de guerre, et des origines de la poudre à canon chez les Arabes, les Persans et les Chinois," in *J. Asiatique*, 1849, xiv, 257–327. See the refs. in ch. I, ref. 221; M. Canard, *J. Asiatique*, 1926, ccviii, 61–121; *Bull. des Études Arabes*, Algiers, 1946, vi, no. 26, pp. 3–7; M. Mercier, *Le Feu Grégeois*, Paris, 1952, 40–91, especially 41, 51, 85.

4. Hime, ref. 1, 64–73.

5. E. Renan, *Averroës et l'Averroisme*, 1882, 80.

6. M. Casiri, *Bibliotheca Arabico-Hispana Escurialensis sive Librorum omnium MSS. quos Arabicè ab auctoribus magnam partem Arabo-Hispanis compositos Bibliotheca Cænobii Escurialensis complectitur. Recensio & Explanatio Operâ & Studio Michaelis Casiri*, 2 vols., fº, Madrid, 1770 (vol. 1 mis-dated 1760), ii, 7. Gibbon, ref. 132, v, 377, says the work, which is very scarce, was distributed privately by the court of Madrid (he had a copy of it). I used the copy in the London Library and there are three copies in Cambridge University Library. See Reinaud and Favé, *Histoire de l'artillerie* 1ʳᵉ *partie. Du Feu Grégeois des feux à guerre et des origines de la poudre à canon d'après des textes nouveau*, Paris, 1845, 66; S. J. von Romocki, *Geschichte der Explosivstoffe*, 1895, i, 78; Hime, ref. 1, 68. C. F. Temler, *Historische Abhandlungen der Königlichen Gesellschaft der Wissenschaften zu Kopenhagen … übersetzt … von V. A. Heinze*, Kiel, Dresden and Leipzig, 1782, i, 168, makes Peter, Bishop of Leon, report the use of cannon in Seville in 1248; see ref. 29.

7. Reinaud and Favé, ref. 3, 281.

8. Quatremère, *J. Asiatique*, 1850, xv, 214–74 (243); B. Laufer, *American Anthropologist*, 1917, xix, 71–5 (still dated it 1249).

9. H. Hallam, *View of the State of Europe during the Middle Ages*, 1868, 229–30 (or undated ed., 280).

10. Ref. 6, i, 79.

11. L. Lalanne, *Mém. div. Sav. Académie des Inscriptions et Belles-Lettres*, II Sér., Paris, 1843, i, 332.

12. Seyffert, Nettleship, and Sandys, *Dictionary of Classical Antiquities*, 2 ed., 1891, 74: onager or scorpio.

13. *Histoire des Berbères et des Dynasties Musulmanes de l'Afrique Septentrionale, Par Ibn-Khaldoun*, Arabic text, 2 vols., Algiers, 1847–51; tr. by W. McGucken de Slane, 4 vols., Algiers, 1852-4-6-6, 1856, iv, 69–70; the passage is quoted by Reinaud and Favé, ref. 6, 75, and they, and Favé, ref. 71, iii, 65–6, say that ibn Khaldūn wrote a century after the event and did not always scrupulously follow authentic documents. They doubt the authenticity of the information.

14. Gerónimo Zurita, *Los Anales de la Corona de Aragon*, 6 vols., fº, Saragossa, 1562–80 (BM 594.k.4–9); revised and extended by his son, 7 vols., fº, Saragossa, 1610–21; 1610, ii, 99 *v.*; J. G. von Hoyer, *Geschichte der Kriegskunst*, Göttingen, 1797, i, 47; Zusätze und Ergänzungen, 5 (who says it was unknown to Temler, ref. 6).

15. G. Köhler, *Die Entwicklung des Kriegswesens und der Kriegsführung*, Breslau, 1887, III, i, 223.

16. Giovanni Andres, *Dell'origine, progressi e statto attuale d'ogni letteratura*, 8 vols., 4º, Parma, 1785, i, 234; M. Berthelot, *Science et Philosophie*, 1905, 131.

17. Ref. 11, 348.

18. Ref. 6, i, 80 (with Arabic text).

19. Ref. 1, 68–71 (with Arabic text).

20. J. A. Conde, *Historia de la Dominacion de los Arabes en España*, 8º, Paris (Baudry), 1840; the work is said to have been first published in 3 vols., fº, Madrid, 1820–21.

21. Ref. 1, 71; parroted by Ayalon, ref. 37, 42.

22. Ref. 20, 593.

23. I. S. Allouche, "Un texte relatif aux premiers canons," in *Hespéris* (Institut des Hautes Études Marocaines, Rabat), Paris, 1945, xxxii, 81–4 (BM Ac.17.d.).

24. See Sarton, *Introduction to the History of Science*, Baltimore, 1948, iii, 1762, who says long extracts from his historical works were quoted and translated by Casiri.

25. Ref. 8, 256.

26. Ref. 6, 73 f.

27. B. Rathgen, *Das Geschütz im Mittelalter*, Berlin, 1928, 693, 703 ("Brandgeschossen").

28. Ref. 20, 604.

29. Pedro Mexia, *Silva de Varia Lecion*, bk. i, ch. 8; 4°, Seville, 1542; 8°, Antwerp, 1603, 41–2.

30. Juan de Mariana, *Historiæ de Rebus Hispaniæ*, f°, Toledo, 1592, 766 (bk. xvi, ch. 11); Spanish tr., *Historia general de España*, 2 vols., f°, Madrid, 1608, ii, 27; English tr. by Capt. John Stephens, *The General History of Spain*, 2 pts., f°, London, 1699, i, 264 (bk. xvi, ch. 5).

31. R. Watson, *Chemical Essays*, 1793, i, 331.

32. W. H. Prescott, *History of the Reign of Ferdinand and Isabella, the Catholic, of Spain*, 3 vols., London, 1849, i, 133, 353, 428.

33. Gardner, *Archæologia*, 1898, lvi, 133.

34. Ref. 6, ii, 8.

35. *Histoire des Mongols de la Perse, écrit en Persan par Reschid-Eldin*, tr. by Quatremère, in *Collection Orientale. Manuscrits Inédits de la Bibliothèque Royale. Traduits et Publiés par ordre*, Paris, 1836, i, 132–7, 284–295 (only one vol. was published).

36. Ref. 15, III, i, 222–5.

37. See the long account of Ibn Khaldūn in Sarton, ref. 24, iii, 1767–79.

38. David Ayalon (Jerusalem), *Gunpowder and Firearms in the Mamluk Kingdom*, London, 1956, 3, 6. Although I differ from Prof. Ayalon on several matters, I wish to record my admiration for his scholarly book.

39. G. Wiet, *Syria*, 1924, v, 16–53; 1925, vi, 150–73; 1926, vii, 44–66 (62–5); Ayalon, ref. 38, 3, omits the words "of red-hot iron."

40. See also Quatremère, ref. 35, 285 f.

41. See ref. 13; Casiri, ref. 6; and Ayalon, ref. 38, 21.

42. Ayalon, ref. 38, 22.

43. Ref. 38, 16, 35.

44. Ref. 38, 17, 38; for the late use of manjanīqs, *ib.*, 44.

45. Ayalon, ref. 38, 41; this is the author quoted by Casiri, ref. 6.

46. Ref. 38, 9 f., 19 f., 23, 31 f.

47. Ref. 38, 9, 14, 24, 30.

48. M. Amari, *Storia dei Musulmani de Sicilia*, 3 vols., Florence, 1854–72; 1872, III, ii, 367; 3 vols., Catania, 1938, III, ii, 373–4; *id.*, "Su i fuochi da guerra usati nel Mediterraneo nell' XI e XII secolo," in *Atti R. Accad. Lincei, Mem. Classe Sci. Morali, Storiche e Filologiche*, Rome, 1876, iii, 3–16; Lippmann, *Abhandlungen und Vorträge*, 1906, i, 134.

49. E. W. Brooks, *J. Hellenic Studies*, 1899, xix, 19–33.

50. H. Ritter, *Islam*, 1929, xviii, 116–54. Cf. Wüstenfeld, "Das Heerwesen der Muhammedaner," in *Abhl. K. Ges. Göttingen*, 1880, xxvi, 69 (naphtha and incendiaries; from a seventeenth-century MS.).

51. C. Cahen, "Un traité d'armurerie composé pour Saladin," in *Bull. d'Études Orientales*, Beyrouth, 1947–8, xii, 104–63 (incorrect ref. in Ayalon, ref. 38). Cahen gives the Arabic text from a Bodleian MS., Hunt 264, with a translation, and notes.

52. Reinaud, *J. Asiatique*, 1848, xii, 193–237; Reinaud and Favé, *ib.*, 1849, xiv, 262–327; Romocki, ref. 6, i, 29. An Arabic MS. in the Bibliothèque Nationale attributed to Aristotle is alchemical, although it quotes an apocryphal Τακτικά of Aristotle; Fabricius, *Bibliotheca Græca*, 1707, ii, 201; it will be remembered (see ch. II) that the *Liber Ignium* of Marcus Græcus says that Aristotle invented an incendiary which burnt for nine years.

53. Reinaud and Favé, ref. 52, 263–78, with Arabic text.

54. Quatremère, ref. 8, proposed many corrections; reply by Reinaud, *J. Asiatique,* 1850, xv, 371-6.

55. Reinaud, ref. 52, 198; Reinaud and Favé, ref. 52, 279.

56. Reinaud and Favé, ref. 52, 264-6.

57. Ref. 8, 261-2.

58. Reinaud and Favé, ref. 52, 273-4.

59. Reinaud and Favé, ref. 52, 267, 276, 281.

60. Reinaud and Favé, ref. 52, 271-2.

61. Reinaud and Favé, ref. 52, 268-9.

62. Reinaud and Favé, ref. 52, 275; Mercier, ref. 3, 60.

63. Mercier, ref. 3, 56-61.

64. Reinaud and Favé, ref. 52, 269-70.

65. Reinaud and Favé, ref. 52, 266.

66. Reinaud and Favé, ref. 52, 276-8.

67. J. von Somogyi, *Islam*, 1937, xxiv, 105 (119). Although Ayalon, ref. 38, 20, thought this referred to a discharge of firearms, I do not agree with him.

68. Reinaud, ref. 52, 198-9; possibly bārūd here means nitrum (soda).

69. Reinaud and Favé, ref. 52, 282. Saltpetre may have been known rather earlier; see ch. VII.

70. Elliot, ref. 2, 1875, vi, 475-80.

71. This is no. 1128 in Reinaud and Favé, ref. 6, 1845, 20-47; Reinaud, ref. 52, 199; M. Jähns, *Geschichte der Kriegswissenschaften* (in *Geschichte der Wissenschaft in Deutschland*, xxi), 1889, i, 179 f., dates it *c.* 1290; Favé, in Napoleon, *Études sur le Passé et l'Avenir de l'Artillérie*, 1862, iii, 20, dates it 1285-95; Romocki, ref. 6, i, 68, and Hime, ref. 1, 19, date it 1275-95; Rathgen, ref. 27, 703, dates it 1285.

72. Favé, ref. 71, iii, 34; Hime, ref. 1, 19.

73. Hime, ref. 1, 20, thought "wood ashes" meant charcoal, which would remove impurities, and that this was an important detail which, since it is not emphasised, was not original. He gives the Arabic text. A full translation is given by Mercier, ref. 22, 116.

74. Ref. 8, 225.

75. See ch. II; Favé, ref. 71, iii, 34; Hime, ref. 1, 21-2; Watson, ref. 31, i, 337-40, who mentioned that Roger Bacon knew some Arabic, thought that he obtained his information from an Arabic source; E. Charles, *Roger Bacon*, 1861, 118, 123-4, says Bacon admitted that he knew Arabic only imperfectly, but he did know some.

76. Ref. 6, i, 68.

77. R. A. Nicholson, *Literary History of the Arabs*, London, 1907 (or 2 ed., Cambridge, 1930), 273.

78. Sarton, ref. 24, 1931, ii, 1036.

79. Reinaud, ref. 52, 217.

80. D'Incarville, *Mém. div. Sav. Acad. Sci.*, 1763, iv, 66-94; Favé, ref. 71, iii, 21 f.; Mercier, ref. 3, 117, says "fleur mourac" should be "mouerreq," giving the more exact sense of "à pétales," but I have failed to find the words in the Arabic dictionaries.

81. Reinaud and Favé, ref. 6, 1845, 47.

82. Favé, ref. 71, iii, 21-4; Jähns, ref. 71, 1889, i, 180.

83. Reinaud and Favé, ref. 52, 314.

84. Favé, ref. 71, iii, 25-30, plate 2.

85. Romocki, ref. 6, i, 71, fig. 14 (with Arabic text).

86. Favé, ref. 71, iii, 25-30, plate 2; Romocki, ref. 6, i, 70-2.

87. Hime, *Gunpowder and Ammunition*, 1904, 229.

88. Hime, ref. 87, 231.

89. Ref. 3, 120-3.

90. Ref. 71, iii, 33.

91. Ref. 1, 72.

92. Reinaud, ref. 52, 193.

93. Reinaud, ref. 52, 203; Reinaud and Favé, ref. 52, 309–25; Favé, ref. 71, iii, 35–43 (on Ghāzān, *ib.*, 43); Quatremère, *Histoire des Sultans Mamlouks de l'Egypte* (tr. of al-Maqrīzī), Paris, 1845, II, ii, 147; Romocki, ref. 6, i, 76; Jähns, ref. 71, i, 181. The date attributed to the MS. by Reinaud and Favé is questioned by O. Guttmann, *Die Industrie der Explosivstoffe, Handbuch der chemischen Technologie*, ed. P. A. Bolley, K. Birnbaum, and C. Engler, Brunswick, 1895, VI, vi, 5–6, 14; and by Rathgen, *Ostasiatische Zeitschrift*, 1925, ii, 11 (26), 196 (Die Pulverwaffe in Indien).

94. Reinaud, ref. 52, 198; Reinaud and Favé, ref. 52, 311–4; Rathgen, ref. 27, 703 (dates 1311) gives 74 saltpetre, 10 sulphur, 13 charcoal; see Quatremère, ref. 93. The MS. was first described by Count de Rzevuski in *Mines de l'Orient*, Vienna, 1809, i, 189, 248 (giving a translation of the gunpowder recipe but differing at the end from Reinaud and Favé's : "set fire to the powder in the chamber of the gun. It will be perforated below by light and if it is perforated still lower it will be a fault, and woe to him who fires it").

95. Reinaud and Favé, ref. 52, 310, with Arabic text; Romocki, ref. 6, i, 77.

96. *Travels of Venetians in Persia*, Hakluyt Society, 1873: *Travels to Tana and Persia*, 37; *Italian Travels in Persia*, 15, 23, 61, 93, 125, 153, 182; E. G. Browne, *A History of Persian Literature under Tartar Dominion*, Cambridge, 1920, 381, 389, 404, 412; *id.*, *Persian Literature in Modern Times*, Cambridge, 1924 75, 93, 105–6.

97. Reinaud, ref. 52, 215–17; Reinaud and Favé, ref. 52, 311; Hime, ref. 1, 3; *id.*, ref. 87, 91–4.

98. Ref. 8, 237.

99. Revised in Favé, ref. 71, iii, 37. The title (in English) was supplied by Dr. E. J. Holmyard.

100. Reinaud and Favé, ref. 52, 321–3; Favé, ref. 71, iii, 41.

101. M. Berthelot, *La Chimie au Moyen Âge*, 1893, ii, 198; E. O. von Lippmann, *Entstehung und Ausbreitung der Alchemie*, Berlin, 1919, i, 394; Hime, ref. 1, 13, 15, says the word tr. "orge," barley, may be the Indian yavakshara, which means impure potash from burnt barley straw and not saltpetre, and that the tr. "pétard" of a word parakeya in the Syriac text on p. 102 in Berthelot is incorrect, petards not being invented until about 1580; the word seems to mean some machine called "the destroyer," but may be barakeya, "the thunderer."

102. Ayalon, ref. 38, 46–59.

103. Ayalon, ref. 38, 59 f., 92 f., 117 f. The Moors in Spain were accused of using poisoned arrows, a piece of linen or cotton cloth dipped in a "distillation of aconite juice" being attached to the tip, in the wars against the Spaniards in 1483; Prescott, ref. 52, i, 432.

104. In Ayalon, ref. 38, 141 f.

105. Q. by Photios; text in Oppert, ref. 106, 55 (who thought σκώληξ "worm" = Sanskrit culukī, crocodile?); J. W. McCrindle, *Ancient India as described by Ktesias the Knidian*, Calcutta, 1882; C. Lassen, *Indische Altertumskunde*, Bonn, 1849, ii, 641, calls it "an Indian fable known to the Persians, perhaps of magic weapons." In Pliny, *H.N.*, ix, 17, the "worms" found in the Ganges are big enough to seize elephants coming to drink and drag them into the river.

106. Ailian, *De Natura Animalium*, bk. v, ch. 3; Elliot, ref. 2, 1875, vi, 478.

107. Philostratos, *Life of Apollonios of Tyana*, (*a*) bk. iii, ch. 1; (*b*) bk. ii, ch. 33.

108. Essay 58: Of Vicissitude of Things (first publ. in 1625 ed.); *Works*, ed. Spedding and Ellis, 1858, vi, 516; Latin tr. publ. by Rowley in 1638 and perhaps made or seen by Bacon (d. 1626). See ref. 107.

109. P. von Bohlen, *Das alte Indien*, Königsberg, 1830, ii, 63–6; Gen. R. Maclagan, "Early Asiatic Fire Weapons," in *J. Asiatic Soc. Bengal*, 1876, xlv, 30–71 (a good general paper).

110. Kauṭilya's *Arthaśāstra*, tr. R. Shamasastry, with intr. by J. F. Fleet, 2nd (revised) ed., Mysore, 1923 (page refs. in text are to this ed.); M. Winternitz, *Geschichte der Indischen Litteratur*, Leipzig, 1922, iii, 36, 281, 509, 518, 523; *id.*, *Some Problems of Indian Literature*, Calcutta, 1925, 82–109; Müller, *Mitteilungen zur Geschichte der Medizin und Naturwissenschaften*, 1927, xxvi, 97; Johnson, *J. Roy. Asiatic Soc.*, 1929, 77; Rāy, in Symposium in *Proc. Nat. Acad. Sci. India*, 1952, xviii, 323 f., 341; Lippmann, *Chem. Ztg.*, 1925, 941; *id.*, *Beiträge zur Geschichte*

der Naturwissenschaften und der Technik, 1953, ii, 19. A. P. C. and P. Rây, *History of Chemistry in Ancient and Medieval India*, Calcutta, 1956, 49 f., omit to mention these parts of the *Arthaśāstra* in their account of it.

111. J. F. Gmelin, *Geschichte der Chemie*, Göttingen, 1797, i, 96–7; the book is in the same series of "Geschichte der Künste und Wissenschaften" as that by Hoyer, ref. 33.

112. Nathaniel Brassey Halhed, *A Code of Gentoo Laws, or, Ordinations of the Pundits, from a Persian translation made from the original, written in the Shanscrit language*, 4°, London, 1776, pp. lii, 53–5 ("Gentoo" = Hindu). The book was a notable one in its time, and was republished in 1777 and 1781. Halhed (1751–1830) was a civil officer in the East India Company in Bengal; on his return to England he became M.P. for Lymington. For some years before his death he was insane.

113. Quintin Craufurd, *Sketches Chiefly Relating to the History, Religion, Learning, and Manners of the Hindoos*, 8°, London, 1790, 293–7 (296) (publ. anonymously); enlarged 2 ed., 2 vols., 8°, London, 1792, ii, 54–8 (57) (pref. signed by Craufurd). The *Nouvelle Biographie Générale*, 1855, xii, 382, says Craufurd (1743–1819) was a Scot who went to India, served in the war between England and Spain, became president of the India Company at Manila, amassed a fortune, and then lived on the Continent, dying in Paris. It says the book was translated into French by Montesquiou, 2 vols., Dresden, 1791, and that Craufurd also wrote *Researches concerning the Laws, Theology, Learning, Commerce of Ancient and Modern India*, 2 vols., Paris, 1817. For Indian rockets, see Craufurd, 1790, 294, and fig. on title-page; 1792, ii, 54 f., and fig. on title-page.

114. W. Y. Carman, *A History of Firearms. From Earliest Times to 1914*, 1955, 59.

115. Elliot, ref. 2, 1875, vi, 471.

116. P. C. Rāy, *A History of Hindu Chemistry*, Calcutta, 1902, i, 97; Oppert, ref. 118, 70 f., thinks the passages in Manu refer to firearms.

117. W. F. Sinclair, *Indian Antiquary*, 1878, vii, 231; criticising Babū Ram Dās Sen, *ib.*, 136, who had quoted (not quite accurately) some of the *Śukranīti* (as *Agni Purāṇa*) and is defended by Oppert, ref. 118, 80; Romocki, ref. 6, i, 36.

118. Gustav Oppert, *On the Weapons, Army Organisation, and Political Maxims of the Ancient Hindus, with Special Reference to Gunpowder and Firearms*, Madras and London, 1880, 162 pp. Oppert was professor of Sanskrit in Presidency College, Madras, and was very well informed (he mentions, p. 49, the Milemete MS. discussed in ch. III); references in the text are to this work. See also Feldhaus, *Die Technik der Antike*, 1931, 73 f.; P. K. Gode, "Use of Guns and Gunpowder in India from A.D. 1400 onwards," in *A Volume of Indian and Iranian Studies. Presented to Sir Denison Ross*, ed. S. M. Katre and P. K. Gode, New Indian Antiquary, Bombay, 1939, Extra Series, ii, 117–24; Gode's paper on firearms in India, 1450–1850, in *K. M. Munshi Diamond Jubilee Volume*, Bombay, 1948, i, was not available to me.

119. *Nature*, 1880, xxii, 580.

120. J. Beckmann, *A History of Inventions*, 1846, ii, 505, quoted "a paper read in the French National Institute" by Langles in "the year 1798," who had said it was forbidden in the Vedas [which go back to about 1000 B.C.].

121. Ref. 116, i, 96 f.; he gives no date for the work.

122. Ref. 1, 74 f., 77; ref. 87, 1904, 109; Romocki, ref. 6, i, 36, also rejected Halhed's and Oppert's views. John Davy, *An Account of the Interior of Ceylon, and of its People*, 4°, 1821, 267, 268, was told by the natives of Ceylon that they first learnt of firearms and gunpowder from the Portuguese, and were completely ignorant of both before. The gunpowder was made from 5 parts saltpetre, 1 part sulphur, and 1 part charcoal from the wood of the Parwatta tree, ground together with a little very weak lime-water and the acrid juice of the wild yam. It was dried in the sun and was a coarse powder or impalpable dust, not bad in quality. The saltpetre was purified with wood ashes.

123. Rájendralála Mitra, *Notices of Sanskrit MSS.*, Calcutta, 1880, v, 135.

124. Hime, ref. 87, 107.

125. Pliny, *H.N.*, xxxi, 10 (46).

126. Hoefer, *Histoire de la Chimie*, 1866, i, 148 : "Les cavernes de l'Asie, appelées Colyces, desquelles on retirait jadis des quantités considérables de nitre." See the note on colligas

by K. C. Bailey, *The Elder Pliny's Chapters on Chemical Subjects*, 1929, i, 173 : "the original word is irretrievably lost, the copyist having written down colligas from colligi in the line above." P. C. Rāy, ref. 116, i, 100, who volunteers the usual statement that "it is well known that saltpetre has been in use from time immemorial as the basis of rocket and other fireworks both in China and India," quotes an undated *Dasakumāracarita* by Dandī as mentioning a yogavartikā (magic wick) and yagacurna (magic powder), "of which saltpetre was probably the basis."

127. Hime, ref. 1, 15, 79–80; ref. 87, 108–11. I can find no authority for this statement.

128. Ref. 2, 1875, vi, 481–2. Accounts of the extraction of saltpetre in India are given by R. Watson, ref. 35, i, 313–26; Rāy, ref. 116, i, 100–1 (from *The Travels of John Albert de Mandelslo from Persia into the East Indies*, London, 1668, 66–7); and A. Marshall, *Explosives*, 1917, i, 57–62 (illus.). It involves lixiviation of saltpetre-bearing earth, evaporation, and crystallisation.

129. Hime, ref. 1, 82 f.

130. B. Rathgen, ref. 27, 564; *id.*, ref. 93 (1925), 196.

131. A. Cunningham, *J. Asiatic Soc. Bengal*, 1848, xvii, pt. II, 241 (244).

132. Gibbon, *Decline and Fall of the Roman Empire*, ch. 65; Everyman ed., vi, 324.

133. Ref. 2, 1875, vi, 457 f.

134. Ref. 1, 79; ref. 87, 112.

135. Maclagan, ref. 109, 64.

136. Elliot, ref. 2, 1875, vi, 462; Lane Poole, *Mediæval India*, 1903, 8 (q. al-Bāladhurī, *c.* A.D. 840), mentions only heavy catapults.

137. Elliot, ref. 2, 1875, vi, 219. It is not correct to say the Hindus fled in panic, since they would follow the general's movements.

138. Ref. 132, ch. 57; Everyman ed., vi, 3.

139. Lt.-Col. [later Gen.] J. Briggs, *History of the Rise of the Mahomedan Power in India, till the Year A.D. 1612, translated from the original Persian of Mahomed Kasim Ferishta*, 4 vols., 1829, i, 47.

140. Ref. 2, 1875, vi, 455–6.

141. Ref. 136, 28.

142. Ref. 2, 1875, vi, 455–82 : "Appendix on the Early Use of Gunpowder in India," giving many literature references.

143. Ref. 1, 67.

144. Ref. 139, 1829, ii, 312; Elliot, ref. 2, 1875, vi, 466.

145. See ref. 192.

146. In Elliot, ref. 2, 1872, iv, 268.

147. Hime, ref. 1, 81, q. Favé and Grewenitz.

148. Elliot, ref. 2, 1875, vi, 464.

149. Elliot, ref. 2, 1872, iv, 117, 119, 465–6; Hime, ref. 1, 81–2.

150. G. Phillips, *J. Roy. Asiatic Soc.*, 1895, 523–35 (531–2); Pelliot, *T'oung Pao*, 1933, xxx, 236–452; Duyvendak, *ib.*, 1939, xxxiv, 341.

151. *Histoire de Timur-Bec . . . par Cherefeddin, natif d'Yezd, traduite par le feu M.* [François] *Petis de la Croix*, bk. v, ch. 47; 4 vols., 12°, Paris, 1722, iii, 94; Gibbon, ref. 138; I have not found Voltaire's statement.

152. Ref. 3, 90 f.

153. Elliot, ref. 2, 1871, iii, 146; Firishta is a reliable author but other histories of the period are often based largely on "the fertile genius of the compiler," dates being transposed or fictitious dynasties interpolated; Elliot, ref. 2, 1875, vi, 178; A. Rey, *La Science Orientale avant les Grecs*, Paris, 1930, 407 (Indian Chronology).

154. Elliot, ref. 2, 1871, iii, 75 (d. 1325); 1875, vi, 465 (d. 1315).

155. Elliot, ref. 2, 1875, vi, 473 : "probably a piece of musketry," q. from a modern Indian scholar (pandit).

156. Elliot, ref. 2, 1871, iii, 202.

157. Elliot, ref. 2, 1871, iii, 80–82.

158. Ref. 2, 1875, vi, 465.

159. *Malfūẓāt-i-Tīmūrī* (*The Autobiography of Tīmūr*), tr. into Persian by Abū Ṭālibal-Ḥuṣainī and dedicated to the Emperor Shāh Jahān, who began to reign in A.D. 1628; in Elliot, ref. 2, 1871, iii, 424.   Browne, ref. 96, 1920, 183, doubts authenticity.

160. Elliot, ref. 2, 1871, iii, 430 f., 439.

161. Elliot, ref. 2, 1871, iii, 498; see ref. 151.

162. Elliot, ref. 2, 1875, vi, 467; Hime, ref. 1, 82 (incorrect ref. to Elliot); Gode, ref. 118. A mention of "guns" in Bengal in 1405-6, ref. 150, needs investigation.

163. Silvestre de Sacy, *Notices et extraits des Manuscrits de la Bibliothèque Nationale*, An VII, iv, 412 (420): "car les peuples de l'Inde n'avait eu jusque là ni canon ni autres pièces d'Artillerie"; Lane Poole, ref. 136, 176, gives the date as 1508.

164. In Robert Kerr, *A General History and Collection of Voyages and Travels*, 18 vols., Edinburgh, 1811-24; 1812, vi, 178.

165. Lane Poole, ref. 136, 197.

166. W. Erskine, *A History of India under the First Two Sovreigns of the House of Taimur, Báber and Humáyun*, 2 vols., London, 1854; Elliot, ref. 2, 1872, iv, 218.

167. Elliot, ref. 2, 1872, iv, 276.

168. Elliot, ref. 2, 1872, iv, 274.

169. Elliot, ref. 2, 1872, iv, 285; 1875, vi, 468; Hime, ref. 87, 121, asks if these gunners were friendly Bengali employed by Bābur or hostile Bengali using guns taken from Portuguese deserters?

170. Elliot, ref. 2, 1872, iv, 286.

171. Erskine, ref. 166, i, 436, 470, 486; Elliot, ref. 2, 1872, iv, 251-6; Lane Poole, ref. 136, 200.

172. Elliot, ref. 2, 1872, iv, 131-2; Lane Poole, *Rulers of India. Bábar*, Oxford, 1899, 81, 140 (casting of cannon, feringhia, in India), 161, 178-81; C. C. Davies, *Ency. of Islam*, 1936, iii, 1025; *id.*, *An Historical Atlas of the Indian Peninsula*, Oxford Univ. Press, 1953, 38.

173. Elliot, ref. 2, 1872, iv, 255, 284; 1875, vi, 468.

174. Elliot, ref. 2, 1872, iv, 268, 270 f.; Lane Poole, ref. 172, 161.

175. Elliot, ref. 2, 1872, iv, 279.

176. Elliot, ref. 2, 1875, vi, 468.

177. Lane Poole, ref. 136, 250.

178. Lane Poole, ref. 136, 211, and plate.

179. *Ā'īn-i Akbarī*, ed. Blochmann, 3 vols., Calcutta, 1873-91-94, i, 112 f.

180. Ref. 166, i, 40; Lane Poole, ref. 136, 247, 256, 280, 283; Wiedemann, *Sitzungsber. Phys.-Med. Soc. Erlangen*, 1911, xliii, 89; 1912, xliv, 210.

181. Lane Poole, ref. 136, 256.

182. H. Goetz, "Das Aufkommen der Feuerwaffen in Indien," in *Ostasiatische Zeitschrift*, 1925, ii, 226-9, with a list of sixteenth-century illustrations of guns.

183. Rathgen, ref. 93, 196.

184. Elliot, ref. 2, 1877, vii, 125.

185. Elliot, ref. 2, 1875, vi, 469.

186. Elliot, ref. 2, 1875, vi, 139.

187. Elliot, ref. 2, 1875, vi, 143, where the account from Faizī Sirhindī gives the fortress as Āsīr, held by Bhādur Khān, and the date A.H. 1006; the account in Lane Poole, ref. 136, 254, suggests that it was Chitor, A.D. 1567.

188. *Roteiro da Viagem que em Descobrimento da India Pelo Cabo da Boa Esperança fez Dom Vasco da Gama em 1497*, ed. D. Kopke and A. da Costa Paiva, Oporto, 1838, 57 (BM 1046.g.6); tr. by F. Denis in *Voyageurs Anciens et Modernes*, ed. E. T. Charton, 4 vols., Paris, 1854-7, 1855, iii, 247 (BM 10027.g.2); Hime, ref. 1, 82; ref. 87, 117; Rathgen, ref. 93 (1925), 11 f.

189. Fernaõ Lopes de Castanheda, *Historia do descobrimento e conquista da India pelos Portuguezes*, 4º, Coimbra, 1551 (BM 582.e.34); 8 vols., sm. 4º, Lisbon, 1833, i, 58 (the text of this ed. reproduces that of the early ones, with the contractions); French tr., fº, 1581 (with other works; 8 bks., BM 182.g.6); English tr. in Kerr, ref. 164, 1811, ii, 292-505 (364). To save a reader's time and patience it may be mentioned that the author is to be found in library catalogues under "Lopez." The first ed. (1551) contained only bk. i, the eight

books being published later. Part of a ninth book was published by Wessels, Hague, 1929. Elliot, ref. 2, 1875, vi, 467. Faria y Souza, ref. 192, tr. Stevens, 1695, i, 101, says: "The Nayres who are their Nobles if they chance to touch any of the Commons, cleanse themselves by washing."

190. Ref. 189, 1833, i, 196, 201, 203; tr. Kerr, ref. 164, 1811, ii, 454; Barthema says the two Italians were afterwards pardoned by the Viceroy but were murdered in Calicut; Barthema, in Kerr, ref. 164, 1811, ii, 513; 1812, vii, 128; Rathgen, ref. 93 (1925), 20. Another account says four Venetians also went to Calicut to make artillery: João Manuel Cordeiro, *Apostamentos para la Historia de Artilheria Portugueza*, 1895, q. by Rathgen, ref. 96 (1925), 16.

191. Giovanni Battista Ramusio, *Navigationi et Viaggi*, 3 vols., Venice, 1550–56–59; 3 ed., fº, Venice, 1563, i, 119D–121C: "Navigatione di Vasco da Gama . . . scritta per un gentilhuomo Fiorētino, che si trouò al tornare della dette armata in Lisbona."

192. Faria y Souza, *Asia Portugueza*, 3 pts., fº, Lisbon, 1666–74–75, I, i, chs. 5, 7; I, ii, ch. 6; i, pp. 55, 61, 146, 148; *The Portuguese Asia; Or, The History of the Discovery and Conquest of India by the Portuguese . . . written in Spanish by Manuel de Faria y Sousa . . . Translated . . . by Cap. John Stevens*, 3 vols., 8º, London, 1695, i, 58, 75, 99, 181, 186; Elliot, ref. 2, 1875, vi, 468; Hime, ref. 87, 78, 100, 117 ("undoubtedly cannon balls"); Kerr, ref. 164, 1812, vi, 70.

193. Ref. 2, 1875, vi, 467.

194. Castanheda, in Kerr, ref. 164, 1811, ii, 418; Rathgen, ref. 93 (1925), 18.

195. Rathgen, ref. 93 (1925).

196. Castanheda, in Kerr, ref. 164, 1811, ii, 459, 476–81 (Faria y Souza said the Zamorin's ships had 382 pieces of ordnance).

197. In Ramusio, ref. 191, i, 133B–145A; 136F, 142D,E, 143A.

198. In Kerr, ref. 164, 1811, ii, 472–86, 502.

199. In Ramusio, ref. 191, i, 147B–173A, 168F–169A,B; in Kerr, ref. 164, 1812, vii, 41 ("Verthema"), 122, 128, 134; 1813, ix, 403.

200. Lane Poole, ref. 136, 177.

201. In Ramusio, ref. 191, i, 288A–323F, 294D, 297A,B, 299C, 318D, 319A. For a Turkish attack on Dīu in 1538, see Kerr, ref. 164, 1812, vi, 267 f.

202. Kerr, ref. 164, 1812, vii, 154.

203. Kerr, ref. 164, 1812, vii, 458.

204. Ref. 96 (1925), 21, 200, 216.

205. J. A. Schmeller, "Valentī Fernandez Alemā und seine Sammlung von Nachrichten über die Entdeckungen und Besitzungen der Portugiesen in Afrika und Asien bis zum Jahre 1508" (a Munich MS. which belonged to Peutinger), in *Abhl. K. Bayr. Akad., philos.-philol. Kl.*, Munich, 1847, iv, pt. 3.

206. A. P. C. and P. Rây, ref. 110, 90–103.

207. R. Mallet, *Trans. Roy. Irish Acad.*, 1856, xxiii, 141–436: "On the Physical Conditions involved in the Construction of Artillery, and on some hitherto unexplained Causes of the Destruction of Cannon." Historical notes, 313 f., illustrated, dealing especially with large Indian guns, although some European are included; Rathgen, ref. 93 (1925), 196 f., and plates. Mallet designed the 42-ton, 36-in. calibre, mortar in Woolwich Arsenal (see p. 143), which has never been fired: Carman, ref. 114, 61. Rây, ref. 206, 214, and plates, figs. 36–8.

208. Rathgen, ref. 93 (1925), 208, plate 9.

209. Oppert, *Contributions to the History of Southern India. Part I, Inscriptions*, Madras and London, 1882, 6–7.

210. Rathgen, ref. 93 (1925), 216, plate 11; the shape of the gun is like Bābur's, ref. 182.

211. Rathgen, ref. 93 (1925), 204, q. Pfister, *Monstergeschütze der Vorzeit*, 1870, 28, on the authority of the "English General Bualo" (?).

212. Mallet, ref. 207, 332.

213. Kerr, ref. 164, 1813, ix, 396, 403.

214. Carman, ref. 114, 65.

215. Elliot, ref. 2, 1875, vi, 94, 100, 144, 410.

216. P. K. Gode, "A History of Fireworks in India between A.D. 1400 and 1900," The Indian Institute of Culture, Basavangudi, Bangalore, 1953, *Transaction* no. 17 (26 pp.).

APPENDIX

Arabic Words

The Arabic alphabet has twenty-eight characters. Some are much alike and some are identical except for diacritical points placed above or below them. The vowels, except for "a" and "u," are denoted by separate signs placed above or below the letters and these are nearly always omitted in printed texts. The vowels "e" and "o" do not occur in the modern English transcription form, but they are much used in French. The transcriptions vary in English, French, and German. The last letter of the alphabet is "y" in English but "j" in French and German, the true "j" sound in Arabic then being given as "dj" in French and "dsch" in German. In a French work, al-Dīnwarī appears as Addainûrî; in a German one, Jābir is Dschābir. Some consonants are denoted by two letters, kh, sh, dh, and th, and these are sometimes underlined to show that they are one; "dh" (one letter) is pronounced rather like "th" in "this," and "nb" (two letters) is more like "mb," so that al-anbīq is pronounced more like "alembic," the second short "a" being pronounced more like "e." There are three letters "s," denoted by s, ṣ, and sh (or š), and two letters "t," denoted by " t " and " ṭ." The consonant which is transliterated "q" in this book is sometimes given as "k" and in French especially it is often "c."

The signs ' (hamza) and ' (ain) are two forms of the letter "a," alif and 'ain, although "a" has another form for the first. The first, often called the glottal stop, is pronounced something like the Cockney's "t" in "bottle."

The letter "l" in the definite article "al" is pronounced like n, r, s or t when it comes before one of these letters in the following word: al-Rāzī is pronounced more like er-Rāzī. In transliterating Arabic words I have tried in most cases to be correct, writing "Muḥammad," for example, instead of "Mahomet," but I have retained some common English forms, such as "Mecca" instead of "Makka," and in the section on India some variants appear, and there are also Persian words.

The Arabic calendar begins with the Hijra (incorrectly "flight") of Muḥammad, and its years are denoted by A.H. (orientalists omit these letters). The Arabs use a lunar year and the calculation of a date A.D. from one A.H. may involve an error running into the following year A.D. Since Arabic historians quite often differ by decades for the same event, this is not serious.

Chapter VI

PYROTECHNICS AND FIREARMS IN CHINA

THE popular conceptions of early Chinese civilisation are full of errors. It is believed that it was of very great antiquity and that the Chinese technical achievements preceded by centuries or millennia those of other nations, even the Egyptians. In actual fact[1] the earliest archæological discoveries in China go back only to about 2500 B.C., when no metal is found in the remains. The Bronze Age appeared only about 1500 B.C., iron is known with any certainty only about 500 B.C., and useful iron (tools and weapons) only about 300 B.C.[2] Yet in what purport to be Hsia period (2000?–1520? B.C.) records, cast iron swords are said to have been in use in 1877 B.C.[3] As in the case of other nations, the Chinese in rather late times attributed the inventions of useful arts to mythical rulers, and such stories have sometimes been taken seriously. These so-called Hsia kingdom documents were composed in the Chou period, when metals were known. As in other cases, the accounts of the earliest periods, which were often compiled much later, contain uncritical material. Recent research on the oracle-bone inscriptions has confirmed some details of the information concerning the Shang dynasty (c. 1500–1050 B.C.) given by Ssu-ma Ch'ien (second century B.C.). There is still not much precise detail but it seems probable that more will be known of earlier Chinese history than was thought to be possible not long ago.

The following table will be useful in fixing the dates of the various dynasties.

Chou	1030–221 B.C.
Ch'in	221–207 B.C.
Han	206 B.C.–A.D. 220
	A.D.
San kuo (Three kingdoms)	222–280
Ch'in	265–419
(Liu) Sung	420–479
Ch'i, Liang, Ch'ên and Sui	479–618
T'ang	618–906
Wu tai (Five dynasties)	907–960
Sung	960–1279
Yuan (Mongol)	1260–1368
Ming	1368–1644
Ch'ing (Manchu)	1644–1911

Mongol invasion on the large scale began in 1210 and lasted intermittently till Kublai Khān established himself on the throne at Peking; Hangchow, the Sung capital, was captured in 1276.

The Chinese dynastic histories which are of interest to us are the following, compiled at the approximate dates A.D. given in brackets: *Liao Shih* (1350), *Chin Shih* (1345), *Sung Shih* (1345), *Yuan Shih* (1370) and *Ming Shih* (1739). They are regarded by sinologists as in general accurate and reliable, but they were compiled by literary scholars and the technical or scientific information in them is not so full or so clear as could be wished. Those of the Yuan and Ming, which are special interest to us, have been said[4] to be less comprehensive than some of the others. As in other fields, the authenticity of the texts and the accuracy of translations of them must always concern those who deal with their scientific aspects. Historical details on scientific discoveries or technological inventions may, in general, be expected to be accurate,[5] but (as in all cases) there may be some exaggeration and the technical details may not always have been understood or appreciated by non-specialist authors. In a country like China, where literary attainments were regarded as more important than practical, this must be expected. Most of the earlier (and some of the more recent) European accounts of gunpowder and firearms were taken from secondary Chinese sources, which are usually accurate, and a word must be said first about some of these.

A continuous Chinese history from 400 B.C. to A.D. 960 was attempted in the *T'ung Chien Kang Mu*. This is based on a work, *Tzu Chih T'ung Chien*, by Ssu-ma Kuang (d. 1086), in 294 books, which was reconstructed and condensed to fifty-nine books by Chu Hsi (*c.* 1189) with the aid of pupils, but was not completed till after his death in 1200,[6] and its present form is dated by Wylie at the end of the fifteenth century. A translation from the Tartar-Manchu version (printed in 1708) was made by the Jesuit missionary Father Joseph Anne Maria de Moyriac de Mailla (1679–1748), who reached Macao in 1703 and went to Canton; it was completed in 1737. It was edited and published[7] (1777–83) by the Jesuit Abbé, Jean Baptiste Gabriel Alexandre Grosier (1743–1823), in conjunction with Le Roux des Hauterayes. Grosier added a thirteenth volume, which was issued with a separate title-page and appeared in two editions in the same year.[8] Grosier does not seem to have visited China. He left a manuscript of a revision of the first work, which has not been published, and he finished as director of the library of the Arsenal.

Another translation from Tartar-Manchu texts was published[9] by the Jesuit Antoine Gaubil (1689–Peking, 1759), who was in China from 1723. Gaubil was an excellent linguist and had (what is rare among sinologists, ancient and modern) a really sound knowledge of science, especially astronomy. His pupils founded the Russian school of sinology. It was noticed that Gaubil's translation often differed from that of de Mailla,[10] and the Jesuit accounts of China were roughly handled by the brilliant but superficial Abbé Cornelius de Pauw (1739–99),[11] a Dutchman who studied in Göttingen and became for a brief time "Private Reader" to Frederick the Great at Potsdam. Although some of his criticism is exaggerated, a good deal of it is sound and is in agreement with the results of modern scholarship.

Joseph Amiot (1718–94), a Jesuit missionary to China, compiled from Chinese sources a history of military arts[12] which was reprinted in the seventh volume of a history of China,[13] and extended in the eighth volume, in which de Pauw's criticisms are taken up. Amiot is said[14] to have had a very accurate knowledge of Chinese, but unfortunately he followed uncritically his Chinese sources, which attributed the invention of gunpowder and firearms to Sun-tzu and Wu-tzu (works ascribed to both have been dated fourth century B.C.), and misled Favé[15] and Romocki.[16] Amiot[17] says some descriptions on the plates in his book (xv and xvi, vol. viii) were unsatisfactory, but he had translated them literally. Most of his material is from the *Wu Pei Chih* (1628).

A revised translation of some Chinese accounts of fireworks was published by the Jesuit, Father d'Incarville,[18] who gave the compositions of some firework mixtures. The full text of d'Incarville is given by Reinaud and Favé.[19]

The *Wu Li Hsiao Shih* (8/26af.) of Fang I-Chih, an encyclopædia which has a section on gunpowder, composed about A.D. 1630 (second edition 1664),[20] says the Emperor Yung-Lo (1403–24) established the Shên-chi brigade; horses had previously been disguised as lions having "divine engines" (shên-chi) attached to their sides, which were tubes filled with inflammable materials. A Ming historian made the Shên-chi brigade use spears and guns (ch'iang p'ao), including large guns on carriages, light artillery and fire-locks, firearms, it is said, being known. Fang I-Chih says "gunpowder came from the outer barbarians" (which need not necessarily mean the Europeans), although fireworks, "fire trees and silver trees," were used in the T'ang dynasty (seventh to tenth centuries), and he thought these contained gunpowder. The *T'ing Shih* of Yo K'o (A.D. 1214), quoted by Fang I-Chih, does not connect the p'ao used to throw stones with the use of gunpowder, although Parker seems to have misunderstood Fang I-Chih as saying that guns were used in the seventh century.[21]

According to Mayers,[22] there is no mention of gunpowder in any work before the T'ang dynasty. Chao I in his *Kai Yü Ts'ung K'ao* (sixteen volumes, 1790) said all the early p'ao were engines for throwing stones by machinery. The first "fire-crackers" (p'ao chu) were not gunpowder crackers, but joints of bamboo thrown on a fire; the use of these crackers was perhaps imported with Buddhism in the sixth century A.D. and they were used in China well into the present (twentieth) century for scaring off devils[23]; they were probably as useful as holy water. The noise of exploding bamboo is graphically described by Marco Polo. He says if the young green canes are put on a camp fire "they burn with such a dreadful noise that it can be heard 10 miles at night, and anyone who was not used to it could easily go into a swoon or even die. Hence the ears are stopped with cotton wool and the clothes drawn over the head, and horses are fettered on all four feet and their ears and eyes covered. For it is the most terrible thing in the world to hear for the first time." Yule (and also Lindsay and Hooker, whom he quotes) says the explosions are like musket shots or salvos of artillery.

The *Wu Yuan* of Lo Ch'i (fourteenth to fifteenth century) says that fireworks of gunpowder (yih i huo yao tsa-hsi) were first used by the Sui Emperor Yang Ti (A.D. 603–17), which is unbelievable. Fang I-Chih and the encyclopædia *Ko Chih Ching Yuan*, quoting from the *Yuan Shu Chi* (T'ang dynasty), mention various fireworks (as yen-huo). Chao I says Wei Hsing (d. A.D. 1164) invented projectile carriages (p'ao-ch'ê) launching "fire-stones" a distance of 400 yards, and in making his "fire-drug," saltpetre, sulphur and willow charcoal were employed; "this was the origin of the pyrotechnics in vogue in modern times" (seventeenth century).[24]   The *T'ai Pai Yin Ching* (A.D. 759) of the Taoist Li Ch'üan mentions fire-arrows (huo shih) and incendiaries only.

The *Wu Pei Chih* (Records of War Preparations) of Mao Yuan-I, who had been a military commander, which was printed in 1628 in eighty volumes of 240 chapters,[25] is an exhaustive treatise, profusely illustrated, and utilising earlier works as well as contemporary material.   It was for long nominally prohibited for general use.   It says (ch. 43) that in the Five Dynasties (A.D. 907–60), a period of war and hostile or friendly relations with the adjoining nations of Central Asia, the sovereign of Wu (the ruler of the south-east kingdom with his capital at Hangchow) sent to Apaoki (the sovereign of the Ch'itan Tartars in the north-east and founder of the Liao dynasty) a "furious fiery oil" (mêng huo yu), blazing more fiercely in contact with water and useful for attacking cities.   Apaoki was very pleased with this gift and assembled a small body of picked troops for the purpose of trying it on his neighbours, but was dissuaded from doing so by his queen Shu Li, who told him that he would only make himself ridiculous by trying to destroy foreign kingdoms with an oil.   In this account it is suggested that the foreign oil was not used. According to Mayers this is the first Chinese reference to the military use of naphtha.   The same account appears in the *Ko Chih Ching Yuan* (Mirror of Scientific and Technical Origins) completed in A.D. 1732, which is usually rather free from legendary material.   Chapters 19 to 52 of its 240 chapters deal with military matters from the earliest times to the Mongol dynasty. Romocki's suggestion that the "fiery oil" contained quicklime is very improbable.

The commentary in the *T'ung Chien Kang Mu* dates this episode in A.D. 917, when the coastal city of Hangchow was full of Arab, Indian, and perhaps Byzantine adventurers.[26]   Just before this time the Byzantine Emperor Leo had introduced fire-tubes ($\sigma\iota\phi\tilde{\omega}\nu\omega\nu$), and a knowledge or rumour of them may have come from Constantinople, where the tradition was much older (see ch. I).   The use of naphtha as an incendiary was also known to the Arabs.   The large Chinese junks sailing in the seventh and eighth centuries in the Persian Gulf were equipped with naphtha as a protection against enemy ships and pirates,[27] and Hirth and Rockhill[28] say that the sea-going ships were Arab till the end of the ninth century.

The *Wu Li Hsiao Shih* of Fang I-Chih (c. 1630, second edition, 1664) says that in A.D. 960 fire-arrows (huo chien) were presented to the Emperor, and that the Sung admiral Yü Yün-wên in the battle at Ts'ai-shih (c. A.D. 1161) used "thunderbolt missiles" (p'i li p'ao) which were paper bags filled with quicklime and sulphur, which when thrown on water burst into flames and leapt upwards, the lime diffusing a sharp smoke which blinded men and horses.   This device also appears in the *Ko Chih Ching Yuan*[29] (A.D. 1735).

Early in the twelfth century the Ch'itan Tartars were in competition with the Jurchen (Nü-chen) Tartars closely related to them, who had combined with the Sung of South China to destroy the kingdom of the Liao.   This was done, but the Jurchen went on to conquer much Sung territory and founded the new Chin dynasty in 1115, which threatened the rest of the Sung. In 1161 the Chin had reached the Yang-tse-kiang and attempted to drive the Sung rear-guard over the river. The Sung admiral, Yü Yün-wên, however, destroyed the Tartar fleet, carrying thousands of men and horses, by the device stated. This device was the "automatic fire" of Julius Africanus (see p. 8), known centuries before in the West. A rumour of it probably reached the Chinese through the Arabs, and it was not a Chinese invention. The *Shui Ching Chu* (*c.* A.D. 500) by Li Tao-Yuan refers to an *I Wu Chih* (there are eight works of this name, the earliest, by Yang Fu, second century A.D.) for the statement that "if water is used to wet [the stones] they are then hot," which may refer to quicklime, but the heat, it is said, can be used for cooking, and there is no mention of incendiary properties.[30] Paper bags of lime and sulphur would not, of course, kill thousands of men and horses.

The *Shou Ch'êng Lu* by Ch'en Kuei, presented A.D. 1170–92, describes the defence of the city Tê-an (near Hankow) by T'ang Tao, in A.D. 1127–32. It says that more than twenty fire-lances (huo ch'iang) were first made then.[31] This is the name given to a weapon which is described in 1233 (see p. 245), and since a kind of gunpowder is described in the *Wu Ching Tsung Yao* in 1044, it seems reasonable to assume that the weapon of *c.* 1130 would be similar to that of 1233.

Undeterred by the unfortunate results of their alliance with the Jurchen, the Sung proceeded to call in the aid of a much more formidable race, the Mongols. These were prepared to assist the Sung in conquering the Chin, after which they intended to swallow the Sung also, and so rule the whole of China. This, however, was not so simple a task as their conquest of Western Asia and Russia. The Chin defended their strong-points stubbornly with all the appliances of war which were known to them. For two years they held out against the Mongols, who had now to develop siege warfare, and finally had the task of capturing two fortresses, one of which was the capital, Pien-king, later known as Kai-fêng-fu.

A translation from the *T'ung Chien Kang Mu* (see p. 258) was made by Stanislas Julien for Reinaud and Favé,[32] from which the following passages are taken:

I. (1232) Deuxième lune. Les Mongols attaquèrent Pien ou Pien-king. Dans la troisième lune, les Mongols assiégèrent Lo-yang; Kiang-chin combattit vaillamment l'ennemie et le repoussat.

Commentaire historique:

Les Mongols établirent des pao pour attaquer Lo-yang . . . Quand les munitions furent épuisées, les assiégés firent des pointes de flèches avec des deniers de cuivre. Dès qu'on avait ramassé une flèche des soldats mongols, on la coupait en quatre morceaux, et on lançait les tronçons à l'aide d'une canne-tube.

Kiang-chin inventa en outre un pao appelé o'pao, c'est-à-dire pao qui arrête . . . (qui lançait) de grosses pierres à plus de cent pas . . . les assiégeants . . . après trois mois d'efforts inutiles, . . . se retirèrent.

There is no mention here of anything beyond arrows and a mangonel throwing large stones.   The short arrows shot from a "canne-tube" were probably discharged from the Chinese cross-bow, consisting of a barrel with a single string which worked in two slits cut in the side of the barrel.   This weapon was used in England under the name of "slur-bow."[33]

II. *Siège de Pien-king.*

Dans le palais . . . on préparait les pierres des pao. . . . Le pao nommé Tsouan-tchou, c'est-à-dire composé de bambous réunis, avait jusqu'à treize angles.   A chaque coin des murailles, les Mongols placèrent une centaine de pao, qui tiraient alternativement, et ne se reposaient ni jour ni nuit.   Au bout de quelques jours, les pierres se trouvèrent de niveau avec les remparts.   Les tours et les guérites placées au haut des murs, et qu'on avait construites avec les plus gross poutres des anciens palais, tombaient en pièces dès qu'elles étaient atteintes.

Les assiégés les recouvraient de fiente de cheval et de paille de blé; de plus, ils les protégeaient avec du feutre et des nattes fortement liées au moyen d'un réseau de grosses cordes.   La partie extérieure des auvents avait été recouverte de peaux de bœuf.   Mais à peine ils étaient atteints par les ho-pao (huó-p'au) ou pao à feu des Mongols, qu'ils s'enflammaient sans qu'il fut possible d'arrêter l'incendie.

The same account is given in the *Wu Pei Chih*.[34]

III.   Alors les assiégés lancèrent un oiseau de papier, sur lequel ils avaient tracé des caractères. . . . Les personnes qui voyaient cela, disaient : Si le général veut repousser l'ennemie à l'aide d'un oiseau, ou d'une lanterne de papier, il aura de la peine.

Paper birds (kites) were used in warfare by the Chinese in much later times, and caused similar amusement to the other side; no doubt they were inscribed with threats or magic characters.   Reinaud and Favé say that kites were flown by the Chinese against the English invasion of the China coasts.[35]

IV.   A cette époque, on faisait usage de ho-pao ou pao à feu, appelés Tchin-tien-louï, ou tonnere qui ébranle le ciel.   On se servait pour cela d'un pot en fer que l'on remplissait de yo.   A peine y avait-on mis le feu, que le pao s'élevait, et que le feu éclatais de toute part.   Son bruit ressemblait à celui du tonnerre, et s'entendait à plus de cent lis [33 miles]; il pouvait répandre l'incendie sur une surface de plus d'un demi-arpent [half a mou, a twelfth of an acre, 400 square yards, or 20 by 20 yards]. Ce feu perçait même les cuirasses de fer auxquelles il s'attachait.

V.   Les Mongols construisirent avec des peaux de bœuf un couloir qui leur permit d'arriver jusqu'au pied des remparts.   Ils se mirent à saper les murs, et y pratiquèrent des cavités, ou l'on pouvait se loger sans avoir rien à craindre des hommes placés en haut. Un des assiégés proposa de suspendre à des chaînes de fer des pao à feu, et de les descendre le long du mur. Arrivés aux endroits qui étaient minés, les pao éclataient et mettaient en pièces les ennemies et les peaux de bœuf, au point même de ne pas en laisser de vestige.

VI. De plus, les assiégés avaient à leur dispositions des flèches à feu volant (Feï-ho-tsiang). On attachait à la flèche une matière susceptible de prendre feu; la flèche partait subitement en ligne droit, et répandait l'incendie sur une largeur de dix pas. Personne n'osait en approcher. Les pao à feu et les flèches à feu volant étaient très redoutés des Mongols.

The protective passage (couloir) was a well-known military device for enabling engineer troops to reach a wall for the purpose of making an excavation or sap. The work on the wall would probably not have been begun until the men were protected from stones, incendiaries, or arrows from the wall above, and we may be reasonably sure that there was no gap left at the end through which such objects could pass. If they passed at all, they must have penetrated the very strong covering.

De Mailla[32] had translated fei huo ch'iang by "javelot of flying fire" or fire-lance, instead of "flèche à feu" or incendiary arrow; ch'iang is a spear. In spite of the use of "thunder shaking the sky" against them, the Mongols captured the city. Since the exact translation of V is important, Dr. Needham very kindly made one for me from the *Chin Shih* written by T'o-T'o and Ouyang Hsüan in A.D. 1345.

Chin Shih 113/19a.

"The Mongol soldiers again made approaches covered with oxhide (niu p'i tung) (not clear whether just mobile mantlets or trench covered with oxhide roof), so as to reach the foot of the wall and dug as it were niches (k'an) large enough to contain a man; so that the defenders on the city wall could not do anything about it. Thereupon someone suggested taking an iron chain (t'ieh shêng) and letting down on it a chen t'ien lei ('heaven-shaking thunder' bomb?). When this reached the place where they were digging, fire burst forth, and the men and the oxhides were all broken to fragments (chieh sui) flying in all directions (pêng) so that no trace was left (wu chi)."

The passage continues with the "flying fire spears" (fire-lances).

The word pao meant a bundle; p'ao at first meant a catapult or trebuchet, *or* the missile thrown by it; huo p'ao meant either a catapult throwing incendiaries, *or* the incendiary object, which later could include some sort of packet or bomb containing an incendiary or explosive composition (e.g. gunpowder); in the sixteenth century huo p'ao meant a cannon. The word yao (yo in the French translation) means "drug" or "chemical"; huo yao (which first occurs about A.D. 1044, see p. 268) is "fire drug" and in later times meant gunpowder.

In IV the "thunder shaking the sky" (chen t'ien lei) is said to be an iron pot filled with a drug (yao). When fire was put to this the following effects ensued : (i) it rose, (ii) it produced a noise like thunder which could be heard 33 miles, (iii) fire burst out of every part, (iv) an incendiary effect extending over 20 by 20 yards was produced, (v) this incendiary could pierce iron armour when it attached itself to it.

Effect (i) suggests a rocket or bomb, but since the object is an iron pot it must be a bomb. Effect (ii) suggests a powerful detonation, heard 33 miles away. This must be an exaggeration and to make sense it will have to be omitted. The largest modern shells filled with high explosive are not heard like thunder when they burst 33 miles away. The meaning of (iii) has been variously given by military experts. Colonel Favé[36] thought it meant jets of fire issuing from holes in an iron pot containing an incendiary. Romocki[16] suggested that the mixture in the pot contained saltpetre and incendiaries like naphtha and resins. This exploded and burst the pot, the fragments of which were scattered, together with unburnt incendiary which then took fire. The incendiary effect extended over a small area only and pierced armour (probably by penetration through chinks in it, as happened with Greek fire, see p. 26). If the exaggerated language is cut away it is possible to grant that some sort of explosive bomb was concerned.

The effects described in V are difficult to understand. A pot lowered on a chain and not exploding could not penetrate oxhides. Romocki thought sui, translated "éclataient" by Julien, meant that an explosive bomb was used (it really means "fragments"), but he was misled by thinking that a seventeenth century bomb described by Amiot (see p. 255) was like one used in 1232. Berthelot[37] and Hime[38] agreed with Favé that an incendiary was used, Hime saying that there was no "explosion" since it is said that the event "left no vestige," and a true explosion would scatter about fragments of oxhides and bodies but not destroy them, which he thought a powerful incendiary would do. De Mailla[32] had translated : "a machine took fire and destroyed the Mongol sappers without leaving a trace."

The Chin Shih reports an event of 1277 in which, as will be seen later (p. 250), it is said that a "fire-p'ao" made a noise like a clap of thunder and 250 men "disappeared without trace." Since this effect is a military impossibility we may suspect that the words "without leaving a trace" should be omitted here also, in which case the explosion of a bomb would be possible. If they are retained, the effect must either have been incendiary or imaginary.

The name huo p'ao used in IV for the chen t'ien lei meant in later times a "cannon," and although Gaubil[9] in 1739 had avoided the mistake of translating huo p'ao in early accounts as "cannon," this has often been done in recent times. Martin[39] said that "cannon" were used against the Mongols in 1232, whilst their first incontestable use in Europe is in 1338 (see, however, ch. III), and that various small arms were used in the Sung dynasty. Goodrich[40] asserts that "in the battle of the Jurchen against the Mongols under Subutai at K'aifêng in 1232-3, bombs or hand-grenades were used against the Mongols," and adds that "cannon" were used by engineers hired by the Mongols in Java in 1293. Davis[41] had correctly said a century before this that real cannon were unknown anywhere at this time, and Reinaud and Favé[32] "did not fear to state" that in 1232 neither the Chinese nor the Mongols were using cannon, and that "they were ignorant of the art of throwing projectiles by the force of gunpowder."

Mayers[42] thought that the chen t'ien lei was perhaps made from two hemispherical iron cups, filled with an incendiary composition and fastened together, like Valturio's bomb (p. 164). The *Ko Chih Ching Yuan* (1735), copying from the *Pai Pien* (1581), says[43] that on the walls of Si-an there was kept from older times an iron p'ao called a chen t'ien lei made of two iron cups, with an opening above, the width of a finger. This form was long out of use but the Chin used it in defending K'ai-fêng-fu (i.e. the event of 1232). From the improbable range of the sound of 33 miles in IV, it was suggested that a rocket[44] or an incendiary arrow[10] (shown in Amiot's figures and called "Chinese arrows" by the Arabs) was meant. Martin[45] quoted the *Ko Chih Ching Yuan* as saying that T'ang Fu in A.D. 998 [1000] introduced "a new kind of rocket" with an iron head; but he quotes other Chinese texts as saying that rockets were used even in the Chou period (1030–221 B.C.), which is impossible.

Průšek[46] gives from the *Chin Shih* (111/8b) compiled in 1345, an account of a river battle in 1232 in which the defeated Chin vessels, armed with huo p'ao called chen t'ien lei, set upon a barricade of Mongol ships:

"Ils les déchargèrent encore une fois et le feu des p'ao brilla vivement. La bravoure des soldats sur les vaisseaux du Nord [the Mongols] fut bientôt sapée, la chaîne des vaisseaux s'ouvrit et les fugitifs purent arriver à T'ung Kuan."

The *Chin Shih* (116/12a and b) says[46] that in 1233 fire-lances (huo ch'iang) were made, which were used on small boats and also by foot-soldiers. In the first case they created great disorder among the Mongols and more than 3500 were drowned.

"The method of making these lances was as follows. The tube (t'ung) was more than 2 ft. long, made of sixteen layers of yellow paper (lai) and was filled with a mixture of willow charcoal, iron powder (t'ieh tzu), powdered porcelain, sulphur, arsenic and other things. At the end of the tube was a cord. Every soldier carried some fire in an iron vessel. When they faced the enemy the soldiers kindled them and afterwards flame shot forwards more than 6 ft. (or 10 ft.). The drug (yao) issued from it but the tube was not damaged. They were used at the time of the siege of Pien and were then used a second time."

Although saltpetre is not specifically mentioned as a component of the mixture, this could not have behaved as described unless it contained it, and with this assumption the account is reasonable.

A Chinese account of an event in A.D. 1259 as translated by Julien[47] reads :

Dans la première année de la période Khai-King [1259] on fabriqua une arme appelée tho-ho-tsiang, c'est a dire, lance à feu impétueux. On introduisait un nid de grains dans un long tube de bambou auquel on mettait le feu. Il en sortait une flamme violente, et ensuite le nid de grains était lancé avec un bruit semblable à celui d'un pao, qui s'entendait à une distance d'environ cent cinquante pas.

Dr. Needham has given me a translation (from the *Sung Shih* 197/15*b*, A.D. 1345) as follows :

Inventions at the arsenal of Shou-Ch'un Fu.

They also made a t'u huo ch'iang (impetuous fire-lance) using a huge bamboo tube as the barrel (t'ung), and inside they put a nest of pellets* (tzu k'o). When ignited a violent blazing flame (yen) came forth and as this was ending (all) the pellets were shot out like trebuchet projectiles (p'ao). The noise could be heard for more than 150 paces (i.e. × 5 ft. = 750 ft. = 250 yd.).

The fire-lance (huo ch'iang) mentioned in *c.* 1130 (p. 264) and described in 1233 as made from paper and having a range of 6 or 10 ft., has in 1259 been constructed of bamboo and the range extended to 250 yards. A modification of this weapon is given in the *Wu Pei Chih* (1628),[48] which incorporates some older material (see Fig. 12, p. 255). A stout bamboo is bored through all the joints except the last. The base is strengthened by a wad of clay above which is a round plate of metal. Above this is a hole in the bamboo to serve as a touch-hole, with 4 or 5 in. of slow-burning match. The powder is put in, then, in an order which is not very clear, a collection of missiles : a lead bullet, fragments of cast iron, and a stone bullet or instead of this an iron coin 0·1 in. thick. At the base of the bamboo tube a wooden handle was inserted, shaped so as to be convenient for carrying, and the whole was tightly bound round with twine (to strengthen the bamboo). The tube was supported by a crotch formed by two stakes 3 ft. long tied together, with a large stone to hold the handle in place and prevent it moving back. The wood cover (inside?) is sealed with oiled paper or persimmon varnish. On firing, the collection of missiles is said to be thrown 700 or 800 pu, which is about 1500 yards, and the effects are alleged to be most terrific and destructive.

Since Mayers could not understand how a range of nearly a mile could be achieved with a bamboo tube (it is reached only by a modern rifle charged with cordite), he erroneously supposed that the apparatus was a rocket. There is little doubt that the distance given in the *Wu Pei Chih* is exaggerated (if it is not, the description does not make sense), and that the 250 yards in the older account is nearer the truth, although even this would seem to be too large.

The contrivance of 1259 was, I think, a large Roman candle projecting pellets by some composition analogous to gunpowder (which was then known in Europe). Lippmann[49] suggested that this "Feuerlanze" threw blazing

---

* Not said to be metal but rather balls of explosive material.

balls ("Brandsatzklümpchen"), in a way similar to the appliance devised by Brock,[50] which projected in succession, with pauses between each, balls of an incendiary containing saltpetre. It is more likely, since the *Wu Pei Chih* account is probably an old one, that the projectiles were pellets.

The next literary record which will be considered is the description of the siege of Hsiang-yang by the Mongols in 1273.

Chingiz Khān (1155–1227) was a Tatar (Tartar) chief with his capital at Qaraqorum in Mongolia; after his death he was given the Chinese imperial title T'ai Tsu (Great Ancestor). His grandsons were Hūlāgū, who ruled in Persia, and Kublai Khān (the Chinese Hou-pi-li, or Shih Tsu, the Great Khan), who started the Mongol Yuan dynasty in China in 1280. The name Mongol (or its Arabic form Mughal) appeared only about 1250. The Persian physician and historian Rashīd al-dīn (1247–1318) reported[51] that in Tartary, north of the Great Wall, there was a people renowned in the art of military fires (des feux de guerre), perhaps because there were petroleum wells in the district. In his expedition of 1254–5, says Rashīd al-dīn, Hūlāgū sent to Khitai (North China) for a thousand families of men skilled in the use of manjanīq (stone-throwing machines) and arbalests, and in throwing naphtha. In 1258, naphtha pots and incendiary arrows were used by the Mongols at Baghdād. In his account of Hūlāgū's campaign in 1260, Rashīd al-dīn does not refer to explosives.[52]

Hūlāgū was interested in science and built an observatory at Marāgha, where Chinese and Persian astronomers worked; although Rashīd al-dīn (who could have found other uses for the money) tartly says Hūlāgū established an alchemical laboratory in which a band of swindlers helped to exhaust his revenues,[53] this may well have been a chemical laboratory in which useful experiments were made, and it would be reasonable to suppose that these covered explosives.

Kublai began his conquest of China in 1268 and the walls of Hsiang-yang were finally breached in 1273. There are four different accounts of the siege.

I. The brothers Niccolò Polo (Marco's father) and Maffeo Polo lived with Marco in Shangtu ("Xanadu") for seventeen years from about 1275, and were in high favour with Kublai Khān; Marco was perhaps for a time governor of Yangchow. His book was partly written, in bad French, in a prison in Genoa in 1298,[55] and was put together by Rustichello of Pisa. It exists in different old versions in French, Italian and Latin. Three of the oldest French manuscripts (called A, B and C), in the Bibliothèque Nationale, were used by Pauthier in his edition.[56] Since Marco Polo says[57] that he and his father and uncle, in their journey to the Mongolian court, heard at Layassa of the election of the new Pope (Gregory X, which occurred on 1st September, 1271) and that their journey lasted three and a half years, they could not have arrived in China until 1275, two years after the siege of Hsiang Yang, as Pauthier pointed out. Nevertheless, Rustichello was a contemporary of Marco, and although he may have made some additions he could hardly

have invented the whole account, which must be based on something which
he was told by Marco. The account has some value, because it agrees with
the others which are given below. We might expect that if Marco Polo
had been told of the use of pyrotechnic devices he would have mentioned
them. At that time gunpowder was known in Europe. The circumstance
that Marco Polo was not present at the siege of Hsiang-yang is not a good
reason for disregarding his account.

In Pauthier's MS. A, the account of the siege of Hsiang-yang is missing.
In MS. B it is said that Niccolò and Maffeo Polo set out in 1260 and reached
the court of Kublai Khān at Khānbaliq, on the site of the present Peiping.
Marco Polo calls Siang-yang "Sa-yan-fu," and says that it had held out in
a siege for three years, the Khān feeling hurt that this was the case. Niccolò
and Maffeo, hearing of this, presented themselves to the Khān saying (ch. cxlv)
that they could make engines which would bring about the surrender of the
city. They made large mangonels of timber which threw large stones, breaking
down the walls. In MS. C, Marco was also present, and he and his father
and uncle were joined by a Nestorian Christian and a German (alemant de
Alemaigne); they made engines throwing stones of 300 lb. weight more than
60 paces (plus de .lx. routes), which made a noise like a great tempest (et
firent moult remour et grant tempest).

Although the Polos were not present at the siege, Marco's account, apart
from unessential details, agrees with the other three.

II. Rashīd al-dīn[58] says that Kublai applied to the Persian court and
asked that "an engineer, who had come back Baalbek[59] and Damascus,
knowing of a large machine (kumga) not previously used in China, should
be sent to him. The sons of this engineer, Abū Bakr, Ibrāhīm and Muḥammad,
together with men who accompanied them, built seven large machines and
attacked the besieged town. The side facing the Mongols was protected by
a strong citadel, a rampart and a deep moat. The stones from the engines
shook and then brought down the towers, whereupon the city surrendered."

Rathgen[60] says the machines were trebuchets, using a sling operated by
a large counterpoise-weight, which he thought were of Italian or German
origin, although they may have been invented independently in China. No
old machines of this type remain, although there was one in Basel in 1800
and one found later in a church in East Prussia was burnt as fire-wood.
Napoleon III had one constructed and tested in Vincennes.[61]

III. The Tartar account of the siege of Hsiang-yang[62] says that a Mongol
general proposed to Kublai in 1271 that engineers from the West who
understood a machine (chi) for throwing stones of 150 lb. weight, capable
of making holes of 7 or 8 ft. in the thickest walls, should be sent for. This
was done, and the machines they constructed in China were used with success.

IV. The Chinese account of the siege refers to Sayan-fu and Fan-ching,
which were two parts of the city of Hsiang-yang Fu separated by the river.
They were joined by a bridge of boats. The translation, made by Stanislas

Julien from the continuation of the *T'ung Chien Kang Mu* (it is in the *Yuan Shih*, A.D. 1370) says[63] that in 1272 a fleet of 100 Chinese ships attempted to force the Han river, each ship being armed with incendiary arrows, fire p'ao, burning coals, large axes and strong bows.   In 1273 :

> A-li-haï-ya, ayant obtenu communication du procédé des nouveaux p'ao[a], présenté par un homme du Si-yu[b], attaqua de nouveau la ville de Fan, et détruisait ses murailles. . . . Liu-wen-huan ayant fait connaître au gouvernement la situation critique de Siang-yang, Hia-sse-tao demanda au prince de marcher vers la frontière; mais, en secret, il engagea les membres du conseil à présenter des suppliques pour qu'on l'obligeât à rester. . . . Peu de temps après, A-li-haï-ya tourna contre Siang-yang les p'ao et autres instruments de guerre qui avaient servi à soumettre la ville de Fan.   Un projectile, lancé par ces p'ao, vint frapper la tour de la ville où était la cloche qui marque les veilles, et produisait un bruit semblable au tonnerre.   Tout la ville fut en émoi, et un grand nombre de généraux escaladèrent les murs pour se rendre. . . . Liu-wen-houan sortit aussi, et, faisant sa soumission, remit au général mongol les clefs de la ville.

[a] This has the stone radical (shih) and this has been supposed to mean that an engine for throwing stones mechanically is meant.
[b] The countries of the West.

The translation by Pauthier[64] says that the chief Mongol general A-li-haï-ya directed all his efforts against Hsiang-yang.   Inside his p'ao so much noise was made by the friction of the wooden parts (ts'ai liao) that it sounded like thunder.   All the generals thought, by the great fear it produced in all parts, that the city would be obliged to surrender.   A-li-haï-ya advanced to the walls and sent the Chinese general, Liu Wên-Huan, a letter from the Mongol sovereign asking him to surrender, which he did.

The Chinese, Tartar, Persian and European accounts of the siege of Hsiang-yang in 1273 all fail to mention the use of any kind of cannon or high explosives.   The fire p'ao used in the river attack in 1272 by the Chinese may have been some device similar to that used in 1232 at Pien-ching, or that used in 1259, or sulphur-lime missiles used in 1161, or (probably) naphtha pots.

The *Yuan Shih*[46] reports that in 1271 large p'ao were made for Kublai, which were used in capturing towns.   In 1274 a Mongol army attacked Sha-yang with "fire p'ao" which set fire to huts inside the town and destroyed them completely.   These "fire p'ao" also demolished the ramparts of Yang-lo.   They were probably engines throwing incendiaries.

Gibbon,[65] in the face of a statement by Gaubil, an eminent sinologist and scientist,[9] that "the use of gunpowder in cannon and bombs appears as a familiar practice" in the siege of Hsiang-yang in 1273, rejected it.   There is more, however, to come.

The first European to explore the Mongol empire, Giovanni del Pian del Càrpine (Joannes de Plano Carpini), who was in Qaraqorum in 1245, says in his *Historia Mongolorum* that the Mongols used a Greek fire containing human fat, which could be extinguished only by wine or beer; and the Persian chronicler Vassaf (al-Waṣṣāf; fl. A.D. 1303–28), a protégé of Rashīd al-dīn, reported that the Persian Khān Uljāi'tū in Syria collected in 1313 an army to fight the Egyptians with stone-throwing machines, armour from Europe, archers from Baghdād, bottles of naphtha, and pyrotechnists from China.[66] Neither Carpini nor Vassaf says anything of the use by the Mongols of explosives or cannon.

After defeating the Kipchak Turks (Cumans), Bulgars and Russians, the Mongol army under Subutai took Cracow and Breslau, and on 9th April, 1241, defeated a German army under Duke Henry of Silesia at Liegnitz. The Mongols under Batu defeated the Hungarians under King Béla IV at Mohi on the Sajó on 11th April, 1241. Prawdin[67] says:

> "there was suddenly raised (so it is reported) 'a bearded human head of hideous aspect, mounted upon a long lance. This sent forth evil-smelling vapours and smoke, which threw Duke Henry's army into confusion, and hid the Tartars from their eyes.' . . . If this . . . was the first gas-attack upon European soil, it has priority over the use of gunpowder, which the Mongols used two days later in the battle beside the Sajo."

Goodrich[68] has improved Prawdin's statement by making it refer to "the first use of cannon in Europe," although it says nothing of cannon. Gibbon[69] had already pointed out that there is no mention in Thwrocz's account of the use of gunpowder by either side. Romocki[70] mentioned that a "flying dragon" (see p. 149) was used at Liegnitz in April, 1241, by the Mongols, and the chroniclers speak of a "feuerspeienden Kopf."

The Polish historian Jan Długosz (1415–80), in his chronicle,[71] says that in the battle of Wahlstatt, near Liegnitz, on 9th April, 1241, the Mongol army had an immense standard on which the sign of a cross was painted, and on the top a picture of a terrifying black head with a bearded chin. As the Mongols were retreating the standard-bearer shook it as violently as he could, whereupon a vapour, smoke and stinking fog came from it (vapor, fumus & nebula tam fœtidissimè exhalauit), which covered the whole Polish army; the terrible and unbearable stench overcame the Poles, choked them, and rendered them almost lifeless and unable to fight, their resistance being completely destroyed.

The *Sung Shih* (451/6a, b) reports[46] that in 1277 Lou Ch'ien-Hsia ordered his men to bring up a huo p'ao. "He lit it and a clap of thunder was heard, the walls crumbled, and smoke covered the sky. Many soldiers outside (en dehors) died of fright. When the fire went out, they went inside and failed to find even the ashes of the 250 defenders; they had disappeared without trace." This is obviously an exaggeration (the effect is impossible) but the small core of fact in the account suggests that a chen t'ien lei like that of 1232 (see p. 244) is concerned.

Pauthier[72] quotes "les Annales chinoises . . . Développement" as reporting that Nayan (Nan-yen), the younger brother of Chingiz Khān, revolted against the Emperor Kublai Khān in 1287, who sent ten soldiers under Li T'ing-Yu by night into Nayan's camp, carrying firearms (pao huo p'ao), the detonations of which produced so much confusion that the enemy ranks fled in all directions. A similar thing had happened in an expedition of Kublai Khān against Japan in 1283–4.[73] What these pao huo p'ao were, whether crackers, rockets, or small firearms, cannot be decided.

An extract from the *Yuan Shih* (A.D. 1370) translated by Pauthier[74] specifies the (moderate) taxes on gold, silver, gems, copper, iron, mercury, cinnabar, certain precious stones, lead (yuan), tin (hsi), alum (fan), "nitre" (hsiao), "salpêtre" (kân, perhaps chien, soda), bamboo, wood, "and other similar materials which the sky and the earth produce naturally."

AMIOT'S ACCOUNT

It will perhaps be interesting at this point to consider the information translated by Amiot[13] on Chinese pyrotechnic devices and different kinds of gunpowder, "known long before its use in Europe." In his translation of the work on *Military Arts* "composed before the Christian era,"[12] which was reprinted in 1782 in the seventh volume of the *Mémoires* (1782, vii, pp. 372 f., 381, plates xix–xxi), Amiot shows and describes matchlocks, muskets, mountain guns, cannon, mortars, a "horse-foot gun" (ma ti p'ao), and a "hundred-ball gun," some of which, old and rusty, he had seen. Of one he says: "cette construction me paroît Européene." In the eighth volume (1782, viii, pp. 326–41, 360 f.) he quotes a Chinese author to the effect that after the beginning of the Christian era, when firearms were introduced by K'ung Ming, they were not used again until the end of the Ming dynasty (*c.* 1600), when cannon were used against Japanese invaders. K'ung Ming is a name for Chuko Liang (Chuko K'ung-Ming), the most famous general of the Warring States period (San Kuo, A.D. 181–234), who is constantly said in Chinese accounts to have invented gunpowder (as Alexander the Great was said to be its inventor by some Arabic authors).

Soon after the use of cannon at the end of the Ming dynasty, continues Amiot, "l'on adopta la manière des Européens; l'on fit de gros & de petits canons, & aujourd'hui notre artillerie est complette." The use of gunpowder and firearms "dès le commencement de l'Ere chrétienne" is, says Amiot, "un fait attesté par tous les Historiens," and those who disbelieve his word must believe theirs. He says:

> "I have expatiated at length on the article on gunpowder only with the design of proving that it was known to the ancient Chinese, since all that I have just reported is taken from their most ancient and authentic books."

In treating of a subject outside his province, he had opened out a field large enough for his critics (including de Pauw[16]) to exert all their efforts, "mais leurs traits seront déjà emoussés quand ils parirendront jusqu'à moi." The Jesuits no doubt felt that the sceptics, who could not read Chinese, had been effectively silenced, since they could not know what the "ancient" books "composed before the Christian era," were. Most of Amiot's material seems to have been taken from the *Wu Pei Chih*, composed about A.D. 1628, but he may have obtained the idea of a pre-Christian date for gunpowder from the *Huo Lung Ching*, dating from A.D. 1411 with some rather earlier parts, certain parts of which are ascribed on their title-pages to Chuko Liang. He may also have seen the *Wu Ching Tsung Yao* (A.D. 1044). It is not suggested that Amiot or any other Jesuit deliberately misled his readers. Chinese textual criticism was then in a rudimentary state. If Amiot had said that the information he found in Chinese works that gunpowder and cannon were known in China in the early centuries of the Christian era, or even before, did not agree with what is known of their use elsewhere, he might have been told that he was arguing in a circle, and that this is no reason for rejecting their use in China.

Amiot followed his Chinese sources too uncritically, and his treatment of de Pauw (who happens to have been right) was too confident even for his time. Amiot had a good knowledge of Chinese[14] and his translations are generally very accurate; some chemical terms he left untranslated. Parts of his text (*Mémoires*, vol. viii) are given here.

"On the subject of the earth-thunder (ti lei) used with success by K'ung-Ming (Chuko Liang) about the second [really the third] century A.D., I observe that the authors who speak of K'ung-Ming do not make him the inventor of this manner of destroying the enemy. They say, on the contrary, that he had drawn upon the works of the ancient militarists, which is a proof without reply that the Chinese knew of gunpowder and its use in warfare long before this knowledge reached Europe. The fêng wo (ruche d'abeilles), another weapon no less murderous than the ti lei and dating from the same time, is a confirmation of this. I pass over in silence what they call the devouring fire (huo yao), the box of fire (huo tûng) and tube of fire (huo t'ung), which come to the same thing, as well as the t'ien huo ch'iu, i.e. the globe containing the fire of heaven. The effects attributed to this fire of heaven recall the idea commonly entertained on the ancient Greek fire. It was used in the Chinese armies in the time of Sun Tzu and Wu Tzu, and even centuries before them, i.e. several centuries B.C. But, says the author I am following, since it is always as dangerous to those employing it as it is to those against whom it is employed, its use has been discontinued."

"The ancient Chinese used powder (chen huo yao) either in combats or to set fire to the camp of the enemy. It is not said how. The powder which goes against the wind, called for this reason ni fêng yao, is one of those which has the greatest force. This powder has a virtue which, it seems to me, could be of great use in our armies, viz. that the smoke goes contrary to the wind."

"The Chinese make their ordinary powder in several ways with the materials we use, viz. saltpetre, sulphur and charcoal. To 3 parts of saltpetre they add 1 of sulphur and 1 of charcoal, or to 4 parts of saltpetre they add 1 of sulphur and 1 of charcoal, and obtain the best powder for artifices of all kinds. The different drugs added to this first composition produce in it different effects. For example, to produce a powder which inflames with a red fire they add t'ao hua p'i. If the five colours are wanted at the same time, instead of t'ao hua p'i, ma nao p'i is added. If white fire is wanted, ch'ao nao (i.e. camphor) is added. A strong explosion requires chen sha, and mercury is also added. A black fire requires hei chiao p'i. For globes of fire, kan ch'i (dry varnish) is added. For fire which goes against the wind, porpoise fat, chiang, and powdered burnt porpoise bones are added. For a powder with the quickest effects, powdered wolf's dung and powdered pan mao are needed. To make a powder producing much smoke, to a pound of saltpetre are added four ounces of sulphur which has been boiled in human urine, three ounces of charcoal, an ounce of ch'ang nao (deer's brain), a tenth of an ounce of ch'ing fan, four ounces of p'i shuang, and a pound of shih huang."

"To make the powder which drives the rocket highest (qui pousse la fusée fort haut), to an ounce of saltpetre, three-tenths of an ounce of sulphur, four-hundredths of an ounce (sic) of mi t'o sêng, and three-tenths of an ounce of charcoal, are required. These kinds of rockets are used for giving signals in the day-time. For signal rockets for night-time, to four ounces of saltpetre, two-tenths of an ounce of sulphur and one ounce of charcoal are added."

"Pour la composition de la poudre qui pousse horizontalement en avant, & en eparpillant, il faut sur deux onces de salpêtre, trois dixiemes de charbon, deux dixiemes & demi de soufre, sept dixiemes & demi de sable tres-fin (ce sable doit avoir été trempé dans l'huile de l'arbre toung, ou dans l'huile appellée pa teou yeou), & trois dixiemes & demi de charbon. Il s'agit de dixiemes d'once." [The "sand" soaked in t'ung oil is iron powder.]

"For petard composition, ten ounces of saltpetre, six ounces of sulphur, three ounces of calabash charcoal, and one ounce of shih huang are used. For powder for large cannon, to sixteen ounces of saltpetre there are added six ounces of sulphur, six and eight-tenths of an ounce of calabash charcoal, or charcoal of béringene or simply willow charcoal. It is essential that the saltpetre should be purified to the highest degree and that it should not leave the least residue (marc). The gunpowder is pounded several times in a stone mortar, wetted with water, and allowed to dry to a paste. This paste is again pounded well in a stone mortar and then taken out to dry. When quite dry it is pounded again, and after wetting a second and third time these operations are repeated. Three ounces of this powder suffice to drive an ordinary bullet."

"Darts (aiguilettes) are made from gunpowder, p'i ts'ao shuang, mallow roots, jujubes with the kernel removed, and saltpetre water; they are shaped on the palm of the hand, dried, and one is put over the gunpowder in the cannon (par un bout dans la lumière du canon jusqu'à la charge), and the other end is covered with ordinary gunpowder, which is set on fire."

"Ordinary powder is made, according to my author, in the following way. A just balance must be provided, the sulphur must be well purified and reduced to an impalpable powder, and similarly the saltpetre, and when they are in proportioned amounts they are mixed and ground on a marble with a stone roller. The powdered charcoal is then added and the whole wetted with water to reduce it to a paste (l'on met le tout dans l'eau pour le délayer et le réduire en pâte). This paste is rolled on the marble as before. When it is well milled and in a state of sufficient consistency, it is removed to dry in the sun. When it is dry, it is put back on the marble and manipulated till it is reduced to powder. The finer this powder, the better it is; it serves for all kinds of uses."

Amiot then gives a recipe for artillery powder: 5 lb. saltpetre, 1 lb. sulphur and 1 lb. charcoal. Various kinds of charcoal are described, but it must not be made from resinous or oily trees. The mixture is ground on a marble 3800 times, wetted, and made into pieces the size of millet seeds, rice grains, or peas. It is good when some grains set on fire on the hand burn without producing any sensation of heat. This is twice as strong as ordinary powder.

Quick matches (étoupilles) are made from 1 lb. of ordinary gunpowder, $3\frac{6}{10}$ oz. of sulphur, powdered willow charcoal $4\frac{3}{10}$ oz., béringue charcoal $\frac{5}{10}$, pe pi $\frac{5}{10}$, camphor $\frac{3}{10}$, worked up as with double-force powder. Other recipes for gunpowder are given, in some of which strong brandy (eau de vie) is used for moistening and the dry powders are passed through sieves. Presumably brandy also was known "several centuries B.C."

The identifications of some of the constituents given without translation by Amiot may be attempted as follows (I am indebted to Dr. Needham for these):

> *t'ao hua p'i*, peach-flower coloured arsenolite?
> *ma nao p'i*, agate-coloured arsenolite?
> *chen sha*, powdered iron filings
> *kan ch'i*, dry lacquer
> *chiang*, ginger?
> *ch'ing fan*, green vitriol (ferrous sulphate)
> *p'i shuang*, arsenious oxide
> *shih huang*, yellow sulphur
> *mi t'o sêng*, litharge or massicot (lead oxide).

The descriptions given to the figures (Figs. 12 and 13) in Amiot, copied from the Chinese sources, are:

Fire Arrows (figs. 71, 72, 74)

"The tube into which the powder is put must be extremely narrow, hardly four inches long, and its extremity two inches from the iron. An arrow so thrown is equivalent to the strongest gun-shot (coup de fusil)."

Three arrows are shown with what looks like a fuse projecting from the lower part of the case, so suggesting a rocket.

It has been emphasised above (p. 2) that the Greek and Roman incendiary arrows had to be shot slowly, otherwise they were extinguished by the strong current of air, and when shot slowly they had only a short range. With a burning composition containing saltpetre this disadvantage would be obviated, since the combustion does not now need the participation of air, and is not

quenched by the cooling effect of a current of air.   But the jet of flame projected forward now tends to retard the flight of the arrow, and someone
must have had the idea of turning the container round, so that the jet issued
at the rear.   An arrow so contrived, when laid in a cross-bow and kindled
before shooting would, if it contained a large and suitably disposed charge of
gunpowder, leave the bow spontaneously and fly through the air.   This
was a rocket.   Incendiary arrows with the flame directed *forwards* were still
in use in France in the seventeenth century and the composition also included
mercury, which produced toxic effects.[75]   The suggestion[46] that the Chinese
incendiary arrows contained a "phosphorus mixture" (un mélange de phos-
phore sans doute) is impossible.

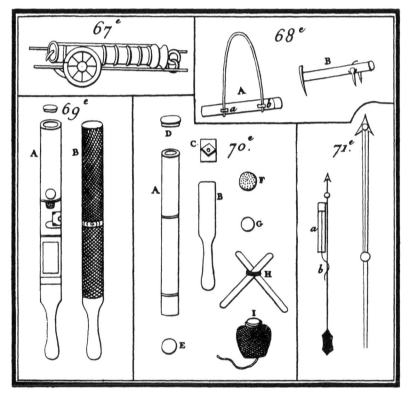

Fig. 12.   Illustrations from the *Wu Pei Chih* (1628) given by Amiot.   The
numbers are those given in the text.   Figs. 69 and 70 show the huo ch'iang
described on p. 246.   Figs. 67 and 68 are guns, also mentioned on p. 279.
Fig. 71, and Figs. 72 and 74 in Fig. 13 are fire arrows and Fig. 73 the
container for them.

Earth Thunder (fig. 75)

"It is an iron globe hollow inside; it should be large enough to contain a
bushel (boisseau) of powder.   This powder is compressed as strongly as possible
by beating it strongly as it is introduced through the opening (par l'ouverture).
As much scrap iron (mitraille) is mixed with the powder as desired, and the

globe is buried a foot or two deep in the place where the enemy is expected
to pass.   As many globes are used as will produce the proposed effect, and
they are placed at some distance from one another.   Sulphured string is
inserted at one end into the globe.   The person who is to fire them, at some
distance, holds the other end.   This cord is concealed in bamboo tubes
buried in the earth and communicating with one another, so that the wick
gives fire at the same time to all the globes.   This device was often used by
K'ung-Ming, who used it particularly against the Tartars, whom he nearly
always conquered.   He was a general at the end of the Han, i.e. about
A.D. 200, and it is added that he had learnt this manner of making war from
the books of the ancients."

Bee-hive (fig. 76)

"Globe of iron filled with powder mixed with bits of iron of all shapes
and strongly compressed.   This globe is buried as described above and is
lit in the same way."

Devouring Fire (fig. 77)

"This artifice is used in a siege or in a naval combat.   A paper globe
covered outside with resin, oil and yellow wax is filled with powder mixed
with resin and metal scrap (mitraille); it is set on fire by a wick and thrown
at the enemy."

Tuyau de Feu (figs. 69–70)

"On choisit parmi les bambous qu'on nomme *mao-chu* (ils son plus forts
que les autres) ceux qui sont les plus rond et ont aux moins deux pieds et
deux dixièmes.   On les lie fortement avec des cordes de chanvre pour empêcher
qu'ils ne se fendent.   On enchasse chaque tuyau dans un manche de bois
fort, au moyen duquel on le tient à la main; le tuyau et le manche pris ensemble
ne doivent pas avoir plus de cinq pieds.   On le charge de plusieurs couches
de poudre diversement composées, et par-dessus l'on met une balle fait avec
une certaine pâte.   Ces balles sont au nombre de cinq.   La portée de ces
balles est d'environ cent pieds, et leur effet est d'embraser."

This is the contrivance described on p. 246.   Fig. 79 seems to show in
perspective the contrivance in operation, the inclined tube appearing shorter
than it actually was.   It was a large Roman candle; Romocki[75] called it "the
first gunpowder weapon" (Pulver-Schusswaffe), and if we knew that it did
contain gunpowder, it would be.   The guns shown in Amiot's figs. 67 and 68
are from the *Wu Pei Chih* (1628).   A very interesting object is the naval mine:

Boîte de Feu (fig. 78)

"(*a*) Couvercle du vase.   (*b*) Bassin où l'on met l'amorce qui doit mettre
le feu.   La capacité du vase contient cent pieces de mitraille, dont on verra
la figure ci-après.   Ces boîtes doivent être d'un bois léger, de la forme qu'on
voit, & proportionnées à l'objet qu'on se propose.   Celle dont se servoit
Tsi-nan-tang, inventeur de cet artifice, contenoit cinq livres de poudre, &
cent pieces de mitraille.   Il mettoit dans le fond une légere couche de sable

fin, ou de terre sablonneuse; sur cette couche, il etendoit avec egalité cinq livres de poudre; qu'il recouvroit de sable ou terre sablonneuse. Il enfonçoit au milieu de cette couche, une tasse de porcelaine grossiere, où il avoit mis du charbon pulvérisé, & sur lequel il mettoit deux ou trois charbons rouges de feu. Il fermoit la caisse avec son couvercle, & la faisoit porter dans l'eau le plus doucement qu'il etoit possible, lorsqu'il vouloit s'en servir. La tasse

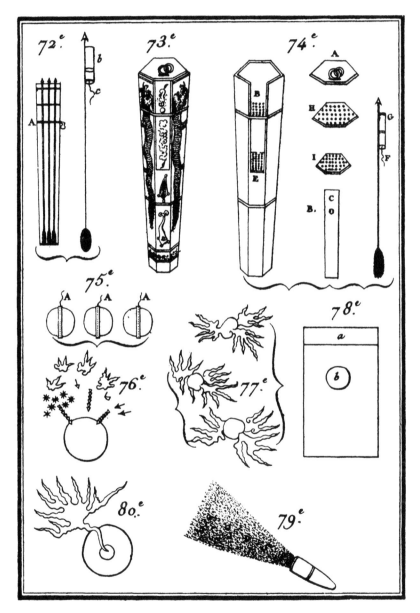

Fig. 13. Illustrations from the *Wu Pei Chih* (1628) given by Amiot. These are described in the text.

qui tenoit lieu de bassin pour contenir l'amorce, n'etoit placée qu'au moment
où on alloit abandonner la boîte ou courant de l'eau. C'est le brulot des
Chinois. Ils ne s'en servent que lorsque leurs barques sont au-dessus du
courant, vis-à-vis les ennemis. Ce brulot prend feu de lui-même, quand il
heurte contre quelque barque ennemie. La charbon pulvérisé, qui a eu le
tems de s'allumer, est renversé par le choc, & enflamme la poudre.''

This mine could be used only on rivers, since it would be rocked and
exploded prematurely on the sea, and it seems to be a Chinese invention,
since I have not found it in European works, except Amiot's.

Romocki[76] quite wrongly supposed that these accounts of incendiary arrows
(or rockets) and bombs were from a Chinese *Firework Book* of 1232, ''im
Original mir wenigstens leider nicht bekannt geworden,'' and that ''the
invention of explosives in China must be taken as the second half of the twelfth
century,'' soon after the use of the sulphur–quicklime mixture. He mentions
that Morrison[77] had attributed the invention of gunpowder to the Sung
commander Wei Hsing (or Wei Shêng) but gave no source for this information.
It is given by Chao I (A.D. 1790),[78] who includes some incredible material.

''The iron bombs shown by Amiot [Fig. 75],'' continues Romocki, ''are
in two hemispherical parts, also used later in Europe (see p. 146). These are
described in the encyclopædia of inventions from the *Pai Pien* of 1581; the
'ruche d'abeilles' was apparently in one piece. Some of the incendiary
arrows were probably rockets; the 'tuyau de feu' was a primitive gun and
only required to be made of metal and properly charged to become a real
gun, but if the Chinese had made this step they afterwards forgot it.''

SOME OPINIONS

Reinaud and Favé,[79] after mentioning that Gaubil[80] had ''not dared to
translate p'ao or huo p'ao as cannon,'' concluded that the Chinese knew of
gunpowder long before the Arabs, who got to know of it from China perhaps
through the Mongols, between 1225 and 1250. Reinaud and Favé had at
first (1845) concluded that cannon were invented in Hungary (see p. 125),
but the contents of the St. Petersburg Arabic manuscript (see p. 204) convinced them that the first propulsive weapons using gunpowder originated
among the Arabs. The knowledge of chemistry among the Arabs (which,
it may be said, Reinaud and Favé exaggerated) enabled them to improve
considerably the purification of saltpetre (wood-ashes were specified by
Ḥasan al-Rammāḥ) and to mix the ingredients of gunpowder more intimately.
The Chinese first introduced saltpetre into incendiaries and fireworks and
invented rockets; the Arabs borrowed this knowledge and extended it to the
throwing of projectiles by gunpowder, at first with a feeble force.

Mayers,[20] in a detailed examination of the material available to him,
concluded that the use of gunpowder, but not of guns, is clearly established
for the battles between the decaying House of Sung and the Nü-chen (Jurchen)
Tartars (who had been active in China from the eleventh century) in the
period about 1150, ''but whether the invention first gained ground among

the Chinese forces or with their Northern enemies is a question the solution of which cannot be ventured upon." Although Chao I attributes the invention of gunpowder to the Sung commander Wei Hsing, he is not a reliable authority. The Chinese made free use of various incendiary mixtures, which were then also known to the Arabs and Europeans, in the thirteenth century, but they made no progress in the development of missiles in the fourteenth century and true gunpowder and cannon appear among them only in the fifteenth century.

Illustrations in the *Wu Pei Chih* show large ballistas (chuang-ch'ê), one requiring forty men to work it and another 100 men (throwing two stones of 36 lb. a distance of 160 yards), very similar to the Greek and Roman engines (from which they were probably derived); and some discharged incendiaries of the Greek type, "the fore-father of all guns (p'ao)," as it says. The latter machine was first used in the defence of K'ai-fêng Fu (A.D. 1127), although Li Kang, says Mayers, lays no stress upon it.

What Mayers thought were bamboo tubes with fire streaming from the forward end, to be thrown from a machine, are really parts of a machine for throwing stones or other projectiles; the *Wu Pei Chih* (1628) calls this machine a "Sung huo p'ao," i.e. a "cannon" of the Sung period (A.D. 960–1280), but incorrectly, since it does not use gunpowder. The "fire streaming from the forward end" was really a set of cords for manipulating a machine for throwing stones.[81]

Mayers said that descriptions of incendiary arrows and similar devices become more frequent in the period from 1150 and throughout the Mongol (Yuan) dynasty, i.e. most of the fourteenth century. The first sovereigns of the Ming dynasty (from 1368) devoted serious attention to the use of gunpowder in warfare, e.g. in battles of 1401 and 1407, when cannon were well known in the West.

Mayers suggested that rockets and cannon were both taken over from their vanquished enemies by the Chinese in the invasion of Tonquin (Indo-China) by Chung Fu in 1407; Chao I says this, adding that large cannon were carefully concealed from observers by order of the Emperor. The *Ming Shih* (1739) says the earliest p'ao were engines for throwing stones; the first firearms used in China came from the West (Hsi Yü) at the end of the Yuan dynasty, but the secret of their manufacture was lost and afterwards they were little used. The envoys of Yung-Lo in the Malay Archipelago, Delhi, Mecca and Herat, says Mayers, probably brought a knowledge of weapons used by the Arabs. Chao I says shên ch'iang (divine guns) were not used in the Chinese army until after Chia-Ch'ing's reign began (A.D. 1422). A knowledge of the propulsive force of gunpowder, and of cannon, said Mayers,[82] probably came to China from the West.

Since Hime[83] relied largely on Mayers, it is not surprising that he reached the same conclusions. He thought the purification of saltpetre was first described by Roger Bacon (see p. 73) and the Chinese followed his method of using willow charcoal, straining the saltpetre solution through straw and

using animal glue and charcoal to remove impurities from it.   Hime's interpretation of Bacon, however, as has been shown (p. 76), is not free from
doubt, and the first known explicit description of the purification of saltpetre
by using wood-ashes is that of Ḥasan al-Rammāḥ (p. 201).[84]   The Chinese,
says Hime, incorporated the ingredients of gunpowder on a marble slab,
as in Marcus Græcus (p. 49), passed rocket powder through a silk sieve,
loaded bombs with the maximum charge and added bits of metal, all suggesting
Western practice (these manipulations are mostly from the *Wu Pei Chih*),
and repaired their built-up bombards with "Western iron" (see, however,
p. 278).   The iron guns called "Franks" seen on Portuguese ships about
1520 were the models for Chinese cannon.

A Western origin for Chinese gunpowder, says Hime, is indicated by the
addition to it of camphor, mercury and varnish, which also occur in the
Arabic recipes of Ḥasan al-Rammāḥ.   But these are just what we should
expect the Chinese, who had early knowledge of camphor, mercury and
lacquer, to use.

Hime gave seventeen arguments to show that "there is no trustworthy
evidence to prove that the Chinese invented gunpowder," and that they
"obtained their first gunpowder and firearms from the West."   The mangonels
used in the siege of Hsiang-yang (1268–73) were Western, and besides the
Polos many Western mechanics and merchants visited China.   Yung-Lo,
the first Emperor to have ch'iang (cannon), had agents in Malaya, Delhi,
Herat and Mecca.[85]   There was communication by land and sea with the
West from the Roman period, which was re-established about 1250 after the
disorders in the ninth century.[86]   There was also contact with the west coast of
India at least as early as 1400–50; 'Abd al-Razzāq said the men of Calicut
were such bold navigators that they were called "sons of China," but when
John Deza destroyed the fleet of the Zamorin of Calicut it was commanded
by a Chinese admiral, Cutiale.[87]

Hime concluded that, although the Chinese used fireworks and incendiaries
from about 1250, "there is nothing in the military history of China in the
thirteenth and fourteenth centuries to lead us to suppose that the Chinese
possessed an explosive during that period."   He quotes Berthelot[88] as saying
that the Chinese were ignorant of cannon in 1621 and were also ignorant
of the propulsive force of gunpowder before it was known in the West; Berthelot
admitted their use of incendiaries from the tenth century (A.D. 969, 1002,
1232, 1259, 1273), but supposed (incorrectly I believe) that the Greek fire
used in Constantinople about A.D. 673 was a similar mixture which contained
saltpetre.

Giles[89] gave the first use of firearms recorded as by the Ming general
Chang Fu in 1407, when the Annamites were defeated; Geil[90] said that
gunpowder was invented in China, was made known in Europe by the Mongols
in the thirteenth century, and in the Ming period (1368–1644) it was used to
throw stones or lumps of metal, but cannon were cast only after foreign
influence.

The suggestion of Schlegel[91] that cannon were used in 1232 was rightly rejected by Pelliot,[92] who thought that, although "explosives" were used in China as early as 1162, firearms came from the West. Laufer[93] put the use of fiery projectiles in warfare in the twelfth century and fire weapons under the Mongols in the thirteenth and fourteenth centuries. Sarton[94] reports that there is no mention of gunpowder or firearms in the extant volumes of the encyclopædia *Yung-Lo Ta Tien*, compiled by the order of the Ming emperor Yung-Lo, being commissioned in 1403 and completed in 1408. It was never printed. Only one copy was made, in 1565, and all but 368 of the 11,095 volumes were destroyed by fire in 1900, so that we do not know whether gunpowder and firearms were described in the missing volumes.

These older views have been given because they bring out some details of importance to us, and are also instructive in their diversity. They have been considerably modified in recent publications, to which we now turn.

THE WU CHING TSUNG YAO

Needham[95] refers to the Sung period (A.D. 960–1279) as "the major focal point" for pure and applied science. By A.D. 1000 at least "explosive grenades" and "bombs" were being fired from catapults. By A.D. 1040 (or 1044) the *Wu Ching Tsung Yao* "stabilised the Chinese name for gunpowder (huo yao) and registered an extremely rapid development of projectile weapons, poisonous and signalling smokes, flame-throwers and other novel inventions. These were used in naval as well as in land warfare." This concise and accurate summary of the conclusions drawn from recent studies of an important Chinese work will now be developed in some detail.

The *Wu Ching Tsung Yao* (Essentials of the Military Classics, *or* Collection of the most important Military Techniques), edited by Tsêng Kung-Liang, was compiled by imperial order and completed in its original form about A.D. 1040, the preface being dated A.D. 1044.

Goodrich and Fêng Chia-sheng,[96] who credit the work to "Tsêng Kung-Liang and others," say that: "The edition used by us is that found in the photographically reproduced *Ssu-k'u ch'üan shu chên pên* series. The original in the National Library of Peiping, unfortunately incomplete, was printed from the wood blocks prepared in 1232, recut in the years 1403–24, and repaired in 1439. How old the illustrations are is uncertain; they may date back to 1232, but we suspect that they are not older than the fifteenth century, possibly later. Chüan 12 is one of the parts missing from the fifteenth-century copy." Shortly after the paper by Goodrich and Fêng Chia-sheng appeared, another independent study by Wang Ling (Wang-Ching Ning) was published.[97]

The *Ssu K'u Ch'üan Shu Tsung Mu T'i Yao* (compiled in 1788) says (vol. 2, p. 2041, in the 1933 reprint) that the work was compiled in the K'ang-Ting period (1040–1) and is a selection of old military methods to the time it was written. It is in two parts, each of twenty chapters, of which fifteen are on regulations, five on frontier defence and the remaining twenty (in the second part) are subdivided into five and fifteen, and are on strategy, tactics, etc. It was written by Tsêng Kung-Liang and others "who favoured peace and were not educated in military affairs." Their information on frontier defence was based on hearsay only and the information on rivers and mountains is incorrect as compared with the present day.

The *Wu Ching Tsung Yao* is not mentioned by Mayers,[20] Wylie,[98] Giles,[89] Romocki,[16] or Hime.[10] No critical examination of its authenticity in a European language is available to me.[99] Chapter 11 deals with attack by fire (huo kung). It describes incendiaries, a fire-pot like that in the *Feuerwerkbuch* (see p. 100), and incendiaries attached to birds and expendable animals. There is a smoke ball (or smoke-screen) (yen ch'iu) and a poisonous smoke ball (tu yao yen ch'iu), flame-throwers, toxic smoke canisters, incendiary whip-arrows, and bombs, all containing huo yao, the recipes for which show that it is a low-nitrate gunpowder. The edition of the work which I have seen contains two illustrations of seventeenth-century European metal cannon, but they are out of place (among descriptions of carriages). Dr. Needham informs me that they are not described in the text, and are not found in the 1439 edition. The edition used by Wang Ling included these illustrations.

Standing between the technical manuals, such as the *Wu Ching Tsung Yao* (1044) and the *Wu Pei Chih* (1628), and the Taoist alchemical works, such as Ko Hung's and many later ones of similar character,[100] we find (as in the West with Kyeser's) a treatise like the *Huo Lung Ching* (Fire Dragon Manual) by Liu Chi (Po-Wên) (1311–75), with a preface dated 1412 by Chiao Yü, who was a practical artillerist in the service of the Ming emperor.[101] It is in three parts. It says there were no fire weapons in the Han dynasty and gives a legend of a Taoist hermit who met Chuko Liang and gave him a book. Chiao Yü says he himself met a Taoist stranger about 1355, who gave him scrolls on fire weapons. Chiao Yü claims to have cast quite a number of cannon (chu huo ch'i shu chien), which he presented to the emperor. They shot projectiles like flying dragons, which penetrated armour (and must, therefore, have been very small guns, otherwise the projectiles would have blown it to pieces). Factories for making standard gunpowder, and arsenals for keeping these magic weapons, were established in the capital. Chiao Yü claims that he received secrets from genii (hsien). The third part of the *Huo Lung Ching* is by Mao Yuan-I, the author of the *Wu Pei Chih* (1628) and the preface is dated 1644. Many parts of the book are found also in the *Wu Pei Chih*, such as the "archaic gunpowder compositions," toxic smokes, "shrapnel bombs" (i.e. with fragments for scattering—not true shrapnel), fire-lances throwing spattering fire and projectiles of the Roman candle type.

The *Wu Ching Tsung Yao* gives the composition of an incendiary powder (huo yao) for use with whip-arrows or incendiary arrows, 5 oz. of it being packed behind the barb, and also for incendiary balls to be thrown by catapults[102]:

(*a*) Grind together with a pestle and sift 14 oz. of Chin-chou sulphur, 7 oz. of K'o sulphur, 2·5 lb. saltpetre. Grind together 1 oz. realgar, 1 oz.

Ting powder (?) and 1 oz. massicot.   Powder 1 oz. of dry lacquer.   Roast to bits or powder 1 oz. hemp roots, 1 oz. bamboo roots.   Boil to a paste 0·5 oz. bee's wax, 0·1 oz. clear oil, 0·5 oz. t'ung oil, 14 oz. pine pitch, 0·1 oz. heavy oil.   Mix all together evenly, wrap in five thicknesses of paper, bind with hemp, smear with pine pitch and throw by a catapult.

It also describes an improved fire ball making a roaring noise when burning :

(b)   Pestle to powder 1·25 lb. sulphur, 2·5 lb. saltpetre, 5 oz. charcoal, 2·5 oz. resin, 2·5 oz. dry lacquer.   Cut into small pieces 1·1 oz. bamboo roots and 1·1 oz. hemp roots.   Melt and mix 2·5 oz. t'ung oil, 2·5 oz. small (?) oil, 2·5 oz. bee's wax.   Melt and mix 12·5 oz. paper, 10 oz. hemp, 1·1 oz. massicot, 8 oz. charcoal powder, 2·5 oz. resin, 2·5 oz. yellow bee's wax. Combine all these and smear over the ball.   Make the ball by wrapping around a piece of bamboo [a mixture of] 30 pieces of thin porcelain [the size of] iron cash and 3 or 4 lb. of incendiary powder.   The two ends of the bamboo protrude about an inch.   Smear powder over the outside.   If the enemy attacks by tunnelling, we dig holes to intercept him.   When the ball is lighted with a hot poker, it will make a noise like a rumbling.   When it is lighted, fan its smoke and flames so as to smoke and burn the enemy.

(c)   A poison smoke ball contains 1 lb. 14 oz. saltpetre, 15 oz. sulphur, 5 oz. aconite tubers, 5 oz. powdered croton beans, 5 oz. wolfsbane, 2½ oz. t'ung oil, 2½ oz. vegetable oil, 5 oz. charcoal, 2½ oz. resin, 2 oz. arsenious oxide, 1 oz. yellow wax, 1 oz. bamboo roots, 1 oz. hemp roots.   These are mixed and smeared over a ball made from 12 ft. of hempen rope.

Advice is given[103] to collect various stores, including lime, iron, charcoal, pine wood, straw, mustard, fatty oil, hemp fibre, felt, thorn tree branches, etc. (but not sulphur or saltpetre), and incendiaries could have been made from some of these.

Goodrich and Fêng Chia-Shêng[104] quote the *Wu Ching Tsung Yao* as saying that huo-ch'iu made of huo-yao ("explosive powder"), and huo-p'ao also made of powder, were thrown from a trebuchet; this may have been a sort of bomb.   The p'i-li huo-ch'iu was a dry bamboo tube, 3 nodes long and 1½ in. diameter, a mixture of "gunpowder," pieces of iron and pottery fragments "was attached to the outside of the bamboo, at the middle, and was lighted with a length of red-hot iron."   It "produced a noise like thunder" and "seems to have been . . . a hand-grenade."   This is Wang Ling's contrivance (b).   The "explosion" of bamboo alone has been dealt with above (p. 239) and the incendiary mixture and mitraille were here put on the *outside* of the bamboo; the heat generated would perhaps have caused it to burst. The correct translation of huo yao is not "explosive powder" but "fire drug."

The authors then give some particulars which we have already dealt with, at least in part :

1126 : The Jurchen attacked the Sung capital, Pien (modern Kaifêng), and a Sung general Yao Yu-Chung defended the city with huo-p'ao, chi-li-p'ao and chin-chih-p'ao thrown by trebuchets or bows. The Jurchen then threw in huo-p'ao from elevated platforms. The literal meaning of chin chih is "gold juice,"[105] but chin also means "metal" and if chih is taken to mean "liquid," chin chih would be liquid metal, say molten lead, and the chin chih p'ao a device for throwing this, probably in pots. Molten lead was used in warfare by the ancient Assyrians (p. 1).

1127 : The Jurchen machines attacking Hopei were destroyed by the Chinese huo-p'ao.

1130 : The Jurchen used chin chih p'ao in attacking Shan-Chou (modern Shensi), destroying everything within reach.

1132 : The huo ch'iang used in the defence of Tê-An was "a long bamboo tube filled with explosive powder." This was the fire-lance and a bamboo tube "filled with explosive powder" would have burst and killed the person discharging it.

1161 : The Chin fleet of 600 vessels on the Yangtze river was destroyed with huo-p'ao and p'i-li-p'ao. The p'i li p'ao were the paper bags containing quicklime and sulphur (see p. 240), although Goodrich and Fêng call them "explosive bombs." Průšek[46] was clear that they did not contain gunpowder.

The huo p'ao may be an apparatus for throwing incendiaries, since in 1206 a Chinese commander cleared a city of inflammable materials and distributed pails of water, and there is no evidence that it contained anything "explosive."

1163 : Wei Shêng made chariots (p'ao ch'ê) which carried stone and fire throwers (huo shih p'ao) with a range of 200 paces, and the emperor issued a decree that chariots should be made everywhere according to his model. Průšek,[46] giving the full passage from the *Sung Shih* (368/15b), said this is the first reference (très peu claire d'ailleurs) to the use of fire p'ao with a range of 200 paces, but, he says, they were probably not firearms.

1221 : The Jurchen used a t'ieh huo p'ao "cast in the shape of a gourd of pig-iron 2 in. thick pierced with a small hole. Its explosion . . . sounded like a thunderbolt." An eye-witness claims to have seen a bomb of this kind on the walls of Si-an about 1490. The *Ko Chih Ching Yüan*, extracting from the *Pai Pien*, printed in 1581 but compiled somewhat earlier, says that such an object was preserved on the city walls of Si-an.[43] The casting of an iron globe 2 in. thick with only a small hole would be difficult, and such a thick and strong container could be burst only by a large charge of good gunpowder.

1231–3 : The yellow paper tube filled with mixture (not said to contain saltpetre) used in the siege of Pien-ching by the Jurchen[106] : this has been considered on p. 245. This, say Goodrich and Fêng, was a "firearm" throwing a "bullet," but it was a fire-lance. Průšek[46] thought saltpetre must be understood as added and that this description shows "avec une certitude presque absolue" that gunpowder "d'aujourd'hui" was known in North China in 1233 and was used in making arms. It shows nothing of the kind.

1236 : The Jurchen collected metals, including gold, silver, bronze and iron, and cast them to form p'ao in order to resist. The word p'ao meant a projectile as well as an instrument for throwing it, not necessarily a gun, although gold and silver guns are reported from India (see p. 226).

1257 : "The surprising, indeed almost incredible information" that more than 100,000 huo p'ao were used against the Mongols is given by Li Tsêng-Po. This is not necessarily "incredible" if they were projectiles.

1259 : The Chinese used a t'u-huo-ch'iang made of bamboo into which "a bullet (or bullets)" (tzu k'o) was inserted. "When the powder was ignited, smoke and fire came forth, followed by the bullet." This is the fire-lance described on p. 246.

1280 : A Mongol was grinding sulphur and set fire to an arsenal which, with all its p'ao, blew up with a resounding bang, "like mountains falling," and caused great devastation. In a similar event, four tigers were killed. The four tigers story is in the *Kuei Hsin Tsa Chih*, which is a reliable book of memoirs; they were perhaps kept in a zoo near the "arsenal."

1281 : A "bursting cannon ball" fired by a Mongol-led army is a curiosity; cannon balls do not burst. From the crude picture given, which includes a man with a bow, it seems to be some kind of incendiary ball, although by this date the explosive properties of gunpowder were known in the West. The original illustration is on a scroll in the Imperial Collection of Japan, painted in 1293 and representing a Mongol attack on Japan in 1281. I have examined a reproduction of the complete scroll without finding anything suggesting the use of "cannon."

Accounts of the use of huo ch'iang and huo p'ao in 1233, 1241 and 1268–73 are supposed by Goodrich and Fêng to be referring to cannon, which is very doubtful. They also thought the "thundering" mangonels used by Muslims were firearms, although this was not the case (see p. 249). Exaggerations like "a hundred claps of thunder" are ingenously interpreted as "explosions" and hui-hui p'ao, which were mangonels, are said to be "metal cannon." From 1356, we are told, there are many records of the use of "explosives and artillery," which is not surprising, since quite big cannon were then in use in the West (see ch. III).

The paper by Goodrich and Fêng Chia-Shêng, who are sinologists of acknowledged competence, published in a journal of high reputation which circulates among historians of science, can hardly fail to mislead readers unfamiliar with military facts. I understand that Fêng Chia-Shêng has corrected some mistakes in a Chinese publication, but this will be inaccessible to European scholars ignorant of Chinese.

TERMINOLOGY

Any modern writer who deals with the history of gunpowder and weapons using it must, if his account is to be of any interest or value, most carefully attend to the meanings attached to technical terms by experts. Great confusion has been introduced by several authors in recent publications by the use of incorrect or misleading names. What could at best be slow-burning incendiaries have been called "gunpowder" or "explosive powder," and their combustion has been called "explosion." The vessels containing them have been called "bombs"; if they contained bits of metal or broken pottery these have been called "bullets" or even "cannon balls," or the whole is a "shrapnel bomb," which is something unknown in military terminology.

It seems possible to draw up a table of combustible materials on some such lines as the following, based on the character of the combustion:

(1) *Slow burning.* Old incendiaries: oils, pitch, sulphur; sometimes used as incendiary arrows.

(2) *Quick burning.* Greek fire; distilled petroleum or naphtha; used in pots or projected by special devices; still incendiary only.

(3) *Deflagration.* Low-nitrate powders containing (*a*) carbonaceous materials and/or (*b*) charcoal as such; essentially incendiary but, on the higher nitrate side, suitable for rockets, or Roman candles projecting incendiary balls or perhaps even fragments of pottery or metal; "proto-gunpowder."

(4) *Explosion.* Higher nitrate mixtures, best with sulphur and charcoal alone as combustibles but sometimes used with additions, e.g. arsenic; giving appreciable noise if fired in a closed space.

(5) *Detonation.* Modern gunpowder, a suitably prepared mixture of saltpetre, sulphur and charcoal in proportions approaching 75 : 15 : 10; making a loud noise if fired in a suitable container, tearing and scattering but leaving débris; suitable for bombs, etc., also as a propellant in a sufficiently strong (e.g. metal) gun.

There may, admittedly, be some difficulty in defining the boundaries between (3), (4) and (5), especially between (3) and (4). I think that anything which can be described as "gunpowder," or better "proto-gunpowder," should contain actual charcoal as well as saltpetre and sulphur, and that "true gunpowder" should be taken as something between (4) and (5) rather than between (3) and (4).

SALTPETRE

The history of Chinese gunpowder is linked with that of saltpetre. The modern Chinese name for saltpetre is hsiao, which is said[107] to occur in the famous *Shih Chi* of Sssu-ma Ch'ien (first century B.C.), and in the *Ch'ieh Yün*, a dictionary compiled in A.D. 605 but several times revised, the present text being that of Ch'en P'êng (A.D. 1011). It must not be assumed that hsiao in such early texts necessarily means saltpetre unless the material it denotes is said to have some characteristic property of saltpetre, such as (i) deflagrating with combustibles, or (ii), less certainly, producing cooling when dissolved in water.

The name hsiao shih ("salt stone," cf. sal petræ) occurs in works attributed to the Taoist alchemist Ko Hung (A.D. 265–317), which some[108] think are later forgeries, or at least interpolated, although they are mostly regarded as genuine.

Ko Hung directs that 1 oz. of hsiao shih and $\frac{1}{2}$ oz. of sulphur be powdered, sifted and mixed. Vinegar is then added and the paste spread over the inside of a small iron box 2 in. thick which is exposed to a strong fire. It is not said what happens. (The 2 in., curiously enough, is also the thickness of the iron "bomb" previously mentioned, p. 264). Wang Ling[97] thought the vinegar would char and the charcoal formed, together with the hsiao shih and sulphur, would form gunpowder, which exploded. Although this is unlikely, the saltpetre and sulphur alone, when heated in a closed vessel, would produce some sort of explosion, but hardly strong enough to burst 2 in. of iron. This may well have been one of the first experiments on saltpetre explosives, although it sounds suspiciously like the mythical one of Berthold Schwartz (p. 91).

Hsiao shih occurs in a Taoist work, the *Chen Yuan Miao Tao Lüeh* (Important Details about the Mysterious Tao) of Chêng Ssu-Yuan. According to Fêng Chia-Shêng[109] the book was written in the later half of the T'ang, after A.D. 760 and before A.D. 906, and probably not before A.D. 900. It is in three chapters, the first entitled "Guide to the Rejection of False Experiments," and the passage reads:

"Some have used sulphur (liu huang) and realgar (hsiung huang) mixed with hsiao shih and honey. Upon ignition this has blazed up and burnt the hands and faces (of people) and set their houses on fire."

This satisfies one criterion that hsiao shih is saltpetre, and the text links up very well with the recipes in the *Wu Ching Tsung Yao* (A.D. 1044), which Goodrich and Fêng and Wang Ling think are the earliest ones for "gunpowder," and also with the passage in Ko Hung.

GUNPOWDER

Wang Ling says that fireworks are mentioned at an early date as yen huo (smoke fire). Poems of A.D. 605–16 speak of a gunpowder play (huo yao tsa hsi) in which flames and sparks shoot forth from the seven branches, flames of fire move round the wheel, peach blossoms spring from the falling branches, clouds of smoke move round the house, and the fiery lake reflects the floating lights. The name huo yao, however, is not in the original text, which Dr. Wang kindly looked up for me.

Wang Ling also refers to a T'ang (A.D. 618–906) work as speaking of huo shu (fire trees) and yin hua (silver sparkle), which are thought to mean fireworks. Since we rejected a similar claim for a mention of fireworks in the poet Claudianus, two centuries earlier (see p. 10), we shall do the same here. Wang Ling refers to a play mentioned in A.D. 1103 in which "crackers" were used and ghost-like figures of various colours appeared. He suggests that "strontium, barium, copper, etc." were used in making the coloured fires (as they are to-day), but strontium and barium compounds were discovered in Europe only at the end of the eighteenth century. In China, red and green fires were made in the seventeenth century (see p. 285); the green contained copper compounds and the red fire mentioned by Amiot (p. 253) contained t'ao hua p'i, which seems to have been an arsenic compound.

In any case, poetry and drama have often been used to "prove" that all kinds of inventions were known in antiquity. The book by Dutens, mentioned previously (see p. 43), although it rejects a passage from Claudianus as proving that the ancients knew of sex in plants, seriously maintains[110] that the legend of Salmoneus, son of Sisyphos, proves that he was the inventor of gunpowder. Pliny (bk. vii, ch. 57) in a list of "the inventors of various things" which has been drawn upon by older authors, says that Pyrodes, son of Cilix, first struck fire from flint, and Prometheus taught how to preserve it in the stalk of the ferula. I have not seen this used to prove that the stalk of the ferula was a cannon and the fire in it was gunpowder.

The almost contemporary historians Zosimos and Sozomenos report[111] that in A.D. 409 some Tuscan magicians promised Pompeianus, the prefect of Rome, that they could extract thunder and lightning from clouds and direct them against the camp of the barbarians. Zosimos says the secret was entrusted to Innocent, bishop of Rome, and Sozomenos that the invention was tried but found unsuccessful. We need to reject all such stories given by historians of repute, whether European or Oriental. We have seen (p. 10) that a recent author has proved to his own satisfaction that gunpowder was invented by Moses, and poets, if they are worthy of their craft, can invent anything.

The literal meaning of huo yao, gunpowder, is "fire drug," or "fire medicine," or "fire chemical," which may suggest a Taoist alchemical origin. In European languages some early names for gunpowder were derived from words meaning "plant," "herb," or (derivatively) "drug" (see p. 313).

In the *Wu Ching Tsung Yao*, huo yao occurs in two recipes along with saltpetre, sulphur and charcoal, and although other things such as vegetable products, arsenic, etc., were added as well (as they were also sometimes in the West, see p. 152), it seems reasonable to assume that huo yao is here a low-nitrate powder, which would at least deflagrate (see p. 266). Dr. Needham informs me that he has not found huo yao used for "incendiary compositions not containing saltpetre, or in fact for anything other than gunpowder or proto-gunpowder," which is important. He has also found a reference to it about contemporary with the *Wu Ching Tsung Yao*, viz. in the *Ch'en Shih* (Objective Reminiscences of the Official World, 1/4b) by Wang Tê-Ch'en, the preface being dated A.D. 1115. This quotes the *Tung Ching Chi* (Records of the Eastern Capital) by Sung Tz'u-Tao (Min-Ch'iu) (A.D. 1019–79) as follows (I have to thank Dr. Needham for permission to use this text) :

"Apart from the eight offices for the government factories there is also the Office of General Siege Train Material, now in two sections, Eastern and Western, both of which are under the authority of the Arsenals Administration. Their work comprises ten departments, viz.: *huo yao*, blue pottery kilns,[a] fierce fire oil,[b] metal,[c] fire,[d] timber large and small, furnaces large and small, oxhides,[e] hemp,[f] and other pottery kilns. All these have their regular specifications and procedures, written down so that those in charge can read them, but it is forbidden to spread about this knowledge."

[a] perhaps for making naphtha 'grenades'
[b] naphtha
[c] for making molten metal? see p. 264
[d] incendiaries
[e] for making covered ways, see p. 243
[f] rope, etc.

The last remark about security precautions is confirmed by the *Sung Shih* (165/23*b*). Apart from these two mentions of huo yao, no other contemporary or earlier use of the name is known, which suggests that it was a technical term applied to some sort of gunpowder.

The earliest fire crackers were simply bamboo thrown on fires. Wang Ling says that under the North Sung (A.D. 960–1126) a new type of cracker is mentioned, p'ao chang: "all of a sudden a clap was heard, it was called p'ao chang," and this suggests that the cracker contained gunpowder. Under the South Sung (A.D. 1127–1279) the p'ao chang had a new shape, connected by a continuous fuse, and "one after another exploded." (This is the toy now called the "rip-rap.") The fuse is called yin hsien ("leading string"). With the fire radical, yin huo hsien, it is the modern name for a fuse, another being tao huo kuan. We are now in the period (*c.* A.D. 1250) when gunpowder crackers were known in Europe to Roger Bacon (p. 77).

Chinese incendiary arrows have already been considered (pp. 254, 277). A new type of fire arrow is mentioned for A.D. 969. The *Wu Ching Tsung Yao* mentions huo yao as used with an arrow, and there is a huo yao pien chien arrow (huo yao whip-arrow), with 5 oz. of huo yao at the end. These arrows were used in A.D. 1206. They no doubt contained a weak gunpowder of type (4), and are described in the *Wu Pei Chih* (1628).[48] In the early Sung a "bursting fire ball" containing bits of porcelain and "something producing fire and smoke" is mentioned.

The mêng huo yü, a "fierce fiery oil" apparatus, an illustration of which is given by Wang Ling,[97] is a very interesting contrivance. It consisted of a rectangular tank filled with oil and fitted with a pump for projecting the oil through a tube. Diagrams of the pump and its connections with the tank are given. The oil issuing from the tube was set on fire by an apparatus at the end of the tube called a "fire tower," which contained gunpowder (huo yao), according to the *Wu Ching Tsung Yao*. If the "oil" were distilled naphtha or light petroleum, a single flash of powder could ignite the jet, which would then continue to burn. The oil projector is mentioned in the *Wu Yüeh Pei Shih* (A.D. 919) and the *Ch'ing Hsiang Tsa Chih* (A.D. 1004), but *without* mention of the huo yao, which suggests that gunpowder was not then in use. The "burning oil" came from Arabia in an iron tank decorated with silver, so that if it was captured the enemy would scrape off the silver and leave the more valuable oil. Although the *Wu Ching Tsung Yao* in A.D. 1044 says this apparatus was used in A.D. 919 (in which case it would be the first use of gunpowder in Chinese armies), the failure to mention huo yao as used in it in the accounts of A.D. 919 and 1004 suggests that this was not, in fact, used in A.D. 919, and if this argument is sound it fixes rather sharply the date of about A.D. 1040 for the introduction of gunpowder in China.

Wang Ling quotes Liang Chang-Chu (A.D. 1280) as saying that gunpowder and firearms were imported from the West (firearms were not known then in Europe), but this need not mean that they came from Europe (see p. 278). Firearms are not mentioned in records of army equipment till the (one year) K'ai Ch'ing reign-period (A.D. 1259). In 1393 gunners are mentioned. Japanese scholars say that gunpowder was introduced into Japan by the Mongols, and some Japanese texts (from A.D. 1274) say that in 1281 there were "huge cannon" on ships. Although ships then were quite large, as we know from what Marco Polo says, it is doubtful whether "huge cannon" could be mounted and fired on them, and the *Wu Pei Chih* (1628) says that the large guns used in naval warfare were mounted on special rafts (see p. 280). The possibility of exaggeration cannot be excluded.

Wang Ling claims that huo p'ao were used in the beginning of the Northern Sung dynasty (A.D. 960–1126), and the "fire barrel" in the Yuan dynasty (A.D. 1260–1368); the name huo p'ao is repeatedly used in the Sung, Chin and Yuan dynasties, when it meant an engine for throwing incendiaries, perhaps gunpowder, but it did not mean a true cannon before the Ming dynasty (1368–1644).

Various types of missiles mentioned in the *Wu Ching Tsung Yao*, which the *Sung Shih* says were invented in A.D. 1000, appear to have contained some kind of gunpowder. Want Ling gives as examples:

(i) A barbed fire ball wrapped in paper with hooks for attachment to thatched roofs, etc.

(ii) An iron-beaked bird with the tail filled with incendiary.

(iii) A bamboo fire-bird with an open-mesh body filled with 1 lb. of "gunpowder" wrapped in paper and a tail of plant stalks.

(iv) A smoke ball weighing 5 lb. with an inner layer composed of sulphur, saltpetre, charcoal powder and dried mint plant, and an outer layer of paper, hemp-fibre and charcoal.

(v) Poison-drug smoke balls filled with "gunpowder," pitch, wax, arsenious oxide and various oils and poisonous plants, or with powdered dried human fæces, arsenic oxide and sulphide, lime, oil and various plants.

(vi) Signal balls, mainly made from bamboo and paper.

Catapults throwing incendiaries are mentioned from 1126 to 1287. They were first used in Sung armies, then adopted by the Chin Tartars and the Yuan Mongols. In 1231 the fei huo ch'iang (fire-lance) came into use among the Chin Tartars. The t'u huo ch'iang (impetuous fire-lance) was used in 1259 (see p. 246); it was made of bamboo. Wang Ling mentions that bamboo bound with twine will support an internal pressure of 80 lb. per sq. in. Even in the last few years the Chinese were using wooden cannon (which were used in Europe in the seventeenth century, see p. 104). Průšek[46] quotes from a book published in 1946 by an eye-witness:

Une bûche de bois dur, d'environ 10 pieds de long, est creusée pour faire un canon qu'on double parfois de métal. L'extérieur est solidement entortillé de fils de teléphone pris aux Japonais, pour empêcher de se fendre. La charge, ne pesant pas moins de cinq livres, davantage si l'on veut, est faite de pots cassées, de morceaux de poêles à frire, de verre, de pierres, de tous les débris que l'on peut avoir sous la main.

Apart from the fact that the old weapons were of bamboo, this, and its charge, is the old t'u huo ch'iang, which Průšek mistakenly thought was a bomb (obus susceptible d'être jeté contre l'ennemi). A wooden gun from Cochin China in the Paris Artillery Museum is described[112] as in two halves hollowed out of wood and bound together with iron hoops.

THE WU PEI CHIH

After Mayers, the *Wu Pei Chih* (1628) was studied by Davies and Ware,[48] who reproduced forty-seven illustrations from it. They say uncritically that "guns were in use in China . . . probably as early as 1236," that "bullets" were fired in 1259, and other things which we have shown are highly improbable, and they remark (as has so often been done) that "from fireworks firearms were undoubtedly derived." They also mention the *Têng T'an Pi Chiu* of Wang Ming-Hao, written soon before 1600, from which nearly all the illustrations were incorporated into 222 single pages of the *Wu Pei Chih*. Some earlier accounts are by authors "whose history is very beclouded."

Davis and Ware give a long list of materials specified in the *Wu Pei Chih* (119/11a–b), including saltpetre, sulphur, various kinds of charcoal, arsenic compounds, mercury compounds and many vegetable and animal materials, among which are aconite, varnish, blister-fly, oils and fats, and snake poisons. They mistakenly supposed that Greek fire contained saltpetre. The idea of combining toxic with incendiary or explosive mixtures is also found in older European works (see ch. IV).

The *Wu Pei Chih* gives several recipes for gunpowder[113] and some material from it, translated by Amiot, has been dealt with previously (pp. 251–258).

(1) *Fire powder* (for fireworks). Saltpetre 5 lb., sulphur 1 lb., egg-plant, willow, or pine-branch charcoal[114] 1 lb. [A good 5 : 1 : 1 mixture.] Grind [pestle ?] together in three batches, 5800 strokes. Mix 3 powder with 1 spirits. Then shape into pellets the size of panicled millet or green peas. The batch is right if a lighted pellet will not burn the palm of the hand. Use in small quantities, for it is very fast (119/10a).

(2) *Lead* [bullet] *gunpowder*. Purified [N.B.] saltpetre 40 oz., sulphur 6 oz., willow, calabash, or egg-plant-stalks charcoal 6–8 oz. Grind separately till very fine, mix with a little water, dry, pestle 1000 times, and dry in the sun. Do this three times. It is useful to add a spider skin. The charge for a gun is 0·25 oz. or 0·3 oz. [therefore a very small gun] (119/21b). The need for good incorporation of the materials was clearly appreciated, even if the spider skin was superfluous.

(3) *Ordinary gunpowder*. Saltpetre 4 oz., sulphur 0·1 oz., charcoal 0·17 oz., blister-fly 0·1 oz. (119/20b). (The sulphur and charcoal as given seem to be much too small and I think they should be 1 oz. and 1·7 oz.)

(4) *Cannon powder*. Saltpetre 10 oz., sulphur 6 oz., calabash or bamboo charcoal 3 oz., orpiment 1 oz., realgar 0·5 oz. (119/21b). This is a toxic powder but could probably be used in cannon.

(5) *Rising powder* (for rockets?), Saltpetre 1 oz., sulphur 0·3 (or 0·03) oz., litharge 0·04 oz., charcoal 0·3 (or 0·05) oz. (119/21a).

(6) *Spattering fire* (this name, not in the original, was introduced by Mayers). 1 lb. saltpetre, 4·5 oz. sulphur, 3·8 oz. pine charcoal, 1·6 oz. cinnabar. This is gunpowder containing mercury sulphide and the mercurial fumes would be poisonous and would cause burns difficult to heal. Davis and Ware report that charred infected wounds were caused by "ancient" incendiary arrows from a disused arsenal at Hami, and thought they were caused by a mercurial mixture. A mercurial mixture was used in Europe in the seventeenth century (see p. 177).

(7) *A powder combining the three fires* (flying, poison and superior). Saltpetre 1 lb., sulphur 6 oz., bamboo, calabash, willow, or pine charcoal 4 oz., cinnabar 0·3 oz., mercury 0·3 oz., ground till the mercury drops disappeared.

(8) *Blue smoke*. Saltpetre 2 oz., birch bark 1 oz., sulphur 0·05 oz., charcoal 0·1 oz., blue dye (from woad) 0·3 oz. (120/5b). (The indigo might give a purple vapour.)

(9) *Violet smoke*. Saltpetre 1 oz., sulphur 0·3 oz., charcoal 0·1 oz., violet powder (cinnabar) 0·5 oz., hemp oil a trifle (120/6a).

The *Wu Li Hsiao Shih* of Fang I-Chih (*c.* A.D. 1630)[115] says that saltpetre mixed with charcoal explodes straight forward and when mixed with sulphur it explodes laterally. Huang fan (ferrous sulphate?) gives showers of sparks, jo p'iao (gourd?) charcoal makes the explosion noiseless, powdered si kuang stone prevents the flame being seen. Iron dust and camphor make coruscations. These are called yen huo (smoke-fire, fireworks). The idea of "noiseless" and (apparently) flameless gunpowder is found in old European works.

The *Wu Pei Chih* describes military equipment of various dates, some without doubt old. This includes smoke pots packed with paper tubes of composition containing mercury oxide, and cast iron bullets, etc.; "spattering fire" weapons; magical fire-serpent staffs for attacking horses; Roman candles; throwers of toxic incendiary pellets; "string-of-100 bullets cannon" (a copper tube containing fa powder, i.e. gunpowder with mercury and arsenic compounds, plants and human sperm); "magic smoke cannon" (a similar device charged with powder containing wolf-dung); a Roman candle throwing crackers; rocket arrows discharged "like a hundred tigers" all at once from a long basket held under the arm, or from boxes with several perforated partitions; and primitive bamboo hand-guns with handles and shooting arrows or used as fire-lances. These are old devices. Later weapons are a "flying-cloud-thunder shell," of iron, "as round as a bowl," shot ten at a time from a cannon (the old chen t'ien lei used as an explosive shell); and iron pot-shaped mortars about 1 ft. high throwing iron shells filled with gunpowder. Very noteworthy is the tool for boring a hollow space in a rocket propelling charge, which is essential for a long trajectory. Last of all, and possibly of European origin, are different types of cannon and small arms, which will be discussed presently.

The gunpowder recipes in the *Wu Ching Tsung Yao* may be reduced to the following compositions (recalculated on a basis of 40 parts of saltpetre):

					Saltpetre.	Sulphur.	"Charcoal."
Recipe *a*	..	..	..	..	40	21	21
Recipe *b*	..	..	..	..	40	20	56½
Recipe *c*	..	..	..	..	40	30	65
Recipe *d*	..	..	..	..	40	30	19½

The last recipe is one given by Wang Ling,[97] a chin being taken as 16 oz. The "charcoal" in all cases includes plant materials, oils, resin, etc., and some mineral constituents.

Gunpowder recipes in the *Wu Pei Chih* include:

					Saltpetre.	Sulphur.	Charcoal.
Recipe 1	..	..	..	..	40	8	8*
Recipe 2	..	..	..	..	40	6	6·8*
Recipe 3	..	..	..	..	40	10 (?)	17 (?)†
Recipe 4	..	..	..	..	40	24	12†
Recipe 5	..	..	..	..	40	12 (?)	12†
Recipe 6	..	..	..	..	40	11¼	9¼‡
Recipe 7	..	..	..	..	40	15	10‡

The compositions in the *Wu Ching Tsung Yao* are (apart from the materials other than saltpetre, sulphur and charcoal, which would probably tend to weaken the effect) like the 40 : 30 : 30 composition given by Roger Bacon, and this was also the composition of a true "slow" gunpowder used in French

* Charcoal alone.
† With arsenic sulphides or diluents, not counted.
‡ With 4 oz. of cinnabar (mercuric sulphide), not counted.

mines,[116] with a relatively small brisance. Some of the compositions in the *Wu Pei Chih* approach modern gunpowder (75 : 10 : 15 or 40 : 5·3 : 8) and would be more powerful than, although some tend towards, the older composition. The variety of compositions for special purposes must have originated from careful experiments. Although the 40 : 30 : 30 composition without additives can truly explode in a suitable container (as Roger Bacon says, see p. 76), the more complicated mixtures in the *Wu Ching Tsung Yao* would probably be more on the deflagrating side (see p. 266).

It could be argued that it is rather improbable that the composition of European gunpowder was known to Mao Yuan-I, the author of the *Wu Pei Chih* (1628). The contacts with the Portuguese in the sixteenth century had not been very numerous, and the few Jesuits had only been in the country a few decades. The development of the compositions given in the *Wu Ching Tsung Yao* (1044) to the modern gunpowder in the *Wu Pei Chih* could have resulted from Chinese experiments rather than from the importation of European information. Against this must be set the fact that Mao Yuan-I describes Portuguese firearms, and the probability that, as in India (see p. 223) and elsewhere, there were always Europeans able and willing to impart military information for suitable reward. The Jesuits at that time were also supervising the casting of cannon in China. In my opinion it would be very unsafe to assume that Mao Yuan-I, a very well-informed man who had been a military commander, was ignorant of the composition of European gunpowder, or of European firearms.

Some analyses of Chinese gunpowders of fairly recent date all show that they had practically the same composition as European gunpowder.[117] Napier in 1788 found for a specimen of large-grained and hard gunpowder from Canton, which was not very strong, 1 oz. 10 dwt. saltpetre, 3 dwt. 14 grains sulphur and 6 dwt. charcoal. There is a deficiency of 10 grains but the proportions give 75·79 : 9·05 : 15·16. An analysis of Chinese gunpowder made by Wilkinson for Davis (who was in China from 1816) gave 75·7 : 9·9 : 14·4. Berthelot in 1883 gave 61·5 : 15·5 : 23·0, which was repeated (perhaps by copying) by Perry in 1922. Davis says that even in 1824 Chinese gunpowder was made on the field by pounding the materials in stone mortars and a spark which was struck once ignited the whole quantity of powder, the explosion killing eleven persons. I have no information on Chinese gunpowder mills.

GUNS IN CHINA

The statement in the *Wu Yuan* that cannon were invented by Lü Wang in the Chou dynasty (1030–221 B.C.)[97] is incredible to me, although I am unable to refute it. The modern European cannon illustrated in the *Wu Ching Tsung Yao* (A.D. 1044) are a later addition (see p. 262), and it seems safe to say that they were unknown in 1044.

A Japanese account says that in 1265 iron p'ao threw out 2000 to 3000 iron balls at a time, which rolled down the hills with a noise like thunder.[96]

In the invasion of Japan in 1274–81 the Mongols are said to have used iron cannon, according to three different accounts, one of which says they used iron cannon balls.[118] These statements need investigation by linguistic and military experts.

Although Davis[119] suggested that the earliest Chinese cannon were probably wrought-iron tubes bound with hoops, the earliest extant specimens appear to be of cast iron. Bishop[120] depicts a Chinese iron bombard in the Shansi museum which is dated 1378. It is in two pieces, almost exactly like the contemporary European guns (see ch. III), and the parts, before firing, must have been forced together by wedges, as in European practice. Goodrich[121] says the date 1378 should be 1377. He points out that this gun was mentioned by Read,[122] and reports and illustrates some other early Chinese iron cannon, viz. of 1356 (500 catties or 666 lb., in Nan-t'ung museum), of 1357 (350 catties or 466 lb., in Nan-t'ung museum) and a doubtful one of 1426. He thinks iron cannon were used earlier in Korea (see below). The guns shown by Goodrich are long tapering muzzle-loaders, not having separate breeches. They have trunnions for mounting on carriages and are like Bābur's guns of about 1530 (p. 220) and those shown in later editions of the *Wu Ching Tsung Yao* (p. 262). These all await examination.

Sarton[123] says the 1356 and 1357 cannon are described as being from a find of 500 of assorted sizes said to have been made for the anti-Mongol rebel Chang Shih-Ch'êng (d. 1367), who set himself up as emperor in 1354–7 under the title Chou T'ien Yu; also that "there are several hundred cannon in the Peking Historical Museum dating from the time of Hung-Wu and later times down to the middle of the following century." (Hung-Wu began to reign as the first Ming Emperor in 1368.)

Goodrich[124] gives as other dated Chinese cannon, a "huge iron piece" of 1372 in the Peking museum; a bronze (copper?) piece 44·6 cm. long and 3·9 cm. muzzle calibre, December 1372–January 1373, in Nanking; and another bronze (copper?) gun 16 in. long, 1379, in Peking. It is not likely that such guns appeared without prototypes and we should expect metal cannon not very different to have been in use soon after 1300.[118] Ibn Baṭṭūṭa,[125] who was in India and China in 1333–47, and visited Zaytūn, Canton and Peking, although he reports the military use of naphtha, including its use on ships to repel Indian and Arab pirates, says nothing of gunpowder or firearms in China; he may not have been interested in them, or may not have seen or heard of them.

Rathgen[126] refers to a small hand-gun from Java dated (he thinks wrongly) 1340 which is in the Darmstadt Museum and has been thought to be Chinese. A small bronze hand-gun, 35 cm. long, in the Museum für Völkerkunde in Berlin, dated by an inscription in the nineteenth year of the Ming Emperor Yung-Lo, i.e. 1421, is remarkable in having a covered touch-hole (Zündpfanne), which is said to be unknown in European specimens of that date.[127] It had a socket for a wooden handle, like those on early fourteenth-century European guns of similar type (p. 108). Rathgen says it is so exactly like the small guns made at Nüremberg in 1420–30, the Hussite guns, and those made at Speyer, of the same period, that it is probably European and reached China by some overland caravan route. In the sixteenth century, he says, information about Portuguese artillery reached China by way of the Arabs in the traffic with India in the Red Sea, the "Franks" mentioned in the Chinese accounts (p. 278) being the Portuguese.

General Rathgen was an expert on artillery but he failed to notice the similarity in shape and mounting to the earlier Chinese bamboo "gun" (p. 246). The *Wu Pei Chih* shows a copper gun of this type, with a touch-hole in the lowest part (apparently misplaced), charged with an arrow.[128] Post[129] disagrees with Rathgen's thesis that artillery is of German origin, since the earliest gun is Milemete's of 1326 (see pp. 98–100), which is similar in shape to a Japanese howitzer of 1830 in the Berlin Zeughaus. Although Post thought from this that the idea for European guns may have come from the Far East before 1326 (an old suggestion),[130] the transmission might equally well have been in the opposite direction. A bottle-shaped bombard shown in the *Wu Pei Chih*[131] may, as will be mentioned later, be based on a European model. The European hand-guns mounted on wooden staves are almost certainly of native origin, developed from the very early ribaudequins, as has been shown (pp. 102, 109, 116).

Some rather primitive information on Korean artillery is given by J. L. Boots.[132] He quotes from a Korean document that at the end of Korai (A.D. 1329) a Chinese merchant visited the general in charge of weapons along the river Imchin, near Songdo (then the capital of Korea). The general told his servant to treat him kindly "and the merchant showed him how to mix saltpetre. This was the first time we had powder in Korea." Another work (the title is given only in Korean) gives a recipe for gunpowder: 1 lb. of saltpetre, 1 yang (38 g.) of sulphur and 5 yang of "ash of willow" [willow charcoal?] are mixed and ground to a flour. The mixture is moistened with water and pressed into a bowl. For use it is dried in the sun. The best powder is pounded 20,000 times (*sic*). If 1 lb. is 453 g., this gives a saltpetre: sulphur: charcoal composition as 66·5 : 5 : 28, which is a low-nitrate mixture with the sulphur-to-charcoal ratio too small.

"In our country they began to use cannon at the end of Korai [1392]; the official studied it from a Chinese nitrate (*sic*) maker." Plate 20 in the paper shows "early types" of heavy muzzle-loading cannon (which have carrying rings); plate 21 a cannon for shooting arrows and a hand-bomb; plate 22 shows guns and mortars, undated except that some illustrations are from the *Book of Five Ceremonies*, 1474, which says the largest cannon throws a 75-lb. stone ball, 3 ft. in circumference (which would be approximately the correct weight), and mentions a "fire tube," 1 ft. 4 in. in circumference, but the barrel only 2 in. bore, for shooting arrows.

Accounts of guns of 1614 and later are given; the guns shown in plate 23 are quite evidently late. A three-barrel signal gun carried by a horseman is shown (pl. 24), also matchlocks (jingals) (pls. 25–7). It is said (p. 23) that only the hand-guns came from Japan, the cannon and powder from China;

they are described in a Korean book of 1567–1608, earlier than the Chinese *Wu Pei Chih* (1628). Sarton[133] reports that "a general bureau of gunpowder artillery (?) was set up in Korea in 1377 and that the gunpowder and weapons were formally inspected in 1381." No source is given.

In 1593 the Ming Emperor is said to have used cannon and rockets against the Japanese in P'ingyang, Korea.[134] European cannon are said to have been used by the Japanese in 1592, and Korean ships were protected by iron plates from the balls from Japanese guns.[135] Boots refers to a cast iron cannon in Tokyo Museum as perhaps (according to Japanese experts) cast in Korea; it is 2 ft. 2 in. long, $1\frac{3}{4}$ in. bore, and is "an early type of breech-loading cannon" (p. 24). A Korean hand-grenade was a hollow round wooden ball (p. 24). The matchlocks (p. 25) are European; it is said the Portuguese gave matchlocks to the Japanese general Yoshimitsu in 1568, and the earliest matchlocks in Japan have Christian designs. A description of making them, composed after 1590, is given (p. 26), and it is said "there are no guns like these in China. We [the Koreans] got them from the Japanese barbarians." The ball would pierce armour. The Korean matchlocks have Chinese characters on the breech end of the barrel, the Japanese have clan emblems. The earliest laminated iron armour and iron helmets in China appear in the Later Han (A.D. 25–220); perhaps copper was used earlier. There is a definite change of Chinese weapons from bronze to iron in A.D. 219. Metal armour perhaps first appeared in Japan in A.D. 780 (pp. 29–30). These Korean guns require an examination by a military expert, but in any case they are very late.

Pauthier[136] quotes Ch'iu Chün (d. 1495) as saying in his *Ta Hsüeh Yen I Pu* (*c.* 1480) that hui-hui p'ao were first made in the time of the Mongols to make a breach in the walls of Hsiang-yang (see p. 247) in 1273. Although Pauthier calls these "cannon," they were counterweighted trebuchets throwing stones. Pauthier quotes the *Chi Yin Mou* (a work I cannot trace) as saying that these p'ao were of the same form as "the p'ao of our times." They were made of copper or iron, were like a tube, and were filled with gunpowder (yao) and round stones. The mouth (embouchure) was closed and a wick communicated with one side. Fire was used to set them off. This account shows that the later Chinese authors confused p'ao with metal cannon when these became known, and their accounts are unreliable.

Pauthier also quotes the *Huang Ch'ao Li Ch'i T'u Shih*, edited by Tung Kao (sixteen volumes, 1759), a title which he translates as: "Modèles figurés des objets de toutes natures conformément aux rites, à l'usage de l'empereur et de la cour." It says that the official Ming history, *Ming Shih* (*Ping Chih*), compiled in 1739, in the section on war, states that all the early p'ao (machines) were for throwing stones. At the beginning of the Yuan (Mongol) dynasty, p'ao of war had been obtained from the West (Hsi Yü). In 1233 in the siege of Ts'ai-chou, a town belonging to the Chin, fire was first used in these p'ao. But the art of making them was not transmitted and they were rarely used in later times. In the Ming dynasty, the Emperor Ching Tsu (Yung-Lo, 1403–24) procured "p'ao ou canons qui furent nommés les p'ao ou canons retentissant à mouvements surnaturels" (shên chi ch'iang p'ao) to conquer the Chiao-Chih (of Cochin China).[137]   They were made with red copper and soft and malleable iron in the interstices, iron of the West[138] being the best. There were large and small p'ao, which were of different kinds.   The large were moved on cars, and were used for the defence of towns.   The small had sometimes wooden forms (stocks), sometimes stakes in the ground, sometimes simply a spike (levier).   The last were very useful to armies on the march.

Průšek[139] says this account in the *Ming Shih* is "devoid of all truth," that the p'ao of the Muslims or those of the West (Turkestan, he thinks) were stone-throwing ballistas and had nothing to do with gunpowder.   The Mongol engineer who made them for the siege of Fan in 1271, Chang Chün-Tso, was probably of Chinese origin, and he, and not the Muslims, made the chin chih p'ao (molten metal engines) which burnt Sha-yang in 1274.   The theory that a knowledge of gunpowder and cannon came to China by way of Central Asia is thus improbable.   The guns described in the *Ming Shih* were probably European.

Pauthier[136] reports that the *Huang Ch'ao Li Ch'i T'u Shih* says that in 1529 there were made p'ao called "Franks" (Fo-Lang-Chi p'ao), which were generally placed on the flanks to protect them.   Fo Lang Chi is the name of a kingdom [Portugal].   At the end of the Ching-Tê period (about 1521) the ships of this kingdom arrived at Canton and a model of their cannon was obtained from them.   From this the same were made of copper, with a length of 5 or 6 ft.   The largest weigh 1000 chin or more (504 kg.) and the smallest 150 chin (75 kg.).   In the Wên-li reign-period (1573) ships from the great Western Ocean (Europe) again arrived at Canton and large cannon called "red barbarians" were obtained from them.   These had a length of 2 ch'ang (7 m.) or more and weighed up to 3000 chin (1514 kg.).   They could pierce a stone wall and throw it down as by a stroke of thunder at a distance of several leagues.

Malacca was besieged and captured in 1511 by Albuquerque, who sent a party to China to explore openings for trade.   In 1514 a Portuguese diplomatic envoy reached China in a Chinese ship, and visited the Ming court at Peking.   The Emperor officially ostracised the newcomers and refused to countenance trade.   In the southern ports, however, unofficial encouragement to trade was given, and by 1541 the Portuguese were trafficking with Canton.   In 1557 they leased territory at the neighbouring port of Macao, which for two centuries was the principal entrepôt for all far-eastern trade.   In 1542 Mendez Pinto landed and was welcomed in Japan, and in a few years Portuguese ships were trading there.

Fernão Mendez Pinto, who was in the East in 1537–58, reports[140] that his ship was fired on by a Chinese pirate junk mounting fifteen guns; in China he saw pots of wild fire used but no guns.   The Chinese were greatly surprised to see arquebuses used in hunting; they had none themselves and no gunpowder, and thought the effects were due to magic.

The date of the arrival of Portuguese ships carrying cannon in Canton is variously given (1517, 1520). The *Wu Pei Chih* (1628) quotes[141] Ku Ying-Hsiang, a contemporary author, as saying that Fo-Lang-Chi is the name of a country but was given to the cannon, which were 6 or 7 ft. long, were broad and had elongated necks, and were bored longitudinally. They had five small chambers loaded with powder, put in succession into the body of the piece, and fired off. They were hooped with timber to prevent bursting. Portuguese guns of this type are depicted in the *Wu Pei Chih* (see p. 282), and the use of chambers inserted into the piece is found in early English guns (see p. 113). Mayers says the Japanese very soon began to make copies of the Portuguese guns, as well as bell-mouthed blunderbusses with wooden stocks.

Wang Chou-Jen (d. 1529) reports that 100 p'ao fo-lang-chi were used by Huang Kuan in 1519 in Fukien. They are said to have been known in 1510, and since the Portuguese arrived in Malacca only in 1511 it is possible that the name fo-lang-chi was used for the gun before the Chinese knew of the Portuguese. Pelliot[140] mentions that Bābur used the Turkish name farangī for cannon just after 1500, and thinks the name may have reached China by way of Malaya, and become fo-lang-chi.

Odoard Barbosa[142] reported that about A.D. 1500 Chinese ships from Malacca carried saltpetre, as did those trading with Java, which carried many other kinds of cargo, such as iron. He does not say they carried guns. The saltpetre may have come from India. Faria y Souza[143] says that when Albuquerque invaded Malacca in 1511 he captured 3000 large cannon out of a total of 8000, which the king of Malacca had obtained from Pahang. In 1518 the Portuguese captured 300 cannon, some of brass, at Maur, near Malacca.[144] But a Chinese fleet which attacked the Portuguese[145] in 1521 does not seem to have had guns.

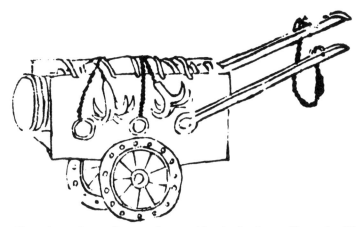

Fig. 14. Kung jung p'ao. Cannon for attacking barbarians. From the *Wu Pei Chih*.

Some of the Chinese guns illustrated in the *Wu Pei Chih* (1628)[146] seem to me to be either European or copies of European guns. The "tiger-crouching-gun", so called from its appearance, has a wooden plug put over the powder charge, weighs 36 lb., has the barrel strengthened by hoops, has a U-shaped piece over the muzzle for holding it to the ground, and a ring with two eyes at the breech for spikes for fixing to the ground. The "attack gun" (fig. 14), made in three sizes (280, 200 and 150–60 lb.), is a hooped gun on a carriage with grapnels to anchor it to the ground. The t'ung (copper, bronze, or brass) siege gun (fig. 15) weighs 500 lb., and discharges at a single

Fig. 15. T'ung fa fan. Siege cannon. From the *Wu Pei Chih*.

shot 100 lead balls each weighing 4 lb. It is "good for shooting up, not down," which suggests that it was used as a howitzer, although it is not shown mounted for this purpose. It is unlikely that such a gun would be strong enough to discharge nearly its own weight of metal. "Small ones are usable at sea, but great precautions must be exercised and the guns ought to be mounted on rafts of their own. Not to be used for defence from a wall"—probably

Fig. 16. Ch'ien tzu lei p'ao. Thousand-bullet thunder cannon. From the *Wu Pei Chih*.

because of the recoil. This gun has a large bulbous breech with trunnions, a muzzle decoration like that on earlier European cannon,[147] and also has a shield. This gun may be the "great commander" (ta chiang-chün) mentioned by Davis and Ware. The cannon mounted on a car (fig. 16) is like a Burgundian cannon; it could fire 1000 shots at once.

A "three-shot matchlock shoulder gun" (fig. 17A) is made of wrought iron, 1·2 ft. long and 0·22 ft. diameter, with a slot in the barrel. It has a wooden stock for carrying. Three cartridges, 0·3 ft. long and 0·17 ft. diameter are inserted into the barrel with the fuses projecting through the slit, each cartridge containing 0·8 oz. of powder and two 0·2 oz. pellets. The gun is put to the shoulder and fired by lighting the fuses with a match. This gun seems to me to be imaginary, since the barrel is open at the top along its length, and hence no propulsive effect could be exerted on a ball.

Fig. 17A. Three-shot match-lock Shoulder Gun.

Fig. 17B. Twenty-shot Gun.

From the *Wu Pei Chih.*

The multiple-barrel guns shown mounted in a circle on a board carried on a horse but fired from some sort of support on the ground, a "cart-wheel cannon," is the old "revolver" described by Kyeser two centuries before. A mobile mortar (No. 68 in fig. 12) which a soldier can carry at his belt, "a hornets' nest of lead shot," is a short gun with a spike at the back for fixing to the ground and a hoop over the front for holding it down. It fires 100 shot 4 or 5 li distance. The muzzle is raised 3 or 4 in. It is "like the barrel of a cock-musket but shorter," and the cock-musket throws only one

Fig. 18. "Changing-seasons" Gun. From the *Wu Pei Chih.*

shot. A narrow iron gun (fig. 17B) 5 ft. long, weighing 15 lb., to be carried, has two hollow ends and a solid middle. The hollow ends are perforated with touch-holes, between which are powder charges and 0·15 oz. pellets, these being separated by paper packing. Ten such charges can be set off one after the other.

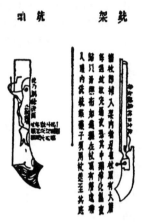

Fig. 19.   Bird-beak gun.
From the *Wu Pei Chih*.

A "changing-seasons" gun, so called because it is fired by four men in succession, is a long thin gun (barrel 2 ft. 2 in. long and stock 2 ft. 9 in. long) weighing 20 lb. It has front and rear sights, the aimer looking along the barrel. The front part is supported in a ring in a stock which is held by another man (fig. 18). With 2 oz. of powder, one large bullet weighing 4 oz. goes 500–600 paces, or 30 small bullets each weighing 0·6 oz. go 300–400 paces.

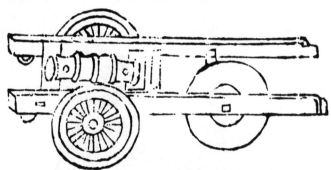

Fig. 20.   Yeh kung shen t'ung ch'ê p'ao.   The Ven. Mr. Yeh's
magic barrel field-gun carriage.   From the *Wu Pei Chih*.

The "bird-beak gun" (niao tsui), a short musket or pistol, is said in the *Wu Pei Chih* to have been captured in a Japanese invasion in 1522, and although diagrams of its parts are given (fig. 19) the author was unable to describe it. Davis and Ware thought it was a Portuguese flintlock (the flintlock was first used in 1520, see p. 168). The heavy gun (fig. 16), the long gun supported in a loop on a stake for firing, the multiple guns, and the match-locks, I would say are European, in spite of Davis and Ware's classification (they show no knowledge of earlier European firearms).

The Fo-lang-chi or Portuguese caliver (ch'ung) of 1517, also shown in the *Têng T'an Pi Chiu* (c. 1600), has a cylindrical barrel and a bulbous breech, behind which is an extension for training the gun (fig. 21). The breech is

Fig. 21. Fo-lang-chi shih. Portuguese-style culverin. From the *Wu Pei Chih*.

fitted with a lid having a hoop for a handle and what seems to be a fuse projecting from the charge. This type of gun, with the loading into an aperture in the body and not into a detachable breech, is like the European gun described in the Warwick manuscript (see p. 113). Ch'i Chi-Kuang (d. 1585) in his *Chi Hsiao Hsin Shu*, quoted by Davis and Ware,[48] says the bullets are slightly larger than the chamber opening and are rammed in for 1 in.; his *Lien Ping Shih Tsa Chi* (5/18b–20a) describes and illustrates a Fo-lang-chi with nine chambers separately loaded with complete rounds of ammunition, each chamber being bolted to the gun. Ch'i Chi-Kuang says that the old guns had a wooden "horse" (plug) and lead shot, which had a low velocity— the old Steinbüchse, in fact. The bullet was rammed in with a flat-headed ramrod, which distorted it, but "to-day we make a concave-ended ram of iron, and ram the bullet into the barrel to a depth of 0·8 in.," so that it remains round and will travel more than 1 li.

For small firearms the Chinese, say Davis and Ware,[48] were indebted to the Japanese, who derived their knowledge of them from Portuguese ships. The bird-beak gun (niao tsui) shown in the *Wu Pei Chih* is a small European flintlock pistol.

The Chinese call narrow tapering guns (6–14 ft. long), carried on the shoulders or set on tripods, jingals, which name was, apparently, introduced by the Manchus in the Ming dynasty (1368–1644) and is said to be derived (although this is doubtful) from the European espringal, used from about 1435 for small copper guns.[148] The Chinese jingals were charged with a handful of gunpowder and a $2\frac{1}{2}$ oz. bullet. In naval warfare grenades consisting of earthenware pots containing 2 to 3 lb. of gunpowder, with a lid cemented on, and a fuse, were used.[149] Staunton[150] thought that some small holes which he saw in the Great Wall were for swivel-guns.

In 1618 large guns were cast in Peking under the supervision of Jesuits.[151] Cannon were sent from Macao (Portuguese) to Peking in 1621. There was an accident when they were discharged, a Portuguese and four Chinese being killed. They were used against the Tartars, who were entirely ignorant of their effects.[152] Other accounts[153] say that the Jesuit Adam Schall was, in 1626, set the task of supervising the casting of twenty cannon, since the Ming Emperor, much impressed by the Portuguese cannon, believed that all Europeans must know the art of making them. Some heavy Chinese guns cast under the supervision of the Belgian Jesuit Father Ferdinand Verbiest, Schall's assistant, who was in China from 1659, were brought from China to Vienna and there is (or was) an older one, of 1641, in Berlin.[154] Verbiest supervised the casting of 300 cannon, blessing them and giving them the names of saints, and he also wrote a treatise on artillery. Pope Innocent XI in 1681 sent Verbiest a letter praising him for having, by his knowledge of profane sciences, contributed to the welfare of the Chinese. On his death the Emperor of China composed an encomium which was read at his funeral.[155]

If the Chinese had made good iron guns at the end of the Yuan dynasty (A.D. 1368) they do not seem to have acquired the art of casting large bronze cannon during a period of nearly three centuries, when the Jesuits appeared with a knowledge of European cannon. They often did fail to develop their technical achievements.

TOXIC MILITARY DEVICES

The *Têng T'an Pi Chiu* (*c.* 1600) and the *Wu Pei Chih* (1628) give compositions containing arsenic and mercury compounds which would produce toxic smokes.[156] Brock[157] mentions that the natives of Brazil generated a smoke by sprinkling red peppers on pans of glowing charcoal, "a no doubt effective weapon against their Spanish invaders," and toxic smokes were used in Europe at least as early as A.D. 1400. The addition of vegetable drugs, green vitriol, borax, etc., to Chinese compositions would tend to produce irritating smokes.[158] Arsenic sulphide or oxide would form a poisonous smoke and it also makes the noise of gunpowder crackers louder and sharper.[159] The Chinese for long had a knowledge of arsenic minerals (realgar and orpiment) and of cinnabar (mercury sulphide) and of their toxic properties, and such additions to pyrotechnic compositions would probably suggest themselves early. They are mentioned in the *Wu Ching Tsung Yao* (1044), as has been explained, as well as in the *Wu Pei Chih* (1628), and were certainly known to the Taoist alchemists. If the latter, as is probable, also discovered the properties of saltpetre, the use of such combinations of lethal agents such as we find in the recipes would be quite natural.

FIREWORKS

It has often been said[160] that gunpowder was first used by the Chinese for fireworks. Davis[161] said the Chinese fireworks of his time were inferior to the European, the rockets being bad but the blue fire usable on European ships, and Brock[162] said that the Chinese had made no progress in pyrotechny. D'Incarville[163] supposed that if the Chinese gunpowder of his time was better than the French, this was due to the purity (goodness) of the ingredients, since the Chinese had made no improvements in the manufacture, still corned it badly, and could not polish it. Even their fireworks were no better than the European; they did not use stars, and their largest rockets were only 5 in. long and 8 lines in internal diameter.

The Jesuit Gabriel de Magalhâes, who was in China from 1640, was, however, favourably impressed by Chinese fireworks, which he describes[164]:

je fus particulierement surpris de la machine suivante. C'estois une treille de raisins rouges, dont tout la menuiserie brûloit sans se consumer, pendant qu'au contraire les seps, les branches, les feuilles, les grappes & les grains se consumoient peu à peu, ensorte toutefois qu'on ne laissoit pas d'y voir toûjours le rouge des grappes, le verd des foüilles & le chatain de la vigne si bien representez, qu'on auroit juré que toutes ces choses estoient naturelles & non pas contrefaites . . . le feu . . . sembloit avoir quitté sa nature pour suivre les preceptes de l'Art.

The red and green colours are noteworthy (see p. 268). Jean Baptiste Tavernier[165] reported that in 1676 he saw in the west of Java Chinese "grenades, fusees, and other things of that kind that run on the water; for the Chinese surpass all the nations of the world in this respect."

The use of powdered cast-iron scale ("iron sand"), presumably the oxide Fe_3O_4, is said by Brock[166] to be "an important contribution to Western pyrotechny," including the coating of the mixture (which otherwise soon spoils by rusting of the particles) by roasting it in t'ung oil, a detail overlooked by d'Incarville which was rediscovered in Europe only after "much trouble." Iron and steel filings, however, are much better than the Chinese "iron sand" and soon replaced it in Europe.

The *Huo Hsi Lüeh* of Chao Hsüeh-Min has been said[167] to be the only Chinese textbook on civil pyrotechnics, although there are earlier works on military pyrotechnics. It was probably written before A.D. 1753 and shows, according to Davis and Chao Yün-Ts'ung, that the art was more advanced than in contemporary Europe. They concluded that it owes nothing to European works, because it does not mention antimony sulphide, although this common constituent of European fireworks occurs abundantly in China in Hunan province, which now supplies the bulk of it used in Europe. Early bronzes from China, Korea and Japan contain antimony.[168] The *Huo Hsi Lüeh* mentions the deleterious effect of common salt in saltpetre (hsiao), which is "the essence of the Moon (yin)," and its refining by boiling with a little glue and white turnips, and crystallising. Charcoal (the soul of fire) of different kinds, iron filings, matches (fuses) and containers (clay, paper and iron cylinders), are mentioned. Colours are made by orpiment, lazuli, cinnabar, calomel and resin. Set-pieces are specified. Fireproof screens are made from paper impregnated with alum, zinc carbonate (lu kan shih), fuller's earth (shih chih fên), or lime. Fireworks for use under water are described, and gunpowder is mixed with lacquer and cast to prevent the effect of moisture (not in use in Europe). Coated and composite grains thrown from rockets, Roman candles, etc., to produce stars, and daylight fireworks and coloured smokes are all mentioned.

Professor Davis was an expert on pyrotechny, but in all his writing on Chinese alchemy and technology he either minimises Western achievements or fails altogether to mention them, and in consequence his statements and conclusions are misleading. In the present case he has underestimated the standard of European books on pyrotechny which are contemporary with or even earlier than the Chinese book.[169] The mention of saltpetre as "the essence of the moon" in an eighteenth-century Chinese work is interesting. The Chinese theory of the explosion of gunpowder is that saltpetre is female (yin) and sulphur is male (yang) and their union produces fire.[170] The semi-rational theory of the explosion as due to the escape of a gas (Dunst) had been proposed in European works about 1400 (see p. 155). It may be that the biological outlook on chemical phenomena, which retarded the development of European alchemy into chemistry, played a similar inauspicious rôle in China; some statements in Chinese scientific works such as the *Pên Ts'ao Kang Mu* (printed in 1596)[171] are no better than those of Paracelsus (d. 1541).

SALTPETRE

Some early Chinese references to saltpetre (hsiao) have been given (p. 276). A Taoist, "Mao K'ua" (A.D. 756), is reported[172] as saying in his *P'ing Lung Jen* (*Recognition of the Recumbent Dragon*) that on heating huo hsiao (saltpetre) the yin of the air is obtained, which combines with sulphur, carbon and metals (except gold), and this has been assumed to mean oxygen. The book is otherwise unknown.

The *Pên Ts'ao Kang Mu* mentions "tooth saltpetre" (ya hsiao), which may be saltpetre (crystallising in needles), but it is fancifully said to be formed from liquid rock salt and a salt called yen.[173] Marco Polo[174] describes the manufacture of "salt" in the town of Chan Glu (Chang-lo in the province of Chih-li) by percolating earth with water, evaporating, and crystallising. The note quotes Grosier as saying it was "nitre" and identifies it with saltpetre (as did Marshall),[175] but it is clear from Grosier's remark that it was used for washing linen that it was soda, as the name (kien, i.e. chien) he gives it also shows.

Staunton[176] says that a "coarse or unpurified nitre" is abundant in the Pei-chih-li province and was used as a substitute for common salt, as it was in parts of India, where it was called shara (see p. 215), and he thought that gunpowder was probably known in China before it was in Europe because of the large deposits of saltpetre which occur there. Davis[177] says that saltpetre "abounds in the alluvial plains near Peking as much as it does in those of Bengal," which would promote an early discovery of gunpowder. Other authors[178] report that saltpetre was made on the large scale in the eighteenth and early nineteenth centuries from efflorescences in the Shanghai province and that it also occurs in North China and Tibet.

Terrien de Lacouperie[179] said that saltpetre (hsiao shih, which he thought was from the Indian name sārāka, Persian shora, see p. 201) is reported in the Fan Tzu Chi Jan [Chi Ni Tzu, attributed to Fan Li (Chi Jan), fourth century B.C.], which Lacouperie says is "of our era," as being imported through the Lung Tao (South-West Kansu). It was brought about 100 B.C. from India to China by the I Chou route, and from Khun-Lun (Turkestan) about 98–87 B.C. It is dug or extracted from an immense cave near 'Kung Tia Pin', north-east of Lake 'Ta Ly Fu'. I think that a knowledge of saltpetre at such an early date is very unlikely, and the hsiao shih mentioned must mean something else.

CONCLUSIONS

I have now given all the information which is known to me on the use of incendiaries, gunpowder and firearms in China, and the following conclusions seem to follow from it.

(i) Saltpetre was known during the Sung dynasty (A.D. 960–1279). There is some evidence that it was known about A.D. 900, but all supposed mentions of it in earlier times are doubtful. It was probably discovered by Taoist alchemists, and may have been known to Ko Hung (A.D. 265–317).

(ii) If the date A.D. 1044 attributed to the original of the Wu Ching Tsung Yao is correct (and I have no reason to doubt it), mixtures containing saltpetre, sulphur and charcoal, usually with other ingredients such as oils, vegetable matter, arsenic compounds, etc., were known at that time. The name huo yao was used for such mixtures. These were deflagrating (see p. 266) and some were explosive, though probably not so powerfully as modern gunpowder. They may be called "proto-gunpowder." They were used in bombs and not as propellants.

(iii) About 1232 a kind of bomb called a chen t'ien lei was used, which had a powerful effect and seems to have contained something approximating to modern gunpowder. About 1130 the fire-lance (huo ch'iang) had been used. In 1233 it is described as a paper tube, in 1259 a bamboo, from which solid fragments were projected by a charge which can be described as a weak gunpowder. This is the prototype of a gun.

(iv) True gunpowder, of more or less the modern composition, was known in the later part of the Mongol Yuan dynasty (A.D. 1260–1368), when it was also known in Europe. It is uncertain whether it was developed by the Chinese or the Mongols, or even if a knowledge of it came from the West. The statement under (ii) suggests that it was discovered in China.

(v) There are extant Chinese cast iron cannon dated from 1356. If the dating is substantiated by expert examination, these are ten or twenty years older than European iron cannon of comparable size, but the information about the Chinese guns is at present unsatisfactory. The statements about cannon in later Chinese historical works are contradictory and of no real value. The question as to whether cannon were invented in China before they were known in Europe cannot be answered with the information available, and dogmatic statements on one side or the other are worthless.

These conclusions are substantially the same as, but are more precise than, those reached by Reinaud and Favé.[180] The opinion[181] that modern gunpowder was known to the Chinese from the first century A.D. would not require to be denied if it had not very recently been made.[182]

REFERENCES

1. Useful surveys of Chinese history and technology will be found in the following recent works : E. T. Williams, *A Short History of China*, New York, 1928; A. Rey, *La Science Orientale ava nt les Grecs*, Paris, 1930, 333 f.; H. G. Creel, *The Birth of China*, 1936 (earliest period); R.P. Hommel, *China at Work*, New York, 1937; C. S. Gardner, *Chinese Traditional Histriography*, Cambridge, Mass., 1938; Tsui Chi, *A Short History of Chinese Civilisation*, 1942; L. C. Goodrich, *A Short History of the Chinese People*, New York, 1943; F. S. Mason, *A History of the Sciences*, 1953; J. Needham, *Science and Civilisation in China*, Cambridge, 1954, i; 1956, ii (to appear in several volumes, when it will be the only work dealing adequately with the subject).   Some useful older books are : J. B. Du Halde, *Description géographique, historique, chronologique, politique et physique de la Chine et de la Tartarie Chinoise*, 4 vols., f°, Paris, 1735 (the Hague ed., 1736, is inferior; the work contains descriptions of technology not found elsewhere); Sir G. L. Staunton, *An Authentic Account of an Embassy from the King of Great Britain to the Emperor of China . . . taken chiefly from the papers of His Excellency the Earl of Macartney*, 2 vols., 4°, and f° vol. of plates, 1797; J. F. Davis, *The Chinese*, 2 vols., 1836; W. A. P. Martin, *The Lore of Cathay*, New York, 1901; Li Ung Bing, *Outlines of Chinese History*, ed. J. Whiteside, Shanghai, 1914; F. E. A. Krause, *Geschichte Ostasiens*, 3 vols., Göttingen, 1925; H. Maspero, *La Chine Antique*, Paris, 1927.   Some of the older works are, of course, unreliable for historical information but are mentioned here because they contain technical information not to be found elsewhere.

2. Dubs, *Isis*, 1947, xxxviii, 82, says the earliest definite mention of iron is in 513 B.C. Goodrich, ref. 1, 27, 29, says that a "somewhat disputed text" mentions all kinds of iron tools and implements as used in the fourth to third century B.C., and that an iron sword appeared in 300 B.C. Needham, *The Development of Iron and Steel Technology in China*, 1958.

3. Hommel, ref. 1, 28.

4. Krause, ref. 1, i, 106, 157, 386.

5. Gardner, ref. 1; Needham, ref. 1, i, 43.

6. A Wylie, *Notes on Chinese Literature*, Shanghai, 1902, 25; cf. G. Sarton, *Introduction to the History of Science*, 1927, i, 778.

7. *Histoire Générale de la Chine, ou Annales de cet Empire; traduite du Tong-kien-kang mou par . . . de Moyriac de Mailla . . . Missionaire à Pékin*, 12 vols., 4°, Paris, 1777–83. (This work must not be confused with that of ref. 13.)

8. *De la Chine, ou description générale de cet empire, redigé d'après les Mémoires de la mission de Pékin*, in 3 pts., 4°, Paris, 1786.

9. A. Gaubil, *Histoire de Gentchiscan et de toute la Dinastie des Mangoux ses successeurs, conquerants de la Chine*, 4°, Paris, 1739. The section of Gaubil of present interest is reproduced in Reinaud and Favé, ref. 19, 188 f.

10. Cf. J. T. Reinaud and I. Favé, *J. Asiatique*, 1849, xiv, 257–327 (258); Hime, *The Origin of Artillery*, 1915, 86 f.

11. *Recherches philosophiques sur les Égyptiens et les Chinois, par M. de P * * **, 2 vols., Berlin, 1773, and Geneva, 1774; tr. by Capt. J. Thomson, *Philosophical Dissertations on the Egyptians and Chinese*, 2 vols., London, 1795. De Pauw says (1795, ii, 312), that Amiot was ignorant of military matters and the prints sent by him to Paris "did not deserve to be engraved." The Chinese work on *Military Art* translated by Amiot was worthless (*ib.*, i, p. xvii); the firelocks were copied from Portuguese and Spanish weapons and were fired by matches and supported by forks, since they could not use flints, yet the Tartar Manchus used very efficient field pieces (*ib.*, i, 329–32).

12. J. Amiot, *Art militaire des Chinois, ou Recueil d'anciens Traités sur la guerre, composés avant l'ère Chrétienne par différents généraux chinois* . . . 4°, Paris, 1772, with 21 plates (BM 147.c.10).

13. *Mémoires concernant l'Histoire, des Sciences, les Arts, les Mœurs, les Usages &c des Chinois par les Missionaires de Pe-kin*, 16 vols., 4°, 1776–1814 (CUL Ll.24.25–; vol. xv, 1791, with portrait of Amiot; vol. xvi, 1814, *Traité de Chronologie Chinois*, is by S. de Sacy). The "Supplément à l'art militaire des Chinois," in vol. viii (1782), pp. 327–75, plates I–XXX, is dated Peking, 1st September, 1778, and begins and ends with references to de Pauw's book, ref. 11.

14. E. V. Zenker, *Geschichte der Chinesischen Philosophie*, Reichenberg, 1926, i, 5: "einer der genausten Kenner der chinesischen Sprache."

15. I. Favé, in Napoleon, *Études sur le Passé et l'Avenir de l'Artillerie*, Paris, 1862, iii, 7, 39

16. S. J. von Romocki, *Geschichte der Explosivstoffe*, 2 vols., Berlin, 1895-6, i, 49–58.

17. Amiot, ref. 13, 1782, VIII, pref., vi.

18. D'Incarville, *Mém. div. Sav. Acad. Sci.*, Paris, 1763, iv, 66–94.

19. J. T. Reinaud and I. Favé, *Histoire de l'artillerie* 1re *Partie. Du Feu Gregéois des feux à guerre et des origines de la poudre à canon d'après des textes nouveaux*, Paris, 1845, 198, 242 f., 267. (There is a separate volume of plates.)

20. W. F. Mayers, "On the Introduction and Use of Gunpowder and Firearms among the Chinese," in *J. North China Branch Roy. Asiatic Soc.*, Shanghai, 1871, vi, 73–104, illustrations; 74–9, 83–4, 93.

21. H. Cordier, in Sir H. Yule, *The Book of Ser Marco Polo*, additional vol., 1920, 95.

22. Ref. 20, 76.

23. Mayers, ref. 20, 76–9, 83; B. Laufer, *American Anthropologist*, 1917, xix, 71–5; Marco Polo, ch. cxiv; Pauthier, ref. 56, 372, and Yule, ref. 21, 1903, ii, 42–6.

24. Mayers, ref. 20, 80–4. Dr. Needham informs me that the statement in the *Wu Yuan* is to be found in the *Shih Wu Chi Yuan* (A.D. 1085) of Kao Chêng.

25. Mayers and Romocki date it 1621; Goodrich, ref. 1, 205, 1628. It contains a set of maps said to be copies of those used by Chêng Ho in the expedition of 1405 (see p. 217); G. Phillips, *J. Roy. Asiatic Soc.*, 1895, 523. The copy I have seen was printed in Kyoto in 1664.

26. Mayers, ref. 20, 85 f.; Romocki, ref. 16, i, 43.

27. J. T. Reinaud, "Mémoire géographique, historique et scientifique sur l'Inde," in *Mém. Acad. Inscriptions*, Paris, 1849, xviii, pt. 2 (399 pp.), 200 (from a work attributed to al-Mas'ūdī, d. A.D. 957).

28. F. Hirth and W. W. Rockhill, *Chau Ju-kua*, St. Petersburg, 1911, 15, 18. Chao Ju-kua's *Chu fan chih* (Description of Barbarians) was composed about 1225 or 1242-58. It was little known in China but was incorporated into encyclopædias without naming the author.

29. Mayers, ref. 20, 74–7, 83–4, 93 f.; Romocki, ref. 16, i, 43–4 (with Chinese text); Martin, ref. 1, 26, q. the *Ko Chih Ching Yuan*, dates the event in A.D. 1131; Průšek, ref. 46, dates the presentation of the fire arrows in A.D. 970 (from the *Sung Shih*).

30. T. T. Read, *Trans. Newcomen Soc.*, 1940, xx, 119 (127).

31. Information from Dr. Needham.

32. Reinaud and Favé, ref. 10, 284–92; reproduced with partial German tr. by Romocki, ref. 16, i, 45–8, with Chinese text; the translation was from the *Hsü T'ung Chien Kang Mu* (A.D. 1476) of Shang Lu, and the reference given to the *T'ung Chien Kang Mu* is misleading, since this work was completed before the event described. Dr. Needham informs me that the original text is in the *Chin Shih*, written not later than 1345 by T'o-t'o and Ouyang Hsüan; section I in our text is in this, 113/12a f., II in 113/18a f., III in 113/19a (biography of Ch'ih-chan Ho-Hsi), and V in 113/19a. There are older translations by de Mailla, ref. 7, ix, 167, and Gaubil, ref. 9, both from Tartar-Manchu texts. See Cordier, in Yule, ref. 21, 1903, i, 342 (dates 1233); Yule, *ib.*, ii, 167. A translation from the *Chin Shih* by Prŭšek, ref. 46, 256–8, 262, 274, agrees with Julien's, except that it is said: "Parmi les armes avec lesquelles les Mongols attaquaient la ville, il y avait même des *huo p'ao*, qu'on appelait *chen t'ien lei*" at the beginning of IV instead of "à cette époque." In V, however, the weapon is used *against* the Mongols. Prŭšek thinks the use *by* the Mongols was inserted, since the *Chin Shih* was compiled under a Mongol editor.

33. Hime, *Gunpowder and Ammunition*, 1904, 168.

34. Mayers, ref. 20, 85, 91.

35. Ref. 10, 288; Feldhaus, ref. 127 (1931), 51, thought the "kites" were "flying dragons" like that described by Kyeser (1405); see Romocki, ref. 16, i, 161, fig. 22.

36. Ref. 32, 291: "*les pao à feu éclataient* s'applique aux éclats de la flamme qui sortait par les ouvertures"; Mayers, ref. 20, translating from the *Wu Pei Chih*, has "the fire burst out from them."

37. M. Berthelot, *Sur la Force des Matières Explosives*, Paris, 1883, ii, 354: "Gaubil a fait observer avec raison que la machine . . . *ho pao* n'est probablement pas le canon, mais plutôt une machine à fronde, lançant des pots à feu dont la flamme s'étendait au loin." The machines used at Hsien yang (see ref. 63) were, he says, of the same type.

38. Hime, ref. 10, 92. A picture of bombs lowered from a wall and throwing men about is shown in a Chinese work of 1726; the bombs, curiously enough, are not bursting; Feldhaus, ref. 127 (1931), 57.

39. Martin, ref. 1, 25.

40. Goodrich, ref. 1, 13, 68; Schlegel, ref. 91, mentions the expedition of 1293, but says only that *p'ao* were used, not "cannon." Mason, ref. 1, 62–3, states that in 1233 the Mongols captured a Chinese "ammunition works," the Mongol general Subutai a few years later leading the Mongol invasion of Europe.

41. Davis, ref. 1, i, 229; see also K. Faulmann, *Illustrierte Culturgeschichte*, Vienna, Pest, and Leipzig, 1881, 300; F. Hirth, *Chinesische Studien*, Munich and Leipzig, 1890, 260.

42. Mayers, ref. 20, 91; Feldhaus, ref. 127 (1931), 58 f., supposed that iron bombs filled with gunpowder were used in China about 1175 but admitted that he had no texts to refer to in support of this. See also Schlegel, ref. 91.

43. Romocki, ref. 16, i, 55.

44. Romocki, ref. 16, i, 49; Hime, ref. 10, 144, who mentions, *ib.*, 140, that incendiary arrows were still used by the Chinese in 1860.

45. Ref. 1, 26; from the *Sung Shih* (A.D. 1345), 197/2a, according to Dr. Needham.

46. J. Prŭšek, "Quelques remarques sur l'emploi de la poudre à canon en Chine," in *Archiv Orientální*, Praha, 1952, xx, 250–277 (scories de fer, t'ie-tsï, probablement au lieu de t'ie-ch'a-tsï).

47. Favé, ref. 15, iii, 39; Romocki, ref. 16, i, 57 (with Chinese text); Cordier, in Yule, ref. 21, i, 342 (dates 1260); Prŭšek, ref. 46, 258, translated: "Quand cela fut allumé et dechargé, il brûla (le bambou?), après quoi la mitraille explosa en faisant du bruit comme un coup de *p'ao*. On put l'entendre à la distance de plus de 150 pas."

48. Amiot, ref. 13, viii, plate, fig. 69; Mayers, ref. 20, 101–3, fig. ix; T. L. Davis and J. R. Ware, "Early Chinese Military Pyrotechnics," in *J. Chemical Education*, 1947, xxiv, 522–37 (533, fig. 33).

49. Lippmann, *Abhandlungen und Vorträge*, Leipzig, 1906, i, 133 (Lippmann is notably weak in his sections on China).

50. Alan St. H. Brock, *A History of Fireworks*, 1949, 232.

51. Reinaud and Favé, ref. 10, 292–5; ref. 19, 195; R. Levy, *J. Roy. Asiatic Soc.*, 1946, 74–78, says the letters of Rashīd al-dīn (not his historical work, the *Jāmi' al-tawārīkh*, used here) are apocryphal and probably of the fifteenth century. Goodrich, ref. 1, 169, gives 1253–8 for the events and speaks of "1000 engineers from China" instead of a thousand "families." Under Ghāzān Maḥmūd (1295–1304), Rashīd al-dīn attained great power as prime minister, but in 1318 he was executed on the charge of having poisoned Uljāi'tū.

52. Hime, ref. 10, 94; Romocki, ref. 16, i, 51; for the ambiguous use of the word *naft* see, however, ch. V.

53. Minorsky, *Encyclopædia of Islam*, 1936, iii, 261 (265).

54. *Histoire des Mongols de la Perse, écrit en Persan par Raschid-Eldin*, tr. by Quatremère, in *Collection Orientale. Manuscrits Inédits de la Bibliothèque Royale. Traduits et Publiés par ordre*, Paris, 1836, i, 401–3 (only one vol. published).

55. C. Langlois, *Histoire littéraire de la France*, 1931, xxv, 232–59; A. L. Moule and P. Pelliot, *Marco Polo, The Description of the World*, 2 vols., London, 1938.

56. Pauthier, *Le Livre de Marco Polo*, 2 vols. (pagination continuous), Paris, 1865, ii, 474; Moule and Pelliot, ref. 55, i, 318–9, follow MS. C; Romocki, ref. 16, i, 64–6; Reinaud and Favé, ref. 10, 303–6; ref. 19, 195; Davis, ref. 1, ii, 229 (no mention of cannon by Polo); Yule, ref. 21, 1903, ii, 167 (for the mangonels, *ib.*, 159). In Marsden's translation the siege of Sa-yan-fu is in ch. lxii of bk. II; Everyman ed., 1918, 280–2. De Pauw, ref. 11, 1795, i, 309, had found it "surprising" that the return of Marco Polo to Venice "was so soon followed by the invention of both powder and cannon in Italy." He does not comment on the accidental resemblance between the names Marco and Marcus Græcus (Græcus was probably Rūmi, European, in an Arabic text).

57. Bk. i, ch. 1, § 3, Everyman ed., 18, 20; Layassa (Laiassus in the Latin text, Giazza in the Italian) is a port on the northern side of the Gulf of Scandaroon or Issus (Asia Minor). In Moule and Pelliot, ref. 55, i, 80, the Polos arrived in Laias in 1260 and there heard of the death of Pope Clement IV, which actually occurred in 1268!

58. Romocki, ref. 16, i, 67; Reinaud and Favé, ref. 10, 300–2 (with Arabic text).

59. Kallinikos, the reputed inventor of Greek fire, also came from Baalbek (or Heliopolis); see ch. I.

60. B. Rathgen, *Das Geschütz im Mittelalter*, Berlin, 1928, 611–13.

61. Ref. 15, 1851, ii, 26 f., plate III; Romocki, ref. 16, i, 72; other reconstructed machines in Feldhaus, *Die Technik der Vorzeit*, Leipzig and Berlin, 1914, 383, figs. 247–52. Romocki points out that such machines were still used when cannon were known; they threw stones weighing some hundredweights *inside* fortified places by using a high trajectory, whilst cannon, with a flat trajectory, were used to batter down walls.

62. Gaubil, ref. 9, 155; Průšek, ref. 46, 259.

63. Reinaud and Favé, ref. 10, 297–300; Romocki, ref. 16, i, 62–3; Průšek, ref. 46, 259 (from the *Sung Shih* 450/2a). De Mailla had translated: "Les bateaux étaient armés de flèches à feux, et de machines qui, au moyen d'une poudre inflammable, lançaient des pierres et des charbons allumés."

64. Ref. 56, 1865, 473 (from *T'ung Chien Kang Mu*, "Développment"); Romocki, ref. 16, i, 63; Průšek, ref. 46, 259 (from the *Yuan Shih*). Průšek, 259–60, gives other statements for 1273–77.

65. *Decline and Fall of the Roman Empire*, ch. lxiv; Everyman ed., vi, 281.

66. Reinaud and Favé, ref. 10, 307–8; Romocki, ref. 16, i, 54, 76.

67. M. Prawdin, *The Mongol Empire, its Rise and Legacy*, tr. from the Russian by E. and C. Paul, London, 1940, 259–60; no authority is quoted.

68. Ref. 1, 148. I have wasted a good deal of time trying to find the source for Goodrich's statement; it may be in Voltaire.

69. Ref. 65, ch. lxiv, Everyman ed., vi, 277, 285; Thwrocz, *Chronica Hungarorum*, in *Scriptores Rerum Hungaricvm*, ed. Schwandtner, Vienna, 1746, i, 150.

70. Ref. 16, i, 76; Mouradja d'Ohsson, *Histoire des Mongols depuis Tchinguiz Khan jusqu'à Timour Bey ou Tamerlan*, 4 vols., Hague, 1834–5, iv, 555, 563; Joseph von Hammer-Purgstall,

Geschichte der Ilchane, das ist, Mongolen in Persien, 2 vols., Darmstadt, 1842–3, ii, 227 f. (ninety ballista, seventy machines with iron hooks, 100 bottles of naphtha); G. Strakosch-Grassmann, *Der Einfall der Mongolen in Mitteleuropa,* Innsbruck, 1893, 85, mentions only "brennende Zeug," and refers to d'Ohsson, ref. 66, ii, 142, 622, who says nothing about the incidents reported by Prawdin.

71. Joannis Długossi sev Longini, *Historiæ Polonicæ Libri XII. Quorum sex posteriores, nondum editi, nunc simul cum prioribus, ex manuscripto rarissimo, in lucem prodeunt,* 2 vols., fº, Leipzig, 1711–12, i, 679 (he says nothing about cannon); J. de Guignes, *Histoire Générale des Huns,* 4º, Paris, 1757, iii, 98; Feldhaus, ref. 127 (1931), 51–2.

72. Pauthier, ref. 56, 238–9; Yule, ref. 21, 1903, i, 342; Průšek, ref. 46, 261, from the *Yuan Shih* 162/8b.

73. Pauthier, ref. 56, 475, 540, 547.

74. Ref. 56, 512; Pauthier, and Lippmann, ref. 49, 1913, ii, 286, gave siao (*hsiao*) as "soda" and kân (*chien*?) as "saltpetre," but I think these identifications should be reversed.

75. Jean Appier, *La Pyrotechnie de Hanzelet,* Pont à Mousson, 1630, 162; Davis and Ware, ref. 48. On Appier, see ch. IV. The suggested sequence in the invention of rockets given in the text is due to Romocki, ref. 16, i, 56.

76. Ref. 16, i, 40, 49, 50–58. He mentions the *Wu Pei Chih* (p. 40) and says (p. 50) that much of Amiot's account is taken from it and also includes some eighteenth-century material.

77. Robert Morrison, *A View of China for Philological Purposes,* 4º, Macao, 1817, 21: "Wei-shing made engines for throwing stones in which he used powder. His powder was made of saltpetre, sulphur and willow charcoal."

78. Mayers, ref. 20, 84, 87; Dr. Needham referred me to the *Sung Shih,* 368/15b (mentioning huo shih p'ao; see the quotation in ref. 77).

79. Ref. 19, 199, 213 f.; Favé, ref. 15, iii, 68; Lalanne, *Mém. div. Sav. Acad. Inscriptions,* II Sér., Paris, 1843, i, 344.

80. Ref. 9, 70 f.

81. Mayers, ref. 20, 88–90; Romocki, ref. 16, i, 41, fig. 1.

82. Mayers, ref. 20, 93–7.

83. Hime, ref. 10, 86 f., 93–100, 111; ref. 33, 135–40.

84. Reinaud and Favé, ref. 19, 181; Romocki, ref. 16, i, 157.

85. Mayers, ref. 20, 95.

86. F. Hirth, *China and the Roman Orient,* Leipzig, 1885, 55 (Ta Shih = Arabs), 63 f., 90; Reinaud and Favé, ref. 19, 201; Tsui Chi, ref. 1, 185–92 (Indian traffic); Needham, ref. 1, i, 191–248; see ref. 28. Hirth's identification of Ta Ch'in with the Roman Orient, Syria, has been criticised (Chavannes, *T'oung Pao,* 1904, v, 163; Yule, *Cathay and the Way Thither,* 1915, i, 44), but it is now, apparently, accepted.

87. H. M. Elliot, *The History of India as told by its own Historians,* ed. J. Dowson, 1872, iv, 103.

88. Ref. 37, ii, 354–5; *Rev. Deux Mondes,* 1891, cvi, 808; *Archéologie et Histoire des Sciences,* 1906, 220 (in *Mém. Acad. Sci.,* Paris, 1906, xlix).

89. H. A. Giles, *A Chinese Biographical Dictionary,* London and Shanghai, 1898, 21 (no. 52); Cordier, in Yule, ref. 21, 1903, ii, 596, says this is a commonly accepted date, but he thinks the publication of Schlegel, ref. 91, shows that they were used earlier.

90. W. E. Geil, *The Great Wall of China,* Shanghai and London, 1909, 82, 85, 151.

91. Schlegel, *T'oung Pao,* 1902, iii, 1 (very careless and uncritical).

92. Pelliot, *T'oung Pao,* 1922, xxi, 399 (432), quoting Mayers, ref. 20, and mostly relying on him; he has no new material of any consequence.

93. Laufer, ref. 23.

94. Sarton, ref. 6, 1947, iii, 830, 1151, 1549; Needham, ref. 1, i, 267 (Chinese census of locations and contents of still existing volumes, published in 1939).

95. Ref. 1, i, 134, 261.

96. L. C. Goodrich and Fêng Chia-shêng, "The Early Development of Firearms in China," in *Isis,* 1946, xxxvi, 114–23, 250.

97. Wang Ling (Wang Ching-Ning), "On the Invention and Use of Gunpowder and Firearms in China," in *Isis*, 1947, xxxvii, 160–78. This is one of the best papers on the subject but needs some revision.

98. A. Wylie, *Notes on Chinese Literature*, Shanghai, 1902, Military Affairs, 89–92 (several forgeries).

99. Průšek, ref. 46, 269; "tout dépend de la véracité du traité chinois [the *Wu Ching Tsung Yao*] fixant la date de la fabrication de la poudre à canon à des fins militaries aux environs de 1040–41, véracité qui est encore à vérifier," but he does not appear to have seen the work. Dr. Needham informs me that the *Wu Ching Tsung Yao* is named in the official bibliography of the *Sung Shih*; more could hardly be expected.

100. List in L. Wieger, *Taoïsme*, 2 vols., Hien-hien, 1911–13, i, *passim*; short bibliography in Partington, *A Short History of Chemistry*, 1957, 33. Dubs, *Isis*, 1947, xxxviii, 62, quoted there, is in error in stating that transmutation by projection does not occur in Greek alchemy, and T. L. Davis is quite uncritical in his treatment of Chinese alchemy.

101. Information from Dr. J. Needham.

102. Davis and Ware, ref. 48, 524, who take 1 *catty* (lb.) = 16 *liang* (oz.) = 10 *ch'ien*; 1 *ch'ien* = 10 *fên*, and say that at present 1 *catty* = 500 g. Mayers, ref. 20, 103, took 1 *catty* = 1⅓ lb. and used *tael* and *mace* for Davis and Ware's *liang* and *ch'ien*. Pauthier, ref. 56, took 1 *catty* = 0·5 kg.; or 1·1 lb.

103. Wang Ling, ref. 97.

104. Ref. 96; Fêng Chia-Sheng, "The Muslims as the Transmitters of Gunpowder from China to Europe," in *Histor. J. Peiping Nat. Acad.*, 1949, VI, i (51 pp. and 7 plates; in Chinese).

105. See the long discussion by Průšek, ref. 46, 255, 270, 276, who quotes a description from the *Sung Shih* of a battle in A.D. 1161 in which machines throwing stones and *huo niu chin chih* are mentioned. He translates *chin chih* as "jus d'or," and *huo niu chin chih* as "jus d'or du bœuf brûlant," concluding that *chin chih* was an incendiary. The interpretation in the text is Dr. Needham's.

106. Goodrich and Fêng quote this item from E. H. Parker, *The China Review*, Hongkong, 1901, xxv, 257, who gives no reference for the statement.

107. Wang Ling, ref. 97.

108. Berthelot, ref. 88 (1906), 181, 203, 221, 263; E. O. von Lippmann, *Entstehung und Ausbreitung der Alchemie*, Berlin, 1919, i, 449 f.; 1931, ii, 17; 1954, iii, 18. For translations of Ko Hung, see Wu and Davis, *Proc. Amer. Acad.*, 1935, lxx, 221; 1941, lxxiv, 287; E. Feifel, *Monvmenta Serica*, Peking, 1941, vi, 113.

109. Fêng Chia-Shêng, "The Discovery of Gunpowder and its Diffusion, I," in *Histor. J. Peiping Nat. Acad.*, 1947, xxix (in Chinese); the passage was independently discovered by Ts'ao T'ien-Chin in a microfilm of the *Tao Tsang* at Caius College, Cambridge. The translation in the text is Dr. Needham's, from the work, p. 3a, *Tao Tsang*, 917.

110. [Louis] Dutens, *An Inquiry into the Origin of the Discoveries attributed to the Moderns; Wherein it is demonstrated That our most celebrated Philosophers have, for the most part, taken what they advance from the Works of the Ancients*, 1769, 262. In spite of its faults the book collects a large amount of interesting information, and since it quotes all the texts on which it relies, a reader can always use his own judgment. The work first appeared (in French) in Paris, 1766, and a fourth ed., with considerable additions, Paris, 1812. Dutens had studied Oriental languages.

111. Zosimos, *Historia nova*, bk. v, ch. 41; ed. L. Mendelssohn, Leipzig, 1887, 269–70; Sozomenos, *Historia Ecclesiastica*, bk. ix, ch. 6; ed. R. Hussey, Oxford, 1860, ii, 895.

112. W. Y. Carman, *A History of Firearms. From Earliest Times to 1914*, 1945, 64. The part dealing with China is based on Hime, ref. 10, and is out of date; there are other errors in the book, which is on the whole very interesting and useful, especially for the illustrations.

113. Mayers, ref. 20, 103 (who incorrectly gave 4·8 oz. of charcoal in Recipe 6 and said the mixture was "the nearest approach to modern gunpowder"); Davis and Ware, ref. 48, 524–7.

114. Smith, *J. North China Branch Roy. Asiat. Soc.*, 1871, vi, 139, says charcoal of Chinese pine or willow; Davis, ref. 1, ii, 230, willow. The text translated by Amiot, ref. 13, says that charcoal from resinous woods is unsuitable. European practice does not lay much stress on the source of the charcoal, the carbonisation temperature being more important.

115. Mayers, ref. 20, 74–9, 83–4, 93.

116. Berthelot, ref. 37, ii, 311 ("poudre de mine dite lente ... aujourd'hui hors d'usage"); A. Marshall, *Explosives*, 3 vols., 1917–32, i, 74 ("French official").

117. Hon. George Napier, *Trans. Roy. Irish Acad.*, 1788, 97 (104–5); Davis, ref. 1, ii, 230–2; Berthelot, ref. 37, 1883, ii, 286; G. H. Perry, in T. E. Thorpe's *Dictionary of Applied Chemistry*, 1922, iii, 2 f.

118. Mason, ref. 1, 62–3.   Needham, ref. 1, i, 142, says metal-barrel cannon were first used in the Ming army in A.D. 1356, "although the Mongols probably used them half a century earlier." But see p. 265.

119. Davis, ref. 1, ii, 231; he was probably thinking of the "tiger" gun in the *Wu Pei Chih*, Davis and Ware, ref. 48, 534, fig. 36.

120. C. W. Bishop, "The Beginnings of Civilisation in Eastern Asia," in *Smithsonian Report*, 1940 (1941), 431, plate 10.

121. Goodrich, *Isis*, 1944, xxxv, 211 (photographs of guns); cf. Sarton, *ib.*, 177.

122. T. T. Read, "The Early Casting of Iron," in *The Geographical Review*, New York, 1934, xxiv, 548.   Some Chinese cast iron contains phosphorus (from the ore), which makes it more fusible : other analysed specimens are said not to contain any appreciable amount of phosphorus.

123. Sarton, ref. 121; *id.*, ref. 6, iii, 1549; Mason, ref. 1, 62–3, mentions a cannon of 1354, and Needham, ref. 1, i, 422, says "The earliest dated Chinese iron bombard is contemporary with the Battle of Crecy (+1346)," but I have found no other notice of these items. In any case the date on a piece is of little interest to an expert on European firearms; some of these, as was seen in ch. III, are dated much earlier than the Chinese cannon.   Dates cannot replace expert examination.

124. *Isis*, 1948, xxxix, 63 (supplementing ref. 121).

125. Pauthier, ref. 56, 656; Yule, ref. 21, 1903, ii, 252.

126. B. Rathgen, *Ostasiatische Zeitschrift*, 1925, ii, 28, fig. 1; *id.*, ref. 60, 674, 705; A. Demmin, *Die Kriegswaffen in ihrer historischen Entwickelung*, 2 ed., Leipzig, 1886, 669 (the general account of firearms, 661 f., richly illustrated, contains some very interesting information).

127. Feldhaus, ref. 61, 424, fig. 281; *id.*, *Die Technik der Antike und des Mittelalters*, 1931, 59 (fig. 65), 347.

128. Davis and Ware, ref. 48, 533, fig. 31.

129. P. Post, *Zeitschrift für historische Waffen- und Kostüm-Kunde*, Dresden, 1938, xv, 137–41.

130. Isaac Voss, *Variarum Observationum Liber*, 4°, London, 1685, 84, 86, 89.

131. See fig. 14 on p. 279.

132. J. L. Boots, D.D.S., "Korean Weapons and Armour," in *Trans. Korea Branch Roy. Asiatic Soc.*, Seoul, 1934, xxiii, pt. 2, 1–37 (read 1931–2 but extended; many plates; the parts of interest are pp. 20 f., as quoted in the text).   The military and technical side is very weak.

133. Ref. 6, 1947, iii, 1549.

134. Tsui Chi, ref. 1, 205.

135. Krause, ref. 1, i, 251.

136. Pauthier, ref. 56, ii, 474; Romocki, ref. 16, i, 60.   F. Hirth, ref. 86, 29, 307, said : "I would advise all Oriental scholars not knowing Chinese not to accept a single sentence out of his [Pauthier's] translations."

137. The *Ming Shih* 92, gives the event under the date 1388.

138. This may mean iron from Szechuan and not European iron.

139. Ref. 46, 254, 275.

140. *The Voyages and Adventures of Ferdinand Mendez Pinto*, tr. by H. Cogan, reprint (somewhat abridged), London, 1891, 120, 245, 263; the original was published as *Perigrinaçao*, 1614; Lisbon, 1762. See Pelliot, *T'oung Pao*, 1948, xxxviii, 86–7, 203–6; A. Kammerer, *ib.*, 1944, xxxix, Suppl. (La découverte de la Chine par les Portugais an XVIème Siècle); C. R.

Boxer, *South China in the Sixteenth Century*, Hakluyt Soc., 1953 (nothing on saltpetre, gunpowder, or cannon). Mendez Pinto does not enjoy a reputation for accuracy, and Dr. Needham thinks there is real doubt whether he was ever in or near China.

141. Mayers, ref. 20, 93–7; Davis and Ware, ref. 48, 536.

142. In Ramusio, *Navigationi et Viaggi*, 3 ed., f°, Venice, 1563, i, 317F, 320E.

143. In R. Kerr, *A General History and Collection of Voyages and Travels*, Edinburgh, 1812, vi, 139, 142.

144. Ref. 143, vi, 171.

145. Ref. 143, vi, 188.

146. Davis and Ware, ref. 48.

147. Favé, ref. 15, 1862, iii, 171, plate 12, fig. 2 (from BN MS. Fonds du Roi 6993, undated but probably 1450–1500); 196, plate 30, fig. 6 (Italian basilisco, *c.* 1450–1500); Burgundian gun on car, *ib.*, 177, plate 16, fig. 5 (1450–1500).

148. Clephan, *Archæological J.*, 1911, lxviii, 57, 95, fig. 8; Feldhaus, ref. 61, 423, fig. 280, from Kyeser's *Bellifortis* (1405).

149. J. Dyer Ball, *Things Chinese*, 1900, 38 (no dates given).

150. Ref. 1, ii, 198; he says that a Chinese claim to an early knowledge of gunpowder "may not be without foundation."

151. Mayers, ref. 20.

152. De Mailla, ref. 7, x, 484.

153. Anon., in *Nouvelle Biographie Générale*, 1864, xliii, 484; Tsui Chi, ref. 1, 195.

154. Rathgen, ref. 126, 1925, 11, 30.

155. Regnard, in ref. 153, 1866, xlv, 1095.

156. See ref. 48.

157. Ref. 50, 265.

158. Mayers, ref. 20, 103.

159. Smith, ref. 114; Prušek, ref. 46, 265, says he was told by Prof. O. Pertold that arsenic was added to modern gunpowder to increase its inflammability.

160. Geil, ref. 90; Davis, ref. 1, ii, 229, says the name *huo yao* ("fire drug") has no reference to guns, and hence gunpowder was first used for fireworks.

161. Ref. 1, i, 304.

162. Ref. 50, 20 f.

163. Ref. 18.

164. Gabriel de Magaillans, *Nouvelle Relation de la Chine, contenant la Description des Particularités les plus remarquables de ce grand Empire*, tr. Bernaut, 4°, Paris, 1688, 128 (an English tr. also appeared in London, 1688); paraphrased by F. Hoefer, *Histoire de la Chimie*, Paris, 1842, i, 10; 2 ed., 1866, i, 12.

165. Tavernier, *Travels in India*, tr. from the 1676 French ed. by V. Ball, 2 vols., 1888, ii, 360.

166. Ref. 50, 23, 155, 189, 231.

167. T. L. Davis and Chao Yün-Ts'ung, *Proc. Amer. Acad.*, 1943, lxxv, 95–107.

168. M. Chikashige, *J. Chem. Soc.*, 1920, cxviii, 917; *id.*, *Oriental Alchemy*, Tokyo, 1936, 67–74.

169. See, for example, the books by Appier (1620, 1630) and Frezier (1706) mentioned in ch. IV.

170. S. Julien and J. Champion, *Industries anciennes et modernes de l'Empire Chinois*, Paris, 1869, 24, 26.

171. Berthelot, ref. 88 (1906), 210–19; F. de Mély and M. H. Courel, *Les Lapidaires Chinois*, Paris, 1896. Hoefer, ref. 164, 1842, i, 9, had mentioned that the *T'ien Kung K'ai Wu* (1637) describes the distillation of zinc and alcohol, and has chapters on saltpetre, sulphur, gunpowder, cannon and firearms, and mines, but he said (which is not correct) that the *Pen Ts'ao Kang Mu* (1596) is of little interest for the history of chemistry; see Smith, ref. 178.

172. J. Klaproth, *Mém. Acad. Imp. des Sciences de St. Pétersbourg*, 1810, ii, 476 (read 1807); mentioned by Murray, *Elements of Chemical Science*, 1818, 103; Duckworth, *Chem. News*, 1886, liii, 250; Jörgensen, *Sammlung chem.- und chem.-techn. Vorträge*, Stuttgart, 1909, xiv, 113;

Partington, *Textbook of Inorganic Chemistry*, 1921, 159; Muccioli, *Archivio de Storia della Scienza*, 1926, vii, 382; Vacca, *J. North China Branch Roy. Asiatic Soc.*, 1930, lxi, 10; Lippmann, *Entstehung und Ausbreitung der Alchemie*, 1931, ii, 66 ("gänzlich unhaltbar").

173. A. von Humboldt, *Asie Centrale*, 1843, i, 326; ii, 45–6.

174. Marco Polo, bk, ii, ch. 50; Everyman ed., 267.

175. Marshall, ref. 116, i, 14.

176. Staunton, ref. 1, ii, 22, 191, 292.

177. Davis, ref. 1, ii, 230.

178. De Pauw, ref. 11, 1795, i, 329; Julien and Champion, ref. 170, 27 f.; F. P. Smith, *Contributions towards the Materia Medica and Natural History of China*, Shanghai, 1871, p. vi; id., "Chinese Chemical Manufactures," in *J. North China Branch Roy. Asiatic Soc.*, 1871, vi, 141 (also reprod. in *Pharmaceutical J.*, 1872, ii, 1008, 1031); D. Hanbury, *Science Papers*, 1876, 217; Berthelot, ref. 88 (1906), 217.

179. Terrien de Lacouperie, *The Western Origin of Early Chinese Civilisation from 2300 B.C. to A.D. 200, or Chapters on the Elements derived from the Old Civilisation of West Asia to the formation of the Ancient Chinese Culture*, London, 1894, 232, 369, 392. As a source he quotes F. Garnier, *Voyage d'Exploration en Indo-Chine effectué pendant les Années* 1866, 1867 *et* 1868 *par une Commission Française . . . publié par les ordres du Ministre de la Marine*, 2 vols., 4º, Paris, 1873, and Atlas. This, i, 521, after describing a stalactite cave says only: "On en retire du salpêtre." Vol. ii contains detailed accounts of geology, mining, metallurgy, botany, agriculture, etc., but as far as I noticed (it has no index) says nothing of saltpetre or gunpowder. Lacouperie's views were fantastic, but his book is a mine of information.

180. Ref. 10, 260.

181. Hoefer, ref. 164, 1842, i, 10; 1866, i, 12, 301; Voss, ref. 130, 83–4: quatenus tamen pulvis iste nitratus bellicis inservit usibus, certum est hæc in parte cedere Europæis, qui in rebus bellicis longe sunt superiores Sinensibus.

182. M. Mercier, *Le Feu Grégeois*, Paris, 1952, 35.

APPENDIX

PRONUNCIATION OF CHINESE WORDS

SINCE many Chinese words used in this chapter are not pronounced as they are spelled, a *rough* indication of the pronunciation is given below. All words not listed are pronounced approximately as they are spelled. The vowels are *e* as in bed, *ee* as in bee, *i* as in kit, *j* as in jet, *oo* as in food, *ow* as in now, *u* as in bun, *ü* French, *y* as in yet.

cha jah	*ching* jing	*hsiang* shi-yang
chang jahng	*ch'ing* ching	*hsiao* shi-ow
chao jow	*chiu* jew	*hsien* shi-en
ch'ao chow	*ch'iu* chee-oo	*hsin* shin
ch'ê cher	*chou* jō	*hsing* shing
chên jun	*ch'ou* chō	*hsieh* shu-e
ch'êng chung	*chu* joo	*hua* hwa
chi jee	*chüan* juan	*huang* hwang
ch'i chee	*ch'üan* chuan	*hui* hway
chia ji-ah	*chuang* jwahng	*huo* hwō
chiang ji-ang	*chun* jun	*i* ee
ch'iang chi-ang	*ch'ün* jüin	*jur* ru
chiao ji-ow	*chung* jung	*kai* gai
ch'ieh chi-eh	*ch'ung* chung	*k'ai* kai
chien ji-en	*fei* fay	*kan* gan
ch'ien chi-en	*fên* fun	*kang* gahng
chih jer	*feng* fung	*kao* gow
chin jin	*hao* how	*kiang* gee-ahng
ch'in chin	*hsi* shee	*ko* gō

k'o kō
kuan gwan
kuang gwahng
k'uei kway
kuo gwo
ku goo
kung gung
k'ung kung
li lee
liang lee-ahng
liu li-oo
leuh lü-eh
mao mow
meng mong
mi mee
mu moo
nao now
niao nee-ow
pai buy
pao bow
p'ao pow
pei bey
pên bun
p'ên pun

pêng bung
p'êng pung
pi bee
p'i pea
pien bi-en
po bō-a
pu bu
sêng sung
shuang shwang
shên shun
shêng shung
shih shee
shu shoe
shui shway
sui sway
ta dah
tai dye
t'ai tie
t'an tan
t'ang tahng
tao daw
t'ao tow
têng dung
ti dee

t'ieh tia
tien di-en
t'ing ting
t'o toe
tsa dzah
tsang dzahng
ts'ao tsow
tsê dzu
tsu dzoo
tsui dzway
tsung dzung
tu doo
t'u too
tung dung
tzu dz-oo
wei way
wên won
wu woo
yao yow
yo yo-a
yüan yü-an
yüeh yü-eh

Chapter VII

SALTPETRE

THE names "nitre" used in the English Bible,[1] nitrum in the Vulgate and νίτρον in the Septuagint, all translate a Hebrew word neter, which means "soda" (sodium carbonate) and is connected with natar, to produce effervescence, i.e. with vinegar.[2] That "effervescence" could include "deflagration," and that neter, νίτρον, and nitrum meant saltpetre[3] are very improbable suggestions. Bochart[4] had read neter as nitrum without comment.

Native soda was used in Egypt from about 3000 B.C. and the name νίτρον may be derived from an Egyptian name ntrj for soda.[5] A Coptic papyrus (ninth to tenth century A.D.) mentions both soda (nitron) and Arabian nitron (sosm narabikon); the old Egyptian name for soda was hosmen or hesmen, and the common Coptic name was qosen, but bsn and bsm also occur in Egyptian texts, as well as a "red nitrum" (hesmen tešr).[6] The usual meaning of neter or ntrj in Egyptian texts is incense.[7] The Hebrew neter may be a Phœnician word.[8]

The form λίτρον used by Herodotos[9] is only the Attic form of νίτρον, and is not an older form of the latter.[10] The Egyptian soda lakes are called Νιτρίαι by Strabo.[11] There is little or no doubt that the Hebrew neter and the Greek νίτρον always meant soda (sodium carbonate) and never saltpetre.

Aristotle[12] says nitron (νίτρον) is fusible and soluble or coagulated by heat like cream. It has an astringent taste[13]; when "foam" is coloured red it is called aphro[nitron].[14] The waters of certain lakes (Ascanius in Bithynia; Pæsa in Palestine) give nitron by evaporation; it can be used for washing clothes but causes them to fall to pieces on standing.[15]

The longest account of nitrum given by a classical author is that of Pliny.[16] His account is confused and is obviously taken from several sources now lost; among the authors given at the end of this book are Theophrastos, Apion, Demokritos and Anaxilaos, who are possible sources. Pliny says nitrum was not properly known to physicians and only Theophrastos had given it great attention. The works of Theophrastos which have come to us, however, mention νίτρον only in passing; it has a bitter taste and is used by the Egyptians in growing cabbages.[17] Pliny, therefore, must have used a lost work of Theophrastos.

Pliny says nitrum occurs near Philippi in Thrace but contaminated with earth, and is called agrion ("wild"), and in small quantities in hot and dried-up valleys in Media, where it is called halmyrax ("salt bursting from the earth"). Bailey[18] thought halmyrax might be saltpetre; also the kind collected in Asia, where, Pliny says, it is found oozing from the soft sides of certain mines and was called colyces; it was then dried in the sun.[19] The best nitrum was from Lydia, which was very light, porous (fistulosum), and purple in

colour, and was imported in tablets. That of Chalastra[20], used instead of
salt in making bread, was probably soda. Hoefer[21] thought fistulosum meant
"prismes allongés et creux de l'azotate de potasse." The use of soda in making
bread is also mentioned in the *Geoponika*[22]: ἄρτον ποιοῦσί τινες ἄνευ ζύμης,
νίτρον ἐμβάλλοντες (those who make bread without leaven add nitron). Pliny
says the only manufacture of nitrum in Egypt was formerly at Naukratis
and Memphis, where it was gathered in heaps which became as hard as stone
from which vessels are made (faciunt ex his vasa). J. Beckmann[23] found this
hard to explain, but the heaps of the salt would aggregate to a mass on
standing and vessels could be cut from it as curiosities, as they still are from
rock salt.

Pliny says nitrum was made in Egypt in nitraria (often incorrectly translated "nitre beds") and the "red nitrum" (nitrum rufum) made was coloured
with the earth of the vicinity. Pliny thought it was made from Nile water,
impregnated with nitrum. It is formed in the nitrum beds which, though
pregnant, are not yet in travail, whilst others concluded that it is formed by
the fermentation of the heaps (alii acervorum fermento gigni existimavere),
which suggests, says Bailey,[24] the formation of saltpetre.

The Egyptian nitraria, which are shallow lakes in the Nile valley (Wādī
al-Naṭrūn) near Memphis from which soda (latroni) is still extracted, are
described by St. Jerome, who visited them about A.D. 385[25]:

nitrum a Nitria provincia, ubi maxime nasci solet, nomen accepit. . . .
Hanc (speciem salis) indigenæ sumentes servant, et ubi opus exstiterit,
pro lomento utuntur. . . . Crepitat autem in aqua quomodo calx viva,
et ipsum quidem disperit, sed aquam lavationi habilem reddit, cujus
natura cui sit apta figuræ, cernens Salamon ait: acetum in nitro, qui
cantat carmina cordi pessimo. Acetum quippe si mittatur in nitrum
protinus ebullit.
Nitrum takes its name from the province of Nitria, where it is especially
found in large quantity. The people living there make use of it for
washing. . . . It decrepitates in water like quicklime and this somewhat
divides it, but it makes the water good for washing; its nature is aptly
described in the proverb of Solomon saying: "as vinegar upon nitrum
so is he that singeth songs to an empty heart." Indeed, vinegar if put
upon nitrum at once bubbles up.

Pliny says the water of the nitrum pits rots shoe-leather, since nitrum is
more corrosive than common salt (maiorem esse acrimoniam nitri; Bailey[26]
has, incorrectly "greater bitterness").

Nitrum, says Pliny, can be used in baths instead of oil (in balneis utuntur
sine oleo) since it removes dirt (ad aliqua sordidum optimum est). It is
used in making glass, in dyeing purple, cooking radishes, and in giving boiled
vegetables a greener colour. Plutarch[27] says it was used in dyeing with
scarlet (κόκκος). Pliny says it was used when burnt as a dentifrice to clean
blackened teeth (nigrescentes crematum dentifricis ad colorem reducit), and
he gives a large number of medical uses; it is good for the eyes, acts as a

sudorific, and when mixed with oil and used as a liniment it softens the body; mixed with turpentine, cypress oil, honey, wine, etc., it is used internally and externally.

Pliny[28] quotes Vergil[29]—who says : semina vidi equidem multos medicare serentis et nitro prius et nigra perfundere amurga (I have seen many sowers artificially prepare their seeds, and steep them first in soda and the black lees of oil), as recommending moistening beans with nitrum and amurca (lees of olive oil) before sowing, to make them grow larger. Pliny[30] also says that in Egypt growing radishes are sprinkled with nitrum, and[31] that nitrum was applied to growing cabbages. These uses have been thought to suggest that this kind of nitrum was saltpetre, which is a fertiliser,[32] and its use as a diuretic (hydropicis cum fico datur)[33] has also been said to favour an identification with this salt.[34] In many cases, however, it must mean soda, as when Pliny says it imparts an extra greenness to cabbages[35] and other vegetables (olera viridiora)[36] : soda is still used in boiling cabbages. Columella,[37] Martial[38] (nitrata viridis brassica fiat aqua) and Apicius[39] (omne olus smaragdinum fiet si cum nitro coquatur) mention this use.

Pliny says nitrum is burnt in a close earthen vessel, as otherwise it would decrepitate (uritur in testa opertum ne exsultet), which might suggest that if burnt in an open fire it would deflagrate, but he goes on to say that, "except in this last case, however, the action of fire does not cause it to decrepitate (alias igni non exsilit nitrum)," which suggests that the first effect was due to decrepitating common salt mixed, as an impurity, with the soda.[40] In two obscure passages Pliny says :

(a) necnon frequenter liquatum cum sulphure coquentes in carbonibus. Ad ea quoque, quæ inveterari volunt, illo nitro utuntur.

(b) sal nitrum sulphuri concoctum in lapidem vertitur.

Bostock and Riley[41] translated (a) :

and very frequently they melt it with sulphur on a charcoal fire. When substances are wanted to keep, they employ this last kind of nitrum.

They mention that Beckmann had pronounced this passage to be one of the darkest parts in the history of nitrum, and thought liver of sulphur (impure sodium sulphide) was produced. Romocki[42] suggested that "in carbonibus" meant "for [the treatment of] carbuncles." Serenus Sammonicus uses carbo for a carbuncle, and in French charbon has this meaning, from ἄνθραξ, Greek coal, but at present anthrax is a carbuncle. Bailey[43] made one of his sparing textual emendations and read in (a) : in corporibus quoque quæ inveteri . . . , "again, nitrum is used for mummifying bodies." The passage is part of a description of Egypt, and the use of nitrum in mummifying is described by Herodotos. (The emendation of quas in the manuscripts to quæ had already been made by de Laet.)[44] It is curious that Pliny nowhere refers to mummies. Bailey does not deal with the text (b), which may be translated : "both salt and nitrum, cooked with sulphur, are changed into stone"; common salt, however, does not react when heated with sulphur.

Pliny says Egyptian nitrum was imported in vessels pitched inside to prevent it from liquefying (by deliquescence) (in vasis picatis, ne liquescat), the vessels having previously been well dried in the sun. This suggests that the alkali was causticised, and Pliny also says the Egyptian nitrum was adulterated with lime, which could be detected by tasting it, since the pure kind dissolves at once but the adulterated has a pungent taste (adulteratur in Ægypto calce: deprehenditur gustu: sincerum enim facile resolvitur: adulteratum pungit). Kind[45] said that caustic alkalis (as well as metallic salts) were used as caustics in ancient medicine, but Celsus,[46] whom he quotes, gives only lime (calx) and burnt tartar (fæx combusta) separately in a long list. Galen[47] called lye κονία. Paul of Ægina (c. A.D. 650)[48] calls lye which has been percolated through quicklime πρωτόστακτον (calx colata). He says the lixivium of ashes is called lye (κονία) but when the ashes have taken something from the lime (τίτανος) caustic lye is produced (κονία τὸ οἷον περίπλυμα τῆς τέφρας ὀνομάζεται. . . . εἰ δὲ προσλάβοι καὶ τιτάνου ἡ τέφρα, καυστικὴν ἐργάζεται τὴν κονίαν).

Isidore of Seville[49] mentions the medical and detergent uses of nitrum (nitrum nascitur in oppido vel regione Ægypti Nitria ex quo et medicinæ fiunt et sordes corporum vestiumque lavantur), and Salmasius[50] quotes Arcadius Grammaticus (c. A.D. 200) as saying that it was used for washing the head.

The accounts of νίτρον in Dioskourides[51] (about Pliny's time), Galen (A.D. 129–99)[52] and later Greek and Latin medical authors add little to Pliny's. Cassius Felix[53] mentions a nitrum vermicarum. Dioskourides, Pliny, and Galen speak of a "foam of nitron" (ἀφρόνιτρον, aphronitrum, spuma nitri), which was probably an alkaline efflorescence of soda.[54] Galen says νίτρον on roasting becomes more like ἀφρόνιτρον, which is bad for the stomach, and Bailey[55] thought this meant a slight caustification of soda on heating, but it probably refers to the loss of carbon dioxide from the native sodium sesquicarbonate to form the normal carbonate, which is much more strongly alkaline. It has been suggested[56] that aphronitron sometimes meant "wall-saltpetre," containing calcium and potassium nitrates, but it probably never meant this in older authors.

Mitscherlich[57] suggested that the Romans first gave the name nitrum (which he connects with νίζω or νίπτω, Sanskrit nēnekti, to wash) to any efflorescent salt, and hence it was later connected with saltpetre. Since νίζειν occurs in Homer, this would be an alternative derivation to the one from neter. No one seems to have suggested a derivation from νίψ, snow (known only in the accusative in Hesiod), although an Arabic name for saltpetre was "snow of China," and saline efflorescences do look like snow.

Dioskourides[58] and Pliny[59] speak of a "flower of salt," ἄνθος ἁλός, flos salis. This is said to feel oily (ὑπολίπαρον; optimum ex eo quod olei quandam pinguitudinem reddit (est enim etiam in sale pinguitudo, quod miremur)), which suggests that it was alkaline, an efflorescence of soda like aphronitron. It was used by the salve and ointment makers on account of its colour (μίγνυται καὶ ἐμπλάστροις καὶ μύροις εἰς χρῶσιν: unguentarii propter colorem eo maxime utantur). Dioskourides says of it: ἔτι τὸ ἀκέραιον [ἁλός ἄνθος] ἐλαίῳ συνανίεται μόνον· τὸ δεδολωμένον δὲ, ἐκ μέρους καὶ ὕδατος (verus ac sincerus non nisi oleo resolvitur: adulteratus nonulla ex parte etiam aqua diluitur); Pliny says "verus non nisi oleo resolvatur."   Kopp[60] thought the passage in Dioskourides should be translated as "otherwise the unadulterated ἄνθος ἁλός alone dissolves in oil, the adulterated also partly in water," since alkali was the only saline substance known to the Greeks which dissolves in oil. Sodium carbonate, however, does not dissolve in oil, and it is possible that the emulsification of oil in water produced by the alkali is meant. The passage, however, is hopelessly obscure.

The genuine kind is yellow, the adulterated is red; and it had a smell of fish-brine. Since Pliny says the "flower of salt" floats on water, Bailey[61] thought it was some kind of fatty material, since it is said to have an oily feel, but it was probably formed as a scum or efflorescence on an alkaline solution and the oily feel may have been due to the alkaline nature. Pliny[62] in describing burnt tartar (fæcis vini) says it has the same nature as nitrum and the same powers, except that it has an oilier feel (cinis ejus nitri naturam habet, easdemque vires, hoc amplius, quo pinguior sentitur). Meyer in 1764 proposed the theory that causticity is due to an "oily acid" (acidum pingue), which is transferred from quicklime to the mild alkalis (carbonates) in making them caustic, and Kopp[63] thought that Pliny's wording might have suggested this.

Pliny[64] says of nitrum, "olei natura intervenit" (its properties include an oiliness), that the Egyptian kind was adulterated with lime, and that aphronitrum came from Egypt in pitched vessels to prevent deliquescence, the vessels being finished by drying in the sun. All this suggests that the caustification of alkali with lime was known in Egypt. It is also true, as Pliny says, that burnt tartar (potassium carbonate) is more strongly alkaline than nitrum (sodium carbonate); it was supposed even in the early nineteenth century that normal potassium carbonate was a basic salt (subcarbonate) and that the bicarbonate ($KHCO_3$) was the true normal salt.

Dioskourides says the best νίτρον is light and reddish (ῥοδωπόν), or white, porous and spongy (κατατετρημένον, οἱονεὶ στρογγῶδές τι), such as that of the Bouni (ἐκ τῶν Βουνῶν).[65] The "red nitrum" (Βερνικάριον, τὸ πυρρὸν νίτρον) was discussed by Salmasius[66]; it may have been coloured by oxide of iron, or with organic matter or bacteria.[67] The Egyptian nitron perhaps contained some ammoniacal salt, since Pliny[68] says that when sprinkled with lime it gave out a strong smell (calce aspersum reddit odorem vehementem).

A sal Indicus is mentioned in the Liber Sacerdotum,[69] translated from Arabic by Renaldo of Cremona and containing many Arabic words. Berthelot[70] thought it was composed in the tenth to eleventh century in Syria (since it is ascribed to a Johannis) and translated in the twelfth to thirteenth century, and that the sal Indicus was perhaps saltpetre. Hime,[71] however, thought it translated a Persian word meaning "salt of bitumen" or "salt of naphtha," i.e. common salt from petroleum brine. A version of the Mappæ Clavicula[72] supposed to have been written by Adelard of Bath (d. c. 1150), containing

Greek and Arabic words, speaks of "nitrum est sal qui nascitur in terra, fiet in laminas in tempore cavatur," which seems to mean saltpetre. A manuscript copied in Paris in 1431 from manuscripts written in Milan and Bologna in 1409–10, mentions saltpetre (sal nitrum, salniterum, salepetre),[73] as does a fifteenth-century Bologna manuscript (salnetrio, salnitrio),[74] in both cases for making aqua regia. But in a manuscript written in 1431, "salpetre, autrement appelle assafetide [= afronitre?]" is, from its use, soda.[75]

It is said that there is no Anglo-Saxon name for saltpetre,[76] but it occurs in earlier Middle English (1100–1500).[77] The definite mention in the *Liber Ignium*, which has some resemblance to that in the *Mappæ Clavicula*, has been dealt with previously (ch. II). A good account of the ancient information on nitrum was given by the Jesuit Bernard Cæsius,[78] who distinguished it fairly well from saltpetre. Ulysses Aldrovandi[79] is less clear, and the account in Athenasius Kircher[80] is even more confused. Casimir Siemienowicz[81] thought the ancient nitrum was an impure saltpetre.

Robert Boyle[82] found that "Egyptian Nitre" was a "lixivial salt" quite different from saltpetre, since it deliquesced in moist air and effervesced with acid, and hence was probably the "nitre" mentioned in the Bible, which was not saltpetre. The specimen was given to him by the British Ambassador at the Ottoman Court. Charles Leigh, of Oxford,[83] examined specimens of Egyptian "nitre" and "water of natron" given him by Robert Plot (the professor of chemistry in Oxford, who was then Secretary of the Royal Society) and showed that the nitrum was not saltpetre by its effervescing with acid, its taste, its failure to detonate on a fire, and other tests. He also distinguished it from sal ammoniac, but wrongly supposed that it was a compound of common salt and "a urinous salt" (ammonia). William Clarke[84] believed that saltpetre was known to the ancients "thousands of years ago," and J. Hardouin in his edition of Pliny[85] refers to Clarke and says the ancient nitrum was the same as the modern saltpetre, but of inferior character (nitrum nostrum, quod Salpetræ nominamus, *le Salpetre*, idem plane est cum veterum nitro, sed aliquanto tamen inferioris notæ). G. C. Schelhamer[86] gave some good philological discussions for the time but left the nature of ancient νίτρον, λίτρον, and nitrum undecided, thinking that the ancient kinds of nitre were unknown in his time, all modern nitre being factitious; his copious references to authorities would form a good foundation for a modern detailed study. His book is also interesting for its long attack on John Mayow's account of nitre in his *Tractatus Quinque Medico-Physici*, Oxford, 1674.

Pliny's account of nitrum was analysed in detail by Johann David Michaelis,[87] who concluded that nitrum sometimes meant alkali and sometimes saltpetre. His arguments are unconvincing. He proposed some alterations of the text of Pliny, e.g. that "non" should be omitted in the passage : "uritur in testa opertum, ne exsulet : alias igni non exsilit nitrum," on the ground that "intelligo de vero nitro." F. G. W. Schröder[88] thought νίτρον or nitrum was either saltpetre, which was known to the ancient Egyptians and used in making mummies, or fixed saltpetre (potassium carbonate); in Herodotos, Strabo and Pliny it meant saltpetre, which Schröder thought was formed by the fixation of sunlight in the soil. J. C. Wiegleb,[89] however, pointed out that Egyptian nitrum is not saltpetre, but soda, as Boyle had shown.

Vincenzo Requeno y Vives, a Spanish Jesuit (1743–1811), devoted nearly the whole of the second volume of a work intended to explain the process of encaustic painting[90] to a discussion of the nature of Punic wax and ancient nitre (Della cera Punica degli Antichi Pittori e dell'antico nitro necessario per fabbricarla). He prints a "Discorso del Cavaliere Lorgna,"[91] denying that Pliny's nitrum was saltpetre, and then refutes it at great length, quoting most of the older authors named in the present section and also works on chemistry by Lemery, Neumann, Boerhaave, Baumé, etc. He concludes[92] that the ancient and Pliny's nitrum was modern saltpetre (il nitro antico, o sia il nitro di Plinio, è il ie nitro de' moderni). The name natron, he says, is not found in Classical authors (it is not in Ducange's *Lexicon*) but is Turkish and late Byzantine (νάτρον), first appearing in Belloni[93] (Pierre Belon, *c.* 1517–64), who[94] says the name natron was used in his time in the East for soda. G. G. Model[95] analysed a native soda from Ochotzk.

The first really critical examination of the question of the identification of νίτρον and nitrum is probably that of Johann Beckmann (1739–1811), professor in Göttingen, and a competent linguist and scientist, who has often been quoted.[96] After reviewing some older authorities he says:

"Before I ascribe to the ancients a knowledge of our saltpetre, I must be shown in their writings properties of their *nitrum* sufficient to satisfy me that it was the same substance."

The only satisfactory property which could be known to them would be the deflagration of the substance with combustible material, such as could be observed when it was thrown on a fire, and as no ancient author ever mentions this property, none of the things they call *nitrum* could be saltpetre. Since Dioskourides and Galen say that nitrum was "burnt" to strengthen its properties:

"it is impossible that the ancients should not in burning it have observed its deflagration, and this property is too remarkable not to have been mentioned. But nothing is to be found that can with any probability be supposed to allude to it."

Beckmann's arguments convinced Kopp[97] that nitrum always meant soda, and Bailey[98] concluded that "though a number of cases has been mentioned, in which nitrum may have referred to nitre, it is possible that the material was soda in every case."

The late Greek name for saltpetre was σαλόνιτρον or σαλονίτριον.[99] It should also be noticed that the adjective νιτρώδης is used for an insipid taste by Galen, Theodoros, and Symeon Seth (A.D. 1075), and by the Persian pharmacist Abū Mansūr Muwaffak (tenth century) this is called "borax-like," būraq meaning soda.[100] Albertus Magnus[101] uses borax for a white or black stone from the head of a toad, baurac for an alkaline salt like nitrum. True borax (from Tibet) was probably not known in the Roman period; the red glaze of the Arretine terra sigillata pottery of the first century B.C. to the first century A.D. contains borates, perhaps obtained from the natural boric acid of the soffioni of Tuscany, but the glaze of older red Etruscan ware does not contain borate.[102]

Since wood ashes containing potash (potassium carbonate) were used in refining and purifying saltpetre, a few words may be said on the early history of potash. Hippokrates[103] mentions mixing burnt tartar (potash) with nitron (soda) in a prescription (τρύγα κατακαίων . . . λίτρον συμμίσγων; fecem vini comburito . . . et pari nitro ammixto). The distinction between potash and soda appears here in a rudimentary form, although the two were afterwards confused by Pliny.

Aristotle[104] says the Umbrians (ἐν Ὀμβρικοῖς) burn reeds and rushes (κάλαμος καὶ σχοῖνος), throw the ashes into water and boil till the water leaves a residue, which on cooling is a mass of salt (ἁλῶν γίνεται πλῆθος); that a salt extracted from cinders (plant ash) is like that of the lakes[105]; and that ashes contain potential heat, and water strained through them contains the heat of the ash.[106]

Pliny[107] says nitrum was formerly made in small quantities by burning oak wood (ex quercu cremata fieri nitrum) but the process had been given up. This would give crude potash, and the product of burning dry wine lees (fæx vini) or tartar, which he says[108] gives nitrum, would also be potash (fæx vini siccata recipit ignes, ac sine alimento per se flagrat. Cinis ejus nitri naturam habet, easdem vires, hoc amplius, quo pinguior senitur : wine lees when dried will take fire and burn without the addition of fuel; the ashes so produced have very much the nature of soda and similar virtues, the more so indeed the oilier they feel). He also says[109] that oak wood when burnt produces an ash like soda (cremati roboris cinerem nitrosum esse certum est). Pliny, therefore, confused potash and soda, but the two alkalis were clearly distinguished by Abū Manṣūr Muwaffak.[100]

Synesios, an alchemical author who may be the same as the famous bishop of Ptolemais (A.D. fourth century), distinguished water of nitron (ὕδωρ νίτρου) and water of tartar (ὕδωρ φέκλης) as solvents.[110] Dioskourides[111] frequently mentions the ashes of the wood of various trees (oak, wild and cultivated fig trees, etc.), and also of oak galls, as used to make lye (κονία), but he does not refer to solid potash. He says[112] that lye is made from the ashes of the vine and burnt tartar (τρὺξ οἴνου); the latter gives a white ash with a burning taste, and must be used without delay since it quickly becomes moist (διαπνεῖται, literally "evaporates") and cannot be kept except in a closed vessel. Celsus[113] also mentions burnt tartar (fæx combusta) as used in medicine. Pliny[114] calls ashes used to make lye cinis lixivius and Columella[115] uses lixivia for lye from ashes; lixivium is used by Palladius.[116] The common mediæval name for potash was cineres clavellati, which may be derived from clavicula, vine tendrils, and later became cineres gravellati (cendre gravelée); the name "pot-ash" seems to have been introduced from the use of pots for boiling down the lixivium.[117]

Varro[118] says the people on the banks of the Rhine, having no sea salt or rock salt, used the saline coal of burnt plants. Pliny[119] reports that in the provinces of Gaul and in Germany, salt is made by pouring salt water over burning wood (ardentibus lignis aquam salsam infundunt), and Tacitus[120] says the same of the Hermanduri and the Catti.   This method of pouring sea water or salt brine gradually over a wood fire, leaving in the end a mixture of salt, charcoal and wood ashes was tried by du Rondeau.[121]   Bostock and Riley[122] report that, according to Townson, this method was still used in Transylvania and Moldavia.   Beckmann[123] says the inhabitants of Zealand, perhaps descendants of the Catti, used no other salt than that made from mud thrown up from the sea, burned and moistened with sea-water, and gives a number of other references to a similar process used in other places, but he is in error in saying that Boerhaave[124] should have said that salt water must be poured over wood ashes before these can be used in place of salt, since Boerhaave gives the references to Pliny, Varro, etc., which make it clear that the ashes alone were used.

SOAP

Before leaving the history of alkalis, a few words may be said about soap, although this does not really concern the subject of the present volume.

In Homer[125] water only is used for washing clothes.   The later Greeks seem to have added wood ashes to water for detergent purposes,[126] and in Rome the fullers used stale urine (containing ammonia) which was collected for the purpose.[127]   Some kind of washing balls (smegmata) were made from the plant osyris (perhaps toad-flax),[128] and also the fuller's weed (struthion; radicula).[129]   The smāma ($\sigma\mu\hat{a}\mu a$) of Theokritos (c. 300 B.C.)[130] and smēma ($\sigma\mu\hat{\eta}\mu a$) of Athenaios,[131] used for washing the hands, was probably some cleansing paste, not soap.   Goguet[132] says the American Indians use fruits, the women of Iceland ashes and urine, and the Persians earths, for washing. Pliny[133] and Galen[134] mention the use of Cimolian, Lemnian, Samian, Selusinian and Chian earths for washing; some of these would resemble fuller's earth.

Aristophanes (445–386 B.C.) mentions[135] an "adulterated soda-lye" as used with Cimolian earth for washing ($\psi\epsilon\upsilon\delta o\lambda i\tau\rho ov\ \tau\epsilon\ \kappa ov i a\varsigma\ \kappa a i\ \kappa\iota\mu\omega\lambda i a\varsigma\ \gamma\hat{\eta}\varsigma$: falsoque nitro mixtum pulverem et Cimoliam terram).   Bochart[136] quotes scholiasts as saying that Cimolian earth was used to make $\nu i\tau\rho ov$ (which is incorrect).   Dioskourides[137] says the best soap was made from nitron and Cimolian earth ($\sigma\grave{v}v\ \nu i\tau\rho\omega\ \mathring{\eta}\ \gamma\hat{\eta}\ \kappa\iota\mu\omega\lambda i a\ \sigma\mu\hat{\eta}\gamma\mu a\ \kappa\rho a\tau i\sigma\tau ov$).   It may be that $\kappa ov i a$ in Aristophanes means quicklime and the soda was causticised.[138]

The first definite mention of soap (sāpo) is in Pliny[139] who says it was an invention of the Gauls, who used it for reddening the hair; it was made from wood ashes (the best being of the yolk-elm and the beech), and that there were two kinds, a hard and a soft.   Among the Germans it was more used by men than by women.   (Prodest et sapo; Gallorum hoc inventum rutilandis capillis ex sevo et cinere.   Optimus fagino et caprino, duobus modis, spissus et liquidus : uterque apud Germanos majore in usu viris quam feminis.) Martial[140] also calls it sapo or caustica spuma, although this reading is doubtful, some manuscripts having Castica and others Chatica instead of caustica.[141]

Martial[141a] also calls it "Dutch pomade" (spuma Batava, comes Teutonicus accendit spuma capillos). Imitating the fashions, art and music of less civilised nations has always had an attraction for some races, and the Roman ladies and dandies thought it smart to have red hair like the Gauls and Germans.[142]

The origin of the name sāpo has been much discussed. Some think it is from the German saipjō, others from the English sepe (still used in Scotland), passing by way of Batavia to Gaul.[143] Blümner[144] says true soap was unknown to the ancients, Pliny's sapo being a pomade made from unsaponified fat and alkali. When true soap was first made by boiling fats or oils with causticised lye seems to be unknown, but the use of causticised lye in making soap (σάπων) is mentioned by Galen,[145] perhaps from Asklepiades junior (c. A.D. 100), who says it is made from the fat of oxen, goats, or wethers, and causticised lye (sapo conficitur ex sevo bulbulo, vel caprino, aut vervecino, et lixivio cum calce). Galen says that the best soap was the German, since it was purest and in some ways the most fatty, that of the Gauls being next best, and that it acted as a medicine and removed all impurity from the body and from clothing. This is the first certain mention of the use of soap as a detergent. Galen[146] says soap is a better detergent than soda (λίτρον). If the mention of Gallic soap in Oreibasios[147] is from Rufus of Ephesus (c. A.D. 100) this would precede Galen's in a Greek writer. Zosimos the alchemist[148] (c. A.D. 250) mentions both soap (σαπώνιον) and soap-making (σαπωναρικὴ τέχνη).

Kopp[149] thought German soap may have been soft soap made from the ashes of land plants (containing potash), the soap of Gaul a hard soap made from the ashes of sea-shore plants (containing soda). Crude soda from sea-shore plants was long made in Narbonne. Pliny[150] says "flower of salt" (flos salis) is used in smegmatis, and although he does not use the name smegma (σμῆγμα) for soap, it is just possible that there is a reference to the use of salt in making hard soap by salting out. It is also possible that he has confused salt with soda.

The medical use of soap is mentioned by Aretaios (second century),[151] and Aïtios (sixth century).[152] Theodorus Priscianus (fourth century)[153] says it was used for washing the head (attamen Gallico sapone caput lavabis) and speaks of a soap-boiler (saponarius). Lippmann[154] speaks of "soap which had been blessed" as sold for ritual washing in the church of St. Thekla in Seleucia, formerly a temple of Athena, as reported by St. Basil, c. A.D. 350. The reference he gives,[155] however, quotes *Acts of Paul and Thekla*[156] attributed to Basil, bishop of Seleucia (c. A.D. 450), in which it is said that "washing materials" (ῥύμματα, purgamentis) were sold not in the church but in booths outside it. Although Liddell-Scott-Jones[157] give as meanings of ῥύμματα, "anything for washing, soap, lye," they quote as using the word Hippokrates, Aristophanes and Plato, in none of whom could it mean "soap." It may well have meant soap in the fifth century *Acts of Paul and Thekla*.

Soap is mentioned by Arabian physicians as used for medical purposes, e.g. by Serapion[158] and Rhases.[159] Soap does not appear in the index of the *Canon* of Avicenna. Soap from fats is mentioned in a Persian work of the late tenth century,[160] although it says saponaceous plants were also used in Persia for washing clothes and carpets.   Shams al-dīn al-Dimashqī (1256–1326) in his *Kitāb Nukhbat* (mentioning events in 1323) reported that a fine soap (raqqī soap, "distilled soap") was made in Nablus in Syria and exported to Mediterranean and other countries.[161]   A "paste of Egyptian soap" was apparently green.[162]   Arabian authors attribute the invention of soap to Hermes, to whom it was the object of a revelation.[163]   The Arabs took over the Greek name σάπων as ṣābūn (which could also mean "lye"[164]).

According to Feldhaus[165] the first certain mention of soap boilers in Germany is about 805 in the *Capitulare de Villis* (43–4, 62) of Charlemagne, and there is an Anglo-Saxon mention of saffiûn of about the same time; soap was principally made in Marseilles from the ninth century and Venice from the fourteenth century but only in small quantities; these two places were the main centres of soap manufacture in the seventeenth century.   The use of soap in England in the Middle Ages for personal washing is uncertain, but either soap or a lye of wood ashes was used for washing clothes.[166] Feldhaus[165] mentions for accounts of soap-making in the fourteenth to fifteenth centuries a work on domestic economy by Anton Tucher of Nürnberg, and a poem of 1492 on Nürnberg saying that washerwomen must not add lime or potash (Weidaschen) to the lye; further sources of the fifteenth century which he quotes refer to the use of ashes and baruth (soda) in washing, the use of good oil, especially Apuleian, and the repeated application of the lye to the clothes by a process called "buiken," "buken," "buchen," or "biuchen," a name which survives as "bucking" in English bleaching terminology.

J. Sheridan Muspratt[167] said: "In the excavations at Pompeii a complete soap-boiling establishment was discovered, containing soap still perfect. . . . The Editor was greatly interested inspecting the factory."   The pieces of supposed soap are in the Museo Nazionale in Naples.[168]   A specimen of it was examined by de Luca,[169] who reported that it blackened when heated on platinum foil, and when it was warmed with dilute hydrochloric acid "a fatty substance of the consistency of butter was set free."   K. B. Hofmann,[170] whilst saying that he did not question the result found by de Luca with his specimen, examined another specimen of reputed soap from Pompeii.   It was insoluble in ether, alcohol and petroleum ether, had only a small part soluble in water, and effervesced with dilute hydrochloric acid, which dissolved only a small part.   The residue was found by qualitative analysis to consist mainly of fuller's earth of medium quality and it dispersed in water like this:

> Die in der Fullonica gefundene, von mir geprüfte Masse ist also nichts als Walkerde.   Was de Luca analysiert hat, weiss ich nicht—jedenfalls auch keine Seife; denn er gibt an, der unlösliche Antheil seien thon- und kalkartige Stoffe gewesen.

Further, said Hofmann, soap was never found among toilet articles in Pompeii, and modern soap does not redden the hair.   There was more interest taken in appearance than cleanliness; Tacitus says the Germans were dirty (sordidus) and, says Hofmann: "Wir haben uns also unsere Vorväter zwar ungewaschen, aber mit pomadierten Köpfen zu denken (We have, therefore, to think of our forefathers as unwashed but with pomaded heads)." A satisfactory history of soap has still to be written.

THE ASSIAN STONE AND BĀRŪD

It is time now to return to the history of saltpetre and to consider a mysterious "Assian stone" which, first mentioned by Pliny, was linked with saltpetre by Arabian authors in the thirteenth century. Pliny[171], quoting Mucianus, says:

In Asso Troadis sarcophagus lapis fissili vena scinditur. Corpora defunctorum condita in eo absumi constat intra xl diem, exceptis dentibus. . . . Eiusdem generis et in Lycia saxa sunt et in Oriente, quæ viventibus quoque adalligata erodunt corpora. . . . Assius gustatu salsus podagras lenit, pedibus in vasa ex eo cavatum inditis. . . . Ejusdem lapidis, flos appellatur, in farinam mollis, ad quædam perinde efficax: est autem similis pumice rufo. . . . Fit et cataplasma ex eo podagricis mixto fabæ lomento.

At Assos in the Troad is found a stone of laminated texture, called sarcophagus [Greek "flesh-eater"]. It is known that dead bodies buried in this are consumed within forty days, except the teeth. Of a similar kind are stones in Lycia and in the East which when applied to the bodies even of the living corrode the flesh. . . . The Assian stone, which has a salty taste, assuages gout, the feet being placed in a vessel hollowed out of it. . . . Flower of this stone [of Assos] is the name of the floury dust [formed from the stone], which resembles red pumice. A plaster of bean-meal and this is applied for gout.

Dioskourides[172] describes the λίθος Ἄσιος, the stone of Asia, as laminated, friable, and in appearance like pumice. The flower (ἄνθος) of the stone is partly white and partly like pumice, verging to dull yellow, and with a biting taste. From the medicinal uses he describes, it was a caustic. It is mentioned by Loukian[173]: ἀνθρώπου κόπρον, ἄλευρα κυάμων, ἄνθος Ἀσσίου λίθου (human dung, bean-meal, flower of the stone of Assios). Lenz[174] thought that it was crude salt from sea-foam or mud, or mud from salt springs, coloured with oxide of iron, as used for mud baths. A favourite hypothesis[175] is that it was quicklime. The first mention of the property of "a stone" (λίθος), of which coffins were made and which reduced bodies to ashes (ἐν ἑαυτῷ τέφραν ποιεῖ), is in Theophrastos,[176] who adds that it acts by virtue of its intrinsic heat (θερμότητι καὶ τῇ φύσει θερμός), a proof of which is τὸ γίνεσθαι κονίαν ἐξ αὐτοῦ, the meaning of which is not clear. Wimmer[177] translated: "argumento est quod ex eo calx fiat" (a proof of which is that lime is made from it), in which case the stone would be limestone. But κονία can also mean "dust" or "an alkaline powder" (see ch. I). Hoefer[178] and Bostock and Riley[179] adopted Ajasson's identification with some variety of alunite (alum-stone). All the above suggestions are unsatisfactory.   There is a mineral gaylussite, sodium calcium carbonate, Na_2CO_3, $CaCO_3$, $5H_2O$, which would probably form an efflorescence of sodium carbonate, and this seems to be the most likely identification.   Pliny (xxxi, 46) says an infusion of nitrum was used in a bath for gout.   Celsus[180] mentions lapis asius, much in favour in Asia, and also refers to the use of vessels made from it as foot-baths for the treatment of gout. This would be a treatment with an alkaline water, which would be beneficial in removing uric acid.

Abū Muḥammad 'Abdallāh ibn Aḥmad Almaliqī ibn al-Bayṭār, usually called al-Baiṭār, was born in Malaga in Spain in 1197 and died at Damascus in 1248.   He travelled in Greece, Egypt and Asia Minor and was well acquainted with Indian and Persian drugs.[181]   His main work is the *Kitāb al-jāmi' fī al-adwiya al-mufrada*, a treatise on simple drugs, mentioning about 1400 items, completed about 1240.   It was translated by J. von Sontheimer,[182] who gives :

i, 42 : Asius, lapis Asius.   Bei den alten ägyptischen Aerzten ist dieses das chinesische Salt.   Die afrikanischen Aerzte und das Volk nennt es unter dem Namen Bârud.

i, 122 : Bârud.   Flores Lapidis Asii.   Dieses Arzneimittel ist die Blume des Lapis Asius, welche man auch den chinesischen Schnee nennt.   Im Buchstabe A ist dieser Stein bereits erwähnt worden.

Leclerc, in a better translation,[183] gives :

xxiii, 71 : asiyūs.   C'est la neige de Chine (thalj al-Ṣīn) chez les anciens médicins d'Égypte.   Le peuple et les médicins du Maghreb lui donnent le nom de bâroud (bārūd).

xxiii, 200 : bārūd.   C'est la fleure de la pierre d'Assious.

xxiii, 333 : thalj Ṣīnī.   C'est le bārūd, généralement connu sous le nom de fleure de la pierre d'Assos.

xxiii, 420 : ḥajar asiyūs.   C'est la pierre dite *bâroud*, dont il a été question à la lettre *ba*.   Les habitans de l'Égypte lui donnent le nom de neige de Chine.

The flower of the stone Asios, which is called the snow of China among the old physicians of Egypt, is called in the West by the physicians and the common people bārūd.

Ibn abī-Uṣaybi'a, who met al-Bayṭār in Damascus about 1235 and was his pupil, says of him[184] : "It excited my surprise that he never mentions a drug unless it is to be found in some work of Dioskourides and Galen, or unless it is numbered among the host of known medicines."   This suggests that if al-Bayṭār came across a drug with an unfamiliar name he would try to identify it with one mentioned by Dioskourides, Galen, or Avicenna (A.D. 980–1037).   Avicenna[185] says :

Asius quid est ?   Est lapis super quem nascitur sal, cuius flos nominatur asius, et videtur, quòd sit eius generatio ex rore maris, & imbre, qui cadet super eum. . . . Cum farina superposita valet podagræ.

What is asius? It is the stone upon which grows the salt, the flower of which is called asius, and it is possible that this may be generated from it by the morning dew and rain which falls upon it. . . . Mixed with flour and applied, it cures gout.

The last words agree with the old medical use of the flower of the Asian stone, and will not be considered further. The Arabic text is translated by Reinaud[186]: "Asius is the stone on which the salt, the flower of which is called asius, is formed"; and by Quatremère[187]: "Asius is the stone on which the salt called flower of asius is formed."

The "Antiqua Expositio Arabicorum Nominum" in the Latin *Canon* of Avicenna[188] simply repeats the definition in the text given above. The new list by Andreas Alpago Bellunensis (sixteenth century) says[189]:

Assius secundum quosdam idem est, quod nix sini, quæ quidem Ebenbitar [ibn al-Baytār] est lapis quidam transparens, sicut salgemma, qui cum exponitur aeri conuertitur in puluerem, sicut est puluis aluminis contriti, & in sapore est acutus, & salsus, & defertur ex India, sed tamen iste sermo non est conformis dictis Auic. 2. can. qui dixit q. assius est lapis super quem generatur sal ex roratione maris cadentis super ipsum. quare &c multi tamen utuntur niue syni loco assius, & affirmāt q. Auicē. ignorauit essentiam assius.

Assios, according to some, is the same as the snow of China, of which al-Baytār says that it is a transparent stone something like rock salt, which when exposed to air is turned into a powder like powdered alum, with a sharp and salty taste and is brought from India. But this account does not agree with what Avicenna says in the second book of the *Canon* [quoted] . . . many use the snow of China in place of assius and say that Avicenna is ignorant of the nature of assius.

Romocki[190] said the statements in al-Baytār about the "flower of the Asian stone" and bārūd comprise the first mention of saltpetre by an Arabic author.

Ibn abī-Uṣaybi'a (1203–70) in his History of Medicine (*'Uyūn al-anbā'*) composed in 1242 but revised later, quotes an otherwise unknown Ibn Bakhtawayh in his *Kitāb al-Muqaddimāt* (Book of Introductions) on the use of saltpetre as a cooling agent[191]:

"Take a pound (riṭl) of best South Arabian alum (shabb Yamānī), powder it finely, put it into a new earthenware pot, and pour over it six pounds of pure water. Put the whole into an oven, which is closed, and allow two-thirds to evaporate. A third remains, which neither increases nor diminishes, since it has become a solid mass. This is put into a bottle, which is well closed. If you wish to make ice, take another vessel (thaljiyya) in which pure water is placed and put into it ten ounces (mithqāls) of the alum water, and after an hour the water will become snow."

The words shabb Yamānī mean literally "alum of Yaman," described by Avicenna as yellowish-white, flaky, astringent and with a sour taste, i.e. a kind of common alum. Von Kremer and Fischer no doubt identified shabb

Yamānī with saltpetre because of the lowering of temperature of water in which it was dissolved, which is a property of saltpetre. It is noteworthy that Ibn abī-Uṣaybi'a does not use any of the names for saltpetre used by al-Bayṭār, and notably not bārūd, which suggests that these names were new at the time. Although Arabic authors generally are rather careless in their use of chemical names, Ibn abī-Uṣaybi'a was a physician and we should expect him to use the right names.

The confused passage seems to refer to the purification of saltpetre by solution in water, evaporation, and crystallisation, and the addition of the crystals or of a paste of them with water, to water, when a cooling effect would, in fact, be produced, and since this would not occur with soda, the passage apparently refers to saltpetre.

Quatremère[192] thought bārūd originally meant a cooling substance, from the use of saltpetre in cooling water, and that the name was first used for saltpetre in North Africa, being afterwards used to mean gunpowder; but Hime[193] says that, since the Hebrew for hailstone is barad, the name bārūd in Arabic first meant a hailstone, and then was used to mean saltpetre, from the crystalline nature of this.

Yūsuf ibn Ismā'il al-Kutubī ("the bookseller"), commonly called al-Juni (c. A.D. 1311), in an abbreviation of al-Bayṭār called *Mā lā yasa'u al-ṭabība jahluhu* ("what a physician cannot afford to ignore"), says[194]:

Bārūd is a name for the flower of asyūs used in North Africa (al-Maghrib). In the common language of the people of 'Irāq [this name] is given to the wall-salt (milḥ al-ḥāyit). This is a salt which creeps on ancient walls; so they collect it. It is sharp. It is stronger than common salt and loosens the bowels. It cleanses the body from internal superfluities. It resembles borax (būraq; perhaps soda?). They use it to make a fire which rises and moves, thus increasing it in lightness and inflammability. It is not otherwise used in therapeutics.

Romocki[195] has:

Die Bewohner Iraks bedienen sich seiner, um das Feuer zu bereiten, welches zu steigen sucht und sich bewegt: er erhöht die Leichtigkeit und die Entzündlichkeit des Feuers.

Reinaud and Favé thought propulsion was meant and the fire might be a rocket; Hime[196] said al-Kutubī knew the effect of the progressive combustion of saltpetre mixtures, but not their explosive combustion, although "the fire which rises and moves" is a rocket.

A "nitrum" which forms an efflorescence on the soil is mentioned in the Salernitan *Circa instans* of Platearius (A.D. 1140–50).[197] Richard Pococke, in his *Description of the East* (1743–5)[198], says the ground near Tadmor and Aleppo is "impregnated with nitre," which was purified and sent to Damascus, Aleppo, etc. Both these efflorescences may have been soda.

Since the "stone of Assios" in classical authors corresponds in some of its properties with quicklime, it is quite possible that Ibn al-Bayṭār, who knew that saltpetre was formed as an efflorescence on old walls, thought that it was produced from the lime in the mortar. It is true that all the properties do not agree with those of quicklime, but its most important one, that of corroding flesh, may have impressed him most, and hence he gave the old name to the new material. Reinaud and Favé[199] said bārūd meant saltpetre in thirteenth-century manuscripts. It is the "flower of the stone of Assios" in ibn al-Bayṭār, who called it "Chinese snow." Quatremère[192] thought "snow" (thalj) should be read "salt" (milḥ), the two words in Arabic script being very similar apart from diacritical points (which Sontheimer said were often missing in the manuscripts), and this would correspond with the Persian "salt of China" (namak Chīnī)[193]; Sontheimer had translated "salt," but not Leclerc. Hime[200] pointed out that the Persians call their own native alkaline salt jamādi-i Chīnī, although it does not come from China, any more than the Jerusalem artichoke comes from Jerusalem (its name is a corruption of the Italian girasole, "sunflower").

From the mention in Ibn al-Bayṭār, Romocki[201] concluded that saltpetre first came to the Arabs from China in 1225–50, and the Arabs then passed on the knowledge of it to Europe, where it was known to Roger Bacon in 1248.

Jābir ibn Ḥayyān (? A.D. 760) is said to mention bārūd among stones as good for facilitating the evacuation of blood and for epilepsy[202] (*Kutub al-Mawāzīn*). According to Ruska[203] the Arabic word used is barad, hail, and not bārūd, and saltpetre was unknown to Jābir. Laufer[204] said the Arabs received saltpetre from China as "Chinese snow (thalj al-Ṣīn)" in the thirteenth century. Stapleton, Azo, and Ḥusain[205] say that Ibn al-Bayṭār equates thalj al-Ṣīn with asiyūs, the stone of Asios, as bārūd, which, apparently they do not think is saltpetre. Yet Reinaud and Favé,[206] Hammer-Purgstall,[207] Berthelot[208] and Romocki[209] all identified thalj al-Ṣīn with bārūd and saltpetre, and I think correctly. Laufer mentioned that the Persian name for saltpetre is shora, derived from the Indian śorāka.

For many centuries gunpowder was called in Germany kraut, i.e. plant; a German-Latin dictionary of 1475 explains nitrum (saltpetre) as krijt (plant).[210] In Danish the name krud for gunpowder has not long been obsolete, and in Dutch kruid is still used.[211] The origin of this name appears to be the transference of the literal meaning "plant" or "herb" to the meaning "drug," at first a vegetable drug, and this can be traced in the Greek φάρμακον and the Arabic dawā'.[212] The Byzantine name for gunpowder was βοτάνη, "herb," and the Chinese name is huo yao, "fire drug." An Arabic name for sugar candy (crystals of sugar) is nabāt, "plant," and this may have been transferred first to the crystalline saltpetre used in making gunpowder and then to gunpowder itself.[213]

In an alchemical manuscript in Arabic written in Syriac characters, of the sixteenth century but probably based on an original of the ninth or tenth century, there is said to be a mention of saltpetre among seven salts.[213a] It refers to a borax of the jewellers (būraq al-ṣāgha) which is white "et ressemble au salpêtre," and as the next salt, "le salpêtre qui se trouve au pied des murs" (al-shīḥa al-lātī takūn fī 'uṣūl al-ḥītān). The word shīḥa, translated "sal-pêtre," is apparently derived from an adjective meaning greyish-white, like the leaves of the plant shīḥ (wormwood).[213b] Mr. A. Z. Iskandar suggested to me that the word read shīḥ should really be sabakh, the two in Arabic being similar except for diacritical points. Sabakh means a saline earth or salt, or an efflorescence on salty soil, and sabikha also means salty soil.

THE MANUFACTURE OF SALTPETRE

Although crude saltpetre occurs in many places in Europe, Africa and Asia, in its natural state it usually contains so little potassium nitrate, and is so contaminated with deliquescent calcium salts, that without refining it is useless for making gunpowder. The first step in the invention of gunpowder must have been the discovery of an efficient process for making purified salt-petre. This can to some extent be done by simple recrystallisation from water, but the removal of the calcium salts is most certainly achieved by adding potassium carbonate in the form of wood ashes, when the calcium is pre-cipitated as carbonate and may be removed by filtration, leaving the chemically equivalent amount of potassium nitrate in solution. Crude Indian and Chinese saltpetre, although richer in potassium nitrate than European, must also first be purified by crystallisation before they are suitable for making gunpowder. The purification of saltpetre by crystallisation is described by Roger Bacon, who may also have used wood ashes (see ch. II), but the first certain reference to the use of the latter is by Ḥasan al-Rammāḥ (see ch. V).

The early accounts of the production of saltpetre and of its sources are very scanty and a comprehensive history of saltpetre is lacking. The usual opinion is that the earliest saltpetre, apart from small amounts made locally, came from India, and that the main trade in it passed through Venice.[214] It was very expensive and was at first made on a small scale in Germany by gun-makers; the first mention of a nitre plantation is in Kyeser (1405) (see ch. IV), but they may have been known before 1398. The cost in Frankfurt in 1381–3 was more than 41 fl. per cwt.; in 1399–1416 it had dropped to 16 fl. In 1388 "abgedrades adensteynes zue pulfer" is mentioned; "adenstein" was "jauchenstein" or manurial stone, and "abgedrad" was "abgerieben, abgeschabt, or abgekratzt," scraped. This crude product cost 8 fl. per cwt.; a little later, refined saltpetre cost 16 fl., and in 1440 only 10 fl., per cwt. Leather sacks for powder are mentioned in 1412. Saltpetre was also made in the South of France. In Nürnberg in 1378 saltpetre made by a gunner cost 52 fl. 20s. per cwt., in 1393, 14 fl. The "purnsteyn" mentioned was not saltpetre ("harnstein") as Rathgen says[215], but amber, as is also the "agsteinspen" or "agstein" in Berne (1383–4), not "abgedrades adenstein" as Rathgen[216] says. In Switzerland (especially Basel) in 1388 saltpetre cost $32\frac{5}{12}$ fl. per cwt., in 1425–6 only $15\frac{1}{2}$ fl.[217] In 1483 saltpetre was exported from Germany to Milan.[218] The Venetian saltpetre was often adulterated. In 1428, it is said, there was a saltpetre monopoly in Milan.[219]

The old method of scraping saltpetre from walls is described in a Vienna manuscript (Hofmuseum 141.H.10) of 1420, which has illustrations of distillation apparatus with glass tubes, furnaces and chemical apparatus[220]; it also shows the digging of saltpetre earth, filtering the solution through a conical bag, and pressing the crystals in a sack. A method of making saltpetre in pots instead of beds was rather vaguely described by Konrad Kyeser.[221] The Archbishop of Magdeburg in 1419, 1460 and 1477 permitted the collection of saltpetre only on payment of a licence.[222] Even the Pope and the Archduke of Bavaria were engaged in manufacturing gunpowder at an early date.[223] Louis XI of France in 1477 appointed commissioners to collect saltpetre wherever it could be found, and with power to force an entry where they suspected it was stored.[224]

A description of a nitre bed is given in the so-called *Mittelalterliche Hausbuch* of *c.* 1480 (?), probably written in South Germany (Heidelberg or Speyer?) and in the possession of the princes of Waldburg-Wolfegg-Waldsee.[225] To "draw (ziehen)" saltpetre, a pit is filled in alternate layers of two fingers' breadth of quicklime, dried straw, and a foot thick of earth. Every day for three weeks urine is poured over the mass. Then the earth is extracted with hot water in a large copper pot (12 ft. diameter) and the solution boiled down. The concentrated solution is poured into wooden vats and allowed to crystallise. The last process is that given for alum, but it is said the same method is used for saltpetre. There is a description of making gunpowder.

A Frankfurt fifteenth-century manuscript (*Rust vnd feuerwerck buych*) shows a saltpetre refinery.[226] G. Agricola[227] adds little to Pliny's account, but distinguishes nitrum from halinitrum (saltpetre), and mentions places in Germany where saltpetre is made. He gives in 1556 a picture of the Egyptian nitrariæ "as I conjecture them to be (tales verò esse conjicio)",[228] in which nitrum was made from water of the Nile percolating through nitrous earth. He is better informed on the preparation of saltpetre (halinitrum) from a dry slightly fatty earth with an acrid salty taste.

This is put into a vat with alternate layers of a mixture of 2 parts of quicklime powder and 3 parts of wood ashes and covered with water, which percolates and is drawn off through a plugged vent into a tub, from which it is ladled into small vats. Three percolations are used, but the second and third liquors are used instead of water with fresh material. The solution is evaporated in a rectangular copper pan to half, run off to a vat with a lid, and allowed to settle. It is reboiled, some caustic alkaline lye being added to prevent the formation of scum, and the common salt which separates at the bottom of the pan is removed with iron ladles. The solution is allowed to crystallise in a vat on rods placed in it. He gives further details of purification by recrystallisation from solution containing lye.

Another method of refining given by Agricola is to melt the saltpetre in a copper pot with a copper lid, and throw on it powdered sulphur, which burns up the thick greasy matter floating on the saltpetre, and the latter is purified. On cooling, the purest saltpetre, looking like white marble, is taken out, the earthly impurities remaining at the bottom. The saltpetre earth, mixed with branches of trees, is exposed in the open and sprinkled with water containing saltpetre, and after five or six years the earth is again ready for treatment.

I have already pointed out[229] that much of Agricola's information was copied from the earlier Italian work of Biringuccio.[230] Lippmann[231] took exception to this statement, but it is again true in the present instance, as Hoover[232] had pointed out. Biringuccio's account is fuller and better than Agricola's, since he mentions the mixing of the earth with manure, especially pig dung, or human urine, and he gives the same two methods of purification in detail. Although he does not give a description of a nitre-bed, he suggests in another place that saltpetre is formed in the manurial earth from a "thickening from moist air, so that this earth becomes impregnated by the air" with saltpetre, and also says that saltpetre is formed in caves, can be extracted from the soil in Tuscany, and forms on walls. He implies that its extraction is a modern discovery.[233]

The sections on saltpetre, gunpowder, mines, bombs and military fireworks in Peter Whitehorne's *Certain Waies for the Orderyng of Souldiers in Battelray*, 1562, are little more than translations of Biringuccio, his illustrations being also pirated.[234] Peter Whitehorne (or Whithorne), who flourished about 1543-63,[235] translated Machiavelli.[236] The original which I have seen is a small quarto.[237] It gives many recipes for gunpowder[238] with ratios saltpetre : sulphur : charcoal of 1 : 1 : 1, 3 : 2 : 1, 10 : 3 : 3, 12 : 3 : 2, 9 : 2 : 3 ("of late dayes for handgunnes"), 4 : 1 : 1, 100 : 20 : 37, 5 : 1 : 1 ("of newer making"), 20 : 4 : 5 ("nowedayes"), etc., etc. Other recipes[239] are 3 : 1 : 2, 5 : 1 : 1½, 10 : 1 : 1, etc., and the corning of powder is described. Whitehorne[240] says :

> Saltpetre is a mixture of manie substaunces, gotten oute with fire and water of drie and durtie grounde, or of that flower, that groweth owte of newe walles in Selars, or of that grownde which is fownde loose within toombes, or desolate caues.[241]

This is taken from Biringuccio[242] :

> Il sal nitro . . . e vn misto composto di piu sustantie estratto con fuocho & acqua di terre aride & letaminose, o di quel fiore che sputanto le murglie nuoue in luochi opachi, ouero di quella terra chi si troua smossa dentro alle tombe, o dishabitate spelunche doue la pioggia non possa entrare.

Biringuccio goes on to say that in his opinion saltpetre is generated in these soils from an airy moisture which is drunk in and absorbed by the earthy dryness. Some physicians, from its salty and sharp taste and its strong biting qualities, think it is hot and dry; but from its production from the air and its becoming inflammable and vaporous, others thank it is hot and moist; yet again, since it is white and transparent, and easily fusible, and heavy, it is watery; since it is cooling and breaks easily, it is thought to be earthy. "Thus it seems that every one of the elemental qualities is predominant in it (talche per concludere d'ogni qualita de elemēto, par che vi sia proprio predominio)." In 1540, therefore, the true nature of saltpetre was entirely unknown.

Lazarus Ercker, a superintendent of the mines of Hungary, Transylvania and Tyrol for the Emperor Rudolph II, composed a treatise on ores and assaying which went through several editions.[243] The fifth book of this deals with saltpetre, "which contains in itself a fast cold fire." Earth containing saltpetre comes from old sheep pens, old walls, the rubble in vaulted cellars, in buildings with dirt floors, stables, garbage dumps, latrines, etc. Ercker gives a picture of scraping saltpetre from long rectangular nitre beds, but does not describe these. He explains how the saltpetre content of earth can be determined by extracting with water and evaporating in a weighed copper dish. The residue should burn on red-hot charcoal and taste sharp and cold.

The leaching of the earth in vats with rush bottoms, filtering through wood ashes, boiling down until the liquid in a small copper dish crystallises on cooling, and crystallising are described. The mother liquor (containing calcium nitrate) is boiled with wood ashes and crystallised. The saltpetre is then refined by boiling with enough water to dissolve it, skimming out grains of common salt which settle out "because salt does not dissolve so easily as saltpetre," removing the black froth, adding some vinegar, then alum, allowing to settle, and crystallising in wood vats, or copper tubs let into the ground. The coarse common salt which has been separated can be purified for use in cooking, and some saltpetre contained in it recovered.

Ercker adds to this description of the customary method some details of his own improved process, which consists in using saltpetre solution for leaching the earth, and determining the concentration quantitatively by evaporation of a sample and weighing. The liquor left in the earth is then washed out with water. The emphasis on the control of the operations by quantitative determinations is very noteworthy for the time. Ercker's book is practical, clear and free from cant.

Joseph Boillot, born in Langres about 1650, was an architect. He was employed as a military engineer by King Henri IV and made director of the Magasin des Poudres et Salpêtres. He was an excellent draughtsman and engraver and his book is illustrated by many interesting copper plates, one of a monk making gunpowder with the Devil on his back. He describes the purification of saltpetre.[244] The saltpetre earth is leached by water, the liquor boiled three days and two nights, then, "à moitié refroidie," is poured into a vessel containing wood ashes, where it remains two or three hours. It is poured off, and after a day and a night the saltpetre "d'une cuit" deposits. This is dissolved by boiling with water and half an ounce of powdered "alun de glace" thrown in. The fire is drawn, the scum taken off, and the saltpetre "de deux cuites" allowed to crystallise in another copper vessel. It is then heated with very little water and stirred with a wooden pestle till reduced to a flour, saltpetre "de trois cuites," suitable for making gunpowder. Rock saltpetre is made by the third operation by melting without water, throwing on a little sulphur, and stirring to remove ordures; half of it is lost. This is similar to the description in Ercker.

Georg Engelhard von Löhneyss, another superintendent of mines for the Duke of Brunswick and Lüneberg, wrote a book[245] of which about a fifth is taken from Ercker without acknowledgment, including the account of the extraction and purification of saltpetre.   Löhneyss's book is also practical and straightforward.

Johann Rudolph Glauber (1604–70) in several works[246] struck a new note. He believed that common salt, alum, etc., could be converted into saltpetre, and he claimed that God had revealed to him all kinds of ways of making it, in particular from wood.   The stars generate salt, and rain brings it down to the earth; by exposing a wooden tub of water to the rays of the sun in summer, it receives the astral influence.   Nitre is formed by heaping together horns, hoofs, bones, feathers, bird dung, hair, scraps of leather, bits of cloth, roots of trees, shells, coral, etc., since it is a universal spirit, as the ancient philosopher Hermes taught.   Glauber, however, would not divulge his wonderful secret of boiling it out of the juices of wood, which would be most useful if any new foreign enemy should invade "our most dear country," a danger which did not seem too remote in the period of the Thirty Years War.   He gives an account of his method in the third part of his *Prosperity of Germany*,[247] but the practical part, as he says, was taken from Ercker.   There is nothing of practical interest about the production of saltpetre, and although Glauber was an excellent practical chemist his writings are disfigured by large excursions into mystical nonsense of the type which Paracelsus had made popular in Germany.

Thomas Henshaw[248] collected several tubs of may-dew, allowed it to putrify and form worms, then filtered it, evaporated it, and obtained an earth, by calcining which, extracting the residue with water, and evaporating, he extracted 2 oz. of a salt, the crystals of which resembled "rochepeter" ("rockpetre" or saltpetre).   Henshaw also reported[249] that much "Peter" was being made in Pegu (Burma), but it was mixed with common salt and not acceptable.   He had got saltpetre from icicles hanging in vaults and cellars, and the air was full of "volatile nitre."   Nothing was more full of "peter" than the earth of churchyards.

He describes the lixiviation of nitrous earth in tubs, and the refining of saltpetre.   Henshaw thought Pliny's nitrum was saltpetre.   He was clear that "the freer ingress the Air hath into a place, is still more of advantage." The presence of saltpetre in earth is found by tasting it.   Before it was imported from India, saltpetre was made by mixing earth with pigeon dung and horse manure and sprinkling with urine.   The richness of the saltpetre liquor could be found by weighing a quart bottle filled with it, then weighing the same bottle filled with common water, when the difference in weights gives "a very near ghess" how much saltpetre will be obtained by boiling.

The saltpetre liquor is purified by boiling down and running it into a tub containing wood ashes, drawing off the settled liquor and boiling it down, removing the scum, drawing it off into a tall narrow tub, when common salt crystallises, and then running out the solution through a tap into deep wooden trays or brass pans to crystallise.   To refine it, the saltpetre is crystallised after adding vinegar or alum, although some add quicklime.

Henshaw said prophetically that he had hopes that "the Spirit of the Volatile Salt" (i.e. ammonia) could "be reduced to Peter or some Nitrous

Salt not much differing from it," and at the present day a large proportion of the nitric acid manufactured comes from the oxidation of ammonia, and ammonium nitrate is a "nitrous salt" not very different from saltpetre. (Ammonium nitrate was discovered by Glauber, who called it nitrum flammans.)

Although Boerhaave[250] and Neumann[251] asserted that saltpetre cannot be made without the artificial addition of potash (wood ashes), Richard Watson[252] found that decayed mortar from an old barn in Ely gave good saltpetre by merely boiling with water, filtration and evaporation, and reported that a lump of saltpetre was dug up 12 or 14 ft. below the surface in gravel near Bury St. Edmunds, which gave good crystals by the same process. The efflorescences on walls, etc., were found by Brownrigg[253] to be salts of soda, but Watson[254] considered that they might sometimes be saltpetre, or if left to themselves might sometimes be converted into such. One efflorescence on brickwork which I analysed was sodium sulphate.

The construction and operation of nitre beds, the scraping of saltpetre from walls of buildings or walls built specially for the purpose of growing saltpetre, and the extraction and refining of the saltpetre, are described in several works[255] and need not be given here in any detail.

Watson[256] reports that a patent was granted in 1625 to Sir John Brooke and Thomas Russel for making saltpetre by a new invention making great use of all sorts of urine, and in 1627 King Charles I issued a proclamation to all persons to save the urine of their families and as much as they could of that of their animals, for collection every day by the patentees or their assigns. Since the invention was not successful, in 1634 an earlier proclamation of 1625 was revived giving the saltpetre-makers permission to dig up the floors of dove-houses, stables, etc., and prohibiting floors to be laid with anything except mellow earth. In 1656 an Act of Parliament was passed prohibiting saltpetre-makers from digging in houses or lands without the consent of the owners. In England the earth impregnated with pigeon dung, cattle urine, etc., formerly belonged to the Crown, and in France before the Revolution earth of stables, cellars, etc., belonged to the King. In Germany the inhabitants were obliged to build walls of fat earth mixed with straw, which when they became impregnated with saltpetre were demolished and worked up.

An edict issued in 1729 by Friedrich Wilhelm I in Berlin specified fines for concealing sources of saltpetre, and for failing to open underground places for inspection, in Magdeburg and Halberstadt. Scraping too hard, so as to remove the "bloom" of the saltpetre was forbidden, and the saltpetre boilers were forbidden to leave their homes and work for foreign masters.[257]

When saltpetre was imported from India, the home industry in England was completely abandoned, and although many projects for reviving it were started they all failed. The Society of Arts proposed premiums for the making of saltpetre in 1756 to 1764 but they were never claimed, and a works established at a cost of over £6000 was abandoned, since the cost of the product was four times that from India.[258]

A summary of the processes used in the extraction of saltpetre in France is given by Berthelot.[259]   An edict of 1540, renewed in 1572, regularised the industry.   The salpêtriers had the right to exploit with hammers, picks, shovels and mattocks any earth containing saltpetre, and materials in stables, cow-sheds, caves, cellars and pigeon-cotes, and also the plaster of demolished buildings.   They were required to make good any excavations. No one could demolish a house or wall without informing them, and they could reserve any part containing saltpetre, wetting which or mixing it with other débris was forbidden.   In 1745, 1779, etc., the salpêtriers were given the right to purchase wood ashes at fixed prices and to seize them.   They had also the right to enter private buildings and to remain in them from 6 a.m. to 6 p.m., to set up their boilers and apparatus in any public market place, private place, or any place which they wished to use, without fee. They visited a town or village every three years or so.   The communes had to provide wood and if necessary lodgings, also transport for taking the salt-petre to the refineries, and the salpêtriers had military exemption and other privileges, their pay being fixed.

Some parts of France were exempt from these regulations, e.g. Touraine (where there were natural deposits of saltpetre), and in large towns a measure of restraint was customary.   In Burgundy, and especially the Franche-Comté, they were exercised rigorously, with all the vexations, abuses and exactions which follow when civil servants have a free run.   There were many lawsuits, which had mainly the effect of enriching the legal profession.   In return, the government fixed the maximum price of sale of saltpetre, of which, it is true, they were the principal purchasers.   In the time of Louis XIII (1601–43) the amount of saltpetre collected annually was $3\frac{1}{2}$ million livres; in 1775 it had fallen to 1·8 million, since much was imported from India, but in 1789 it rose to 3 million livres again.   In 1777 Turgot put a check on the rights of the salpêtriers; they were restricted to demolitions, and forbidden to enter private dwellings or wine cellars.

In 1775 the Paris Académie des Sciences offered a prize of 4000 livres for a process for securing an abundant supply of saltpetre, to be submitted by 1778.[260]   Altogether sixty-six papers were received, the period was extended to 1782, and the prize was finally awarded to the two Thouvenels, who gave a full account of nitre plantations.[261]   Arising out of this two publications appeared :

I.   Recueil de Memoires et d'Observations Sur la Formation & sur la fabrication du Salpêtre.   Par les Commissaires nommes par l'Academie pour le jugement du Prix du Salpêtre.   A Paris, Chez Lacombe, Libraire rue Christine.   M.DCC.LXXVI, 8⁰, 622 pp., 3 plates[262].

This volume, which is scarce, contains an introduction (pp. 1–48), table of contents (pp. 49–55), extracts from Glauber (pp. 1–42), Stahl (pp. 43–65), Lemery (pp. 66–101, 102–43), Petit (pp. 144–61, on the separation of common salt, published 1729), and Pietsch (pp. 161–214 : prize essay of Berlin Academy), additions (pp. 215–35), instructions of the War Department, Stockholm (pp.

236–83), abridgements of memoirs, etc., by Bertrand (pp. 284–93), Gruner (pp. 294–380), Neuhaus (pp. 381–6), Vannes of Bescançon (1766) (pp. 387–402), Granit of Åbo (1771) (pp. 403–56), de Milly (pp. 457–74), Ducoudray (pp. 475–91), Desmazis (pp. 492–513), Simon of Dresden (1771) (pp. 514–78), Clouet (saltpetre in Asia) (pp. 579–86), Bowle (pp. 586–97), an American method (1775) (pp. 597–600), Lavoisier's memoir (pp. 601–17) and a statement by d'Incarville (p. 618).

II. Recueil des Mémoires sur la Formation & la Fabrication du Salpêtre, published in 1786 in the *Histoire* and *Mémoires* of the Academy.[263]

The *Histoire* contains papers by Macquer, Darcy (*sic*), Lavoisier, Sage, and Baumé, on saltpetre (p. 11); by de la Rochefoucauld, Clouet, and Lavoisier, on saltpetre earth (p. 192); and an anonymous report on experiments [Cavendish's] made in England on the composition of nitric acid (p. 197). The *Mémoires* contain papers by Cornette (p. 1), Thouvenel and Thouvenel (p. 55), Le Lorgna (p. 167), Gavinet and Chevrand (p. 268), de Beunie (p. 371), Romme (p. 421), Clouet and Lavoisier (pp. 503, 571), de Rochefoucauld (p. 610), and Lavoisier on the decomposition of nitre by carbon (p. 625).

In the anonymous announcement of the offer of the prize in 1775 there is a long summary of the views which had been proposed to account for the formation of saltpetre, and it is said that air is necessary:

> il est possible que l'air entre lui-même, comme partie constituante, dans la composition de cet acide [nitric acid], ou qu'il fournisse quelque substance *gazeuse*, ou outre, qui, sans être de l'acide nitreux, se trouveroit cependant un des ingrédiens nécessaires à sa mixtion.

This is probably written by Lavoisier,[264] and as the Thouvenels found that air is necessary in the formation of saltpetre, the award of the prize to them may have been connected with this, since Lavoisier was chairman of the commission which considered the essays submitted. It has been mentioned that the presence of air in the formation of saltpetre had been said to be necessary by Henshaw in 1667,[249] so that Lavoisier's suggestion was less original than he seems to have imagined. Among a collection of books purchased by Lavoisier in 1767 was *Pietsch Sur le Nitre*,[265] a detailed work on the production of saltpetre, *Abhandlung von Erzeugung des Salpeters nebst Gedanken von Vermehrung desselben*, 4⁰, Berlin, 1750[266], of which there is a French translation in the publication of 1786.[263]

A full account of the method of extraction of saltpetre is given by Krünitz[255] and Thiele.[267] The average yield from plantations was $\frac{3}{4}$ to 1 lb. of saltpetre for 100 lb. of earth, although appreciably higher yields could often be obtained, e.g. 2 to $2\frac{1}{2}$ lb. in France. From wall scrapings the average was $\frac{1}{4}$ to $\frac{1}{2}$ lb. per 100 lb. of earth; a works near Mühlhausen in 1813 obtained in about six months $\frac{1}{2}$ lb. per 100 cu. ft. of earth from walls made from 2500 cu. ft. of earth. The crude brown saltpetre from the first boiling contained about 88 per cent. of potassium nitrate; the refined salt for making gunpowder was the result of at least three crystallisations, as was explained above.

An account of the refining of saltpetre, published by the administration of the national factories in France,[268] says the refining of saltpetre of second crystallisation was carried out by percolating it in special vats with perforated bottoms with 20, 10 and 5 per cent. of cold water in succession, leaving it to drain for six to seven days in the vat, and then drying in air or by warming in a pan. The product was suitable for gunpowder. The trade in Indian saltpetre was in the hands of the English and Dutch. Russia obtained its saltpetre from Poland.[269]

In 1792 France was blockaded and the external supplies of saltpetre cut off. A decree of the year II of the Republic (1793) invited all citizens to collect saltpetre in their cellars, stables, etc., and the municipalities were also invited to set up refineries, instructions being published and supervisors appointed. The price of saltpetre was fixed at 24 sous per livre, but in some places it reached 200 livres. The scheme was completely successful. In Paris there were 600 new factories, and the old "salpêtriers" worked alongside the state organisation. In the whole of France there were 6000 factories, producing 16 million livres in one year and 5 million next year. In one year, 1793, 12,000 iron and 7000 bronze guns were made, and the gunpowder for all the French artillery was made with the national saltpetre. In the year V of the Republic (1796) all the saltpetre industry was united under the Administration des poudres et salpêtres.[270] In 1796 also, an attempt was made in France to manufacture saltpetre from nitric acid obtained by the oxidation of ammonia gas by passing it over red-hot manganese dioxide,[271] a reaction which had been described by the Rev. Isaac Milner,[272] of Cambridge, in 1788.

Saltpetre is contained in the soil of Hungary. Rückert[273] mentioned a "Salpeterflöz" in Lower Hungary, 30–36 miles long, 12–15 miles wide, which was worked for saltpetre costing 3 gulden per cwt.

An account of making saltpetre in India was given by Thevenot[274] and further details, with statistics, by Watson.[275] John Albert von Mandelslo (1616–44), a German who visited India in 1636–40, gives an account of the manufacture of saltpetre on a commercial scale in Ajmer, 60 leagues from Agra. He says[276]:

"The blackest and fattest ground yields most of it, though other lands afford some, and it is made thus : they make certain trenches which they fill with their saltpetrous earth, and let into them small rivulets, as much as will serve for its soaking, which may be the more effectually done, they make use of their feet, treading it till it becomes a brooth [broth]. When the water has drawn out all the saltpetre which was in the earth, they take the clearest part of it, and dispose it into another trench, where it grows thick, and then they boil it like salt, continually scumming it, and then they put into earthen pots, wherein the remainder of the dregs goes to the bottom; and when the water begins to thicken, they take it out of these pots, to set it adrying in the sun, where it grows hard, and is reduced into that form wherein it is brought to Europe."

John Davy[277] described the earth and "rock" of the nitre caves of Doombera in Ceylon. The nitre was wrongly supposed to have been formed from the dung of bats. The earth, in 100 parts, contained only 3·3 of potassium nitrate, 3·5 of calcium nitrate, and the rest inert. The rock ("very compounded") contained in 100 parts 2·4 of potassium nitrate, 0·7 of magnesium nitrate, and the rest inert. The nitre earth of Bengal, in India, contained in 100 parts 8·3 of potassium nitrate, 3·7 of calcium nitrate, and the rest calcium carbonate, earth, water, etc. The method of manufacture of saltpetre in Ceylon[278] was to lixiviate the earth in water, add wood ashes, and crystallise. The work is better than the Indian. Whether the addition of wood ashes was a native invention or was learnt from the Portuguese they did not know. The manufacture of saltpetre was prohibited under English rule, but Davy thought that, as the island was quiet, it could be permitted again.

In the absence of soluble calcium and magnesium salts, saltpetre can be separated from common salt (sodium chloride) by simple recrystallisation from water, the product not containing more than 0·2 per cent. of sodium chloride.[279] The nitre caves of Kentucky were worked early in the nineteenth century.[280]

COMPOSITIONS OF GUNPOWDER

The following table gives the compositions of European gunpowder (with one or two Oriental compositions) from the earliest times to about 1900. Chinese, Arabic and Indian gunpowders have been dealt with separately, and the compositions given by Roger Bacon and Marcus Græcus have already been considered.

John Arderne of Newark (c. 1307–77 or later), who practised as a surgeon before 1350, left a number of manuscripts now in the Sloane collection in the British Museum.[281] The Sloane MSS. 56, 357 and 795 contain recipes for Greek fire (fewes Grégeois) and a flying fire (fewe volant), which is an oily mixture which, when put into a tube and kindled by a match, will fly in all directions, as a marginal picture shows. A recipe is given for another fewe volant, which is gunpowder[282]:

> pernez j.li. de souffre vif; de charbones de saux (i. weloghe) ij li.; de saltpetre vj li. Si les fetez bien et sotelment moudre sur un pierre de marbre, puis bultez le poudre parmy vn sotille coverchief; cest poudre vault à gettere pelottes de fer, ou de plom, ou d'areyne, ove vn instrument qe l'em appelle gonne.
>
> Take 1 lb. of native sulphur, 2 lb. of willow charcoal, and 6 lb. of saltpetre. Grind and mix them to a fine powder on a marble, then sift the powder through a fine handkerchief. This powder is used to throw pellets of iron, lead, or bronze from an instrument which is called a gun.

Brackenbury[283] thought Arderne's recipe for a 6 : 1 : 2 mixture was from a laboratory experiment only; Arderne had probably read the *Liber Ignium* of Marcus Græcus, since the recipe as far as the word "marbre" is literally translated from § 13 of this (see ch. II), and other extracts from Arderne given by Hewitt are also from this work. The sifting of the powder through a fine cloth, however, is a new feature not in the *Liber Ignium*.

Another recipe given by Arderne is for burning turpentine thrown from an iron or copper tube to set fire to houses of an enemy[284]:

> Si volueris domos inimicorum tempore guerre cremare, fac unum instrumentum concavum interiùs de ferro vel ere, ad modum fistule, et impleatur de aquâ terbentine; et illud instrumentum ligetur uni sagitte vel querule, et, igne accensum, cum arcu vel balistâ mittatur ubicunque volueris malefacere.

> If you wish to burn the house of an enemy in time of war, make a hollow instrument of iron or copper like a tube and fill it with water of turpentine. Then bind this instrument to an arrow or cross-bow quarrel and having set it on fire throw it with a bow or balista wherever you wish to cause damage.

He also says the incendiary may be attached to a bird (vel cum aliquâ ave portatum).

In the following table, compiled from various sources, the proportions of the components are given in the usual order, saltpetre : sulphur : charcoal. The kind of charcoal (willow, alder, dog-wood, etc.) is sometimes given in the reference, but it is said[285] that the particular wood used is not very important, provided that it is fully charred. The sulphur is often given as sulphur vivum in early recipes (e.g. Cologne, 1373–6, etc.) in Latin; in a Berne account of 1383 this is given in German as lepswebel.[286] Rathgen[287] misunderstood some Rothenburg recipes which specify melting 1 lb. of saltpetre and "ein klein vierdung schwefelz werfen," as mixing saltpetre separately with sulphur and with charcoal and then mixing the two, whereas the first process was a purification of the saltpetre (see p. 313) by throwing sulphur on it when melted. It is said that the powder should be charged at once into the gun to avoid dampness (which does not suggest that the saltpetre was very pure), rammed, and an air space left.

Table of Compositions of Gunpowder[288]

1. R. Bacon, *c*. 1260
   7   5   5
2. Albertus Magnus, *c*. 1275 ?
   6   1   2
3. Marcus Græcus, *c*. 1300
   6   1   2
4. French, *c*. 1338
   2   1   ?
5. English (Arderne), 1350
   6   1   2
6. German, *c*. 1350*
   4   1   1

7. Rothenburg, 1377–80†
   4   1   1
8. Nürnberg, 1382
   4   1   1
9. Montauban, *c*. 1400
   22   4   5
10. German, *c*. 1400
    22   4   5
11. Amiens, 1417‡
    25   24   43
12. Dijon, 1478 (900 lb.)

* In lb., plus 1 oz. salpetri (sal practica ?—see p. 155), 1 oz. sal armoniac, $\frac{1}{12}$ camphor.
† Plus mercury.
‡ Plus 6 fine amber and 2 arsenic.

Table of Compositions of Gunpowder—continued.

13.	German, 1546				24.	French, 1650			
	Large guns:	50	33·3	16·7			75·6	10·8	13·6
	Medium guns:	66·7	20	13·3	25.	English, 1670			
	Mortars:	83·4	8·3	8·3			71·4	14·3	14·3
14.	German, 1555				26.	French, 1686			
		6	2	1			76	12	12
15.	Swedish, 1560				27.	French, 1696			
		4	1	1			6	1	1
16.	Whitehorne's recipe, 1562				28.	Swedish, 1697, 1726			
		5	1	2			73	10	17
17.	German, 1595				29.	English, 1742			
		52·2	21·7	26·1			6	1	1
18.	French (Boillot), 1598				30.	Swedish, 1770			
		6	1	1			75	16	9
19.	Danish, 1608				31.	Prussian, 1774			
		68·3	8·5	23·2		Coarse:	74·4	12·3	13·3
20.	French (de Bry), 1619					Fine:	80	10	10
		6	1	1	32.	English, 1781			
21.	German (Furtenbach)						75	10	15
	Large guns:	69·0	14·5	16·5	33.	Chinese, 1788			
	Small guns:	72·4	13·1	14·5			75·79	9·05	15·16
	Small arms:	75·7	11·3	13·0	34.	French, 1794			
22.	English (Nye), 1647						76	9	15
		4	1	1	35.	Prussian, 1800			
23.	German, 1649						75	10	15
	Large guns:				36.	Prussian, nineteenth century			
	66·8–70·0	16·6–14·0	16·6–16·0				75	11·5	13·5
	Muskets:				37.	Swiss, 1800, 1808			
	72·5–75·5	13·0–11·2	14·5–13·3				76	10	14
	Pistols:				38.	Swedish, 1827			
	78·7–85·6	9·4– 8·5	11·9– 5·9				75	15	10

In France six different powders were formerly made, the reason for the nicety of distinction not being very obvious (the ratios are saltpetre : sulphur : charcoal) :

	Cannon.	Musket.	Pistol.
Strong	100 : 25 : 25	100 : 18 : 20	100 : 12 : 15
Weak	100 : 20 : 24	100 : 15 : 18	100 : 10 : 18

whilst in England the composition, 75 : 10 : 15, was the same for all arms, the difference being only in the size of the grains.[289] Many trials were made in 1756 at the French Royal Factory at Essone, near Paris, to determine the best proportions of the ingredients. The powder was tested by an eprouvette, consisting of a short mortar, 7 in. (French) in calibre, containing 3 oz. of powder and throwing a 60-lb. copper ball. No powder which did not throw the ball 300 ft. was admitted.[290] The common eprouvette was a small barrel, the force being measured by the action on a strong spring or a large weight. At Essone, a powder made from saltpetre and charcoal was strongest with the ratio 16 : 4, giving with the eprouvette a power of 9. All three ingredients, 16 : 1 : 4 (saltpetre : sulphur : charcoal), gave an eprouvette power of 15, a less and greater quantity of sulphur producing a smaller effect. The highest

power, 17, was produced with a ratio 16 : 1 : 3. The powder without sulphur
was tried in the mortar eprouvette. A 2-oz. charge threw the 60-lb. ball
213 ft., the strongest powder containing sulphur threw it 249 ft., but with a
3-oz. charge the effects were reversed, the first powder throwing the ball
475 ft. and the second powder only 472 ft.[291]

In 1883 Berthelot[292] gave the following compositions as in use (KNO_3 =
potassium nitrate or saltpetre, S = sulphur, C = charcoal) :

	KNO_3	S	C
French cannon powder	75	12·5	12·5
Coarse-grain powder "ancienne"	75	10	15
Musket powder B	74	10·5	15·5
„ F	77	8	15
Austria	75·5	10	14·5
Switzerland, U.S.A.	76	10	14
Holland..	70	14	16
China	61·5	15·5	23
Prussia	74	10	16
England, Italy, Russia, Sweden	75	10	15

In 1895 the following compositions of gunpowder (KNO_3 : S : C) were
in use[293] :

I. *Small arms gunpowder (Gewehrpulver)* :

	KNO_3	S	C
China, England, France, Italy, Austria–Hungary, Russia, Sweden, Turkey, U.S.A.	75	10	15
Belgium, Persia, Spain	75·5	12·0	12·5
Germany	74	10	16
Holland..	70	14	16
Portugal	75·7	10·7	13·6
Switzerland	75	11	14

II. *Cannon powder (Geschützpulver)* :

England, France, Switzerland	75	10	15
Austria–Hungary, Germany	74	10	16

III. *Sporting powder*

England, France	75	10	15
Austria–Hungary	75·95	9·43	14·62
Germany	78	10	12
Switzerland	78	9	13

In sporting powder a so-called "red charcoal," made by incomplete carbonisation at a lower temperature, has been used, since it gives a faster-burning
powder.

IV. *Blasting powder*

	KNO_3	S	C
Austria–Hungary	60·19	18·45	21·36
England	75	10	15
France	72	13	15
Germany	70	14	16
Russia	66·6	16·7	16·7

A lower saltpetre content gives a slower-burning powder, less likely to splinter or disintegrate the coal, and also safer, but such powder gives a larger proportion of the poisonous carbon monoxide in the combustion gases, so that in Germany a powder containing up to 78 per cent. of saltpetre has been used. A very low-nitrate powder used in French mines[294] was 40 : 30 : 30.

The so-called "brown powder" or "cocoa powder," invented by Castner about 1882, had the composition 78 : 3 : 19, the charcoal, made from rye-straw, being reddish-brown in colour. It is said to have been very good.[295] The compositions of blasting powders in 1874[296] were: France 60 : 18 : 20, Germany 66 : 12·5 : 21·5, Italy 70 : 18 : 12. A slow powder containing sodium nitrate instead of potassium nitrate and sawdust instead of charcoal was used in mines,[296a] but has the disadvantage that the sodium nitrate absorbs moisture from the atmosphere very easily. The following tables[297] give the compositions of some recent gunpowders (KNO_3 = saltpetre, S = sulphur, C = charcoal; all compositions are for dry materials) :

Military gunpowder

	KNO_3	S	C
England, Russia, Sweden, Italy, Turkey, U.S.A.	75	10	15
France, Belgium, Spain, Persia	75	12·5	12·5
Germany	74	10	16
Austria–Hungary	75·5	10	14·5
Portugal	75·7	10·7	13·6
Switzerland	76	10	14
China	61·5	15·5	23

Sporting powder

England	75	10	15
France	78	10	12
Germany	78	10	12
Austria–Hungary	76	9·5	14·5
Switzerland	78	9	13

Mining and blasting powder

England	75	10	15
France	72	13	15
Germany	70	14	16
Austria–Hungary	60	18·5	21·5
Italy	70	18	12
Russia	66·6	16·7	16·7

The blasting powder burns more slowly the less saltpetre it contains. The use of ordinary gunpowder in coal mines is unsafe, since the flame is hot enough to set fire to the firedamp (methane) which accumulates in "fiery" mines and causes explosions. A "safety black powder" which was at one time used in coal mines was "Bobbinite," which contained 64 saltpetre, 2 sulphur, 19 charcoal and 15 of a mixture of ammonium and copper sulphates, the salts reducing the flame temperature. The salts were later replaced by a mixture of 8 of starch and 3 of paraffin wax, the other ingredients being correspondingly increased. Bobbinite was later found not to be really safe— no gunpowder ever is, and gunpowder is not a permitted explosive in coal mines in which firedamp may be found.

Records are said to speak of gunpowder works in Augsburg in 1340, Spandau in 1344, Liegnitz in 1348, and Hildesheim in 1421.[298] The story of the explosion of gunpowder in the town hall of Lübeck in 1360 has been dealt with in ch. IV. G. Köhler[299] said there was no question of powder mills until 1431, a statement that Michael Behaim set up one in Röthenbach in 1405, in which the powder was ground between millstones (?), being unsupported by a reference to the source. The earliest method of mixing was by a roller on a stone slab ("marble"). Then a pestle and mortar was used, the pestle later being worked by horse or water power. Such a stamp mill, with heavy lignum vitæ pestles, was still in use about 1800 at Battle, in Sussex, only a few pounds of very fine powder being made at a time.[300] Later, rolling mills were introduced in which two vertical stones attached to a cross-piece and vertical axle were rolled over the moistened ingredients on a bed-stone or trough. Not more than 40–50 lb. of powder was worked at one time, the mill being worked by horse or water power. The process used at Woolwich was described by Napier[301] and that at Waltham Abbey by Coleman.[302]

The inflammability of gunpowder is not greatly affected by the mixture ratio. The propulsive force depends mainly on the burning rate and the volume of gas, both of which depend on the mixture ratio. The right mixture for military gunpowder was found only after many trials over a considerable period of time. Even to-day more stress is laid on the method of manufacture than on the mixture ratio.

An anonymous article on "The History of Making Gunpowder"[303] says "there is so great a Latitude, that provided the Materials be perfectly mixt, you may make good Powder with any of the proportions above mention'd; but the more Peter you allow it, it will still be the better, till you come to observe Eight parts," to one of sulphur and two of charcoal.

Berthollet (1777–8), by experiment, found the best composition to be 80 saltpetre, 5 sulphur and 15 charcoal.[304] Berthelot,[305] from theoretical considerations, concluded that it was 74·84, 11·84 and 13·32, but theoretical considerations are not finally decisive, since good and intimate mixture is necessary for complete combustion. Experience shows that an increase in sulphur makes the powder keep better, whereas the charcoal is never pure, the amount of moisture varies, and the mass of powder may change in composition during the various operations in making it.[306]

The composition 75:12:13 corresponds approximately with $2KNO_3$: S : 3C, and the following equation has been given for the reaction on explosion :

$$2KNO_3 + S + 3C = K_2S + N_2 + 3CO_2.$$

The gaseous product contains carbon monoxide as well as carbon dioxide and nitrogen, and the solid product contains potassium carbonate and sulphate as well as sulphide.   R. Bunsen and L. Schischkoff[307] found that many other products exist in the gases and solid, and in a classical research Capt. Noble and F. A. Abel[308] found that the decomposition which an average gunpowder undergoes when fired in an enclosed space cannot be represented even by a comparatively complicated chemical equation.

The ballistic pendulum as a means of comparing the qualities of kinds of gunpowder was invented by the mathematician Benjamin Robins.   He was born in Bath in 1707, the son of poor Quaker parents and was self-educated. He was befriended by Pemberton, a physician and mathematician who had assisted Newton in reprinting the *Principia* in 1726 and himself wrote on mathematics and physiology.   Robins at the age of eighteen established himself in London, where he taught mathematics and also interested himself in navigation, fortification, architecture and gunnery.   He replied to Bishop Berkeley's book *The Analyst*, which attacked the ideas of the differential calculus, by publishing a book on the subject in 1735 in which he used the method of limits, afterwards developed by Brook Taylor, Thomas Simpson, and Colin Maclaurin.   Robins died in Fort St. David, Pondicherry, on 29th July, 1751.

In 1742 Robins published his *New Principles of Gunnery*[309] in which he describes his ballistic pendulum.   This consists of a heavy body which is suspended so that it can swing to-and-fro.   If a ball is fired into the body of the pendulum its velocity of impact can be deduced from the measured recoil of the pendulum according to the principle of momentum.   Robins also showed that Newton's formula for the resistance exerted by the air on a moving missile holds only for small velocities; for larger velocities the resistance is much greater than the formula predicted.

The *New Principles of Gunnery* was translated by the great German mathematician Leonard Euler, who added some notes which mainly serve to show that he did not understand it[310], and his version was re-translated into English by Hugh Brown.[311]   The French translation has been mentioned previously (ch. III, ref. 52).   In the collected works of Robins,[312] the first volume contains some previously unprinted articles on gunnery in which he shows that much smaller charges of gunpowder were at least as, and often more, effective than those used at the time.

REFERENCES

1.   Prov. xxv, 20 : "as vinegar upon nitre" (Luther had translated "chalk"); Jerem. ii, 22 : "wash thee with nitre"; in Ferrar Fenton, *The Complete Bible in Modern English*, 1938, 493, it is still "nitre," but *The Bible in Basic English*, Cambridge, 1949, 552, has "soda," the first correct translation in an English Bible known to me.

2.   F. Hoefer, *Histoire de la Chimie*, 1866, i, 58; Partington, *Origins and Development of Applied Chemistry*, 1935, 498.

3.   M. Mercier, *Le Feu Grégeois*, 1952, 115.

4.   *Geographia Sacra, seu Phaleg et Canaan*; in *Opera Omnia*, Leyden, 1707, i, 419.

5.   Partington, ref. 2, 144; Lippmann, *Naturwiss.*, 1937, xxv, 592; E. Boisacq, *Dictionnaire Étymologique de la Langue Grecque*, 3 ed., Paris, 1938, 671.

6.   Partington, ref. 2, 144, 193.

7. Partington, ref. 2, 165.

8. C. Salmasius, *Pliniana Exercitationes*, Utrecht, 1689, 760; F. Lenormant, *Les Premières Civilisations*, 1874, ii, 425.

9. Hist., ii, 86.

10. Stephanus, *Thesaurus Græcæ Linguæ*, ed. Hase and Dindorf, Paris, 1842–6, v, 338, 1527; O. Schrader, *Reallexikon der Indogermanischen Altertumskunde*, Strassburg, 1901, 781; Liddell-Scott-Jones, *Greek–English Lexicon*, 1177, quote from Sappho, sixth to fifth century B.C., and Hippokrates, De Aere, fifth century B.C.

11. Geogr., XVII, i, 23, p. 803C.

12. Meteor., iv, 9, 10; 385, 389.

13. Prob., i, 38.

14. Colours, iv.

15. De mirab., 53; Prob., xxiii, 40; Meteor., ii, 3; of the works quoted only the Meteorologica is by Aristotle, the others being by members of his school, although fairly early.

16. *H.N.*, xxxi, 46; K. C. Bailey, *The Elder Pliny's Chapters on Chemical Subjects*, 1929, i, 49–55; all references in the text to Pliny are to bk. xxxi, ch. 46, unless otherwise stated.

17. Theophrastos, Caus. plant., II, v, 3; VI, ii, 4; Odor., 65.

18. Ref. 16, i, 169.

19. Bostock and Riley, *Pliny's Natural History*, 1856, v, 515, said Hardouin had derived the name from κόλικας, round cakes; Sillig had suggested, from Photios, that it should be scolecas, "worm-like."

20. Bostock and Riley, ref. 19, say that this was near the mouth of the River Axius, modern Vardar, near modern Kulakia, Gulf of Salonika.

21. Ref. 2, 1860, i, 148.

22. Geoponika, ii, 33; ed. Niclas, 1781, i, 175.

23. *A History of Inventions*, 1846, ii, 502.

24. Ref. 16, i, 169.

25. Commentary on Prov. xxv. 20; in *Opera*, f°, Cologne, 1616, viii, 109, col. 1; and in full in Beckmann, ref. 22, ii, 487; on native soda, see S. Parkes, *Chemical Essays*, 2 ed., 1823, ii, 37.

26. Ref. 16, i, 55.

27. De orac. defect., § 41; *Moralia*, Paris, 1841, i, 526.

28. *H.N.*, xviii, 45.

29. Georgics, i, 192–3.

30. *H.N.*, xix, 26.

31. *H.N.*, xix, 41.

32. Bailey, ref. 16, i, 170.

33. Pliny, *H.N.*, xxxi, 46.

34. Hoefer, ref. 2, i, 148.

35. *H.N.*, xix, 41.

36. *H.N.*, xxxi, 46.

37. Rei rust., xi, 3.

38. Epigr., xiii, 17.

39. De arte coquinaria, iii, 1.

40. Beckmann, ref. 23, ii, 492; H. Kopp, *Geschichte der Chemie*, Brunswick, 1847, iv, 205.

41. Ref. 19, v, 514, 519.

42. *Geschichte der Explosivstoffe*, 1895, i, 5.

43. Ref. 16, i, 53, 172.

44. Pliny, *H.N.*, Leyden (Elzevir), 1635, iii, 285.

45. In Pauly-Wissowa, *Real-Encyklopädie*, xi, 99.

46. De Medicina, v, 8.

47. De Simpl. Med. fac., i, 14; ed. Kühn, xii, 222.

48. Paulus Ægineta, De Re Medica, vii, 3 (κονία); ed. Heiberg, *Corpus Medicorum Græcorum*, 1924, ii, 227; *Works*, tr. Adams, 1846, ii, 265, 395.

49. Orig., XVI, ii, 7.

50. *Liber de Pallio*, Paris, 1622, 403.

51. Dioskourides, Mat. Med., v, 130, 131; ed. Saracenus, f°, Frankfurt, 1598, 378.

52. Ed. Kühn, xii, 212, 225.

53. De Medicina, ch. 48.

54. Salmasius, ref. 8, 760; H. Blümner, *Technologie und Terminologie der Gewerbe und Künste bei Griechen und Römen*, 1887, iv, 500.

55. Ref. 16, i, 170.

56. L. Israelson, *Die Materia Medica des Kl. Galenos*, Jurjew, 1894, 165.

57. *Lehrbuch der Chemie*, 1835, II, i, 51. (Mitscherlich was a competent Oriental linguist).

58. Mat. Med., v, 129.

59. *H.N.*, xxxi, 42.

60. Ref. 40, 1847, iv, 27.

61. Ref. 16, i, 168.

62. *H.N.*, xiv, 26.

63. Ref. 40, 1845, iii, 35; 1847, iv, 25–6.

64. *H.N.*, xxxi, 46.

65. Saracenus in ed. Dioskourides, 1598, 378, read βουνῶν as "tumulis, vel ex Bunis," and said Oreibasios had read νήσων; Pliny, iii, 25, said the Buni were a people of Illyria.

66. Ref. 8, 778.

67. Baas-Becking, *Scientific Monthly*, 1931, 434.

68. *H.N.*, xxxi, 46.

69. Berthelot, *La Chimie au Moyen Âge*, 1893, i, 179–228 (199); from BN Latin MS. 6514, *c.* A.D. 1300. A revised text is given in Corbett, *Catalogue des Manuscrits Alchimiques Latins*, I Paris MSS., Brussels, 1939.

70. Ref. 69, i, 308: "salpêtre?"

71. Hime, *Origin of Artillery*, 1915, 14 f.

72. Phillips, *Archæologia*, 1847, xxxii, 183–244; § cxcii.

73. Mrs. Merrifield, *Original Treatises on the Arts of Painting*, 1849, i, 16, 47, 65, 85.

74. Merrifield, ref. 73, ii, 397, 403.

75. Merrifield, ref. 73, i, 290, 317.

76. Hime, ref. 71, 15.

77. Schöffler, *Lexikalische Studien zur mittelenglischen Medizin*, Halle, 1913, 103; q. by Lippmann, *Entstehung und Ausbreitung der Alchemie*, 1931, ii, 187.

78. *Mineralogia, sive Natvralis Philosophiæ Thesavri*, f°, Lyons, 1636, 327–34.

79. *Mvsævm Metallicvm*, f°, Bologna, 1648, 321 f.

80. *Mundus Subterraneus*, f°, Amsterdam, 1668, i, 304 f.

81. *Artis Magnæ Artilleriæ*, bk. ii, ch. 1: de originis Salisnitri ejus natura & operationibus, la. 4°, Amsterdam, 1650 (BM 64.f.1), 61 f.; *Volkommene Geschütz-, Feuerwerck- und Büchsenmeistery-Kunst*, tr. from the Latin by T. L. Beeren, 2 pts., f°, Frankfurt, 1676 (BM 717.l.9); *The Great Art of Artillery*, tr. from French by George Shelvocke, Junr., Gent., f°, London, 1729 (BM 8825.f.27), 83 f.

82. *Experiments and Notes about the Prodvcibleness of Chymicall Principles* (appendix to 2nd ed. of the *Sceptical Chymist*), Oxford, 1680, 29.

83. *Phil. Trans.*, 1684, xiv, 609–19, no. 160; Abridged ed., 1809, iii, 50, and notes.

84. *The Natural History of Nitre: or, A Philosophical Discourse of the Nature, Generation, Place, and Artificial Extraction of Nitre, with its Vertues and Uses*, London, 1670 (viii ll., 93 pp.), 12 f., 36: "the Nitre of the Ancients is the same with ours"; J. Ferguson, *Bibliotheca Chemica*, Glasgow, 1906, i, 161, mentions a Latin tr., *Naturalis Historia Nitri*, publ. in London, Frankfurt, and Hamburg, 1675, and that long before Clarke's a tract on nitre was published by Sir Thomas Chaloner, Junr., *A Shorte Discourse of the most rare and excellent Vertue of Nitre*, sm. 4°, London, G. Dewes, 1584, ff. ii, 22 (BM 778.e.56).

85. *Historia Naturalis*, 5 vols., 4°, Paris, 1685; iv, 816.

86. *De Nitro cum Veterum tum Nostro Commentatio*, Amsterdam, 1709, xxviii, 243 pp.; C. G. Jöcher, *Allgemeines Gelehrten-Lexicon*, Leipzig, 1751, iv, 240, says Günther Christoph Schelhammer (*sic*) was born in Jena in 1649, professor of medicine in Helmstädt, Jena, and Kiel, wrote several medical works, and left a MS. *Schediasma de chalcantho, alumine, et atramentis.* He died in 1716. His life, with letters to him from scholars, was published in 1727 by C. S. Scheffel. The book on nitrum is not mentioned by Kopp or Hoefer, and J. F. Gmelin, who mentions some of Schelhamer's medical works and the book by Scheffel, does not refer to it (J. F. Gmelin, *Geschichte der Chemie*, Göttingen, 1798, ii, 485; C. S. Scheffel, *Virorum Clarissimorum ad G. C. Schelhammerum Epistolæ selectiores*, Vismariæ et Sundii, 1727 (BM 1165.f.18), and with new title-page and preface, Leipzig, 1740 (BM 246.k.29)).

87. *Commentationes Societati Regiæ Scientiarvm Gættingensi per annos* 1758, 1759, 1760, 1761 *et* 1762. Oblatæ a Ioanne Davide Michaelis, eivs Societatis Directore, 4°, Bremen, 1763, Commentatio V, pp. 134–50 (prælecta die 14. Novembris, 1761) : "De Nitro Plinii."

88. *Geschichte der ältesten Chemie und Philosophie*, Marburg, 1775, 222, 266.

89. *Historisch-kritische Untersuchung der Alchemie*, Weimar, 1777, 102.

90. *Saggi sul Ristabilimento dell' antica arte de' Greci e Romani Pittori, Seconda Edizione* . . . *accresciuta notabilimente*, 2 vols., 8°, Parma, 1787 (BM 7856.d.25, uncut till I used it).

91. Ref. 90, ii, 65 f.

92. Ref. 90, 95 f., 131 f., 182 f.

93. Ref. 90, 113 f. : antichi non conobbero nessum nitro col titolo de natron; Requeno, *ib.*, 155, says Schelhamer was a pupil of N. Lemery.

94. Petri Belloni Cenomani, *De Admirabili Opervm Antiqvorvm et rerum suscipiendarum præstantia*, sm. 4°, Paris, 1553, bk. iii, ch. 8, f. 49 *v.* : Vulgari nomine Natron dicitur; cf. bk. iii, ch. 8, De flore Asiæ petræ, quem malè quidam pro nitro sumunt, f. 53; with an account of mummification, bk. ii, ch. 1 f., De Medicato Funere; f. 20 *v.*, mumiæ. He says in bk. iii, ch. 8, that nitrum was not saltpetre.

95. *Chymische Nebenstunden*, St. Petersburg, 1762, 151.

96. Ref. 23, ii, 482; F. W. Gibbs, *Annals of Science*, 1938, iii, 213–6.

97. Ref. 40, 1845, iii, 219.

98. Ref. 16, i, 170.

99. Berthelot, *Collection des anciens Alchimistes Grecs*, 1888, ii, 326, 335, 445; M. Stephanides, *Rev. des Études Grecques*, 1922, xxxv, 296–320 (311).

100. Die pharmakologischen Grundsätze des Abu Mansur übersetzt und mit Erklärungen versehen von Abdul-Chalig Achundow, in Kobert's *Historische Studien aus dem pharmakologischen Institut der Kaiserlichen Universität Dorpat*, Halle, 1893, iii, 153, 162-3, 174, 245, 280, 316, 322, 325; summary in Lippmann, *Abhandlungen und Vorträge*, 1906, i, 81-96. Abū Manṣūr clearly distinguished potash and soda. Our name borax is derived from the Arabic būraq by substitution of the Spanish letter x for the Arabic q. See also Lippmann, *Geschichte der Rübe*, 1925, 25, 28–9.

101. *De Mineralibus*, Cologne, 1569, 124, 135, 386.

102. R. Nasini, *Chem. Ztg.*, 1930, liv, 985.

103. De morbis, ii; *Opera*, ed. Vander Linden, Leyden, 1665, ii, 45; ed. Kühn, 1826, ii, 224.

104. Meteor., ii, 3, 359; Pliny, xxxi, 40, quotes this as from Theophrastos.

105. Meteor., ii, 3, 359*b*.

106. Meteor., iv, 11, 389*b*.

107. *H.N.*, xvi, 11; xxxi, 40, 46; xxxvi, 27.

108. *H.N.*, xiv, 26.

109. *H.N.*, xvi, 11.

110. Berthelot, ref. 99, 1888, ii, 59.

111. Mat. Med., i, 115, 116, 136, 137, 146, 149, 178, 186.

112. Mat. Med., v, 132, 135.

113. De Med., v, 8.

114. *H.N.*, xv, 18; xxviii, 75.

115. De Rei Rust., xii, 16, 50.

116. Opus agriculturæ, xii, 7, 13; lexiuo or lexiuio in MSS.

117. Lippmann, *Abhandlungen und Vorträge*, Leipzig, 1913, ii, 318, 326; Speter, *Archiv f. d. Geschichte der Naturwiss. u.d. Technik*, 1910, ii, 201.

118. Rei Rust., i, 7.

119. *H.N.*, xxxi, 39.

120. Annal., xiii, 57.

121. "Mémoire sur la Nature du Sel Commun, Dont les anciens Belges & Germains faisoient usage," in *Mém. de l'Académie Impériale et Royale des Sciences et Belles-Lettres de Bruxelles*, Brussels, 1780, i, 357.

122. Ref. 19, v, 503.

123. Ref. 23, ii, 493.

124. *Elementa Chemiae*, Leyden, 1732, i, 767.

125. Od., vi, 92, 199: στεῖβον δ᾽ἐν βόθροισι, treading in pits.

126. T. Thomson, *History of Chemistry*, i, 94.

127. W. Smith and Wayte, in W. Smith and Marinden, *Dictionary of Greek and Roman Antiquities*, 1914, i, 881.

128. Pliny, *H.N.*, xxvii, 88.

129. Pliny, *H.N.*, xix, 18; xxiv, 58; Beckmann, ref. 23, ii, 98.

130. Idylls, xv, 30; ed. Cholmely, 1919, 296.

131. Deipnos., ix, 77.

132. *L'Origine des Lois, des Arts et des Sciences*, 1809, i, 144.

133. *H.N.*, xxxv, 57.

134. Ed. Kühn, xii, 170, 180.

135. Ranae, 709; Liddell-Scott-Jones, *Greek Lexicon*, 2020: "lye or soap made from adulterated soda"; H. B. Dixon, *Mem. Manchester Lit. and Phil. Soc.*, 1924–5, lxix, no. 1: "adulterated soap-balls"; Aristophanes, Lysistrata, 470: λουτρὸν . . . καὶ ταῦτ᾽ ἄνευ κονίας.

136. Ref. 4, 1707, i, 419.

137. Mat. Med., ii, 96.

138. Liddell-Scott-Jones, ref. 135, 977, give as meanings of κονία: sand, ashes, pearl-ash (potash), lye, "soap-powder," alkaline fluid for washing, plaster or stucco, and quicklime.

139. *H.N.*, xxviii, 51.

140. Epigr., XIV, xxvi–xxvii.

141. Marquardt, *Das Privatleben der Römer*, in *Handbuch der Röm. Altert.*, Leipzig, 1879, vii, 764; Postgate, *Corpvs Poetarvm Latinorvm*, 1905, ii, 483, 525.

141a. VII, xxxiii, 20.

142. Martial, XIV, xxv–xxvii; Ovid, Artis amatoriæ, iii, 163.

143. T. Thomson, *History of Chemistry*, 1830, i, 95; Beckmann, ref. 23, ii, 92; Mowat, *Alphita*, Oxford, 1887, 159; Schrader, ref. 10, 760–1.

144. Pauly-Wissowa, ref. 45, II Reihe, 1921, iii, 1112.

145. De compos. med. sec. loc., ii; Kühn, xii, 586.

146. Method. medend., vii, 4; Kühn, x, 569.

147. Synopseos, iii; in Stephanus, *Medicæ artis principes*, 1567, 53.

148. Berthelot, ref. 99, 1888, ii, 142.3, 143.7.

149. Ref. 40, iv, 383.

150. *H.N.*, xxxi, 42.

151. De diuturn. morbis, ii, 13.

152. Tetrabiblos, II, iv, 6; III, iii, 14.

153. De crementis capillorum, in Rerum Medicarum, i, 3; *Medici Antiqvi Omnes*, f°, Venice, 1547, 291 *v*.

154. *Entstehung und Ausbreitung der Alchemie*, 1919, i, 662.

155. E. Lucius, *Die Anfänge des Heiligenkults in der christlichen Kirche*, Tübingen, 1904, 206, 211.

156. Acts of Paul and Thekla, Miracle 28, in Migne, *PG*, lxxxv, 613; on Thekla, see C. Schlau, *Die Akten des Paulus und der Thecla*, Leipzig, 1877; Gwynn, in W. Smith and Wace, *Dict. of Christian Biography*, 1887, iii, 822.

157. Ref. 135, 1576.

158. De Temperamentis Simplicium, ch. 358 (de sapone); ed. Brunfels, f°, Strassburg, 1531, 245; tr. Andreas Alpagus Bellunensis, la. 8°, Venice, 1550, 183 *v*.

159. De Simplicibus, in Serapion, ed. Brunfels, 1531, 373–97 (misnumb. 348) : sapo, calidus existit, qui hulcerans corpus, in ipso fortem efficit, abstersionem.

160. Abū Manṣūr Muwaffak ibn 'Ali al-Harawī, Ref. 100, 252.

161. Al-Dimashqī, *Manuel de la Cosmographie du Moyen Âge*, tr. A. F. Mehren, Copenhagen, 1874, 270.

162. Ref. 161, 77.

163. E. Wiedemann, *Sitzungsber. Phys.-Med. Soc. Erlangen*, 1916, xlviii, 286 (316–17).

164. Wiedemann, ref. 163; A. von Kremer, *Culturgeschichte des Orients unter den Chalifen*, Vienna, 1877, ii, 225, says the name first appears in the dictionary of Ibn Durayd (d. A.D. 933 in Baghdād), and with it the art of the barber appeared.

165. *Die Technik der Vorzeit*, 1914, 1290, see also F. Sherwood Taylor, *A History of Industrial Chemistry*, 1957, 130 f.

166. Salzmann, *English Life in the Middle Ages*, Oxford, 1926, 106.

167. *Chemistry applied to Arts and Manufactures*, Glasgow, n.d. (1857–60), Division vi, 868.

168. Feldhaus, ref. 165, 1289.

169. *Rendiconti dell'Accademia delle Scienze Fisiche e Mathematiche*, Naples, 1877, xvi, 74 : Sopra una materia grassa, ricavata da talune terre rinvenute a Pompeji.

170. "Ueber vermeintliche antike Seife," in *Wiener Studien. Z. f. classische Philologie*, Vienna, 1882, iv, 263–70; Günther, in Iwan Müller, *Handbuch der klassischen Altertumswissenschaft*, V, i, 63, gives this title as "Graz, 1885," which may be another publication by Hofmann.

171. *H.N.*, xxxvi, 27–8; Licinius Mucianus, consul in A.D. 52, 70, and 74, published three books of *Epistles*, and eleven books of a *History* which seems to have treated mainly of eastern affairs. Pliny, *H.N.*, vii, 3, quotes him as giving a case of change of sex of a woman into a man. After squandering his property, he lived in retirement in Asia, probably Lycia, but was recalled by Nero. He was instrumental in the elevation of Vespasian to emperor. Apart from fragments, his works are lost.

172. Mat. Med., v, 142.

173. Tragopodagra, l. 161; *Opera*, ed. Jacobitz, Leipzig, Teubner, 1886, iii, 448; Helm, Pauly-Wissowa, ref. 45, xiii, 1729, says the work is spurious.

174. *Mineralogie der alten Griechen und Römer*, Gotha, 1861, 149.

175. E. Hiller, *Arch. Gesch. Math., Naturwiss. u. d. Technik*, 1931, xiii, 358 (393); *id.*, *Quellen u. Studien z. Gesch. Naturwiss. u. Med.*, 1941–2, viii, 166; Bailey, ref. 16, 1932, ii, 251–3; cf. Salmasius, ref. 8, ii, 847.

176. De Igne, 46.

177. Theophrasti *Opera*, Paris, 1866, 358.

178. Ref. 2, i, 149.

179. Ref. 19, 1857, vi, 357.

180. De Medicina, iv, 24 : lapis etiam, qui carnem exedit, quem σαρκοφάγον Græci vocant . . . ex quo in Asia lapidi asio gratia est.

181. E. A. W. Budge, *The Divine Origin of the Craft of the Herbalist*, 1928, 77; Leclerc, *Histoire de la Médicine Arabe*, 1876, ii, 225; E. H. F. Meyer, *Geschichte der Botanik*, Königsberg, 1856, iii, 227; Carra de Vaux, *Penseurs de l'Islam*, 1921, ii, 291; Ruska, in *Encyclopædia of Islam*, 1927, ii, 366.

182. *Grosse Zusammenstellung über die Kräfte der bekannten einfachen Heil- und Nahrungsmittel*, 2 vols., Stuttgart, 1840–2. Sontheiner also quotes Dioskourides, Galen, and Ridwān (d. 1223-33), a commentator of Avicenna.

183. "Traité des Simples," in *Notices et extraits des manuscrits*, 1877, xxiii, 71, 200, 333, 420; Romocki, ref. 42, i, 37–8, with Arabic text; Mercier, ref. 3, 114. *K. al- jāmi'*, I, 83, l.23; II, 12, ll. 27–8.

184. In Sontheimer, ref. 182, i, p. x.

185. *Liber Canonis*, tr. Gerard of Cremona, revised by Andreas Alpego Bellunensis, f°, Venice, 1557, 101 : lib. II, tract. ii, cap. 30.

186. *J. Asiat.*, 1849, xiv, 262.

187. *J. Asiat.*, 1850, xv, 214.

188. Ref. 185, appendix, 15 *v*.

189. Ref. 185, appendix, 7.

190. Ref. 42, i, 37.

191. Arabic text, ed. Müller, Königsberg, 1884, i, 82; the passage was translated by A. von Kremer, *Culturgeschichte des Orients unter den Chalifen*, Vienna, 1876, ii, 453, who gave "saltpetre" but added the Arabic shabb, "alum," in brackets. A translation made by Prof. A. Fischer for Lippmann, ref. 117, 1906, i, 123 (who says the passage was "previously not sufficiently attended to," but does not mention von Kremer) simply gives "saltpetre." I am indebted to Mr. A. Z. Iskandar, Oxford, for a translation of the Arabic text, which is difficult and contains unfamiliar words.

192. Ref. 187, 214, 237.

193. Ref. 71, 3, 16. Ruska, *Z.f. Assyriologie*, 1927, xxxvii, 282, says the Arabic bārūd, saltpetre, and barad, hail, have nothing to do with one another, and "saltpetre does not lie about like hail." It does seem to me, however, that both words come from a common root, and that the name for saltpetre, bārūd, was derived from the cooling it produced when dissolved in water. The Turkish name, according to Vullers, *Lexicon persico-latinum*, Bonn, 1855, i, 170, is bārūt, meaning both saltpetre and gunpowder. Vullers also gives namak shūra Chīnī (salt of Chinese salt-marsh?) as a Persian name for saltpetre. Serapion, *Practica*, bk. vii, ch. 33, Venice, 1550, f.109, uses the name burut for cooling lotions for the eyes, but none contains saltpetre.

194. Reinaud and Favé, *Du Feu Grégeois*, 1845, 77–8 (from BN MS. Anc. fonds no. 1072); Favé, in Napoleon, *Études sur le Passé et l'Avenir de l'Artillerie*, 1862, iii, 20.

195. Ref. 42, i, 75, with Arabic text.

196. Ref. 71, 72.

197. *Liber de simplici medicina secundum Platearium dictus Circa instans*, f°, Lyons, 1524, 245 (BM 546.i.10(2)).

198. In Pinkerton, *Voyages*, 1811, x, 571.

199. Ref. 194, 1845, 13–15, 197.

200. Ref. 71, 3, 16.

201. Ref. 42, i, 38–9.

202. Berthelot, ref. 69, 1893, iii, 155 (Arabic text, 122). Works attributed to Jābir, perhaps written in the ninth century, mention poisoning fruit trees (see Ch. IV) and the "automatic fire" of ψ-Julius Africanus (Ch. I); Kraus, *Mém. de l'Inst. d'Égypte*, Cairo, 1942, xlv, 76, 142; 1943, xliv, 86; Plessner, *Islamica*, 1931, iv, 525 (553).

203. Ref. 193. Ruska, *Archiv für die Geschichte der Medizin*, 1923, xv, 62, says the K. al–Mawāzīn is apocryphal and not earlier than the tenth century.

204. *Sino-Iranica*, Field Museum of Natural History Publ. 201, *Anthropological Series*, Chicago, 1919, xv, no. 3, 503, 555; G. Jacob, "Oriental Elements of Culture in the Occident," in *Smithsonian Report*, 1902, Washington, 1903, 509–20, relies entirely on secondary sources.

205. *Mem. Asiatic Soc. of Bengal*, 1927, viii, 333.

206. Ref. 194, 1845, 177, 180.

207. *Jahrbücher der Literatur*, Vienna, 1846, cxiv, 163.

208. *Sur la Force des Matières Explosives*, 1883, ii, 354; *Rev. Deux Mondes*, 1891, cvi, 807.

209. Ref. 42, i, 37.

210. M. Jähns, *Handbuch einer Geschichte der Kriegswissenschaft*, Leipzig, 1880, 771.

211. Hime, ref. 71, 1915, 3, 99.

212. M. Jähns, *Geschichte der Kriegswissenschaft in Deutschland*, Munich and Leipzig, 1889, i, 224.

213. Lippmann, *Geschichte des Zuckers*, 1929, 168.

213a. R. Duval, in Berthelot, ref. 69, ii, 145, 164; *id.*, *Journal Asiatique*, 1893, ii, 290 (308, 348–9).

213b. Dozy, *Supplément au Dictionnaires Arabes*, Leiden, 1881, i, 809.

214. Jähns, ref. 210, 1880, 801; B. Rathgen, Z. *Naturwiss.*, 1925, 87; *id.*, *Das Geschütz im Mittelalter*, 1928, 95 f., 99, 700.

215. Ref. 214 (1928), 100; cf. *ib.*, 107, *n.* 26.

216. Ref. 214 (1928), 107.

217. Rathgen, ref. 214 (1928), 106.

218. Rathgen, ref. 214 (1928), 108 : salnitri.

219. Rathgen, ref. 214 (1928), 577.

220. Feldhaus, *Geschichtsblätter der Technik*, ii, 54; Guttmann, *Monumenta Pulveris Pyrii*, 1906, 10, plates 17–20.

221. Bellifortis, *c.* 1405; Romocki, ref. 42, i, 163.

222. Beckmann, ref. 23, 1846, ii, 509.

223. Clarke, ref. 84, 21; Clarke also quotes from some source : aperte enim salem hunc, qui in cavernis sua sponte in rupium superficies erumpebat, florem et spumam nitri, salemque petrosum vel petræ nominat.

224. Favé, in Napoleon, ref. 194, 1862, iii, 205.

225. *Das Mittelalterliche Hausbuch*, ed. Bossert and Storck, Leipzig, 1912, 29; Lippmann, *Beiträge zur Geschichte der Naturwissenschaften und der Technik*, 1923, i, 200–10 (good summary).

226. Guttmann, ref. 220, 10, plate 21.

227. *De Natura Fossilium*, iii; *Opera*, Basel, 1657, 586 f.; tr. M. C. and A. Bandy, *Geological Soc. America, Spec. Paper* 63, 1955, 43.

228. *De Re Metallica*, xii; *Opera*, Basel, 1657, 453 f.

229. *Isis*, 1935, xxiv, 121.

230. *Pirotechnia*, x, i; Venice, 1540, 149 *v.*–152 *r.* : Della natvra del sal nitro e del modo che a farlo si procede; tr. C. S. Smith and M. T. Gnudi, *The Pirotechnia of Vanoccio Biringuccio*, New York, 1942, 404.

231. Ref. 225, 1953, ii, 202.

232. Agricola, *De Re Metallica*, 1912, 562.

233. *Pirotechnia*, ii, 8; 1540, 35 *v.*

234. C. S. Smith and M. T. Gnudi, ref. 229, xxiii, who give the wrong title and date of Whitehorne's book.

235. A. F. Pollard, *Dict. Nat. Biogr.*, 1909, xxi, 137.

236. *The Arte of Warre. Set forthe in Englishe by P. Whitehorne. Certaine waies of Orderyng Souldiers in battelray*, 2 pts., 4º, 1560–62; the reprint in *Tudor Translations*, 1905, xxxix, does not include the powder recipes quoted here.

237. Peter Whitehorne, *Certain Waies / for the orderyng of Souldiers in battelray . . . and more / ouer, howe to / make Saltpeter, / Gunpowder, / and diuers / sortes / of Fireworkes or wilde Fyre*, sm. 4º, Printed in London by Ihon Kingston : for Nicolas Englande, April, 1562. CUL SSS.24.4 (second part only).

238. Ref. 237, ff. 31 *r.*–32 *v.*

239. Ref. 237, ff. 27 *v.*–28.

240. Ref. 237, f. 21 *v.*

241. Incorrect quotation and wrong date in W. Y. Carman, *A History of Firearms. From Earliest Times to 1914*, 1955, 161.

242. *Pirotechnia*, 1540, 149 *v.*

243. *Beschreibung Allerfürnemsten Mineralischen Ertzt vnnd Berckwercksarten . . .* fº, Prague, n.d. [1574]; Frankfurt, 1580, 1598, 1629; enlarged as *Aula Subterranea Domina Dominantium Subditorum. Das ist : Unterirdische Hofhaltung . . . oder Gründliche Beschreibung derjenigen Sachen so in der Tieffe der Erde wachsen . . .* 4º, Frankfurt, 1672, 1684; fº, 1703, 1736; tr. by Sir John Pettus, *Fleta Minor. The Laws of Art and Nature, in Knowing, Judging, Assaying, Fining, Refining and Inlarging the Bodies of Confin'd Metals*, fº, London, 1683, 1686; *Lazarus Ercker's Treatise on Ores and Assaying, translated from the German Edition of* 1580, by A. G. Sisco and C. S. Smith, Chicago, 1951, bk. v, On Salpetre, 291–313.

244. Boillot, *Modeles d'Artifices du Fev et Divers Instrvmens de Gverre avec les Moyés de s'en Prevaloir*, sm. f° in 6's, A. Chaumōt en Bassio, 1598 (BM 534.f.5); it is said in the *Nouvelle Biographie Générale*, 1853, vi, 422, to have been republished with a German tr. by Brantz, f°, Strasbourg, 1603; Favé, in Napoleon, ref. 194, 1862, iii, 293, says Boillot was manager of a saltpetre refinery and gunpowder warehouse in Langres. In it, p. 164, he has a picture of a bursting bomb; the part on saltpetre is in pp. 86 f.

245. *Bericht vom Bergwerck, wie man dieselben bawen und in güten wolstande bringen sol, sampt allen dazu gehörigen arbeiten, ording und Restlichen processen beschrieben*, f°, engr. title-page only, s.l.e.a. [Zellerfeld, 1617]; 2 ed., engr. title-page and title-page, *Gründlicher und aussführlichen Bericht von Bergwercken* . . . f°, Leipzig, 1690; on saltpetre, 1st ed., 333–40.

246. Transl. by Christopher Packe, *The Works of the Highly Experienced and Famous Chymist, John Rudolph Glauber: containing Great Variety of Choice Secrets in Medicine and Alchymy . . . Also, Various Cheap and Easy Ways of making Salt-petre, and Improving of Barren-Land, and the Fruits of the Earth*, f°, 1689, i, 186 f., 259, 309 f., 335 f., 409, 416; ii, 41.

247. Ref. 246, i, 338–50.

248. *Phil. Trans.*, 1665, i, 33, no. 3.

249. In T. Sprat, *The History of the Royal Society*, 1667, 260–76.

250. *Elementa Chemiae*, Leyden, 1732, ii, 386.

251. *Chemical Works*, tr. Lewis, 1759, 197.

252. *Chemical Essays*, 1793, i, 293–6.

253. *Phil. Trans.*, 1774, lxiv, 481.

254. Ref. 252, i, 298.

255. L. Lemery, *Mém. Acad. Sci.*, Paris, 1717, 29 Hist., 31, 122 Mém.; H. Boerhaave, *A New Method of Chemistry*, tr. Shaw, 1741, i, 107; Zedler, *Grosses Universal-lexicon aller Wissenschaften und Künste*, Halle and Leipzig, 1742, xxxiii, 1128–1170 (Salpeter); Caspar Neumann, *Chemical Works*, tr. Lewis, 1759, 197–201; *Instruction sur l'établissement des Nitrières et sur la Fabrication du Salpêtre. Publiée par ordre du Roi par les Régisseurs généraux des Poudres et Salpêtres*, De l'Imprimerie royale, Paris, 1777 (this was written by Lavoisier, *Œuvres*, 1892, v, 391–460); C. M. Cornette, *Mémoir sur la Formation du Salpêtre*, Paris, 1779; J. C. Wiegleb, *Handbuch der allgemeinen Chemie*, Berlin and Stettin, 1781, ii, 102–110; *id.*, *Geschichte des Wachsthums und der Erfindungen in der Chemie in der neuern Zeit*, Berlin and Stettin, 1791, ii, 158–9; R. Watson, *Chemical Essays*, 6th ed., 1793, i, 283–326; Hon. George Napier, "Observations on Gunpowder," in *Trans. Roy. Irish Acad.*, 1788, 97–117 (incl. saltpetre refining); P. Sangiorgio, *Opuscoli sulla Formazione del Nitro e lo Stabilimento delle Nitrerie Artificiali*, Milan, 1805 (190 pp.); J. J. A. Bottée and J. R. D. Riffault, *L'Art du Salpêtrier*, Paris, 1811 (Poggendorff, *Handwörterbuch*, 1863, i, 251, credits this to Bottée and dates it 1813; in *ib.*, ii, 644, he credits it to Riffault); C. F. Becker, *Theoretisch-praktische Anleitung zur künstlichen Erzeugung und Gewinnung von Salpeter nach eigenen und den in Frankreich gemachten Erfahrungen*, Brunswick, 1814; J. G. Krünitz, *Oekonomisch-technologische Encyclopädie*, 242 vols., Berlin, 1773–1858; 1822, cxxxi, 253–548 (Salpeter und Salpetersiederey), bibliogr. 543–8, quoting 95 items; O. Thiele, "Salpeterwirtschaft und Salpeterpolitik," in *Zeitschr. für die gesamte Staatswissenschaft*, Tübingen, 1905, Ergänzungsheft xv, 1–237 (BM P.P.1423ha); K. W. Jurisch, *Salpeter und sein Ersatz*, Leipzig, 1908; Lenoir, *Historique et Legislation du Salpêtre*, Paris, 1923; H. Nicol, *Chemistry and Industry*, 1932, li, 971; J. Massey, A Treatise on Saltpetre, in *Mem. Manchester Lit. and Phil. Soc.*, 1785, i, 184–223.

256. Ref. 252, 1793, i, 286 f.

257. Thiele, ref. 255, 237; F. Hildebrandt, *Z. angew. Chem.*, 1926, xxxix, 90–2; Lippmann, *ib.*, 401; Krünitz, ref. 255, 488 f.

258. Watson, ref. 252, 1793, i, 291–2.

259. Ref. 208 (1883), i, 345–55; Krünitz, ref. 255, 481 f.

260. *Obs. Phys.*, 1775, vi, 329–46 (October, 1775).

261. *Obs. Phys.*, 1786, xxix, 264–72; note by de la Metherie, 272–5.

262. See Partington, *Annals of Science*, 1953, ix, 96–8.

263. *Mém. div. Sav. Acad. Sci.*, 1786, xi.

264. The announcement is printed in Lavoisier's *Œuvres*, 1892, v, 461–472.

265.　*Œuvres de Lavoisier, Correspondance*, ed. R. Fric, 1955, i, 97.

266.　J. F. Gmelin, *Geschichte der Chemie*, Göttingen, 1799, iii, 24, with names of other works on saltpetre, etc.

267.　Ref. 255, 13 (pits), 20 (plantations), 25 (walls), 30 (extraction), 38 (refining), 41 f. (social and economic aspects), 215 f. (documents).

268.　*Annales de Chimie*, 1797, xx, 356–69; the same volume contains a paper by C. A. Prieur, *ib.*, 298–307; and an anonymous paper (by Chaptal?) on nitre beds, *ib.*, 308–55 (tr. by McLachlan, *J. Soc. Chem. Ind.*, 1936, lv, 803R); J. A. Chaptal, *Élémens de Chymie*, 1796, i, 213–81, 281–316, gave detailed accounts of saltpetre and gunpowder.

269.　Krünitz, ref. 255, 509.

270.　Berthelot, ref. 208 (1883), i, 352–3.

271.　J. Black, *Lectures on the Elements of Chemistry*, Edinburgh, 1803, ii, 245, 455; Partington, *The Alkali Industry*, 1925, 196.

272.　Milner, *Phil. Trans.*, 1789, lxxix, 300; see Henshaw, ref. 249.

273.　Crell's *Annalen*, 1793, I, 224–7.

274.　*Phil. Trans.*, 1665, i, 103, no. 6.

275.　Ref. 252, 1793, i, 313–26.

276.　*The Voyages and Travels of J. Albert de Mandelslo from Persia into the East Indies*, tr. J. Davies, 2 ed., 4°, 1669, 66–7 (usually bound as part of A. Olearius, *The Voyages and Travels*); P. C. Rây, *A History of Hindu Chemistry*, Calcutta, 1902, i, 100; P. C. and P. Rây, *History of Chemistry in Ancient and Medieval India*, Calcutta, 1956, 229.

277.　*An Account of the Interior of Ceylon, and of its People*, 4°, 1821, 30–4; *Ann. Chim.*, 1824, xxv, 209.

278.　Davy, ref. 277 (1821), 265 f.

279.　B. G. Rao, J. J. Sudborough, and H. E. Watson, *Proc. 9th Indian Science Congress*, in *J. and Proc. Asiatic Soc. Bengal*, 1922, xviii, 75.

280.　Samuel Brown, "A Description of a Cave on Crooked Creek, with Remarks and Observations on Nitre and Gunpowder," in *Trans. American Phil. Soc.*, 1809, vi, 235–47; R. N. Maxson, *J. Chem. Education*, 1932, ix, 1847.

281.　J. Freind, *History of Physick*, 1726, ii, 325–32; *id.*, *Opera*, 1733, 565; J. F. Payne, *Dict. Nat. Biogr.*, Oxford, 1950, i, 548; G. Sarton, *Introduction to the History of Science*, 1947, iii, 1700.

282.　J. Hewitt, *Ancient Armour and Weapons in Europe*, 2 vols., and Supplement, Oxford and London, 1855–60–60; 1855, ii, 292–3; Burtt, *Archæological J.*, 1862, xix, 68; Jähns, ref. 212, 1889, i, 804; Hime, ref. 71, 62, 115 f., 149 f.; *id.*, *Gunpowder and Ammunition*, 1904, 177.

283.　*Proc. Roy. Artillery Inst. Woolwich*, 1865, iv, 289.

284.　Hewitt, ref. 282, ii, 284; from Sloane MS. 56.

285.　A. and C. R. Aikin, *A Dictionary of Chemistry and Mineralogy*, 1807, i, 542.

286.　Rathgen, ref. 214 (1928), 108.

287.　Ref. 214 (1928), 102 f., 105; q. Kerler, *Anzeiger für Kunde der teutschen Vorzeit*, 1866, xii, 246, not available to me.

288.　Items 1, 4, 5, 15, 16, 17, 19, 22, 24, 25 and 28 are from Hime, ref. 71 (1915), 168–9, and ref. 282 (1904), 197–8, who gives references for them; see also Lacabane, *Bibl. École des Chartes*, 1844, i, 28 f.; Favé, in Napoleon, ref. 194, 1862, iii, 73–4 (for item 4); Favé, in Napoleon, ref. 194, 1862, iii, 329 (for item 24; hazel-nut sized grains); R. Watson, ref. 252, 1781, ii, 16 (where other compositions are given). Items 1, 13, 14 (from Fronsperger), 18, 20, 26, 27, 28, 30, 34, 37 and 38 are given by O. Guttmann, *Die Industrie der Explosivstoffe*, in P. A. Bolley, K. Birnbaum, and C. Engler, *Handbuch der chemischen Technologie*, Brunswick, 1895, VI, vi, 160–3. Item 6 is from Jähns, ref. 212, 1889, i, 229–30; item 10 from *ib.*, 1889, i, 156; item 11 from Favé, ref. 194, iii, 122; item 21 from Fürtenbach, *Halinitro Pyrbolia*, 1627, 6; items 7, 8 and 12 from Rathgen, ref. 214 (1928), 102, 100 and 562, respectively. Chinese and Indian gunpowders are dealt with in the appropriate chapters.

289.　Aikin, ref. 285, i, 544. R. Brandes, Liebig's *Annalen*, 1832, iii, 345–9, gave the composition of English gunpowder as $KNO_3:S:C = 75.40:10.75:13·00$. The charcoal was brown and part was soluble in alcohol. Brandes says it was made by charring in cylinders, a process first used in England.   H. W. Dickinson and E. Straker, *Trans. Newcomen Soc.*, 1938, xviii, 61–6, think cylinder charring was probably first introduced by Bishop Watson, who says he suggested it in 1786 for making charcoal for gunpowder, the quality of the gunpowder being very much improved and a saving of several thousand pounds a year resulting, although Watson received no award from the Government; *Anecdotes of the Life of Richard Watson, Bishop of Landaff; written by Himself at Different Intervals*, 1817, 149 (on Watson see Partington, *Chemistry and Industry*, 1937, lvi, 819).   See also "The Rise and Progress of the British Explosives Industry", in *VIIth International Congress of Applied Chemistry, Explosives Section*, London, 1902.

290.   Eprouvette de M. de la Chaumette (1715), in *Machines et Inventions approuvées par l'Academie Royale des Sciences*, Paris, 1735, iii, 59, no. 162.

291.   Aikin, ref. 285, i, 545.

292.   Ref. 208, 1883, ii, 286.

293.   O. Guttmann, ref. 288, 1895, VI, vi, 160–3.

294.   Berthelot, ref. 208, ii, 311 : poudre de mine dite lente, aujourd'hui hors d'usage; Marshall, *Explosives*, 1917, i, 74 : "French official."

295.   Romocki, ref. 42, ii, 31; Guttmann, ref. 281, 1895, 160 f.

296.   J. Upmann, *Das Schiesspulver*, in P. Bolley and K. Birnbaum, *Handbuch der chemischen Technologie*, Brunswick, 1874, VI, i, 62.

296a.   Berthelot, ref. 204, 1833, ii, 314.

297.   G. H. Perry, in *Thorpe's Dictionary of Applied Chemistry*, 1922, iii, 2 f.

298.   Berthelot, *Science et Philosophie*, 1905, 130; Lippmann, ref. 117, 1906, i, 149, gives Liegnitz, 1344, and Spandau, 1348; Rathgen, ref. 214 (1928), 301 (Hildesheim).

299.   *Die Entwickelung des Kriegswesens und der Kriegführung in der Ritterzeit*, Breslau, 1887, III, i, 336.

300.   Aikin, ref. 285, i, 542.

301.   Hon. George Napier, *Trans. Roy. Irish Acad.*, 1788, 97–117.

302.   R. Coleman, *Phil. Mag.*, 1801, ix, 355–65.

303.   In Sprat, ref. 249, 1667, 277–83.

304.   Guttmann, ref. 288, 15, 160–3.

305.   Ref. 208, 1883, ii, 287.

306.   Guttmann, ref. 288.

307.   *Phil. Mag.*, 1858, xv, 489–512.

308.   *Phil. Trans.*, 1875, clxv, 49–155; Berthelot, ref. 208, 1883, ii, 288–313.

309.   I have not seen this first edition.

310.   *Neue Grundsätze der Artillerie*, Berlin, 1745.

311.   *The True Principles of Gunnery Investigated and Explained. Comprehending Translations of Professor Euler's Observations upon the new Principles of Gunnery published by the late Mr. Benjamin Robins . . .* 4º, London, 1777.

312.   *Mathematical Tracts of the late Benjamin Robins, Esq.: Fellow of the Royal Society, and Engineer General to the Honourable the East India Company*, published by James Wilson, M.D., 2 vols., 8º, London, 1761.   Vol. i (pp. 1–153) contains the *New Principles of Gunnery*.

INDEX OF NAMES

H

Hadrian 2, 6
Ḥājjī Khalīfa 204
Halhed 211, 232
Ḥalil Bassa 125
Hall 129
Hallam 190-1
Halliwell [-Phillips] 182
Hammer-Purgstall 42, 126, 142, 291, 313
Hanbury 296
Hannibal 2
Hansjakob 91, 94-5, 129
Hanzelet 176, 185, 292
Harikavi 221
Hallam 228
Hardouin 303, 330
Hārūn 13
Hasan 187; see Rammāḥ
Hase 330
Haskins 87, 89
Hassenstein 96, 107, 130, 133, 135-6, 138, 141, 152, 154, 156-7, 179-80
Hatton 110
Hay 130
Ḥaydar Mirzā 219
Heiberg 330
Heines 41
Heinze 135
Helm 334
Helmont 155, 180
Helvicus 158
Hemmerlin 91-4
Hendrie 61
Hennebert 2
Henntz 119
Henri IV 317
Henry VI 112; VII 137; VIII 37, 104, 112, 136, 183; of Silesia 250
Henshaw 318, 321, 338
Herakles 209
Herakleios 14
Hermes 58-9, 70, 83, 308, 313
Herodianos 4
Herodotos 1, 5, 171, 183, 298, 300, 303
Heron of Alexandria 13, 16, 33, 37, 163; of Byzantium 2
Hesiod 1, 301
Hesychios 16, 29, 37
Hewitt 128, 143, 324, 338
Hildebrandt 337
Hiller 334
Hime 7, 10, 12, 15, 17, 32-4, 40-5, 56, 59, 61-3, 73-6, 78, 80-1, 97, 100, 129, 132-5, 138, 142-3, 150, 164, 166, 174, 178-9, 182, 184, 190-2, 204, 213, 216, 227-8, 231-4, 244, 259-60, 262, 289-91, 293, 302, 312-3, 331, 335, 338
Hippokrates 43, 67, 305, 307, 330
Hippolytos 4, 6, 59, 72, 149, 178
Hirth 240, 289, 290, 292, 294
Hitti 36, 39

Hody 75
Hoefer 33-5, 43-5, 51, 62, 79, 80, 214, 232, 295-6, 309, 329-30, 332
Hofmann 308
Hoge 112
Hogel 122
Hogge 112
Holinshead 112
Holmyard 199, 204, 206, 231
Homer 7, 301, 306
Hommel 288
Hooker 239
Hope 137
Hopkins 183
Horowitz 178
Hortulan 70
Hou-pi-li 247
Hoyer 129, 138, 141, 179, 191, 228, 232
Hugenin 133
Hūlāgū 188, 247
Hultsch 51
Humāyūn 220
Humboldt 296
Hung-Wu 275
Hunter 93, 131, 136
Hunyadi 188
Ḥusain 313
Hussey 293

I

Ideler 33
Igor the Russian 18
Incarville, d' 230, 239, 285, 289, 321
Innocent, Bishop of Rome 268; XI (Pope) 284
Isaac Angelus 11, 38
Isabella 123, 189
Isidore of Seville 37, 301
Iskandar 314, 335
Ismael 205
Ismāʿīl Ṣafawī 205
Israelson 331
Ivins 181
Ibn abī-Uṣaybiʿa 310-2; al-ʿAmid 189; Baktawayh 311; Baṭṭūṭa 275; Durayd 334; Faḍallāh al-ʿUmarī 196; Iyās 195, 208; Khaldūn 191, 195-6, 228-9; al-Khatīb 191; Sīnā—see Avicenna
Ibrāhīm 219, 248
Ibrāhīm ibn-Sallām 200

J

Jābir ibn Ḥayyān 59, 197, 313, 335
Jacob 335
Jacobs 118, 138, 141
Jacques of Majorca 113-4; of Paris 113-4; de Vitry 29
Jahāngīr 226

INDEX OF PLACES AND NATIONALITIES

SUBJECT INDEX

A

INDEX OF GREEK WORDS

Milton Keynes UK
Ingram Content Group UK Ltd.
UKHW052224040923
428043UK00010B/1144